Embryo development and hatchery practice in poultry production

It is widely recognised that agriculture is a significant contributor to global warming and climate change. Agriculture needs to reduce its environmental impact and adapt to current climate change whilst still feeding a growing population, i.e. become more 'climate-smart'. Burleigh Dodds Science Publishing is playing its part in achieving this by bringing together key research on making the production of the world's most important crops and livestock products more sustainable.

Based on extensive research, our publications specifically target the challenge of climate-smart agriculture. In this way we are using 'smart publishing' to help achieve climate-smart agriculture.

Burleigh Dodds Science Publishing is an independent and innovative publisher delivering high quality customer-focused agricultural science content in both print and online formats for the academic and research communities. Our aim is to build a foundation of knowledge on which researchers can build to meet the challenge of climate-smart agriculture.

For more information about Burleigh Dodds Science Publishing simply call us on +44 (0) 1223 839365, email info@bdspublishing.com or alternatively please visit our website at www.bdspublishing.com.

Related titles:

Achieving sustainable production of eggs Volume 2: Animal welfare and sustainability
Print (ISBN 978-1-78676-080-7); Online (ISBN 978-1-78676-082-1, 978-1-78676-083-3)

Achieving sustainable production of poultry meat Volume 2: Breeding and nutrition
Print (ISBN 978-1-78676-068-5); Online (ISBN 978-1-78676-070-8, 978-1-78676-071-5)

Advances in poultry genetics and genomics
Print (ISBN 978-1-78676-324-2); Online (ISBN 978-1-78676-326-6, 978-1-78676-327-3)

Understanding the behaviour and improving the welfare of chickens
Print (ISBN 978-1-78676-422-5); Online (ISBN 978-1-78676-424-9, 978-1-78676-425-6)

Chapters are available individually from our online bookshop:
https://shop.bdspublishing.com

climate•SMART•publishing
IN AGRICULTURAL SCIENCE

BURLEIGH DODDS SERIES IN AGRICULTURAL SCIENCE
NUMBER 134

Embryo development and hatchery practice in poultry production

Edited by Dr Nick French

burleigh dodds
SCIENCE PUBLISHING

Published by Burleigh Dodds Science Publishing Limited
82 High Street, Sawston, Cambridge CB22 3HJ, UK
www.bdspublishing.com

Burleigh Dodds Science Publishing, 1518 Walnut Street, Suite 900, Philadelphia, PA 19102-3406, USA

First published 2023 by Burleigh Dodds Science Publishing Limited

Library of Congress Control Number: 2022949222

British Library Cataloguing in Publication Data
A catalogue record for this book is available from the British Library

ISBN 978-1-80146-252-5 (print)
ISBN 978-1-80146-254-9 (PDF)
ISBN 978-1-80146-253-2 (ePub)
ISSN 2059-6936 (print)
ISSN 2059-6944 (online)

DOI: 10.19103/AS.2022.0118

Typeset by Deanta Global Publishing Services, Dublin, Ireland

Contents

Part 2 Incubation

Series list

Title	Series number
Achieving sustainable cultivation of maize - Vol 1 From improved varieties to local applications Edited by: Dr Dave Watson, CGIAR Maize Research Program Manager, CIMMYT, Mexico	001
Achieving sustainable cultivation of maize - Vol 2 Cultivation techniques, pest and disease control Edited by: Dr Dave Watson, CGIAR Maize Research Program Manager, CIMMYT, Mexico	002
Achieving sustainable cultivation of rice - Vol 1 Breeding for higher yield and quality Edited by: Prof. Takuji Sasaki, Tokyo University of Agriculture, Japan	003
Achieving sustainable cultivation of rice - Vol 2 Cultivation, pest and disease management Edited by: Prof. Takuji Sasaki, Tokyo University of Agriculture, Japan	004
Achieving sustainable cultivation of wheat - Vol 1 Breeding, quality traits, pests and diseases Edited by: Prof. Peter Langridge, The University of Adelaide, Australia	005
Achieving sustainable cultivation of wheat - Vol 2 Cultivation techniques Edited by: Prof. Peter Langridge, The University of Adelaide, Australia	006
Achieving sustainable cultivation of tomatoes Edited by: Dr Autar Mattoo, USDA-ARS, USA and Prof. Avtar Handa, Purdue University, USA	007
Achieving sustainable production of milk - Vol 1 Milk composition, genetics and breeding Edited by: Dr Nico van Belzen, International Dairy Federation (IDF), Belgium	008
Achieving sustainable production of milk - Vol 2 Safety, quality and sustainability Edited by: Dr Nico van Belzen, International Dairy Federation (IDF), Belgium	009
Achieving sustainable production of milk - Vol 3 Dairy herd management and welfare Edited by: Prof. John Webster, University of Bristol, UK	010
Ensuring safety and quality in the production of beef - Vol 1 Safety Edited by: Prof. Gary Acuff, Texas A&M University, USA and Prof. James Dickson, Iowa State University, USA	011
Ensuring safety and quality in the production of beef - Vol 2 Quality Edited by: Prof. Michael Dikeman, Kansas State University, USA	012
Achieving sustainable production of poultry meat - Vol 1 Safety, quality and sustainability Edited by: Prof. Steven C. Ricke, University of Arkansas, USA	013
Achieving sustainable production of poultry meat - Vol 2 Breeding and nutrition Edited by: Prof. Todd Applegate, University of Georgia, USA	014

Introduction

It is widely recognised that developments in the chick embryo have a profound effect on hatchability and the subsequent health and productive capacity of poultry. With a stronger focus on welfare, prevention of disease and more sustainable poultry production, understanding these early stages has never been more crucial. This volume addresses the key stages and factors in embryo development in poultry to optimise hatchability and chick health.

The book is split into three parts: chapters in Part 1 examine parental influences and embryo development, focusing on areas such as the use of genetic selection to improve reproductive traits and management of breeder poultry flocks to optimise hatchability and chick health. Chapters on the development of defences against pathogens in the chicken egg, assessing poultry semen quality, the key stages of embryo development, chicken egg storage and transport, in ovo sexing in poultry chicks and in ovo vaccination of chicken eggs are also included. Part 2 chapters focus on incubation, specifically drawing attention to incubator design, the effects of incubator temperature, light as well as humidity and air composition on chick development. Chapters also address the role of egg turning in embryo development as well as the importance of incubation conditions on embryo skeletal development. Chapters in Part 3 focus on managing the hatching stage, highlighting the use of alternative hatching systems, the importance of managing the poultry hatcher environment as well as emphasising current key issues in transportation of broiler and layer chicks.

Part 1 Parental influences and embryo development

Part 1 opens with a chapter that reviews genetic selection to improve reproductive traits in chickens. Chapter 1 begins by first focusing on the layer breeding programme structure, then moves on to discuss areas such as egg production, liveability, resilience and adaptability. It also examines how genetic selection can affect egg quality, nesting behaviour and various reproduction traits as well as hatchability, fertility and sperm quality. A section on the molecular tools used is also included, before the chapter concludes with an overview of potential future research trends in the area as well as a section that emphasises the importance of improving reproductive traits to further improve animal health.

The subject of Chapter 2 is managing breeder poultry flocks to optimise hatchability and chick health. The chapter begins by providing an overview of different forms of mating behaviour, such as natural mating behaviour and mating behaviour in a broiler breeding house. It moves on to review housing systems and the methods that can be used to improve hatchability and chick health. A section on male-to-female ratio is also provided, focusing specifically on common male-to-female ratio practice and adjusted male-to-female ratio. The chapter reviews the use of spiking methods such as spiking young males and intra-spiking. A section on improving broiler breeder nutrition is also provided, followed by an overview of feed restriction and feeding programs. The importance of bodyweight management in female and male broiler breeders is also described, before concluding with an overview of why good feather cover quality is important during the production period.

The next chapter focuses on advances in understanding the development of defences against pathogens in the chicken egg. Chapter 3 starts by providing an overview of the basic components of the egg, followed by a breakdown of the key events that occur during the egg's formation. The chapter provides a summary of how antimicrobial proteins and peptides contribute to the egg's defence, followed by an overview of how each egg component can also contribute to the egg's defence with a special focus on the cuticle through the inclusion of a case study. The chapter also describes the embryo derived structures that are involved in the egg's defence.

Chapter 4 discusses assessing poultry semen quality by examining some of the semen evaluation techniques and fertility determination procedures currently used on breeder farms. The chapter begins by examining the biology of fertilisation, focusing on fertilisation versus fertility and how eggs are formed during fertilisation. It moves on to analyse the importance of semen evaluation and provides an analysis of how semen evaluation techniques can be categorised into subjective and objective categories based on how the data is obtained. The chapter also describes the current techniques used for determining fertility, before concluding with an overview of how on-farm semen evaluation remains crucial in poultry breeding management.

Chapter 5 draws attention to the key stages of embryo development in poultry. The chapter first uses nomenclature to describe the different phases of embryonic development, which is then followed by an overview of fertilisation. The chapter also covers intra-uterine development, focusing on zygotic gene activation and the formation of primordial germ cells and the role of these primordial germ cells in reproduction. A section on embryo development during the incubation period is also provided, followed by a discussion of endocrine control systems and maternal thyroids. The chapter also describes growth and development of the embryo E10-E15, the E16-E19 plateau phase and how extra embryonic fluids can influence compartment development. The chapter also provides an overview of the hatching process, drawing specific attention to synchronous hatching and physiology including the endocrinology of hatching.

The next chapter focuses on chicken egg storage and transport. Chapter 6 first describes how both egg quality and embryo development can influence the effect of egg storage. The chapter also considers practical aspects of using storage temperature, humidity and egg turning to keep eggs in good condition, with some reference to particular aspects of egg hygiene and egg transport. A case study describes the work at Aviagen developing a robust method to deliver short periods of incubation during egg storage to improve the hatchability of stored eggs.

The subject of Chapter 7 is in ovo sexing in poultry chicks. The chapter begins by first highlighting the key issues and challenges of in ovo sexing, then moves on to provide three methods for in ovo sexing through the inclusion of case studies. The case studies section highlights the use of hyperspectral analysis, how endocrinology is crucial for analysing certain components in the allantoic fluid of a chicken embryo, the use of DNA analysis and also mass spectrometry. The chapter also provides an overview of research projects whilst also indicating methods that need to be developed further, such as vibrational spectroscopic methods, magnetic resonance imaging, gene editing, avian egg odour detection and protein quantification.

The final chapter of Part 1 expands on topics previously discussed in Chapter 7 by addressing in ovo vaccination of chicken eggs. Chapter 8 first discusses the history and evolution of in ovo vaccination, then goes on to highlight the various interacting factors of in ovo vaccination success. A section on additional interactions and considerations are also provided, focusing specifically on egg candling for viability, chick handling and welfare, labour efficiencies and quality assurance. The chapter concludes by providing an overview of the challenges that need to be overcome in order to develop more efficient vaccines for in ovo use.

Part 2 Incubation

Part 2 opens with a chapter that focuses on the principles and techniques of incubator design for poultry eggs. Chapter 9 begins by first reviewing the key factors that need to be controlled during the incubation process, individual sections are provided for factors such as temperature, heat production and transfer, positioning and turning of eggs as well as the importance of maintaining moisture, carbon dioxide, oxygen and heat levels through ventilation. The chapter also includes an overview of cooling techniques, then moves on to discuss incubator machine design issues. A section on potential hatcher design issues is also provided, followed by an overview of how the hatching process can be managed and what hatchers should look like for a successful incubation process.

Chapter 10 draws attention to the effects of incubator temperature on embryo and post-hatch chick development. The chapter first compares single stage incubation to multiple stage incubation, then moves on to highlight embryonic heat requirements. It moves on to review the effect of temperature on hatchling quality and growth, then discusses how thermal manipulation can be used to improve embryo development. The influence of embryonic temperature on organ development is also addressed, drawing specific attention to bone and heart development, endocrine development and the use of thermoregulation. The chapter also discusses how embryonic temperature can affect muscle and adipose tissue development, intestine development and its effect on the immune system. A section on incubation temperature effects on incidence of ascites is also provided, followed by an overview of temperature effects on the integument system.

Chapter 11 focuses on understanding the effects of light on embryo and post-hatch chick development. It begins by assessing the role of monochromatic light on the growth of meat-type birds, focusing specifically on the effect of post-hatch and in ovo monochromatic photostimulation on broilers. The chapter moves on to review the effect of in ovo green light photostimulation on somatotrophic axis activity in embryos and the post-hatch period. The chapter also suggests that embryos are highly sensitive to green light photostimulation, a phenomenon that can be a tool for acceleration of muscle growth in meat type birds. It concludes by highlighting the importance of improving poultry meat production through different external manipulations.

The next chapter examines the effects of humidity/air composition on embryo and post-hatch chick development. Chapter 12 describes the physical function of the eggshell as a respiratory organ for the developing avian embryo, focusing on eggshell porosity, water vapour conductance and incubational egg weight loss and the influence of incubational temperature and airflow. The chapter moves on to review humidity, discussing aspects such as relative humidity and its effects on hatchability, hatchling quality, embryo metabolism and embryonic and post-hatch chick development. It also addresses the influences of breeder age on the effects of relative humidity on embryonic development. A section on vital gases is also provided, specifically focusing on oxygen and carbon dioxide.

Chapter 13 examines the role of egg turning in embryo development. The chapter identifies the four main aspects of turning deemed crucial to successful integration. It first focuses on turning during the egg storage period, then moves on to highlight critical periods during the incubation period. A section on turning frequency during incubation is also included, followed by a discussion on the influence of the turning angle during incubation. A section on signs of turning failure is also provided.

The final chapter of Part 2 reviews poultry embryo development and skeletal growth. Chapter 14 first provides an overview of embryo bone development, drawing specific attention to early development, the growth of long bones, hormonal regulation and developmental fluctuating asymmetry. The chapter moves on to assess incubation factors that can effect skeletal development, such as temperature, oxygen, carbon dioxide and ventilation, light, egg storage and SPIDES effects on bone and how incubation can effect vertebral development and chondronecrosis with osteomyelitis. Stressful conditions during transportation to the farm is also touched upon.

Part 3 Managing the hatching stage

The final part of the book begins with a chapter that examines alternative hatching systems for broilers. Chapter 15 introduces the current alternative hatching systems in use, then uses this introduction to compare alternative hatching systems to conventional ones. The chapter highlights the environmental factors that can influence both types of hatching systems, such as climate, egg position, light, noise as well as dust and the use of disinfectants. It then moves on to review the management factors that can influence conventional and alternative systems, focusing specifically on chicken handling and transport, the removal of second-grade chickens and time to feed and water access. A section on the effects of alternative hatching systems on health, welfare and poultry production is also provided.

Chapter 16 focuses on managing the poultry hatcher environment. The chapter first looks at the embryonic phase, focusing on development, physiology and metabolism of the chicken embryo and how the hatching position, hatch time and hatch window can effect embryonic mortality. Environmental settings within the hatching phase are also examined, such as temperature settings - both high and low - and relative humidity settings in the hatcher as well. Carbon dioxide concentration settings in the hatcher are also examined. The chapter concludes by providing an overview of why monitoring these settings is crucial in developing and maturing the chicken embryo.

The final chapter of the book discusses the key issues in the transportation of broiler and layer chickens. Chapter 17 focuses first on the transportation of newly hatched chicks, then goes on to look at welfare issues that can affect them during pre-transportation and the transportation process itself. The chapter then reviews how these welfare issues can be assessed which is followed by a discussion on how journey times and periods without feed and water can impact newly hatched chicks. A section on thermal stress in transit is also provided, where the use of thermal comfort zones for chicks is assessed, followed by the use of thermal imaging for surface temperature measurement and the use of thermal micro-environments and ventilation. Stocking density is also examined and exposure of chicks to acceleration and vibration is analysed. The chapter also discusses transport pathologies and provides guidance on good practice for animal transportation as well.

Part 1

Parental influences and embryo development

Chapter 1

Genetic selection to improve reproductive traits in chickens

David Cavero and Maurice Raccoursier, H&N International, Germany

1 Introduction

Global egg production increased from 51.13 million metric tons (Mt) in 2000 to 86.67 Mt in 2020 – an average annual growth of almost 3.5% (FAO, 2022). The number of hens worldwide also increased, albeit at a lower rate, from 4.97 to 7.90 billion hens (2.9% yearly) in the same time frame, which reflects higher productivity on the part of hens.

Disease control, modern housing systems and the use of balanced compound feed have contributed to highly improved performance in the layer industry. However, probably the most significant contribution has come from genetic improvement of birds, which has had a positive and cumulative effect over time. Genetic progress has been documented in experiments designed

http://dx.doi.org/10.19103/AS.2022.0118.01

especially for broilers and turkeys (Havenstein, 2006), in layers (Jones et al., 2001; Pelletier et al., 2014) and across different farm animal species (Hill, 2008).

The layer breeding industry has substantially contributed to the success of the poultry industry with the development of genetically improved birds. Consistent and continuous improvements in multiple traits such as increased egg production, improved feed efficiency, adaptation of egg quality to consumer preferences and improved robustness and liveability have led to high-performance birds with extended single productive cycles over 100 weeks.

Whilst the focus remains on maximising genetic potential for producing high-quality protein at a competitive cost, additional requirements of the egg industry, changing consumer habits and public opinion related to animal welfare and environmental load are being taken into consideration in order to guarantee the sustainability of egg production. The concept of sustainable development was described by the United Nations Brundtland Commission Report (Brundtland, 1987) as 'meeting the needs of current generations without compromising the ability of future generations to meet their own needs'. As Neeteson et al. (2020) explained, breeding contributes to sustainability by delivering genetic potential to match environmental (through lower resource utilisation and carbon footprint), economic (through increased profitability of the wider industry) and welfare requirements (through meeting societal requirements).

Breeding companies must look beyond current requirements and anticipate future needs and opportunities in the medium and long term. Breeding goals are broad and holistic, oriented towards optimising the use of resources, reducing the production of waste and carbon footprint, increasing food security and food quality, and enhancing the economic and social well-being of producers, consumers and society, as well as the welfare and well-being of the birds.

This chapter reviews these developments in the layer breeding industry, focusing on the current performance of commercial strains as well as selection criteria that might be significant in the future. Special attention is given to reproduction traits and the impact of other important traits on hatchability results. Additionally, the incorporation of new traits, the application of new evaluation methods and future improvements in breeding programmes are discussed.

2 Layer breeding programme structure

Since World War II, poultry production has developed from small scale, local enterprises to an economically important branch of agriculture (Fröhlich et al., 2012). Prior to the introduction of modern breeding, laying hens seldom produced more than 150 eggs per year, whilst now more than 320 eggs per year are commonly expected.

Breeding chickens for egg production started during the first half of the twentieth century, based on trap-nesting in single sire pens. Poultry were first selected as purebreds, but around 1940 crossbreeding began to be implemented by corn breeders since these hybrid birds have an advantage in terms of production and viability traits compared to parental lines. This effect is called heterosis.

An antagonism between egg production and weight gain in chickens was recognised early on. Koerhuis and Mckay (1996) showed a negative genetic correlation between juvenile body weight and hen day egg production ranging from −0.12 to −0.20. This led to the development in the middle of the last century of different breeding programmes with highly efficient lines specialised in egg production (layers) or meat production (broilers), making optimal use of resources.

In the late 1960s, more than ten different egg strain crosses competed on the international market, each offered by a different primary breeder. Typically, each breeding company marketed one well-known commercial cross. Since the late 1970s, the number of independent primary breeders represented in the world market has shrunk and can now be counted on the fingers of one hand. Each breeding company offers a choice of several white-egg, brown-egg and tinted-egg crosses, depending on regional preferences.

Geneticists identify individual animals within the breeding pools that carry the best combinations of genes to meet the selection objective. Thus, generation after generation there is an accumulation of alleles with positive effects on the desired traits within the population. Once the birds have been ranked and the best selected, a mating strategy is required to avoid inbreeding. The size of the nucleus needs to be large enough to maintain an adequate effective population size with sufficient genetic diversity within the lines to guarantee further genetic improvements in the future. Inbreeding in relatively small breeding populations reduces fertility and hatchability (Gowe et al., 1993). In commercial layer breeding programmes, a negative effect of inbreeding has been found on hatchability and to a lesser extent on production traits (Flock et al., 1991). Useful variation can be assumed to remain, providing population sizes do not become too small, and therefore continued response can be expected within populations, providing that selection is practised in the direction of market needs and deleterious side effects are eliminated or at least kept at bay (Hill et al., 2016). Kranis et al. (2013) re-sequenced 243 chickens from 24 chicken lines representing experimental, commercial broiler and layer breeds and detected 139 million SNPs (single nucleotide polymorphisms). In total, 78 M SNPs segregated in at least one line and on average 8-9 M segregating SNPs were detected in each commercial line, providing evidence of sufficient genetic variation. Using the best line combination and maintaining an optimal population size are key to success and set the basis of a breeding programme (Sitzenstock et al., 2013).

When poultry breeding is initiated, individual birds with the highest phenotypic values (own performance) are selected to become parents of the next generation, a strategy known as mass selection. This method is not effective for traits with low heritability and cannot be used for birds without own performance records, for instance egg production in males. Acceleration of the genetic progress is achieved by application of quantitative genetic theory combining own and relatives' performance in large closed pure-line populations whilst correcting for known non-genetic effects (Best Linear Unbiased Predictions - BLUP), with clearly defined breeding goals and systematic crossbreeding at subsequent multiplication stages.

Furthermore, advances in molecular biology have contributed to the development of new and more powerful techniques for selection. Currently, custom-designed SNP chips with thousands of SNPs distributed over the whole genome are used. This additional tool identifies individuals with the greatest genetic merit early in life with higher accuracy, which accelerates the genetic improvement of the elite lines.

Genetic selection is performed in relatively small gene pools under optimal conditions at the breeding farm. These genetic improvements are subsequently transferred through the different levels of the multiplication pyramid (pure lines - grandparents - parents - commercial birds) until they reach the level of commercial farms. Today, an extensive collection of phenotypic data is performed on pure-line birds with known pedigree housed in single and group cages as well as in floor pens at breeding farms in different locations with minimal disease pressure and optimal management conditions. An important part of genetic improvement is to collect accurate individual performance data under conditions representative of what the commercial layers will encounter in practice. Therefore, further data is collected on the performance of crosses (half-sibs of the pure-line birds with pedigree information) housed in different countries under different field conditions on commercial farms with the chance of natural exposure to field infections and variable feed quality.

These phenotypes from several generations of both pure lines at the breeding farm and crosses under field conditions are later combined with the pedigree and the molecular information of the birds by means of sophisticated statistical tools, allowing the geneticist to predict the merit or breeding value of each of the birds with high precision.

Heritabilities of all relevant traits and genetic correlations between traits are calculated in each line prior to selection, and estimates are used to calculate breeding values and a composite selection index.

Undoubtedly, the increase in size and a better structure of the populations of pure lines, together with the application of new data collection techniques and the implementation of new computational technologies, with powerful statistical tools for estimating the genetic merit of animals, have crucially contributed to optimising the use of existing genetic variance and maximising genetic improvement (Preisinger, 2018).

Better feed efficiency leads to the production of greater egg mass from fewer resources, thereby reducing the carbon footprint of egg production, which is critical in terms of sustainability. With climate change high on the agenda over the last decade, traits related directly to the emission of greenhouse gases from livestock have become a new focus. Moreover, other aspects such as genetic diversity, food safety, product quality, extended production cycles and the quality of animal well-being and fitness have become a large part of breeding objectives and breeding strategies. Hence, selection goals must be well-designed for future circumstances. A good balanced and broad breeding objective, which takes a holistic approach and includes all relevant traits, is more sustainable in the long term.

It is not surprising that some of the traits are antagonistic and correlate negatively with each other. Geneticists must handle these negative correlated traits by estimating the breeding values and genetic correlations with simultaneous evaluations of multiple traits, balancing the weight of the different traits in the index to achieve the breeding goal, using the variations that exist within the population, and selecting the right animals to obtain improvements in antagonistic traits simultaneously. Whilst it is possible for production traits to have undesirable consequences on bird fitness (Hocking, 2014), these negative consequences have been mitigated through balanced selection where health and welfare traits are included as equally important selection criteria (McKay, 2009; Neeteson, 2010; Avendaño et al., 2017).

The breeding goal in layer breeding programmes can be summarised in broad strokes as 'increasing the genetic potential of the birds to produce the maximum number of saleable eggs of the best quality per bird housed, whilst maintaining good feed efficiency, excellent adaptability to different climate and management conditions and good bird health and fitness'. Good health, robustness and bird viability is a *conditio sine qua non*, as the target is set per hen housed. Furthermore, since all breeding companies are active worldwide, it is necessary to adjust breeding goals according to regional requirements and changing consumer preferences.

3 Egg production

Following the re-discovery of Mendel's laws of inheritance in the early twentieth century, egg production was one of the first areas to illustrate the effects of selection on quantitative traits. Since the early twentieth century, layers have been through significant changes. The genetic performance of the birds has improved substantially during this time, as can be demonstrated by the increase in egg production in the last 50 years, as illustrated in Fig. 1. The number of eggs laid by a hen is determined by its age at sexual maturity, the peak of production and the persistency of lay.

Flock (1977) estimated heritabilities for egg production over different periods. Heritabilities were high for sexual maturity (age at first egg) and egg number during the first 8 weeks (h^2 = 0.54–0.58), lowest around peak of production (h^2 = 0.21–0.23) and increased towards the end of the testing period (h^2 = 0.25). Based on these results, it is clear that greater genetic progress in annual egg production can be achieved without extending the generation interval by putting more emphasis on the rate of lay at the time of selection, combined with full-year records of the parent generation.

Since this publication, more emphasis has been put on persistency at the end of production and pure lines are now tested over 100 weeks of age without moulting. Since flocks are continuously improving their productivity, getting closer to the biological limit of 100% (one egg per day), heritability for the rate of lay at peak of production in current populations is low (h^2 ~ 0.10). However, there is still sufficient genetic variation and a higher heritability (h^2 ~ 0.25) from 60 weeks onwards, which is exploited by selection to increase the genetic potential for more saleable eggs in extended laying periods. Genetic correlations between consecutive part records of egg production are

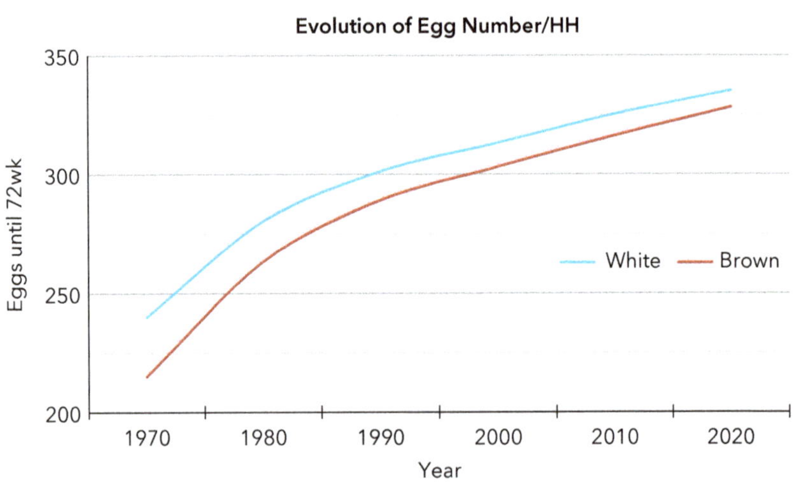

Figure 1 Genetic improvement in egg numbers/hens housed in white and brown-egg layers based on published management guides from Nick Chick and Brown Nick, data from Random Sample Tests summarised in Flock and Heil (2002) and in Anderson et al. (2013), as well as the authors' own data from commercial flocks (H&N International).

generally high and tend to be reduced as the distance between part periods increases.

The reported estimates of heritability for egg number vary substantially depending on the studied period, the population and the method used for the analysis, ranging from 0.1 to 0.5, both in layers (Anang et al., 2002; Nurgiartiningsih et al., 2004; Wolc et al., 2007; Cavero et al., 2011) and broilers (Koerhuis and McKay, 1996; Luo et al., 2007).

From 1950 to 1990, average egg production in Germany increased from 120 to 260 eggs per hen (Hartmann, 1992). Fairfull et al. (1998) reported an increase in egg numbers from 270 to over 340 between 1950 and 1993, with a simultaneous increase in egg mass of 42% during this period. In a summary of 53 years of random sample testing at the North Carolina Station, involving 37 different strains, Anderson et al. (2013) reported an increase of 67 and 64 eggs per hen housed during the first production cycle for white and brown varieties, respectively. According to these authors, mortality decreased from 16.6% to 5.5% in white-egg strains and from 10.9% to 6.4% in brown-egg strains. Moreover, Flock and Heil (2002) analysed a large set of data from 6 white-egg and six brown-egg strains over 24 years (1974–1999) of random sample testing in 5 German stations. Both white-egg and brown-egg strains followed similar trends - egg numbers increased by more than 60 eggs (from 245 to more than 310) with an increase in egg mass from 15 kg to almost 20 kg per hen housed. An annual increase of about 2–3 eggs per year in a 13-month production cycle can be expected (Preisinger, 2018).

In the past, selection was based on eggs laid until a specific age, putting special emphasis on early sexual maturity. A significant decrease in age at 50% of production is reported in the literature over these years, going from about 170 to 145 days of age (Anderson et al., 2013; Flock and Heil, 2002). Since early production and persistency tend to be negatively correlated, there is little interest in selecting further for earlier maturity and this trait has not changed over the recent last years. The goal of genetic selection for annual egg production is sexual maturity at an optimal age, rapid increase in egg weight, a high peak and persistent rate of lay after reaching the peak. Due to the long history of selection for persistent egg production, today's hybrid layers can lay an egg a day for several months, without pause. Analyses of 'clutch' length have shown that many hens lay for more than 100 days without pause. In fact, the average rate of lay in flocks is so high and the difference between birds so small that there is not much room for improvement in this phase. The biological limit is one egg with a normal shell in 24 h, and this limit has almost been achieved during peak production. Therefore, selection for life-time egg production is focused on persistency beyond current laying periods on commercial farms.

Saleable egg production per hen housed will continue to be the most important trait under selection. However, over the years the focus has moved from the start of the laying period in the first few decades to persistency and extending the production life of the birds in a single cycle without moulting in more recent years. These improvements also apply to parent stock, where flocks nowadays achieve more than 320 eggs per hen housed until 75 weeks of age.

Icken et al. (2012) presented moderate heritability for the oviposition time ($h^2 = 0.23$). Higher heritabilities have been calculated by Becot et al. (2021) ($h^2 \sim 0.49$ to 0.68). Good heritability estimates would theoretically allow to select for greater variability in oviposition time within a flock to make better use of the nest.

The time interval between ovipositions measured using electronic nests was on average 24 h, 10 min in both white-egg and brown-egg layers (Icken et al., 2008; Becot et al., 2021). Icken et al. (2008) also showed that hens with a short time interval between ovipositions of 24 h and 24 h and 15 min had the highest laying performance. Icken et al. (2008) estimated a heritability for time intervals between ovipositions of $h^2 = 0.09$ and a negative genetic correlation between laying rate in the nest and time between ovipositions of $rg = -0.29$. In agreement, but with higher genetic parameters, Becot et al. (2021) estimated moderate heritabilities for time between ovipositions ($h^2 \sim 0.34$ to 0.45) as well as a strong negative genetic correlation between laying rate in the nest and time between ovipositions ($rg \sim -0.61$ to -0.87).

Furthermore, Icken et al. (2012) presented moderate heritability for the clutch size ($h^2 = 0.25$). Wolc et al. (2019) published an average for the maximum clutch size of 81.8 eggs in a RIR line and 57.3 eggs in a WL line. Furthermore, they presented moderate estimates of heritability for average clutch size ($h^2 \sim 0.3$), maximum clutch size ($h^2 \sim 0.25$), number of clutches ($h^2 \sim 0.4$) and total number of saleable eggs ($h^2 \sim 0.3$). As expected, the total number of saleable eggs was positively correlated with average and maximum clutch size ($r_g \sim 0.6$) and negatively correlated with the number of clutches ($r_g \sim -0.5$) (Icken et al., 2012; Wolc et al., 2019). The negative genetic correlation estimated by Becot et al. (2021) between laying rate in the nests and number of clutches was even greater ($rg \sim -0.74$ to -0.92). In accordance with this, Chen and Tixier-Boichard (2003) observed an increase in egg number as a correlated response to selection for increased average clutch length. Under a typical 24 h rhythm lighting programme, ovulation cycles start with the laying of an egg (oviposition) early in the day. Further eggs are laid on successive days, generally at a later time, until the sequence or clutch ends with generally one day off, a pause without oviposition. Single clutches extend over several months, as can be seen in the longest clutch from a hen within a flock of pure lines (Fig. 2).

It can be concluded that time intervals between oviposition, number of clutches and clutch size are helpful parameters to further improve birds' laying capacity. They show moderate heritabilities and positive correlation to the total number of saleable eggs.

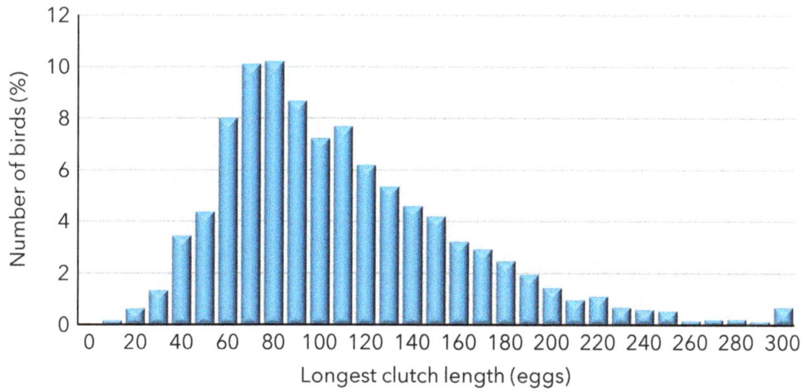

Figure 2 Histogram of the maximum clutch size per hen in a white-egg pure-line flock (generated by authors, H&N International).

4 Liveability

Egg production is expressed per hen housed, which includes liveability. On breeding farms of primary breeders, the best biosecurity standards and vaccination against common diseases assure low mortality in pure-line pedigree populations. Under these conditions, the heritability of liveability is close to 0 and genetic improvement impossible. To select for liveability, pedigreed crossline chickens are tested on commercial farms with a history of increased mortality.

Depending on the frequency of particular causes of death or the relevance of specific diseases, breeding companies may decide on single trait selection in certain lines to learn more about the disease, whilst poultry virologists develop vaccines. However, the exact causes of death are often difficult to determine and time-consuming to deduce under commercial conditions. Breeding for resistance to specific diseases requires birds to be exposed in a controlled and standardised way to these disease challenges, which raises concerns from an animal welfare point of view. Due to these concerns, the fact that disease-causing microorganisms are evolving continuously, and that new vaccines are rapidly developed and available on the market, the general approach of breeding companies is to select for general liveability. Although the heritability of this trait is low - generally in the range of 0.05–0.10 - giving liveability sufficient weight in the selection goal will have substantial cumulative effects on this trait in the long term.

Finding genetic markers associated with disease resistance has always been a goal for animal breeders. The MHC is one of the best-studied genomic regions related to disease resistance in chickens. Early studies in the serology of chicken blood groups were successful in finding various alleles of the B blood group responsible for higher susceptibility to Marek Disease (Bacon and Witter, 1993).

In addition to general liveability, more and more attention is being focused on animal welfare and well-being. Selection to enhance bone stability, increase resilience, reduce the incidence of feather pecking and aggression, as well as improving feather cover are gaining importance in the selection goal.

5 Resilience and adaptability

Resilience is defined as the capacity of an animal to be minimally affected by a disturbance or to rapidly recover to its previous state before a disturbance (Berghof et al., 2018). Bedere et al. (2022) studied resilience indicators derived from daily egg production. Especially interesting was the natural logarithm of the variance of the individual weekly egg production deviation from the average, albeit showing a low to moderate heritability ($h^2 \sim 0.1$ to 0.2). They also found a favourable genetic correlation with egg production (ranging from −0.83 to −0.54). The authors concluded that this trait could be considered in the selection process to improve the resilience of birds in breeding programmes for layers. Furthermore, literature on other species shows that the natural logarithm of variance is favourably associated with general immunity and health (Berghof et al., 2018).

To minimise the effects of genotype x environment interactions that can reduce the response to selection at the commercial level, crossline relatives with known pedigrees are tested under a wide range of commercial conditions in different countries. Heritabilities are consistently higher based on pure-line records from single cages than based on crossbred data from commercial multiple bird cages (Cavero et al., 2011; Bedere et al., 2022). Cavero et al. (2011) estimated genetic correlations between purebred and crossbred performances, which were high for egg quality traits ($r_g \sim +0.73$ to +0.90), moderate to high at sexual maturity ($r_g = +0.63$ and +0.83), whilst the correlations decreased to moderate to low for rate of lay at peak and end of production (rg between +0.1 and +0.5). Bedere et al. (2022) found similar genetic correlations for egg production in pure lines and cross-breeds, ranging from +0.31 to +0.52 in white-egg and brown-egg lines, respectively. Furthermore, in a review, Calus et al. (2020) reported positive genetic correlations between purebred and crossbred animals, ranging from 0.61 to 0.86 in layers. Genetic correlation that deviates from 1 could be explained by genotype by environment interactions, which can contribute to the same trait having a different genetic architecture in different environments. Purebreds and crossbreds are raised and housed for production in very different environments, with different climate, management, feed, housing systems and sanitary status. An additional explanation is that records from purebreds come from hens in single cages, whereas crossbreds are housed in group cages where the effects of social interaction with cage mates may lead to different traits. Breeding programmes benefit from considering both purebred and crossbred data to increase the accuracy of the estimated breeding values of candidate birds in the pure lines for crossbred performance, thus improving response to selection for crossbred performance in commercial environments (Wei and van der Werf, 1994; Dekkers, 2007; Cavero et al., 2011).

Geneticists are seeking to increase birds' productive capacity whilst maintaining appropriate feed intake capacity, in accordance with their productive level. Selecting for low intake capacity could compromise the performance of birds in a hot climate or in poor feed quality conditions.

Feed efficiency has a huge impact on profitability. Sustained improvements in feed conversion ratio (FCR) have a positive impact on the industry through the requirement for fewer feed resources, which reduces production costs. It should be borne in mind that feed accounts for more than 60–70% of production costs in most countries. Feed conversion (kilogram of feed intake per kilogram of egg output) improved from 3.41 to 1.98 from 1960 to 2010 in the United States, according to Pelletier et al. (2014). In the last 20 years, feed conversion has improved by at least 0.45–0.50 kg feed per kg egg mass produced (Preisinger, 2021).

6 Egg quality

Egg quality determines producer income, profit margins and consumer satisfaction. Several external and internal egg quality criteria are routinely measured in breeding programmes and used during the selection process to continuously improve the characteristics of the egg. External egg quality depends mainly on egg weight, shell strength, shell colour and cleanliness as well as shells free of any kind of defect. Internal egg quality is based on yolk percentage, Haugh units, yolk colour, freedom from inclusions and off-flavour. The heritabilities of most of these traits are moderate, as shown in Fig. 3. The routine control and improvement of egg quality in commercial breeding programmes start with the collection of eggs - only 'saleable' eggs with intact shells are counted.

Egg weight has always been a critical trait in layer breeding programmes. In the early years of selection, egg weight was increased significantly to accommodate consumer requirements. Since different countries have differing requirements regarding egg size, it has become impossible to cover the needs of all countries with a single hen. One strategy implemented by breeding companies was to increase the layer portfolio and to offer egg producers a choice of breeds with different egg profiles. Average egg weight has a high heritability ($h^2 \sim 0.6$–0.7) and is negatively correlated with the rate of lay, breaking strength and feed conversion. In breeding programmes, egg weight is measured several times during the life of the hens. The focus is to obtain a rapid increase during the first part of the laying period, progressively reducing

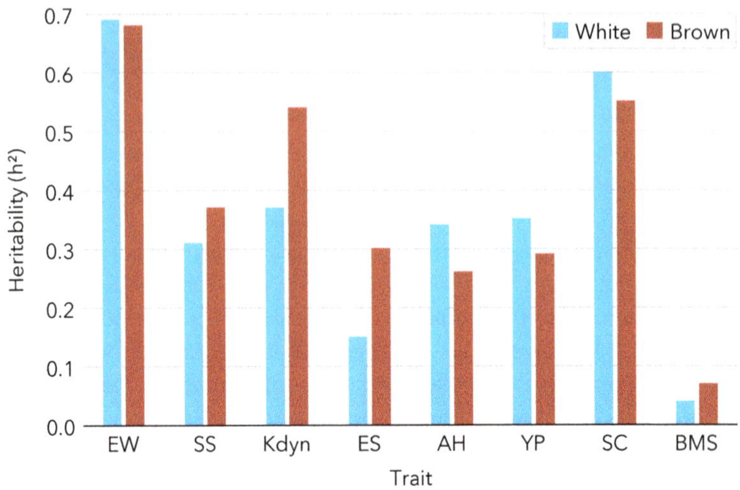

Figure 3 Heritability values for different traits in layers (EW: Egg Weight, SS: Shell Strength, Kdyn: Dynamic Stiffness, ES: Egg Shape, AH: Albumen Height, YP: Yolk Proportion, SC: Shell Colour, BMS: Blood and Meat Spots, SC: Shell Colour, BMS: Blood and Meat Spots/ inclusions).

egg weight increase so that fewer oversized eggs are produced at advance ages and consequently eggshell quality problems are reduced.

Furthermore, egg weight is an important trait for broilers, due to the positive correlation between egg size and chick weight (chick weight at hatch is on average 67% of the egg weight). Hartmann et al. (2003) found that the maternal heritability for chick weight potentially mediated via egg size was 0.5, whilst the direct heritability was close to zero.

Optimum egg size for hatching eggs ranges between 52 g and 68 g since both smaller and larger eggs show lower hatchability. Negative genetic correlations between egg weight and hatchability (r_g ~ −0.2) has been reported in broilers (Wolc et al., 2010) and layers with r_g ranging from −0.2 to −0.5 (Förster et al., 1992; Sewalem and Johansson, 2000; Hartmann et al., 2002; Cavero et al., 2011). This may be explained by the impact of egg weight on eggshell quality and embryo needs during incubation. Moreover, larger eggs tend to have proportionally lower surface area to volume ratio, which can obstruct embryonic exchange of heat and gases (Narushin and Romanov, 2002), making it more difficult to provide an optimal incubation environment. A further explanation could be the decline in yolk proportion with increased egg size at a given age, which could have a negative impact on the development of the embryo. Therefore, breeding programmes aim to control egg size within an optimal range to prevent drops in hatchability as the flock ages. This has been the focus of the genetic programme of layer breeders, especially in the female lines (Cavero et al., 2011) but not in broiler breeders where chick size is critical and of greater economic importance. In general, layer breeders tend to have a relatively constant and uniform egg size throughout the life of the flock and, as a consequence, better hatchability and eggshell quality than broiler breeders where the focus of selection is on weight gain (Fig. 4). In layer breeding programmes, more emphasis on reproductive traits is given in the female lines, whilst the male lines can be selected for higher egg weight. In

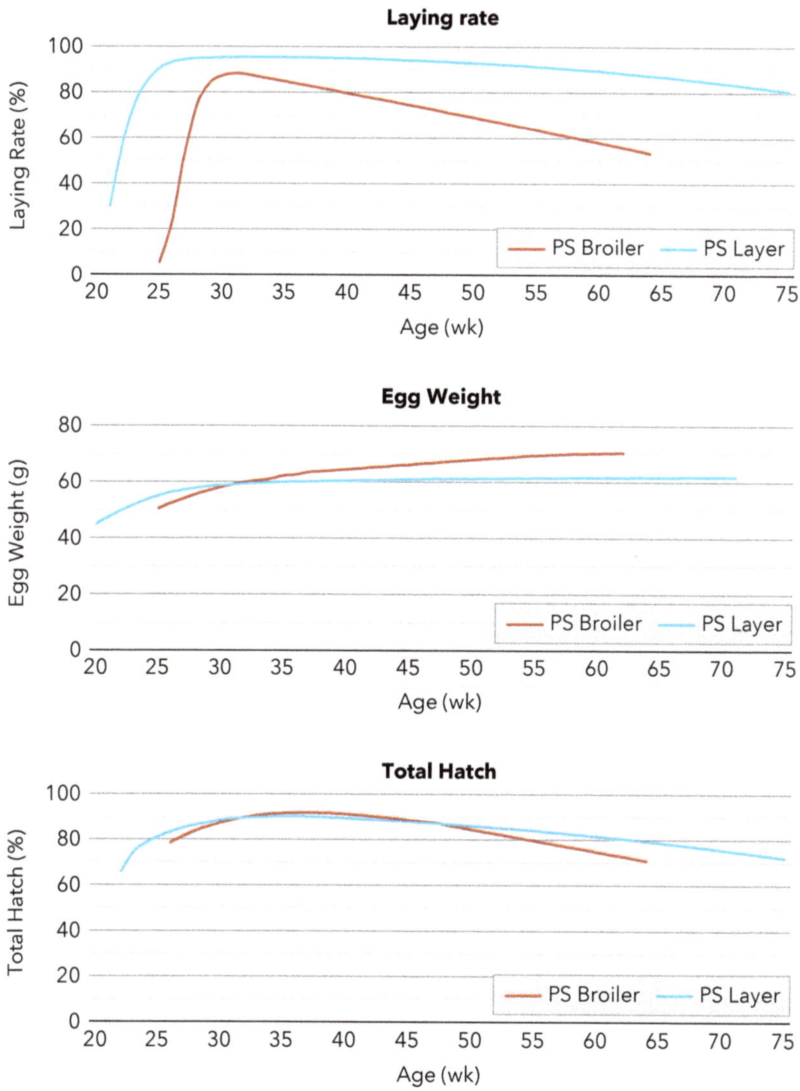

Figure 4 Comparison of laying rate, egg weight and hatchability of layer breeders vs broiler breeders (data from management guides: broiler - Ross 308, white layer - Nick Chick).

this way, the final commercial layer-cross complies with market requirements in terms of egg size whilst it is possible to overcome the negative correlations on hatchability and maintain a balanced performance of breeder flocks.

Selection for higher yolk percentage as an indirect trait to improve the percentage of solids in the egg is of great relevance to the egg-breaking industry in North America and Europe, which generally prefers white eggs. Furthermore, yolk percentage plays an important role, not only for the egg-processing industry but also for appropriate embryonic development, optimal hatchability and chick quality. Cavero et al. (2011) found positive correlations between yolk proportion with hatchability (rg ~ +0.10 to +0.39). Similar results in a White Leghorn line were also reported by Hartmann et al. (2002), who found positive correlations between hatch of fertile eggs and yolk weight, yolk proportion, and albumen dry matter of rg = +0.28, +0.52 and +0.26, respectively.

Cracked or damaged eggshells not only cause financial loss but constitute a risk in terms of food safety due to the easier ingress of microorganisms (Bain et al., 2006). These authors concluded that the dynamic stiffness measurement (Kdyn) can predict the probability of an egg cracking in the field with high precision. The heritability of Kdyn is high (0.35–0.55) and is combined with the conventional breaking strength measurement in layer breeding programmes to improve the stability of the eggshell. The main supply of calcium for eggshell formation is provided by the feed, with the remainder derived from the mobilisation of bone reserves, mostly coming from the medullary bones. Therefore, it is critical to deliver calcium in the feed at the right time, of the right amount and of the right granulometry, to fulfil the requirements of the hen.

Shell strength is of special interest at the end of the laying period as shell quality tends to decrease after peak production and eventually determines the end of the laying period. Several methods have been used to measure shell quality at an earlier age (shell thickness, shell weight, puncture score, breaking strength, crack detection by sound) to predict late shell quality. Crack detectors are used on a large scale in packing stations, and the system has been adapted to test eggs from pedigree hens before the shell is damaged for other measurements. The heritability of most shell quality criteria is intermediate (h^2 = 0.3 to 0.5) and much lower for visual counts of shell defects due to the low incidence at the time of assessment.

Blanco et al. (2014) found a high genetic correlation between breaking strength and dynamic stiffness (rg ~ +0.40 to +0.61). Shell thickness was also positively correlated with breaking strength (rg ~ +0.50 to +0.63) and dynamic stiffness (rg ~ +0.28 to +0.69). These moderate relationships demonstrate that the strength of an egg not only relies on shell thickness but also on the quality and uniformity of eggshell construction. Dynamic stiffness might be preferred for breeding purposes due to its lower negative genetic correlation with egg weight and its higher heritability (h^2 = 0.35 to 0.70) compared with breaking strength (h^2 = 0.23 to 0.35). Breaking strength and Kdyn are positively correlated with shape index, which confirms that round eggs have greater shell stability.

Moreover, egg quality traits are of utmost importance in layer breeding programmes due to their effect on hatchability and the production of quality chicks. Aside from the conditions provided by the setters and incubators of a hatchery, the probability of an embryo to hatch (hatchability of fertile eggs) is determined mostly by the environment provided by the egg. The eggshell performs a double function during embryo development. It must be strong enough to protect the embryo from external influences and penetration of pathogens, but it should have sufficient porosity to allow gas exchange for embryonic development (Narushin and Romanov, 2002). These authors also indicated that hatchability tends to increase with increasing shell thickness and length-to-width ratio. The positive effect of eggshell quality on hatchability has been confirmed in several studies (i.e. Bennet, 1992; Barnett et al., 2004). Furthermore, positive genetic correlations were calculated between hatchability and breaking strength of r_g ~ +0.19 to +0.29 in layers (Cavero et al., 2011) and between hatchability and specific gravity in broilers, r_g = +0.43 (Wolc et al., 2010). In addition, Rozempolska-Rucinska et al. (2011) observed positive phenotypic correlation values ranging from +0.47 to +0.66 between those traits. Eggs with higher specific gravity tend to have lower early embryonic mortality and higher hatch of fertile.

Egg weight loss, which also depends on egg weight and eggshell quality, is generally used as an indicator of incubation conditions to adjust humidity and temperature. Optimum egg weight loss during incubation for embryo survival is about 12% of fresh egg weight. Wolc et al. (2010) estimated a moderate heritability for egg weight loss (h^2 = 0.38) and found that high weight loss is genetically associated with lower mean hatchability and lower persistency of hatch. Both too-low and too-high weight loss at transfer after 18 days in the hatcheries (< 10% and >14%) has been associated with poorer chick quality and decreased hatchability.

The Haugh unit is a measure of albumen consistency and has low heritability. Haugh unit has a negative correlation with both egg storage and hen age. In a review by Narushin and Romanov (2002), the relationship between hatchability and internal quality showed a negative correlation between hatchability and liquid content of the egg. This agrees with the study by Wolc et al. (2010), in which an increase in Haugh units was associated with decreased average hatchability.

Considering the geometry of the egg, shape is very important to detect where the pointed end pole is, in order to set it downwards during storage and the incubation period. If eggs are rounded, the percentage of eggs set upside-down will be higher and their hatchability reduced. In addition, normal-shaped eggs (oval) have better hatchability than eggs that are round or abnormal (Cavero et al., 2011).

7 Nesting behaviour

The trend of moving layers from cages to alternative systems is gaining momentum, especially in the EU following the ban on conventional cages, but also in other parts of the world such as the United States, Canada and Australia. According to the market situation in the eggs sector presented by the European Commission (2022), the number of laying hens in the EU in 2021 totalled 376.0 M, of which 55.1% were housed in cage-free systems. The number of layers in the United States in 2021 was 323.0 M, of which 27.8% were in cage-free systems (Ibarburu, 2022). With the transition from cages to cage-free systems, hens can enjoy greater freedom to move and express locomotive activity, scratching and dustbathing, nesting and perching behaviour. In non-cage systems, new issues are arising for which new management strategies must be adopted. Birds must be reared in cage-free systems similar to the systems in which the birds will be in production. Additionally, birds need to learn to fly and to move around the system in order to find water and feed.

In cage-free environments, laying eggs in the nest becomes a key trait, since only eggs laid in the nest are suitable for human consumption or useable as a hatching egg in the case of parent stock flocks, due to risk of contamination of floor eggs. Cooper and Appleby (1996) studied the nest behaviour of hens and found differences between individual birds in the use of nests. Furthermore, these authors argued that this behaviour can be learned, as could be evidenced in the reduction of floor eggs as the flock aged. The genetic background for the percentage of eggs laid off-nest was presented by Settar et al. (2006) in brown-egg layers raised in floor pens, with moderate heritability estimates ranging from h^2 0.39 to 0.44. Special systems for testing individual performance in floor systems have been developed, using an adapted automatic trap-nest and transponder technology to obtain individual information on egg production and nest acceptance (Icken et al., 2012; Becot et al., 2021). This data is used to improve the 'number of saleable nest eggs', penalising birds with poor nest acceptance, which tend to produce floor eggs. Icken et al. (2012) estimated high genetic correlations (r_g ranging from 0.56 to 0.97) at the beginning of lay and low to moderate genetic correlations at peak of production (r_g ranging from 0.18 to 0.44) between the number of eggs recorded in individual cages and with the electronic trap-nest in floor pens.

As reported by Icken et al. (2013), the heritability of 'nest eggs' produced during early and peak production was moderate and varied substantially between the 12 flocks studied in this floor system. On average, the heritability was 0.26, indicating the possibility to use this trait for selection. Becot et al.

(2021) obtained similar estimates for heritability in White Leghorns, whilst they were lower (h^2 = 0.04 and 0.10 for peak and middle of egg production) in Rhode Island Red birds. Nevertheless, egg producers should not expect miracles from genetic selection and must pay proper attention to rearing conditions, timely transfer to the production house and optimal nest arrangements to minimise the number of displaced eggs. Also critical are an adequate lighting regime, adjusted feed formulation and feeding.

With different strategies and performance testing in cage-free systems, breeders are reducing the number of eggs laid on the floor, favouring the hygiene and quality of eggs from birds that are housed in alternative systems. It should not be forgotten that most breeder hens are housed on the floor and that eggs laid on the floor are not suitable for incubation. Avoiding floor-laid eggs is therefore not only advantageous for poultry farmers using alternative systems but is also very important for incubator companies. Hatchery companies will consequently obtain a greater number of hatching eggs and thus a greater number of chicks per breeder.

8 Reproduction traits

The avian egg is a biological system intended to ensure the well-being of the embryo and its successful hatching into a fully developed chick (Narushin and Romanov, 2002). Predictable reproductive performance of parent stock is essential for every commercial hatchery to produce a high number of first-quality chicks at a competitive price. Healthy chicks with good vitality and good uniformity within a flock set the basis for successful rearing and subsequent production. If conditions in the setters and hatchers are not optimised, the uniformity of chicks will suffer, and the hatching time will spread out. Many factors can affect hatchability, especially egg size and age of breeders, season, nutrition, egg handling and storage, as well as temperature and humidity throughout the incubation and hatching period (Wilson, 1997). Reproductive traits usually reach their peak close to the peak of production and decline with increasing age towards the end of the laying cycle (Fig. 4). The decrease in hatchability in older flocks is well-known and may be explained as a result of lower fertility of males and females. Potentially more important is the reduction in eggshell quality and increased egg weight.

Modern poultry breeding programmes strive to optimise the overall reproductive efficiency of the birds, which is defined as the number of viable day-old chicks produced per breeder hen. The reproductive characteristics include egg production in combination with fertility and hatchability, as well as good chick quality. Reproductive traits usually benefit from heterosis in crossline parents (Flock et al., 1991) and this effect increases with advancing age (Ledur et al., 2000). Furthermore, natural selection will always favour families with above-average reproductive rates as they contribute more progeny to the next generation. In addition, reproductive traits can be improved by genetic selection. However, they belong to the complex of fitness traits, which frequently exhibit low heritability owing to their complex nature (Falconer and Mackay, 1996), suggesting that optimisation of breeder farms and hatchery management is the most promising route for improvement. However, low heritability does not exclude improvement by selection; it just takes longer to see measurable results. One approach could be to have a selection threshold and to not select families with low fitness (Gowe et al., 1993). On the other hand, selecting birds with improved fitness based on their breeding values generates a higher genetic progress than simply eliminating animals with poor fitness based on their phenotypic value or relying on natural selection due to more precise evaluation and efficient selection.

Reproductive fitness has been negatively affected by selection for increased body weight in broilers (Siegel and Dunnington, 1985) although this effect has been counteracted by management practices as well as genetic selection to achieve increases in breeder performance whilst maintaining improved broiler characteristics (Laughlin, 2009). In layers, fertility and hatchability are maintained at a high level and improved every generation, since most of the traits included in the selection goal correlate positively to fitness traits and additionally a direct selection to improve them is implemented. The differences between egg production and hatchability evolution at different ages are shown for broilers and layers in Fig. 4. The higher increase in egg weight with age in the broiler breeders, with a consequent decrease in eggshell stability, might be at least partly responsible for the lower hatchability seen in broiler breeders at advanced ages.

Furthermore, reproductive traits might be decreased in the summer months due to higher ambient temperatures, which cause a decline in reproductive efficiency in both male and female birds. The logistics for egg handling is also more difficult to achieve, to ensure that the egg is collected at the right time and stored under good conditions, avoiding significant temperature changes and making a smooth transition from farm to storage temperature to avoid the eggs sweating.

Fertility depends on both parents, being a consequence of the interaction between a male and a female gamete to produce a viable zygote. In contrast, hatchability is determined by correct embryonic development, which at the same time depends on the characteristics provided by the maternal effect on the egg. Therefore, it is important to evaluate the hatch of fertile eggs because this parameter helps to isolate the impact of infertility and therefore evaluate how the incubation conditions and the egg quality traits impact hatchability.

9 Hatchability

Hatchability follows a pattern similar to egg production, with an initial increase from the start of lay to a peak that is sustained for several months in the middle of production, followed by a decline as the birds age (see Fig. 4). An important effect of age is that eggs laid by older breeders often present higher infertility and total embryo mortality, resulting in a lower hatching percentage (Almeida et al., 2008). Duncan et al. (1990) reported that mating activity is high in young male and female broiler breeders and declines with age. In addition to infrequent mating, lower hatchability of eggs from older females can be associated with changes in shell quality. The shell allows for gaseous exchange and loss of moisture from the egg, and poor shell quality results in excessive loss of relative humidity during incubation (Peebles et al., 2001). Egg quality has a significant effect on hatchability, as conditions in the microenvironment during the storage process and early incubation modify both the external and internal quality of eggs (Narushin and Romanov, 2002).

Research has identified that hatchability can be influenced by several animal-related factors, such as breeder age, egg weight and egg uniformity, as well as environmental factors such as egg storage duration, season and egg weight loss during incubation (Grochowska et al., 2019). In addition, the genetics of the embryo plays an important role in hatchability. It has been established that polygenic factors, as well as single lethal genes and chromosomal aberrations, lead to malformations, malpositions and metabolic disorders during different stages of embryonic development and lead consequently to embryonic death (Liptoi and Hidas, 2006).

Bouba et al. (2021) showed that in breeders older than 45 weeks, hardly any effect of egg weight uniformity on hatchability was found, but for breeders between 30 and 45 weeks of age, higher egg uniformity (> 85%) resulted in higher hatchability. There is a positive correlation between egg weight and incubation time, but the incubation time is affected much more by pre-incubation storage time than by initial egg weight (Wilson, 1991). As a rule of thumb, the lengthening of the incubation time with increasing storage time is estimated to be 0.5–1 h of incubation for each additional storage day.

Prolonged egg storage duration has a negative impact on hatchability results due to modifications to the blastoderm. The best results are obtained with eggs that are no older than one week and from breeders in middle age (Damaziak et al., 2021). One method to reduce the negative effect of long-term storage has been to pre-incubate eggs for short periods before storage (Meijerhof, 1992). Nicholson et al. (2013) reported in a summary of experiments and field tests using short periods of incubation during egg storage (SPIDES) that it is possible to recover 60–70% of the hatchability lost when storage must be prolonged for more than a week.

It is widely known that eggshell plays an important role during embryonic development, isolating the embryo from the external environment whilst allowing sufficient gas exchange through the shell. Bennet (1992) compared thin and thick shells based on specific gravity measurements and reported a reduction in hatchability of 3-9%, which he attributed to increased cracks, moisture loss and bacterial contamination of eggs with thin shells.

To support hatchery management in producing high-quality chicks within a reduced hatch window, even from older flocks and/or prolonged egg storage, primary breeders select families with persistent hatchability under suboptimal conditions (Förster, 1994).

It is well documented that improvement in growth rate in broilers may have a negative impact on reproductive traits such as hatchability and fertility (Chambers, 1990) due to the existence of an antagonistic genetic relationship. In order to overcome this negative effect, breeding companies use multivariate models that predict more accurate breeding values for growth and reproductive traits to allow improved production traits, to prevent or reduce negative impacts on animal welfare and reproduction (Laughlin, 2009; Avendaño et al., 2017).

Estimates of heritability for hatch of fertile eggs in the literature range from 0.05 to 0.25 (i.e. Förster et al., 1992; Beaumont et al., 1997; Szwaczkowski et al., 2000; Sapp et al., 2004; Bennewitz et al., 2007; Rozempolska-Rucinska et al., 2009; Wolc et al., 2010; Cavero et al., 2011; Makanjuola et al., 2021) depending on the lines and model used. By using a multi-trait model it has been shown that the correlation at early and late ages of fertility (Wolc et al., 2009) and hatchability (Makanjuola et al., 2021) is very high (rg ~ +0.70 to +0.99). Based on these results, geneticists can capture information at early ages and use this information to make breeding decisions to improve reproduction traits, which will have a positive effect at all ages. Beaumont et al. (1997) and Wolc et al. (2010) found in their research a higher genetic contribution of the dam than to the sire to hatch of fertile. This is logical since the impact of the male is limited once the egg is fertilised and the embryo starts to develop, whilst the influence of the dam is much more significant due to the characteristics of the egg.

10 Fertility

Assersohn et al. (2021) showed that fertility of females has received comparatively less attention than that of males within poultry production. Of the 718 papers on avian fertility that were published between 1985 and 2019, 42% focused on both sexes, followed by 37% focused on male fertility and 20% focused on female fertility. This might be explained by the determining role of the male in the diffusion of genetic progress since males usually inseminate several females (normally 5 to 12).

Fertility is fundamental to reproductive success, but not all copulation attempts result in a fertilised embryo. Female fertility is the product of not only the fertilisation rate but also the number of eggs produced that are capable of being fertilised. Furthermore, studies on fertility must discriminate between real infertility and embryonic mortality as the cause of hatching failure of eggs not transferred to the hatcher.

Fertility in poultry is traditionally regarded as an independent trait either of the male or the female, but genetic and non-genetic factors originating from both the male and female affect egg fertilisation and embryo development (Brillard, 2003). There are several factors affecting fertility, including body weight of both males and females, leg quality of the males, male-to-female ratio and nutrition. Fertility of an individual egg is also a function of the genotype of the embryo, to which both parents contribute. Therefore, both paternal and maternal components should be accounted for simultaneously when analysing fertility (Wolc et al., 2009).

According to Brillard (2003), hen fertility depends on the ability of females to ovulate, mate, store sperm and provide an appropriate environment for the formation and development of the egg. Wishart (2009) explained that around 1% of the inseminated dose of sperm enters the sperm storage tubules at the uterovaginal junction. He further described that these tubules provide a reservoir from which sperm are released at a rate of around 30% per day in chickens and 10% per day in turkeys. Released spermatozoa migrate up the hen's oviduct to fertilise the newly ovulated ovum at the upper end of the oviduct. However, he concluded that if the oviducal sperm reservoir is not replenished by further inseminations, there will eventually be insufficient sperm to fertilise the ova, which occurs after about 2 weeks in chickens. Brillard (2003) stated that the decline in fertility at an advanced age is a consequence of a faster release of spermatozoa from the sperm storage tubules rather than impaired storage of spermatozoa in the oviduct.

Meanwhile, Brillard (2003) and Mohan et al. (2018) reported that both the quality and quantity of semen deposited by the male are important to obtain good fertility. Thus, fertility in the case of the male will depend on two main points - his mating behaviour and the production of a high number of good-quality sperm.

Several studies have estimated heritabilities for fertility in broilers and layers using different models. As expected, for fitness traits heritability is generally low, with values ranging from 0.06 to 0.21 (Beaumont et al., 1997; Ledur et al., 2000; Sapp et al., 2004; Bennewitz et al., 2007; Rozempolska-Rucinska et al., 2009; Wolc et al., 2009; Cavero et al., 2011; Wolc et al., 2019). The heritability for fertility is usually lower than for hatchability. Wolc et al. (2019) concluded that only a limited genetic component was captured by markers and no QTL of large effects were detected for reproductive performance of White Leghorn males in their study.

Wolc et al. (2009) found that the genetic correlation between male and female fertility was positive but moderate and non-significant (0.15), indicating that improvement of one sex would be slightly beneficial, and certainly not harmful to the other, but that selection on records of both sexes would be needed for maximum improvement of overall flock performance.

Rozempolska-Rucinska et al. (2009) and Beaumont et al. (1997) observed a negative correlation between fertilisation efficiency and the number of chicks that are malformed, weak and unable to survive. Beaumont et al. (1997) found most genetic correlations favourable between fertility and embryonic mortality at three different phases. Moreover, highly positive genetic correlation between fertility, hatch of set and hatch of fertile are reported in the literature (Rozempolska-Rucinska et al., 2009; Wolc et al., 2010; Cavero et al., 2011; Wolc et al., 2019).

Wen et al. (2022) studied five different traits related to duration of fertility in layers with the focus of interest set on the females and using inseminations over two consecutive days with pool ejaculates from seven roosters. They estimated moderate heritabilities ($h^2 \sim 0.11–0.17$) and moderate to high positive genetic correlation between the five traits ($r_g \sim +0.32$ to $+0.94$). Their results showed that laying performance and duration of fertility are not correlated.

11 Sperm quality

Semen quality is also an important factor determining the breeding value of males as it influences the fertilisation rate of the eggs for hatching as well as the reproductive efficiency of their progeny (McGary et al., 2002). Different approaches can be used to assess sperm quality to predict the fertilising ability of the males and to use this information in breeding programmes to improve fertility. It is well documented that, in poultry, sperm quality tests are significantly correlated with fertilising ability, although there is considerable variation in the proportion of fertile eggs laid by hens inseminated with aliquots of the same sample of semen (Wishart, 2009). Some traits related to sperm quality have been shown to be predictive for male fertility (Froman et al., 1999). Aside from sperm volume, the most straightforward parameters of semen quality are its appearance (good-quality semen should have a thick, pearly white appearance) and sperm concentration. This last parameter can be measured directly by haemocytometer counts viewed through a microscope or by indirect methods such as packed cell volumes, spectrophotometry, colorimetry or turbidity (Bakst, 2014). Furthermore, two methods have been developed based on sperm interaction with the egg's outer and inner perivitelline layers (Wishart,

2009). It would appear that all quality parameters maintain a high correlation and are interrelated (Wishart, 2009).

Two of the main characteristics of semen are ejaculated volume (which ranges between 0.2–0.8 and 0.3–1.5 ml in layers and broilers, respectively) and sperm content (ranging from 1–4 to 3–10 billion/ml in white-egg layers and broilers, respectively). However, some authors argue that fertilising capacity should be measured directly by inseminating 10–20 hens with one ejaculate and observing the fertility rate of these eggs. Brillard et al. (1989) found that, independent of hen age, the mean duration of fertility (but not maximum percent fertility) was improved in females inseminated with the highest doses (2 × 120 mill spermatozoa) compared with the lower doses (2 × 30 and 2 × 60 mill spermatozoa).

Fecundity rate and sperm motility appear to follow a saturation curve, which means that fecundity rate is poor at low motility, but no effect or improvement is seen from certain thresholds onwards (De Reviers, 1992). Motility tests could be used to eliminate sperms with low motility that are less suitable for fertilisation.

Different studies have reported moderate heritabilities for most sperm quality traits (Bongalhardo et al., 2000; Wolc et al., 2010; Hu et al., 2013) such as sperm concentration ($h^2 \sim 0.12$–0.34), sperm motility ($h^2 \sim 0.08$–0.34), sperm volume ($h^2 \sim 0.27$–0.32), sperm colour ($h^2 \sim 0.19$), sperm viability ($h^2 \sim 0.52$) and sperm deformities percentage ($h^2 \sim 0.60$).

Based on results by Hu et al. (2013), sperm colour had a positive genetic correlation with sperm concentration ($r_g = +0.88$) and motility ($r_g = +0.28$). However, it is unfavourably correlated with sperm viability ($r_g = -0.29$) and sperm deformities (r_g=0.57). They further estimated a moderate heritability for sperm volume ($h^2 = 0.28$) and a positive genetic correlation with sperm concentration ($r_g = +0.68$) as well as sperm deformations (rg = +0.47). Information regarding sperm volume and colour could be easily and quickly recorded in the farming environment, and the moderate heritability estimated for these traits, as well as the positive correlation with sperm concentration, which could recommend the use of these traits for the selection of good-quality sperm. However, special caution must be taken as regards the potential decrease in sperm viability and increased incidence of sperm deformities.

The results of genetic selection for sperm quality have been disappointing (Wishart, 2009). Furthermore, whilst sperm quality could be used in breeding programmes that use artificial insemination, since most of the (grand)parent stocks in chickens are under natural mating this trait may be not so relevant since there are other more important parameters involved such as mating behaviour. A good review of the complexity of mating behaviour can be found in Duncan (2009).

12 Molecular tools

Genomic selection uses information derived from high-throughput genotyping of SNPs in the genetic evaluation by simultaneous estimation of marker-associated effects across the whole genome (Meuwissen et al., 2001). The idea is to use a dense set of markers distributed over the whole genome, independent of their effect and linking these markers with phenotypic information to estimate breeding values more accurately.

Hill et al. (2016) explain that when assessing individuals without their own or progeny records, information from ancestors does the effect of the unique DNA an individual gets from the sire and dam (Mendelian sampling). Genomics information allows prediction of the mendelian sampling term in the absence of own phenotype. Genomic selection has been successfully implemented in many commercial breeding programmes from several livestock species (Meuwissen et al., 2016). Breeding values can be estimated much more accurately at a younger age and for traits that cannot be easily measured in breeding animals. Selection accuracy has increased by 10–60% with genomic selection compared to conventional pedigree estimations (Neeteson et al., 2020; Sitzenstock et al., 2013; Wolc et al., 2016).

DNA can be analysed in both sexes, which is particularly helpful for sex-limited traits such as egg production. Furthermore, the DNA can be analysed early in life, which helps to shorten the generation interval. Both factors have a positive effect on genetic progress.

SNP markers panels with varying density have been developed and are used for large-scale genotyping in poultry (Kranis et al., 2013). Genomic selection is a recent paradigm shift for animal and plant breeding, a science that has undergone exponential advancement in recent years. Currently, custom-designed medium-density SNP chips are used to shorten the generation interval and to estimate breeding values more accurately compared to pedigree-based analysis, particularly in traits with moderate to high heritabilities. The molecular information also allows selection between full-sib males, which used to have identical breeding values, based on female relatives. Genomic selection provides tools that can be applied to the selection of birds during the growing period before their full- and half-sisters generate new information.

13 Future trends

The availability of less costly sequencing technologies increases the power of genome-wide association studies (GWAS), which together with better annotation information will increase the resolution of mapping QTLs or find causative variants (Kranis and Maniatis, 2020), which could increase the accuracy of genomic predictions in breeding populations. For the last 2 decades, genotyping and sequencing technologies have outpaced the most optimistic predictions, allowing industry-scale application of genomic selection (Pérez-Enciso and Steibel, 2021). However, simulation and empirical results have shown that only marginal or no gains in genetic prediction accuracy can be expected by genome sequencing data as opposed to high-density genotyping data (Ni et al., 2017; Pérez-Enciso and Steibel, 2021).

Moreover, the implementation of automatic digital data recording, where data is reliably recorded, easily presented, available to all and in real time, can help to make better decisions at the right time. Artificial intelligence can learn from different sources of data, not only by the farmer but automatically recorded by sensors and cameras. These systems not only include warnings for current events on farms but are also capable of producing predictions that can be automatically readjusted when new data emerges, which can assist decision-making.

Furthermore, more and more traits are being included in the breeding goals that are not just related to production characteristics. In fact, complex biological and physiological processes, as well as individual behaviours within a group, now play an important role in breeding programmes. Therefore, there is a greater need to develop novel methods to effectively collect and accurately measure all these traits. Chakraborty et al. (2022) stated that it is now possible to collect and access large and complex datasets comprising different genomics, transcriptomics, proteomics, metabolomics and phonemics data as well as animal-level data (such as longevity, behaviour, adaptation, etc.), which provide tools for rapid and effective detection of subtle phenotypic changes, dietary responses and innate phenotypic propensities in animals as well as new opportunities to better understand the mechanisms regulating animals' actual performance. However, all these omics technologies generate huge amounts of complex data (Big Data) that are difficult to handle by traditional data management tools and it is not easy to extract valuable information from them. Proper handling of omics data and a multidisciplinary team with advanced knowledge of statistics, computer science and bioinformatics are prerequisites for the adequate utilisation and interpretation of omics technology (Riggs et al., 2017).

14 Conclusion

Animal breeding programmes are extremely powerful in improving the genetic resource in animal production systems, thereby contributing to long-term sustainability in the sector. Defining clear breeding objectives relevant to a long-term sustainability goal is a key factor, in which improving productivity is still a major component, especially for a longer productive life. Breeding goals can be further fine-tuned, including more emphasis on traits related to animal health and their efficiency and environmental impact. New technologies such as genomics and reproductive technologies can be useful, but only where there is already an existing breeding structure to successfully achieve genetic improvement.

Genetic improvements will continue to be a major contributor to the future development of the poultry industry and will focus on the health and welfare of the birds as well as the requirements of producers, retailers, consumers and legislation. This will produce maximum benefits for a high amount of good-quality food at a reasonable price, better food safety, animal health and welfare, whilst making efficient utilisation of natural resources and reducing the environmental impact of the poultry industry.

Market demands are mainly differentiated by the requirements for adaptability to different local conditions, as well as egg size and eggshell colour. Moreover, breeding companies are in constant communication with egg producers and hatchery companies to understand the realities of their environment in order to recognise potential changes in the market and to be well prepared to deliver future solutions. Great efforts have been made to develop well-dimensioned and internationally distributed production hubs. This production network is critical to ensure the supply of genetic material anytime and anywhere in the world, in the event of a possible crisis that restricts or hinders trading between countries (e.g. outbreaks of avian influenza or events such as the COVID-19 crisis). Nowadays, a broad multidimensional breeding goal is applied, with increased emphasis on biological efficiency, reproduction, adaptability, welfare and robustness that focuses on sustainability, product quality and health (Neeteson et al., 2013). Breeding companies should support producers with the latest management and technical advice and demonstrate corporate responsibility whilst meeting global food demands.

Maintaining fluid communication with clients is paramount as it not only directs current selection objectives but also helps predict potential emerging market niches.

15 Where to look for further information

There are several scientific meetings that provide the appropriate forums to discuss animal production and genetic improvement such as the World's Congress on Genetics Applied to Animal Production (WCGALP - http://www.wcgalp.org).

The World's Poultry Science Association (WPSA) holds regular meetings, including the European Poultry Conference and the World Poultry Congress. Furthermore, several Working Groups (WG) of the WPSA such as WG3 'Breeding and Genetics' and WG6 'Incubation and Fertility Research Group' organise additional meetings and are a good source of information and novel research on poultry genetics and incubation.

Further readings reviewing poultry breeding, genetics and incubation can be found in the following books:

Hocking, P. M. (Ed.) (2009). Biology of breeding poultry. DOI: 10.1079/9781845933753.0000.

Aggrey, S. E., Zhou, H., M. Tixier-Boichard and Rhoads, D. D. (Ed.) (2020). Advances in poultry genetics and genomics. Burleigh Dodds Science Publishing Limited. DOI: 10.19103/AS.2020.0065.

16 References

Almeida, J. G., Vieira, S. L., Reis, R. N., Berres, J., Barros, R., Ferreira, A. K. and Furtado, F. V. F. (2008). Hatching distribution and embryo mortality of eggs laid by broiler breeders of different ages. *Rev. Bras. Cienc. Avic.* 10(2): 89–96.

Anang, A., Mielenz, N. and Schüler, L. (2002). Monthly model for genetic evaluation of laying hens. II. Random regression. *Br. Poult. Sci.* 43(3): 384–390.

Anderson, K. E., Havenstein, G. B., Jenkins, P. K. and Osborne, J. (2013). Changes in commercial laying stock performance, 1958–2011: Thirty-seven flocks of the North Carolina random sample and sequential layer performance and management test. *Worlds Poult. Sci. J.* 69(3): 489–514.

Assersohn, K., Brekke, P. and Hemmings, N. (2021). Physiological factors influencing female fertility in birds. *R. Soc. Open Sci.* 8(7): 202274.

Avendaño, S., Neeteson, A. M. and Fancher, B. (2017). Broiler breeding for sustainability and welfare – Are there trade-offs? In: *Proceedings of the New Zealand Poultry Beyond 2023 Conference*, Queenstown.

Bacon, L. D. and Witter, R. L. (1993). Influence of B-haplotype on the relative efficacy of Marek's disease vaccines of different serotypes. *Avian Dis.* 37(1): 53–59.

Bain, M. M., Dunn, I. C., Wilson, P. W., Joseph, N., De Ketelaere, B., De Baerdemaeker, J. and Waddington, D. (2006). Probability of an egg cracking during packing can be predicted using a simple non-destructive acoustic test. *Br. Poult. Sci.* 47(4): 462–469.

Blanco, A. E., Icken, W., Ould-Ali, D., Cavero, D. and Schmutz, M. (2014). Genetic parameters of egg quality traits on different pedigree layers with special focus on dynamic stiffness. *Poult. Sci.* 93(10): 2457–2463.

Bakst, M. R. (2014). Chapter 2: Determination of sperm concentration. In: Bakst, M. R. and Long, J. A. (Eds.), *Techniques for Semen Evaluation, Semen Storage, and Fertility Determination* (2nd edn.). Buffalo, MN: The Midwest Poultry Federation.

Barnett, D. M., Kumpula, B. L., Petryk, R. L., Robinson, N. A., Renema, R. A. and Robinson, F. E. (2004). Hatchability and early chick growth potential of broiler breeder eggs with hairline cracks. *J. Appl. Poult. Res.* 13(1): 65–70.

Beaumont, C., Millet, N., Le Bihan-Duval, E., Kipi, A. and Dupuy, V. (1997). Genetic parameters of survival to the different stages of embryonic death in laying hens. *Poult. Sci.* 76(9): 1193–1196.

Becot, L., Bedere, N., Burlot, T., Coton, J. and Le Roy, P. (2021). Nest acceptance, clutch, and oviposition traits are promising selection criteria to improve egg production in cage-free system. *PLoS ONE* 16(5): e0251037.

Bedere, N., Berghof, T. V. L., Peeters, K., Pinard-van der Laan, M. H., Visscher, J., David, I. and Mulder, H. A. (2022). Using egg production longitudinal recording to study the genetic background of resilience in purebred and crossbred laying hens. *Genet. Sel. Evol.* 54(1): 26.

Bennet, C. D. (1992). The influence of shell thickness on hatchability in commercial broiler breeder flocks. *J. Appl. Poult. Res.* 1(1): 61–65.

Bennewitz, J., Morgades, O., Preisinger, R., Thaller, G. and Kalm, E. (2007). Variance component and breeding value estimation for reproductive traits in laying hens using a Bayesian threshold model. *Poult. Sci.* 86(5): 823–828.
Berghof, T. V. L., Poppe, M. and Mulder, H. A. (2018). Opportunities to improve resilience in animal breeding programs. *Front. Genet.* 9: 692.
Bongalhardo, D., Dionello, N. and Ledur, M. (2000). Genetic parameters for semen traits in a white Leghorn strain. 1. Heritabilities and correlations. *Rev. Bras. Zootec.* 29: 1320–1326.
Bouba, I., Visser, B., Kemp, B., Rodenburg, T. B. and van den Brand, H. (2021). Predicting hatchability of layer breeders and identifying effects of animal related and environmental factors. *Poult. Sci.* 100(10): 101394.
Brillard, J. P., McDaniel, G. R., De Reviers, M. and Drane, J. W. (1989). Expression of several traits of fertility in young and old dwarf broiler breeder hens inseminated with duplicate doses of semen. *Poult. Sci.* 68(4): 558–563.
Brillard, J. P. (2003). Practical aspects of fertility in poultry. *Worlds Poult. Sci. J.* 59(4): 441–446.
Brundtland, G. H. (1987). *Our Common Future: Report of the World Commission on Environment and Development. Geneva, UN-Document A/42/427.* Oxford, UK: Oxford University Press.
Calus, M. P. L., Bos, J. A., Duenk, P. and Wientjes, Y. C. J. (2020). The purebred-crossbred correlation in broilers and layers: A review. In: *Proceedings of the of the 71st Annual Meeting of the European Federation of Animal Science (EAAP), Online Virtual Meeting*, p. 425.
Cavero, D., Schmutz, M. and Preisinger, R. (2011). Genetic evaluation of pure-line and cross-line performance in layers. *Lohmann Inf.* 45: 18–21.
Chakraborty, D., Sharma, N., Kour, S., Sodhi, S. S., Gupta, M. K., Lee, S. J. and Son, Y. O. (2022). Applications of omics technology for livestock selection and improvement. *Front. Genet.* 13: 774113.
Chambers, J. R. (1990). Chapter 25: Genetics of growth and meat production in chickens. In: Crawford, R. D. (Ed.), *Poultry Breeding and Genetics*. Amsterdam, The Netherlands: Elsevier, pp. 599–643.
Chen, C. F. and Tixier-Boichard, M. (2003). Correlated responses to long-term selection for clutch length in dwarf brown-egg layers carrying or not carrying the naked neck gene. *Poult. Sci.* 82(5): 709–720.
Cooper, J. J. and Appleby, M. C. (1996). Individual variation in prelaying behaviour and the incidence of floor eggs. *Br. Poult. Sci.* 37(2): 245–253.
Damaziak, K., Koznaka-Lipka, M., Gozdowski, D., Goebiowska, A. and Kedziorek, E. K. (2021). Effects of broiler breeder strain, age, and eggs preheating profile in single-stage systems on the hatchability of eggs and quality of chicks. *Animal* 15(1): 100057.
Dekkers, J. C. M. (2007). Marker-assisted selection for commercial crossbred performance. *J. Anim. Sci.* 85(9): 2104–2114.
De Reviers, M. (1992). Chapter 7: Aparato genital masculino y producción de espermatozoides. In: Sauveur, B. and De Reviers, M. (Eds.), *Reproducción de las Aves*. Translated by C. Buxadé Carbo. Madrid, España: Mundi Prensa, pp. 91–238.
Duncan, I. J. H. (2009). Chaper 8: Mating behaviour and fertility. In: Hocking, P. M. (Ed.), *Biology of Breeding Poultry*. Poultry Science Symposium Series Vol. 29. Wallingford, UK: CABI Publishing, pp. 111–132.

Duncan, I. J. H., Hocking, P. M. and Seawright, E. (1990). Sexual behaviour and fertility in broiler breeder domestic fowl. *Appl. Anim. Behav. Sci.* 26: 201–213.

European Commission. (2022). EU market situation for Eggs from 19 May 2022. Available at: https://ec.europa.eu/info/food-farming-fisheries/animals-and-animal-products/animal-products/eggs_en. Accessed on 14 June 2022.

Fairfull, R. W., McMillan, L. and Muir, W. M. (1998). Poultry breeding: Progress and prospects for genetic improvement of egg and meat production. In: *Proceedings of the of the World Congress on Genetics Applied to Livestock Production*, Armidale Australia, pp. 271–278.

Falconer, D. S. and Mackay, T. F. C. (1996). *Introduction to Quantitative Genetics* (4th edn.). Essex, UK: Longman.

FAO. (2022). *FAO Database*. Available at: www.faostat.org. Accessed on 24 May 2022.

Flock, D. K. (1977). Genetic analysis of part-period egg production in a population of White Leghorns under long-term RRS. *Z. Tierzuecht. Züchtungsbiol.* 94: 89–103.

Flock, D. K., Ameli, H. and Glodek, P. (1991). Inbreeding and heterosis effects on quantitative traits in a white leghorn population under long-term reciprocal recurrent selection. *Br. Poult. Sci.* 32(3): 451–462.

Flock, D. K. and Heil, G. (2002). A long-term analysis of time trends in the performance profile of white-egg and brown-egg hybrid laying strains based on results of official German random sample tests from 1974/75 to 1997/9. *Archiv Geflügelk* 66: 1–20.

Förster, A., Kalm, E. and Flock, D. K. (1992). Schlupffähigkeit aus züchterischer Sicht. *Lohmann Inf.* (Sept./Okt.): 13–16.

Förster, A. (1994). Optimierung der Schlupfrate und Kükenqualität - biologische Zusammenhänge und Bruttechnik. *Lohmann Inf.* (Mai/August): 11–16.

Fröhlich, E. K. F., Niebuhr, K., Schrader, L. and Oester, H. (2012). Chapter 1: What are alternative systems for poultry? In: Sandilands, V. and Hocking, P. (Eds.), *Alternative Systems for Poultry - Health, Welfare and Productivity.* Wallingford, UK: CABI Publishing, pp. 1–22.

Froman, D. P., Feltmann, A. J., Rhoads, M. L. and Kirby, J. D. (1999). Sperm mobility: A primary determinant of fertility in the domestic fowl (*Gallus domesticus*). *Biol. Reprod.* 61(2): 400–405.

Gowe, R. S., Fairfull, R. W., Mcmillan, I. and Schmidt, G. S. (1993). A strategy for maintaining high fertility and hatchability in a multiple-trait egg stock selection program. *Poult. Sci.* 72(8): 1433–1448.

Grochowska, E., Kinal, A., Sobek, Z., Siatkowski, I. and Bednarczyk, M. (2019). Field study on the factors affecting egg weight loss, early embryonic mortality, hatchability, and chick mortality with the use of classification tree technique. *Poult. Sci.* 98(9): 3626–3636.

Hartmann, W. (1992). Evaluation of potentials of new scientific developments for commercial poultry breeding. *Worlds Poult. Sci. J.* 48(1): 17–27.

Hartmann, C., Strandberg, E., Rydhmer, L. and Johansson, K. (2002). Genetic relations between reproduction, chick weight and maternal egg composition in a white leghorn line. *Acta Agric. Scand. A.* 52(2): 91–101.

Hartmann, C., Johansson, K., Strandberg, E. and Rydhmer, L. (2003). Genetic correlations between the maternal genetic effect on chick weight and the direct genetic effects on egg composition traits in a White Leghorn line. *Poult. Sci.* 82(1): 1–8.

Havenstein, G. B. (2006). Performance changes in poultry and livestock following 50 years of genetic selection. *Lohmann Inf.* 41: 30-37.

Hill, W. G. (2008). Estimation, effectiveness and opportunities of long-term genetic improvement in animals and maize. *Lohmann Inf.* 43: 3-20.

Hill, W. G., Wolc, A., O'Sullivan, N. P. and Avendaño, S. (2016). Chapter 11: Breeding for sustainability: Maintaining and enhancing multi-trait genetic improvement. In: Europe, E. B., Gatcliffe, J., O'Neill, H. M. and Scholey, D. (Eds.), *Sustainable Poultry Production*. Wallingford, UK: CABI Publishing, pp. 193-213.

Hocking, P. M. (2014). Unexpected consequences of genetic selection in broilers and turkeys: Problems and solutions. *Br. Poult. Sci.* 55(1): 1-12.

Hu, J., Chen, J. L., Wen, J., Zhao, G. P., Zheng, M. Q., Liu, R. R., Liu, W. P., Zhao, L. H., Liu, G. F. and Wang, Z. W. (2013). Estimation of the genetic parameters of semen quality in Beijing-You chickens. *Poult. Sci.* 92(10): 2606-2612.

Ibarburu, M. (2022). *U. S. Flock Trends and Projection Report From 7 June 2022*. Egg Industry Center.

Icken, W., Cavero, D., Schmutz, M., Thurner, S., Wendl, G. and Preisinger, R. (2008). Analysis of time interval within laying sequences in a transponder nest. In: *Proceedings of the XXII World Poultry Congress*, Brisbane, Australia.

Icken, W., Cavero, D., Schmutz, M. and Preisinger, R. (2012). New phenotypes for new breeding goals in layers. *Worlds Poult. Sci. J.* 68(3): 387-400.

Icken, W., Thurner, S., Heinrich, A., Kaiser, A., Cavero, D., Wendl, G., Fries, R., Schmutz, M. and Preisinger, R. (2013). Higher precision level at individual laying performance tests in non-cage housing systems. *Poult. Sci.* 92(9): 2276-2282.

Jones, D. R., Andersson, K. E. and Davis, G. S. (2001). The effects of genetic selection on production parameters of single comb White Leghorn hens. *Poult. Sci.* 80(8): 1139-1143.

Koerhuis, A. N. M. and McKay, J. C. (1996). Restricted maximum likelihood estimation of genetic parameters for egg production traits in relation to juvenile body weight in broiler chickens. *Livest. Prod. Sci.* 46(2): 117-127.

Kranis, A., Gheyas, A. A., Boschiero, C., Turner, F., Yu, L., Smith, S., Talbot, R., Pirani, A., Brew, F., Kaiser, P., Hocking, P. M., Fife, M., Salmon, N., Fulton, J., Strom, T. M., Haberer, G., Weigend, S., Preisinger, R., Gholami, M., Qanbari, S., Simianer, H., Watson, K. A., Woolliams, J. A. and Burt, D. W. (2013). Development of a high density 600K SNP genotyping array for chicken. *BMC Genomics.* 14: 59.

Kranis, A. and Maniatis, G. (2020). Chapter 14: Application of genomic selection (GS) in breeding commercial meat-type chickens. In: Aggrey, S. E., Zhou, H., Tixier-Boichard, M. and Rhoads, D. (Eds.), *Advances in Poultry Genetics and Genomics*. Cambridge, UK: Burleigh Dodds Science Publishing Limited, pp. 385-402.

Laughlin, K. F. (2009). Chapter 2: Breeder management: How did we get here? In: Hocking, P. (Ed.), *Biology of Breeding Poultry*. Wallingford, UK: CABI Publishing, pp. 10-25.

Ledur, M. C., Fairfull, R. W., McMillan, I., Gowe, R. S. and Asseltine, L. (2000). Genetic effects of ageing on fertility and hatchability in the first laying cycle of three White Leghorn strains and their two-way crosses. *Br. Poult. Sci.* 41(5): 552-561.

Liptoi, K. and Hidas, A. (2006). Investigations of possible genetic background of early embryonic mortality in poultry. *Worlds Poult. Sci. J.* 62(2): 326-337.

Luo, P. T., Yang, R. Q. and Yang, N. (2007). Estimation of genetic parameters for cumulative egg numbers in a broiler dam line by using a random regression model. *Poult. Sci.* 86(1): 30-36.

Makanjuola, B. O., Olori, V. E. and Mrode, R. A. (2021). Modeling genetic components of hatch of fertile in broiler breeders. *Poult. Sci.* 100(5): 101062.

McGary, S., Estevez, I., Bakst, M. R. and Pollock, D. L. (2002). Phenotypic traits as reliable indicators of fertility in male broiler breeders. *Poult. Sci.* 81(1): 102-111.

McKay, J. C. (2009). Chapter 1: The genetics of modern commercial poultry. In: Hocking, P. (Ed.), *Biology of Breeding Poultry*. Wallingford, UK: CABI Publishing, pp. 3-9.

Meijerhof, R. (1992). Pre-incubation holding of hatching eggs. *Worlds Poult. Sci. J.* 48(1): 57-68.

Meuwissen, T. H. E., Hayes, B. J. and Goddard, M. E. (2001). Prediction of total genetic value using genome-wide dense marker maps. *Genetics* 157(4): 1819-1829.

Meuwissen, T. H. E., Hayes, B. J. and Goddard, M. E. (2016). Genomic selection: A paradigm shift in animal breeding. *Anim. Front.* 6(1): 6-14.

Mohan, J., Sharma, S., Kolluri, G. and Dhama, K. (2018). History of artificial insemination in poultry, its components and significance. *Worlds Poult. Sci. J.* 74: 1-14.

Narushin, V. G. and Romanov, M. N. (2002). Egg physical characteristics and hatchability. *Worlds Poult. Sci. J.* 58(3): 297-303.

Neeteson, A. M. (2010). Striking the balance. *J. Anim. Breed. Genet.* 127(2): 85-86.

Neeteson, A. M., Knap, P. and Avendaño, S. (2013). The role of sustainable commercial pig and poultry breeding for food security. *Anim. Front.* 3(1): 52-57.

Neeteson, A. M., Avendaño, S. and Koerhuis, A. (2020). Chapter 6: Poultry breeding for sustainability and welfare. In: Ahmadi, B. V., Moran, D. and D'Eath, R. (Eds.), *The Economics of Farm Animal Welfare: Theory, Evidence and Policy*. Wallingford, UK: CABI Publishing, pp. 117-146.

Ni, G., Cavero, D., Fangmann, A., Erbe, M. and Simianer, H. (2017). Whole-genome sequence-based genomic prediction in laying chickens with different genomic relationship matrices to account for genetic architecture. *Genet. Sel. Evol.* 49(1): 8.

Nicholson, D., French, N., Tullet, S., van Lierde, E. and Jun, G. (2013). Short periods of incubation during egg storage - SPIDES. *Lohmann Inf.* 48: 51-61.

Nurgiartiningsih, V. M. A., Mielenz, N., Preisinger, R., Schmutz, M. and Schüler, L. (2004). Estimation of genetic parameters based on individual and group mean records in laying hens. *Br. Poult. Sci.* 45(5): 604-610.

Peebles, E. D., Doyle, S. M., Zumwalt, C. D., Gerard, P. D., Latour, M. A., Boyle, C. R. and Smith, T. W. (2001). Breeder age influences embryogenesis in broiler hatching eggs. *Poult. Sci.* 80(3): 272-277.

Pelletier, N., Ibarburu, M. and Xin, H. (2014). Comparison of the environmental footprint of the egg industry in the United States in 1960 and 2020. *Poult. Sci.* 93(2): 241-255.

Pérez-Enciso, M. and Steibel, J. P. (2021). Phenomes: The current frontier in animal breeding. *Genet. Sel. Evol.* 53(1): 22.

Preisinger, R. (2018). Innovative layer genetics to handle global challenges in egg production. *Br. Poult. Sci.* 59(1): 1-6.

Preisinger, R. (2021). Commercial layer breeding: Review and forecast. *Züchtungskunde.* 93: 210-228.

Riggs, P. K., Tedeschi, L. O., Turner, N. D., Braga-Neto, U. and Jayaraman, A. (2017). The role of "omics" technologies for livestock sustainability. *Arch. Prod. Anim.* 25(3-4): 149-155.

Rozempolska-Rucinska, I., Zieba, G. and Lukaszewicz, M. (2009). Hatchability traits as selection criteria in breeding of laying hens. *Arch. Geflügelk.* 73: 263-267.
Rozempolska-Rucinska, I., Zieba, G., Lukaszewicz, M., Ciechonska, M., Witkowski, A. and Slaska, B. (2011). Egg specific gravity in improvement of hatchability in laying hens. *J. Anim. Feed Sci.* 20(1): 84–92.
Sapp, R. L., Rekaya, R., Misztal, I. and Wing, T. (2004). Male and female fertility and hatchability in chickens: A longitudinal mixed model approach. *Poult. Sci.* 83(8): 1253–1259.
Settar, P., Arango, J. and Arthur, J. (2006). Evidence of genetic variability for floor and nest egg laying behavior in floor pens. In: *Proceedings of the of the 12th European Poultry Conference in Verona*, Italy, pp. 1-3.
Sewalem, A. and Johansson, K. (2000). Egg weight and reproduction traits in laying hens: Estimation of direct and maternal genetic effects using Bayesian approach via Gibbs sampling. *Anim. Sci.* 70(1): 9–16.
Siegel, P. B. and Dunnington, E. A. (1985). Chapter 6: Reproductive complications associated with selection for broiler growth. In: Hill, W. G., Manson, J. H. and Hewitt, D. (Eds.), *Poultry Genetics and Breeding*. Harlow, UK: British Poultry Science, pp. 59–72.
Sitzenstock, F., Ytournel, F., Sharifi, A. R., Cavero, D., Täubert, H., Preisinger, R. and Simianer, H. (2013). Efficiency of genomic selection in an established commercial layer breeding program. *Genet. Sel. Evol.* 45(1): 29.
Szwaczkowski, T., Wezyk, S., Piotrowski, P. and Cywa-Benko, K. (2000). Direct and maternal genetic and environmental effects for fertility and hatchability in laying hens. *Arch. Geflügelk.* 64: 115–120.
Wei, M. and van der Werf, J. H. J. (1994). Maximizing genetic response in crossbreds using both purebred and crossbred information. *Anim. Sci.* 59(3): 401–413.
Wen, C., Mai, C., Cai, R., Gou, Q., Zhang, B., Li, J., Sun, C. and Yang, N. (2022). Inheritance of the duration of fertility in chickens and its correlation with laying performance. *Genet. Sel. Evol.* 54(1): 41.
Wishart, G. L. (2009). Chaper 10: Semen quality and semen storage. In: Hocking, P. M. (Ed.), *Biology of Breeding Poultry*. Poultry Science Symposium Series Vol. 29. Oxfordshire, UK: CABI Publishing, pp151–178.
Wilson, H. R. (1991). Interrelationships of egg size, chick size, posthatching growth and hatchability. *Worlds Poult. Sci. J.* 47(1): 5–20.
Wilson, H. R. (1997). Effects of maternal nutrition on hatchability. *Poult. Sci.* 76(1): 134–143.
Wolc, A., Lisowski, M. and Szwaczkowski, T. (2007). Genetic evaluation of laying hens under fixed regression animal models. *Arch. Anim. Breed.* 50(3): 279-287.
Wolc, A., White, I. M. S., Olori, V. E. and Hill, W. G. (2009). Inheritance of fertility in broiler chickens. *Genet. Sel. Evol.* 41(1): 47.
Wolc, A., White, I. M., Hill, W. G. and Olori, V. E. (2010). Inheritance of hatchability in broiler chickens and its relationship to egg quality traits. *Poult. Sci.* 89(11): 2334–2340.
Wolc, A., Kranis, A., Arango, J., Settar, P., Fulton, J. E., O'Sullivan, N. P., Avendano, A., Watson, K. A., Hickey, J. M., de los Campos, G., Fernando, R. L., Garrick, D. J. and Dekkers, J. C. M. (2016). Implementation of genomic selection in the poultry industry. *Anim. Front.* 6(1): 23–31.
Wolc, A., Jankowski, T., Arango, J., Settar, P., Fulton,J. E., O'Sullivan, N. P. and Dekkers, J. C. M. (2019). Investigating the genetic determination of clutch traits in laying hens. *Poult. Sci.* 98(1): 39–45.

Chapter 2

Managing breeder poultry flocks to optimise hatchability and chick health

R. A. van Emous, Wageningen Livestock Research, The Netherlands

1 Introduction

The main goal of broiler breeder management is to produce fertile, good quality and clean eggs (van Emous et al., 2020). A non-fertile broiler breeder egg is basically a table egg that holds little value for the breeder manager. Fertility declines during the second half of the laying phase. This is not just due to males; hens are also responsible for this phenomenon (Fontana et al., 1992; Bramwell et al., 1996; Chotesangasa, 2001). A maximum number of hatching eggs in combination with high fertility (and hatchability) are the most important traits for the economic profitability of a breeder flock. An improvement of 10 chicks per hen housed on a farm with 20 000 broiler breeders and a hatching egg price of €0.20 means: €40 000 increase in profit per cycle (van Emous et al., 2020). Egg fertility is primarily influenced by optimal mating behaviour between males and females (van Emous, 2010). Mating behaviour and egg fertility are in turn affected by housing system, male-to-female ratio, nutrition, (intra)spiking males, female and male body weight (BW), grading males, litter quality and feather cover (van Emous, 2010). While maximum first-class fertile

http://dx.doi.org/10.19103/AS.2022.0118.02

egg production is very important for breeder farmers, the potential impact of egg quality on chick quality and subsequent broiler performance and health can have a much larger economic impact (Decuypere and Bruggeman, 2007). A good quality broiler chicken starts with a first-grade hatching egg (Tona et al., 2005). A wide range of factors influences the fertility and hatchability of breeder eggs (Heier and Jarp, 2001; van Emous, 2007; Yassin et al., 2008) and many of these factors also affect the quality and health of the offspring (Yassin et al., 2009). First-week mortality in broilers is affected by egg storage, season, breed, feed company, year, breeder age and hatchery (Yassin et al., 2009). Because of the relationship between embryonic mortality and chick quality (Reijrink et al., 2010; Molenaar et al., 2011; van Emous et al., 2015), a major part of this chapter describes the different management issues that affect fertility and hatchability.

Because of the importance of mating behaviour on fertility and hatchability, the chapter begins with a description of natural mating behaviour and mating behaviour in broiler breeder houses. The chapter then describes the different issues affecting mating behaviour, fertility, hatchability and chick quality.

2 Mating behaviour

2.1 Natural mating behaviour

Domestic chickens and the Red Jungle Fowl (the wild progenitor of the domesticated chicken) have a harem polygynous mating system with one dominant male maintaining a territory (Collias and Collias, 1996). In dense vegetation Red Jungle Fowl form small harems up to four hens; however, they form bigger groups in a more open area (Collias et al., 1966; Collias and Collias, 1967). A dominant male among feral domestic fowl has the monopoly on matings in the group of 4–12 females throughout the mating season (McBride et al., 1969). Because males and females under natural conditions have separate social hierarchies, and males dominate females passively, aggressive behaviour to hens by mature males is uncommon (Wood-Gush, 1958; Kruijt, 1964; Rushen, 1983). Under natural circumstances, a female Red Jungle Fowl mates with multiple males (Ligon and Zwartjes, 1995). Females use a wide range of physical characteristics such as comb size and colour, eye colour and spur length for male selection prior to approach and copulation (Zuk et al., 1990, 1995). The size and colour of the comb seem to be the most important ornament for the females for selection. Comb size and colour are signals of male fitness (Zuk et al., 1990), dominance (Johnsen et al., 2001), high testosterone content (Ligon et al., 1990), and fertility and testicular weight (McGary et al., 2002). In addition to male choice by females, female choice by males is also an important mechanism in the sexual behaviour of fowls (Cornwallis and Birkhead, 2007). Researchers observed in different studies that mate choice by the male is driven by body size and body mass of the female (Verrell, 1985; Bonduriansky, 2001; Byrne and Rice, 2006; Cornwallis and Birkhead, 2007). Moreover, Cornwallis and Birkhead (2007) found that female ornamentation (especially combs) influences male mating decisions. They observed that females with larger ornaments than others were more popular with higher quality males and obtained more sperm. They hypothesized that males may benefit by investing more sperm in females with larger ornaments as they were in better condition and produced heavier eggs.

When the preferred male is selected, different courtship behaviour containing a spectrum of stimulus-response patterns between males and females takes place (Kruijt, 1964; Fischer, 1975). Courtship behaviour of males is comprehensive with obtrusive postures (waltzing and high-step advance), feather spreading (feather ruffling and wing flapping) and vocalizations (tidbitting and crowing) (Kruijt, 1964). After courtship behaviour, copulation activity occurs. The female voluntarily crouches and the male approaches her from behind with his neck stretched upward and mounts the hen. After mounting, the male bends his legs and holds the neck feathers of the hen in his beak or pecks gently at the head of the female. The male tramples with his left and right foot in alternation and synchronically moves backwards and lowers his tail with sideways rapid movements until cloacal contact. The male has a small phallus that becomes engorged with lymph to form a copulatory organ with practically no penetration during mating (Kruijt, 1964). The hen everts her vagina during copulation, which helps to transfer the semen into the oviduct. After the final contact, the male releases the female's head feathers and dismounts (Kruijt, 1964).

2.2 Mating behaviour in a broiler breeder house

Under commercial conditions it has been shown that male broiler breeders may demonstrate high levels of aggression towards females, mainly during the performance of mating behaviour, whereas courtship behaviour was virtually absent before mating (Millman et al., 2000; Jones and Prescott, 2000; Millman and Duncan, 2000a,b; Jones et al., 2001; Duncan, 2009; de Jong et al., 2009). In a relatively old but still relevant study, sexual behaviour (mating, mounting and waltzing) of individual marked males was observed in a large pen (31 males and 300 females) and in a small arena (1 male and 6 females) (Mols, 1978; van Grunsven, 1978). The same males performed less mating (45% vs. 17%), less mounting (45% vs. 14%) and more waltzing (69% vs. 10%) in the arena compared to the pen (Mols, 1978; van Grunsven, 1978). They concluded that the same male in a bigger group showed much more mating and less courtship behaviour because they experience male-male competition.

Males and females become sexually mature between 18 weeks and 23 weeks of age and it is important that males and females are equally mature to prevent problems with over-mating (de Jong and Swalander, 2012; de Jong and van Emous, 2017). Over-mating and aggressive behaviour of males lead to fearful females which is possibly one of the major reasons why females tend to remain on the slatted area or hide in the nests instead of being in the litter area (where the majority of males are present) at the end of the light period (Millman et al., 2000). Birds showed sometimes severe wounds on the back, thigh, the back of the head and along the torso beneath the wings which was related to aggressive mating behaviour (Millman et al., 2000). Sexual behaviour of male broiler breeders has been described as rough, the males pecking or chasing females and forcing copulations (Millman et al., 2000; Jones and Prescott, 2000; Jones et al., 2001). This phenomenon of forced copulation is also reported in the study of Leone and Estévez (2008). They found that females stayed on the slatted area instead of the litter area, and found injured female broiler breeders during lay. They suggested that the cause of rough mating behaviour may be that males reach sexual maturity at a younger age than females. It is highly plausible that both aggressive or rough mating behaviour and too early maturation of males play an important role in injuries, fear and stress of females (de Jong and Guemene, 2011). A field study on 8 commercial broiler breeder farms focused on the sexual behaviour of males and females between 20 weeks and 28 weeks of age (de Jong et al., 2009). It was concluded that male behaviour towards females could be described as 'rough' but that male courtship behaviour also appeared to be incomplete. It was remarkable that females in general did not show crouching behaviour in response to approach by a male. Females often showed struggling behaviour during mating or tried to escape from the male. This incomplete mating behaviour of males and females may explain the low percentage of successful matings (44%) and the high percentage of forced matings (80%) (de Jong et al., 2009). They also found that courtship behaviour was almost absent before mating, confirming earlier experiments (Millman et al., 2000; Jones et al., 2001). Because females do not recognise male signals, the behaviour of the females may thus seem to be incomplete. In contrast, it has been shown (in different female broiler breeder lines) that female behaviour can evoke aggressive male behaviour. Feed restriction did not play a role in rough mating behaviour (Millman and Duncan, 2000a) but genetics may have an effect (Millman and Duncan, 2000b). It has been suggested that factors like separate rearing of males and females, large group sizes and high stocking density may also play a role in the development and recognition of courtship behaviour and proper responses to signals of the other sex (Kruijt, 1964; de Jong et al., 2009).

3 Housing system

The transition from the rearing to the production period involves transportation from the rearing to the production house, usually at a separate production farm (de Jong and Swalander, 2012). Males and females are reared separately and mixed at the beginning of the production period and, in Europe, usually arrive on the same day at the production farm (EFSA, 2010). The production period starts between 18 weeks and 22 weeks and lasts until 60–65 weeks, dependent on the performance of a breeder flock. Although broiler breeders are kept in cages during lay in some countries, in the vast majority of cases laying birds are on a littered floor. Floor house systems consist of a a litter area and a percentage of slatted flooring from which nests can be accessed. The proportion of littered flooring (30-60%) versus raised slatted flooring can differ between countries and regions, whilst materials and layout (e.g. wire with plastic coating, plastic or wooden flooring) may also differ.

Housing systems can be divided roughly into two different layouts: (1) breeder houses with a central litter floor or (2) houses with a central slatted floor. In general, 100% of the breeder houses in Europe are equipped with a central slatted floor with an automatic community nest in the middle of the slatted floor. In the rest of the world the majority (approx. 90%) of breeder houses are equipped with a central litter floor (between 30% and 40%) with manual or automatic nests on both slatted floors at the side walls. At the present time there are no studies that show any difference in performance between the two breeder house layouts. However, van Emous (2010) have postulated that there may be benefits of a central slatted layout due to the larger litter area split into two sections resulting in less male-to-male interactions disrupting mating behaviour

To increase fertility in central litter breeder houses, vertical panels (70 × 70 cm) were placed in commercial houses in a staggered pattern every 4.5 m in the central litter area (Leone and Estévez, 2008). In breeder houses with access to cover panels, egg production was improved by 2.1%, while maintaining better hatchability and fertility throughout the breeding cycle leads to higher chicken production (+4.5 chicks/female). The authors postulated that providing enrichment in the form of cover panels improved reproductive performance, most likely by increasing male mating opportunities and reducing female stress. These panels attracted females to the litter floor, thereby decreasing the competition for females.

The effect of flock size and house dimensions has been investigated in commercial breeder farms (van Emous, 2007). The author found that a relatively small breeder house (≤ 12.5 m width and ≤ 70 m length) separated in the middle of the house by nest boxes (flock size: 3.250 breeders) produced a 2% higher egg hatchability compared to larger houses. Research by Brake (1998), however, did not find any effect of group size between 2000 breeders and 8000 breeders on reproduction.

In a study by de Jong et al. (2011) broiler breeders were housed at a high (8.8 birds/m^2) or low (5.2 birds/m^2) stocking density till 60 weeks of age. During the rearing phase the reduction of stocking density had a small but significant effect on the general behaviour of the birds. At a lower stocking density, more hens were foraging and fewer hens were standing, and more males were walking. More successful matings were observed at a lower stocking density during production. Mating behaviour was improved at a lower stocking density: more matings were preceded by courtship behaviour, more matings were performed in which the hens showed crouching behaviour (thus less matings were forced), and less hens struggled during mating. During production, hens at a reduced stocking density produced more hatching eggs. In addition, fertility was higher in eggs from hens housed at a reduced stocking density and the number of day-old chicks per hen was higher for hens at a reduced stocking density. Due to the high investment costs of broiler breeder houses, in Northwest Europe, a relatively high stocking density is used to optimize the profitably per square meter.

A new housing system for broiler breeders called the Quality Time® Concept (QTC) was developed to improve sexual behaviour and reproduction (van Emous and Gunnink, 2011). In total four different experiments were carried out to investigate the effects of sex separation on reproduction, behaviour and feather cover. The idea of the QT house is that males are separated from females for 5 h a day using a separate feeding system and a moving fence to avoid aggressive male-to-female behaviour during the day and improve mating behaviour in the late afternoon. After a pilot experiment, two consecutive experiments were carried out in one commercial broiler breeder house with 15 000 birds. The house was divided into six compartments in the QTC compartments, 10% more voluntary and 20% more successful matings were observed (van Emous and Gunnink, 2011). This improved appropriate sexual behaviour resulted in better feather coverage between 37 weeks and 48 weeks of age. Separating males from females did not increase aggressive behaviour between the males themselves during the separation period. In the first flock no effect on fertility was found; however, in the second flock, average fertility was improved by 1.5% (van Emous and Gunnink, 2011). The first version (QT 1.0) was followed by an on-farm experiment (QT 2.0) with 9900 breeders with a separated male compartment at the rear of the house. The QT 2.0 system resulted in a very good separation of the females (less than 0.1% females in the male compartment) and increased hatchability by approx. 1–2% (R.A. van Emous, non-published data). Overall advantages of the QT house are that males can be easily locked up in the male pen for 2–3 weeks after transfer for excellent separated sex feeding. Besides this, grading second-grade males is easier when males are separated in one small compartment, while separating males and females at the end of lay is easier (van Emous, 2010), After these on-farm studies, an experiment was carried out to investigate the effects of the Quality Time 3.0 or so-called 'skip-a-day male-female mixing' system (van Emous and Mens, 2020). In this system, males and females were mixed every other day followed by a day without mixing. The study showed that skip-a-day male-female contact has no effect on reproduction results and behaviour. When the system is applied in practice, it can improve economic performance by 5-10% due to the higher number of females and lower number of males.

During the production period there is no manure removal from the litter area, which means that a manure-litter mixture remains for the whole laying period (Hocking, 2004). The majority (90%) of matings are performed in the litter area, making it the most important area of the house for sexual behaviour (van Emous, 2010). The more dominant males are located on the litter during daylight (van Emous, 2020, personal observation) which implies that litter quality is very important for optimal leg and footpad health (Kaukonen et al., 2016). The latter observed an increased prevalence of footpad lesions (also known as bumble foot) in aging broiler breeders, with severe lesion occurrence reaching a maximum of 64% at slaughter age, potentially resulting in impaired mating behaviour. It is concluded by Barbato (1999) and Wolanski et al. (2004) that body weight (BW) has a greater effect on footpad condition than age of the breeders. This was confirmed by Chung (2010) who observed a positive correlation between BW of broiler breeder females and foot pad quality.

3.1 Floor eggs prevention

First-class eggs are clean, not broken, without cracks and collected from the nest boxes and not from the floor (de Reu, 2006; Khabisi et al., 2012; van den Brand et al., 2016). In contrast, most floor eggs are dirty (Berrang et al., 1997), contain a surplus of bacteria on the eggshell (Berrang et al., 1997; de Reu et al., 2011; van den Brand et al., 2016), and are more often broken or cracked than clean nest eggs (de Reu, 2006). Compared to clean nest eggs, floor eggs with more bacteria on the eggshell are more at risk of bacterial penetration before and during the incubation process which affects chick health (Smeltzer et al., 1979). There is a clear relationship between the increased microbial infection rate of the yolk sac in floor eggs and higher embryonic and offspring mortality (Deeming et al., 2002). Moreover, eggs contaminated with bacteria can be a critical factor in Salmonella infection of broiler chicks, with negative effects on broiler and human health (Cox et al., 2000). van de Brand et al. (2016) ran two consecutive experiments to investigate the effect of floor eggs, washed floor eggs and clean nest eggs on incubation characteristics and performance in later life of broiler chickens. They concluded that floor or washed floor eggs resulted in lower fertility and hatchability compared to clean nest eggs. No

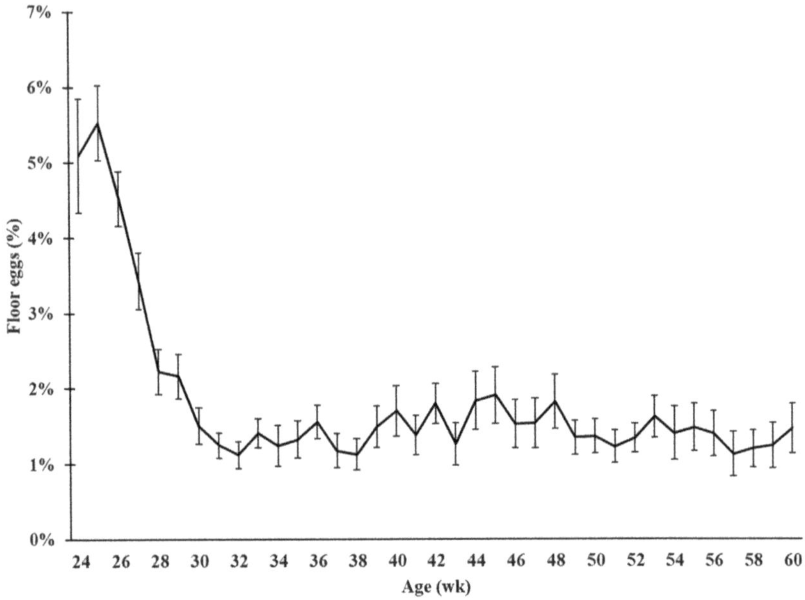

Figure 1 Effect of age on floor eggs (%) (van Emous, unpublished data).

effect was found on first-grade chicks; and performance of offspring was also not affected by using floor and washed floor eggs.

During the first 5 weeks of the laying period, the number of floor eggs can increase by 20% (Cooper and Appleby, 1996; Sheppard and Duncan, 2011). Due to improved management and housing systems, the percentage of floor eggs is between 2% and 6% during initial lay in commercial breeder houses; this decreases to between 1% and 2% from 30 weeks onwards (Fig. 1).

Incubating a relatively low percentage of floor eggs together with clean nest eggs can result in higher antibiotic use in the offspring due to increased infection amongst the day-old chicks (Lourens and Steentjes, 2008). Many management factors during the rearing and laying period have been identified to prevent floor eggs at the breeder farm (Aviagen, 2015). The three key management factors to prevent breeders from laying eggs on the floor are: (1) training: teach birds to use the nests, (2) environment: keep birds in the nests and (3) egg laying: removing alternative nesting sites, especially dark corners in the house.

4 Male-to-female ratio

4.1 Common male-to-female ratio practice

Breeder companies recommend the production period starts with 9–10% males (Aviagen, 2018; Cobb, 2018). This percentage decreases during the laying period due to grading (e.g. leg problems, BW loss) and mortality (de Jong and van Emous, 2017). Around 24 weeks of age 9% sexually mature active males are necessary for a good initial fertility, resulting in 8-8.5% first-grade males around 30 weeks of age (Aviagen, 2018; Cobb, 2018). From 40 weeks to 45 weeks of age, spiking, which is replacing 1-2% inactive males by young active males around 25 weeks of age, is often applied (Leeson and Summers, 2009). Despite spiking, the overall male percentage declines roughly from 9% at the start to 7% males at the end of the production period. Objectively, this decline in physical and physiological characteristics of breeder males is surprising (van Emous, 2010). This means that, besides the decreased male-to-female ratio, the overall quality of males declines with potentially far-reaching effects on fertility and hatchability. It is known from experiments that male aging decreases mating activity (Duncan et al., 1990; Casonovas, 2000), percentage of successful matings (Duncan et al., 1990), ejaculate quality (e.g. Casanovas, 2002; Zhang et al., 1999), sperm penetration ability (e.g. Barna et al., 2007) and testicular size (Sarabia Fragoso et al., 2013).

Number of matings per male per hour was approx. two and half times lower at 58 compared to 28 weeks of age and successful matings decreased in the same period from 40% to 35% (Duncan et al., 1990). Mating activity decreased with aging, caused by the higher male BW (McDaniel and Craig, 1959; Hocking and Duff, 1989) and decreased testosterone levels (Sexton et al., 1989; Weil et al., 1999; Hocking and Bernard, 2000; Chung, 2010; Sarabia Fragoso et al., 2013). It is postulated that the decline in testosterone level is caused by testicular regression (Wolanski et al., 2004; Sarabia Fragoso et al., 2013) and reduced libido of the males (Duncan et al., 1990). The decrease in libido is caused by the correlation between testosterone level and courtship-sexual behaviour (Brougher et al., 2005) and completed matings (Jones and Mench, 1991).

Some experiments observed no effect of male age on semen volume (Wilson et al., 1979), sperm motility (Bilcik et al., 2005), sperm concentration (Wilson et al., 1979), or dead sperm percentage (Wilson et al., 1979; Sexton et al., 1989). Other studies however, found a decrease in sperm production (Wilson et al., 1979; Hocking and Bernard, 1997a; Zhang et al., 1999; Sarabia Fragoso et al., 2013), sperm concentration (Sexton et al., 1989; Hocking and Bernard, 1997a; Zhang et al., 1999; Casanovas, 2002), as well as on semen quality (Hocking, 1990; Casanovas, 2002) with age. Zhang et al. (1999) measured a 50% decrease in sperm volume and 8 times lower sperm concentration between 28 and 46 weeks of age. It has been shown that age is negatively correlated with sperm penetration ability (Bramwell et al., 1996; McGary et al., 2002; Barna et al., 2007). A study by McGary et al. (2002) showed that sperm number declined after peak production and that this was correlated with the decrease in fertility.

4.2 Adjusted male-to-female ratio

As described earlier, mating activity and success rate, ejaculation and sperm quality decrease as males age (Duncan et al., 1990; Zhang et al., 1999; Casonovas, 2000, 2002). For practical reasons the male-to-female ratio decreases during the laying phase (de Jong and van Emous, 2017) potentially resulting in a mutually reinforcing fertility decreasing effect. On the other hand, the relatively high number of males at the start of the laying period will lead to over-mating (van Emous, 2010; de Jong and van Emous, 2017; Aviagen, 2018). It has been estimated that males in a broiler breeder house mate daily with each hen which is up to ten times higher than chickens housed under natural conditions (van Emous, 2010). From a biological point of view, this excessive mating frequency is redundant because a large ejaculation results in females having sufficient sperm to fertilize eggs for 14 days (Warren and Kilpatrick, 1929; Etches, 1996; Pizzari et al., 2004). This excessive mating behaviour is probably counterproductive because successive matings result in semen accumulation in the hen's storage tubes close to the cloaca, resulting in increasing resistance to mating (Løvlie and Pizari, 2007). The relatively high frequency of mating may affect the relationship between the males and females, resulting in females avoiding males as hypothesized by Fontana et al. (1992) and van Emous (2010). This phenomenon is reinforced by the aggressive behaviour of the males towards females, mainly during the performance of mating behaviour, with courtship behaviour virtually absent before mating (e.g. Millman and Duncan, 2000a,b; Millman et al., 2000; de Jong et al., 2009). This incomplete mating behaviour of males and females may explain the low percentage of successful matings (44%) and 80% forced matings between 20 weeks and 28 weeks of age, as found in the study of de Jong et al. (2009).

Male-to-female ratio in practice normally decreases from about 9–10% to 7–8% to the end of the production cycle with once or twice with (intra)spiking once or twice between 40 weeks and 50 weeks of age. An adjusted male-female ratio therefore been suggested by van Emous (2007) and Renema (2011). Starting with a low male-to-female ration (6-7%), and adding extra males at several points during the laying period (e.g. +1% around 30 weeks, 40 weeks and 50 weeks of age), resulted in 7.5–8.5% males around 50 weeks of age (Renema, 2011). van Emous (2007) has suggested starting with an even lower male-to-female ratio (4-5%) and, by spiking three times, increasing to around 8% at 50 weeks of age. He advised transferring a 8-9% male-to-female ratio to the breeder farm , placing 4% in a separate male house and using these as reserve males, placing them in a 4-5 week period in the main flock after 30 weeks of age.

Experiments with different male-to-female ratios in poultry are very rare. A study with native Thai chickens was performed with different male-to-female ratios (from 1-7 to 1-16) and no effects on fertility were observed (Chotesangasa, 2001). Recently, an on-farm experiment in a Veranda cage system (660 females per cage) has started to investigate the effects of a decreasing male-to-female ratio (from 7.5% to 6.0% at 60 weeks of age) compared to an increased male-to-female ratio (from 4.0% to 7.0% at 50 weeks of age). Preliminarily results show that mating behaviour and feather cover improved (van Emous, 2021, unpublished data). Moreover, fertility was higher from 44 weeks onwards in cages with increased male percentage compared to cages with decreasing male-to-female ratio.

5 Spiking

5.1 Spiking young males

Fertility of hatching eggs peaks around 35 weeks of age (Aviagen, 2018; Cobb, 2018); however, fertility rapidly declines after roughly 45 weeks of age (Casanovas, 2000). To prevent this significant decrease in fertility during the second phase of the production period, farmers use different 'spiking' methods (Végi et al., 2013). It is common practice in many countries to remove a certain percentage of old and inactive males from the flock and replace them with younger (25–28 weeks of age) and more active males (Casanovas, 2002; Wilson, 2002; Leeson and Summers, 2009). Spiking involves a high risk of introducing pathogens, and moreover, it may be stressful to the birds because male aggression may increase (EFSA, 2010). In practice, breeder flocks are transferred at around 20 weeks of age from the rearing to the breeder farm, and the surplus males are transferred to another breeder farm to be used as spike males (de Jong and van Emous, 2017). A special compartment or separate house for these spike males is necessary on a breeder farm to raise these young males to maturity from 20 weeks to 25–28 weeks of age with a minimum BW of 4 kg (van Emous, 2010). This male house must be equipped with the same feed system, water system, slats and light system as the main breeder houses to minimize the stress for the males when they are moved from the male to the breeder house (van Emous, 2010). The number of males to be replaced is dependent on the number of graded males due to illness, leg problems or poor condition but is between 2% and 33% of the old males (Casanovas, 2002).

Two spiking experiments were carried out with breeders in small pens (60 weeks of age) with 20%, 25%, and 33% spiking males and in a field trial (41 weeks of age) with 25% spiking males (28 weeks of age) (Casanovas, 2000). Both experiments showed that, after spiking, the old males showed higher mating activity. Moreover, they observed that the old males performed 87% of all matings even though they represented only 75% of the males present. The fertility in the experiment with small pens increased from 74% to 89% in only

15 days and the fertility in the field experiment stayed constant at around 95% during the next 9 weeks. In the first week after spiking they observed an increase in aggression and sexual interference between males. Mortality in males was 2.5 times higher after compared to before spiking, reducing to normal levels after 9 weeks post-spiking. A more extreme spiking program was applied by Ordas et al. (2015), replacing at 44 weeks of age 100% of the old males with young (30 weeks of age) males in a Ross 308 breeder flock. From 44 weeks of age onwards a constant hatchability level of around 84.5% was observed during the consecutive 6 weeks. The breeders without male replacement showed a decline in hatchability in the same period from 86.5% to less than 81%. In other studies, no effect on fertility/hatchability was found in replacing 10–100% of the old males by spiking males at 44 weeks of age (Nazarpak and Fallah-Khair, 2011; Chung, 2010; Végi et al., 2013).

5.2 Intra-spiking

Besides spiking younger males, intra-spiking is used in some countries as a technique to increase male activity and improve fertility with fewer biosecurity risks (Casanovas, 2000). Intra-spiking involves replacing 25-45% of males between 40 weeks and 45 weeks of age by moving them from one compartment or house to another (Casanovas, 2000; Mphepya et al., 2019). The abrupt change to a new environment, and introducing new individuals to a stable social group, can increase sexual behaviour and compensate for loss in libido and mating interest of the older roosters (Casanovas, 2002). It is postulated by Casanovas (2000) that intra-spiking is even or more effective than regular spiking in increasing mating activity of old males. This is probably due to the decreased propensity with aging of males to copulate with familiar females; however, this is renewed when males have access to unfamiliar females (Pizzari et al., 2003).

In the study of Casanovas (2000), two different on-farm experiments with Cobb PS were carried out with single intra-spiking at 45 weeks of age and double intra-spiking at 42 weeks and 50 weeks of age with 25% males at the different ages. He found an approximate 4% higher fertility in the 8 weeks after intra-spiking due to increased mating activity. He also observed increased aggression and sexual interference in the first 2 weeks after intra-spiking; however, mortality was not increased for both the original or spiked males. Mortality in the 50 weeks of age males in the double intra-spiking protocol showed higher mortality due to handling and aggression.

In another study with double intra-spiking of 25% males at 42 weeks and 52 weeks of age, an increased fertility level was found 6 weeks after the first intra-spiking; however, no effect was found after the second intra-spiking (Chung, 2010). She postulated that this was caused by the heat-stressed environment that was part of the experiment setup causing an extra environmental stressor which affected the results. It has been shown recently that the best reproduction results and lowest mortality rate can be achieved with the highest (45% vs. 35% and 25%) percentage of double (40 weeks and 48 weeks of age) intra-spiked males (Mphepya et al., 2019). Average fertility was 94.5%, 95.1% and 95.9% and average hatchability was 85.8%, 86.3% and 87.1%, respectively, for the 25%, 35% and 45% treatments. Male mortality was the lowest (0.23%) for the 45% group compared to the 35% (0.40%) and 25% (0.44%) group.

6 Nutrition

6.1 Crude protein/AA level rearing

Feeding high-yield breeders high levels of amino acids (e.g. lysine) will lead to more breast muscle and this extra muscle tissue requires more energy to maintain (de Beer, 2009). Therefore, during the last decade, several researchers have reported that broiler breeders need a certain (higher) proportion of body fat at the onset of lay for subsequent reproductive performance (Sun and Coon, 2005; de Beer, 2009; Mba et al., 2010; van Emous, 2015). On average, 4.6% lower daily crude protein (CP) intake and 11.4% higher daily energy intake during the rearing phase in the study reported by van Emous et al. (2013, 2015) resulted in 2.1% (16.9% vs 19.0%) lower breast meat and more than doubled (1.00% vs. 0.46%) abdominal fat pad content at onset of lay (20 weeks and 22 weeks of age, respectively). This was in close agreement with Mba et al. (2010) who found the same effects of a 12.5% reduction in CP content of the diet (14% vs. 16%) on body composition at 23 weeks of age. In the study of van Emous et al. (2015) the increased abdominal fat pad and decreased breast meat of the pullets fed the low protein diet resulted in a 1.3% and 1.1% increased hatchability of set and fertile eggs, respectively when compared to the control diet. The improved hatch was primarily due to a reduction in embryo mortality between day 10 and day 21 of incubation. An explanation for this phenomenon may be that egg composition might be affected by the differences in abdominal fat pad content of the body (Hocking et al. 2002). However, the first experiment of van Emous et al. (2013) did not find any effects on incubation traits when pullets received different dietary protein levels during rearing. Egg fertility in the study of van Emous et al. (2015) was not affected by different dietary protein levels during rearing. This is contrary to the findings of Walsh and Brake (1997, 1999) who found that a low CP intake during rearing could reduce fertility during lay. These studies, however, were conducted with broiler breeders two decades ago and may therefore not be comparable to the breeders in the study of van Emous et al. (2013). Current broiler breeders may show smaller effects on fertility due to differences in body compositions at start of lay.

6.2 Crude protein level males rearing

It is suggested by de Reviers and Seigneurin (1990), Brake and Peak (1999), and Romero-Sanchez et al. (2004) that broiler breeder males may need a minimum cumulative nutrient intake during rearing for optimal reproduction. Protein intake during the rearing phase seems to be an important macronutrient that affects male reproduction during the laying phase. Zhang et al. (1999) found a numerically higher semen concentration in caged males that were fed a 12% CP diet during the rearing phase compared with those that were fed a 16% CP diet. An experiment was performed to evaluate the carry-over effects on fertility of different male BW profiles (Concave or Sigmoid) and dietary CP levels (12% vs. 17%) during rearing (between 2 weeks and 26 weeks of age) (Romero-Sanchez et al., 2007). A 2.9% (91.8% vs 88.9%) higher fertility was found for the breeders fed the 12% compared to the 17% CP level during the rearing phase.

6.3 Crude protein level females laying phase

A number of authors reported negative effects of a high compared to a low daily dietary CP intake (27 vs. 23, 26 vs. 21, and 25 vs. 16 g/d) during the entire laying phase on fertility and hatchability of eggs (Pearson and Herron, 1982; Whitehead et al., 1985; Lopez and Leeson, 1995, respectively). Moreover, Pearson and Herron (1982) reported that a high compared to a low daily protein intake (27 vs. 23 g/d) during the laying phase, resulted in an increased mortality and malformation of embryos. van Emous et al. (2015) found a 1.5% higher hatchability of fertile eggs caused by 1.3% lower embryonic mortality when breeders were fed 19.4 versus 21.5 g CP per day, between 45 weeks and 60 weeks of age. The effect of improved hatchability and decreased embryonic mortality in the study of van Emous et al. (2015), due to the low daily protein intake, was underlined by 0.6% fewer second-grade chicks. The relationship between low embryonic mortality and less second-grade chicks was previously found in studies of Reijrink et al. (2010) and Molenaar et al. (2011).

A negative effect of higher levels of especially dietary lysine and isoleucine at peak production on fertility was found by Ekmay et al. (2013) in a dose-response study. They speculated that these amino acids affect the microenvironment of the sperm storage tubules which are located close to the utero-vaginal junction. The high dietary lysine and isoleucine levels results in a nitrogen excess which can be very toxic and must be expelled from the body. This process of breakdown and excretion most likely leads to differences in urine pH which directly affects the sperm host tubules environment (de Beer, 2009). On the contrary, no effect on fertility of feeding a high daily protein intake, during the first as well the second phase of the laying phase was found by Whitehead et al. (1985) and Mejia et al. (2012).

6.4 Vitamins and trace minerals laying phase

Deficiencies of vitamins and trace minerals in breeder diets have negative effects on chick quality and offspring performance (Kidd, 2003; Calini and Sirri, 2007; Hocking, 2007). In particular, deficiencies in fat-soluble vitamins D and E and the trace elements selenium, zinc and manganese have an important effect on chick quality, chick viability and skeletal development (Chang et al., 2016). A mistake by a premixes producer, who delivered feed premix with zero vitamin D (cholecalciferol) to several feed companies, resulted in a significant drop in egg production, soft-shelled eggs, eating floor eggs and reduced hatchability (A. Steentjes, 2020, personal communication). One broiler breeder flock received several weeks of feed with this particular premix which resulted in very poor hatchability and day-old chicks unable to open the egg with their (too soft) egg-tooth. Providing additional vitamin D via the drinking system, however, solved the problem and hatchability improved within a week.

For optimal progeny development, viability and carcass development, sufficient vitamin D3 intake in breeder feed is very important (Sunde et al., 1978). Atencio et al. (2005a) found a lower incidence of tibial dyschondroplasia in progeny of young (30 weeks and 37 weeks of age) receiving 4000 IU vitamin D3/kg instead of 2000 IU vitamin D3/kg; however, this was not observed in older breeders (45 weeks and 47 weeks of age). Supplementing a better biological value of vitamin D3 (25-OH-D3) to the breeder diets results in lower embryonic mortality and higher bone ash in progeny (Atencio et al., 2005b).

The immunity of offspring can be influenced by vitamin E by improving the adaptive antibody transfer from the breeder to the offspring (Boa-Amponsem et al., 2001). This effect of vitamin E in breeder diets on maternal immunity of the offspring was previously found by Haq et al. (1996) and Hossain et al. (1998). The breeders in a dose-response study of Hossain et al. (1998) received experimental diets containing 4 levels of supplementary vitamin E (25 mg/kg, 50 mg/kg, 75 mg/kg and 100 mg/kg) between 24 weeks and 54 weeks of age. They observed that antibody titers against Newcastle disease were increased with increasing levels of Vitamin E supplements in the diets. They concluded that increased Vitamin E levels in breeder diets results in an improved chick immune response. This was confirmed by Johnson (2013), who found that maternal dietary replacement of synthetic vitamin E with a natural vitamin E increased in-vitro E. coli bactericidal capacity of offspring from older (58 weeks of age) breeders, and oxidative burst in offspring from 46 weeks and 58 weeks of age breeders. Supplementing a broader range of vitamins (B,

E, and K at 20% above Aviagen breeder recommendations of 2007) resulted in a 2.2% lower mortality in the offspring probably caused by an improved immune system (de Beer, 2009). This led to recommended vitamin levels being increased between 25% and 100% in nutrition specifications for parent stock between 2007 and 2021 (Aviagen, 2007, 2021).

Breeders were fed, in a 2 × 2 factorial design, 2 vitamin E levels (30 mg/kg and 120 mg/kg) and 2 selenium sources (sodium selenite and zinc-L-selenomethionine) (Urso et al., 2015). They found improved offspring quality expressed in lower embryonic mortality and higher chick weight when diets were supplemented with zinc-L- selenomethionine and vitamin E (120 mg/kg feed). In an earlier study, Surai (2000) investigated the combination of vitamin E (40 mg/kg to 200 mg/kg α-tocopherol) and selenium (200 μg/kg and 400 μg/kg) levels in breeder diets. They found that the liver antioxidant system in 1- and 5-day-old chicks was improved and they concluded that this has potential benefits for chick health.

Selenium (Se) is very important as an antioxidant defence of the embryo which optimizes flock fertility and hatchability (Surai and Fisinin, 2014). The organic form of Se has been identified as an important antioxidant. Several experiments have shown that supplementing Se to breeder diets increased Se concentrations in both eggs and offspring (Surai, 2000; Kemp et al., 2001; Pappas et al., 2005, 2006; Jlali et al., 2013; Couloigner et al., 2015). Breeders fed an additional 500 μg/kg of organic Se increased the antioxidant status of eggs and decreased embryonic mortality which results in a 4% higher hatchability (Rajashree et al., 2014).

Higher zinc concentrations in breeder diets improved cellular immune function and decreased first-day mortality of the offspring (Kidd, 2003). Supplementation of organic zinc in combination with manganese decreased offspring mortality and improved some immune parameters from day 0 to day 34 (Virden et al., 2003, 2004). Moreover, substantially higher maternal dietary concentrations of selenium, zinc and manganese than recommended may improve immune function and liveability of offspring (Hocking, 2007).

6.5 Special male diet

The influence of one individual male broiler breeder on egg fertility is of greater importance than the contribution of one female (Wolc et al., 2009). This is due to the mating ratios where a number of females received sperm from each male, usually in a ratio of 1 male to 11–12 females. Egg fertility will also depend on the fertilizing ability of the spermatozoa produced by each male, with only fecund sperm occupying the sperm storage tubules (Bakst et al., 1994).

No difference in semen concentration or testis weight at 65 weeks of age was reported for males fed either 12% or 14% CP to achieve either 90% or 100% of the recommended BW target, respectively (Fontana et al., 1990). Semen volume was reported to be higher in males fed 12% vs. 16% CP from 28 weeks to 36 weeks of age, but sperm concentration was not affected (Zhang et al, 1999). In contrast, no effects on semen volume or metabolic activity of the spermatozoa were observed in males fed different CP diets from 80 g CP/kg to 400 g CP/kg (Hocking, 1989). No differences in semen volume, sperm concentration, total sperm yield and the proportion of males producing semen were observed from males fed either 8% or 12% CP (Revington et al., 1991). No effects of feeding 9%, 12% or 15% protein were observed on semen volume, concentration, number of spermatozoa ejaculated or spermatogenic activity (assessed by means of histological evaluation of the seminiferous epithelial area, tubule diameter and epithelial height) (Wilson et al., 1987). There was also no effect of the level of protein (11% or 16% CP) observed on sexual behaviour (Duncan et al., 1990). However, with extremely high dietary CP levels (210 g/kg, 290 g/kg and 400 g/kg) the proportion of males producing semen declined to around 10% (Hocking, 1989). Males on 11% CP diets had higher fertility of hatching eggs than those on 16% CP diets (Hocking, 1990), and males fed 16% CP diets as opposed to 12% CP diets had a reduced semen concentration (Hocking and Bernard, 1997a).

A dose-response study was performed with Cobb-500 males fed five different daily crude protein levels (12.0, 14.2, 16.4, 18.6, and 20.8 g CP/day) (Borges et al., 2006a). It was concluded that the intake of 16.9 g CP/male/day gave the best results for semen volume, sperm concentration, motility and fertility. An experiment was performed between 26 weeks and 41 weeks of age to determine the effect of 3 (10.5%, 12.6% and 15.0%) dietary CP levels on Ross 788 broiler breeder male fertility (Tyler and Bekker, 2012). Over all ages, on average, sperm from males fed the 12.6% CP diet resulted in a longer fertile period after artificial insemination than males fed the 10.5% and 15.0% CP diet. The fertile period was assessed by determining the number of sperm in the outer perivitelline layer of eggs and calculated to be 7.0 days, 14.5 days and 8.6 days from for the males fed the 10.5%, 12.6% and 15.0% CP diets, respectively. No treatment effects were found on the measures of normal motility and morphology of live sperm. In a later study by the same research group even lower dietary CP levels (9.7%, 11.7% and 13.0%) were applied to Ross 308 males (Graaf et al., 2018). They did not find a response to dietary CP level on sperm concentration or sperm mobility at different ages; however, a positive effect of increased CP level was observed at the end of the laying phase. They concluded that feeding males 11.7% CP diet throughout the entire production period had minimal effects on sperm quality; however, higher CP intake at the end of the laying phase may positively affect fertility.

Besides dietary protein level, it is suggested by Borges et al. (2006b) in a dose-response study that the optimum energy level for males is around 2660 kcal ME/kg. Too high an energy intake will lead to an increased body fat content, increased BW and decreased fertility and hatchability (Borges et al., 2006a,b).

Two different treatments were applied to breeders in an on-farm study with 26 houses with each containing 7000 females and 840 males (Cobb 500) (Silveira et al., 2014). To half of the houses, a standard female diet (2830 kcal. ME/kg and 15.0% CP) was provided to the females and males, whereas in the other half of the houses, females were fed female diets and males received a special male diet (2750 kcal ME/kg and 13.5% CP). Males in both treatments were fed the same daily amount of feed. Males fed the special male diet had a 6.1% lower BW at 65 weeks of age and a 4.2% higher hatchability. Overall, it seems that a lower dietary CP level is more effective in improving fertility and hatchability in modern broiler breeders males which is probably due to differences in breeds over recent decades.

Because males don't produce eggs, calcium level in female diets is approx. four times too high for male needs. Breeder companies therefore recommend between 0.70% and 0.95% calcium for male diet compared to between 3.0% and 3.5% calcium for female diets (Aviagen, 2018; Cobb, 2018). Feeding female diets to males results in excessive calcium intake which can cause metabolic kidney damage (Leeson and Summers, 2005; Moyle et al., 2011). The latter authors found that broiler breeder males fed high levels of calcium develop kidney asymmetry and urolithiasis, which can contribute to high mortality (Moyle et al., 2011). Moreover, a lower sperm concentration was found at 42 weeks, 57 weeks and 69 weeks of age in males fed diets with high calcium (Tyler et al., 2021).

Over the last decade breeder companies have recommended increasing Vitamin E because of its critical role in sperm quality (Aviagen, 2018). Vitamin E is described as a very important element in the antioxidant defence network of semen and therefore fertility (Surai et al., 2019). Feeding Ross breeder males a diet supplemented with 200 mg of Vit E/kg enhanced semen volume, sperm concentration and sperm motility (Zanussi et al., 2019).

A higher Se level in male and female diets is implemented for two main reasons (Surai and Fisinin, 2014). The first reason is the important role of Se in the maintenance of semen quality which is correlated to the fertility of the hatching eggs (Surai, 2006). The second reason is the important role of Se in the maintenance of the antioxidant system of the growing embryo in the egg which can increase hatchability (Surai, 2006; Fisinin et al., 2008).

More recently a study was performed to investigate the effects of 3 basal diets supplemented with 20 mg/kg, 40 mg/kg or 60 mg/kg organic Zn (zinc-methionine) in 45-week-old females and males (Prabakar et al., 2021). Males fed the 60 mg/kg Zn-Met had higher semen volume, sperm motility, concentration, and live sperm count. The highest testosterone concentration, fertility and hatchability were observed in the males fed the 40 mg/kg Zn-Met. They concluded that the additional supplementation of organic Zn (zinc-methionine) at 40 mg/kg to the basal diet improved the reproductive performance in broiler breeders.

7 Feed restriction and feeding programs

7.1 Feed restriction

The consequence of ad libitum feeding of breeder pullets is excessive BW with impaired effects on reproduction performance (Katanbaf et al., 1989; Savory et al., 1993; Heck et al., 2004). The study by Heck et al. (2004) showed that ad libitum feeding results in 5 times higher mortality (31.1% vs. 6.1%), impaired egg production (33.5 eggs vs. 50.3 eggs) and a higher feed conversion ratio (5.80 g feed /g egg vs. 3.33 g feed /g egg) between 32 weeks and 40 weeks of age. Feed restriction is therefore necessary to prevent these dramatically negative effects on the performance of breeders (Mench, 2002; D'Eath et al., 2009). As early as the 1990s researchers described the effects of feed restriction on behavioural signs of stress and frustration (e.g. Hocking et al., 1993; Savory et al., 1993, 1996). Researchers also observed physiological indicators of (chronic) stress. They found increased plasma corticosterone levels and increased heterophil: lymphocyte ratios in the blood (Hocking et al., 1996; Savory and Mann, 1997; de Jong et al., 2003; Ekmay et al., 2010). Increased levels of adrenocorticotropin (ACTH) and glucocorticoids, as observed in feed-restricted pullets, can have suppressive effects on the immune system (Montiel, 2016). Moreover, higher levels of ACTH and corticosterone can have a negative effect on innate or induced immunity (e.g. Shini and Kaiser, 2009; Moe et al., 2010). Stress in birds can affect innate immunity which decreases the efficacy of pathogen control in birds (Latshaw, 1991; Montiel, 2016). The effect of feed restriction of breeder females on growth rate and immune responses in offspring has recently been investigated by Bowling et al. (2018). The breeders were fed restricted from 24 weeks of age and fed to 3 different BWs between 31 weeks and 43 weeks of age and maintained different target weights: 3.4 kg, 3.6 kg and 4.0 kg. Eggs over a 2-week period were incubated and chicks were hatched, IBV vaccinated at days 7 and 16, and were given an immune challenge of lipopolysaccharide at days 18 and 20. They found that antibodies to the IBV vaccination were increased for the offspring from the high BW breeders compared to offspring from the low BW breeders and postulated that the immune response of the broilers was improved.

7.2 Feeding programs

Breeder pullets are fed a restricted amount of feed, once a day in the morning with feed clean-up time between 15 min and 60 min, depending on the level of feed restriction (de Jong and van Emous, 2017). Due to this, it is a real challenge to optimize nutrient provision to all individual pullets for optimal BW with an optimal flock BW uniformity (Zuidhof et al., 2015). Because of the short feed clean-up time, in the past feeding programs (skip-a-day, skip-another-day, 5/2, 4/3) were developed to optimize BW uniformity (Leeson and Summers, 2005). A skip-a-day program means that the amount of feed every second day is doubled which increased feed clean-up time and improved feed distribution and uniformity (de Beer and Coon, 2007). Using restricted feeding programs is prohibited in most countries in North West Europe because of the negative effects on welfare. Ekmay et al. (2010) found higher stress levels, evidenced by higher corticosterone and insulin-like factor plasma levels, in skip-a-day fed pullets compared to everyday-fed pullets. These higher levels of corticosterone (and ACTH) can affect the innate or induced immunity with potential negative effects on offspring health as mentioned earlier (e.g. Shini and Kaiser, 2009; Moe et al., 2010).

8 Body weight for females and males

Optimal BW for both males and females is necessary because of the negative correlation between breast meat content and reproductive performance (McDaniel et al., 1981; Hocking, 1990; Dunnington et al., 1990; Fontana et al., 1990; Casanovas, 2002; McGary et al., 2002; Bilcik et al., 2005). BW management of broiler breeder females and males is achieved through feed restriction and nutrient levels of the diets (e.g. Zhang et al., 1999; Casanovas, 2002; Bilcik et al., 2005). In the past, obesity in the second phase of the laying phase was a major problem in broiler breeder flocks and resulted often in decreased egg production and fertility (Bornstein et al., 1984; Leclercq et al., 1985; Cahanar et al., 1986; Robinson et al., 1993). Moreover, heavy breeder females show a higher frequency of embryonic mortality during the incubation period resulting in lower hatchability of the flock (Dunnington et al., 1990). Overweight hens have sperm storage problems (due to the fat deposition in the sperm storage glands) and physical problems during the cloacal contact during natural mating (McDaniel et al., 1981). Body composition of breeders, however, has changed dramatically during the last five to six decades (Eitan et al., 2014; van Emous, 2015; Zuidhof et al., 2015). In modern broiler breeders, obesity is not an issue anymore, due to the selection of strains with increased breast muscle and decreased fat pad deposition characteristics as shown by Eitan et al. (2014). They compared a 1980 to a 2000 breeder strain and observed a 42% higher breast meat content (21.2% vs. 14.9% of BW) and 50% lower abdominal fat pad content (2.7% vs. 5.4% of BW) for the 2000 compared to the 1980 strain. Despite this shift in body composition in breeder females, BW control is still necessary for good general health, leg health and vitality (Aviagen, 2018).

There seems to be a clear link between the BW of the male, testes weight and fertility (Powley, 2008). In general, heavy breeder males had heavier testes, although in modern broiler breeders bodyweight is not the only important trait. Under commercial circumstances, over-fleshed (large breast muscle) males with high BW sometimes show lower testes weight (Powley, 2008). Males with large breast muscles will have, due to their BW and balance issues, more problems in mounting and therefore an incomplete mating without cloacal contact with the female and failed semen insemination (Hocking and Bernard, 1997b; Wolanski et al., 2004; Bilcik et al., 2005; Chung, 2010). Furthermore, heavier males show lower mating activity (Bilcik et al., 2005) reflecting lower sexual behaviour and a lower percentage of successful matings (Hocking, 1990). On the other hand, BW loss in males is associated with impaired mating activity, sperm production and fertility (Duncan et al., 1990; Casanova, 2002). This is because a decrease in BW is directly correlated with a decrease in testes weight and testosterone levels (McGovern, 2002; Sarabia Fragoso et al., 2013). Testicular shrinkage in combination with a lower sperm production and testosterone levels in broiler breeder males was found in studies of Vizcarra et al. (2010) and Sarabia Fragoso et al. (2013). Sarabia Fragoso et al. (2013) observed infertility or very low fertility, low levels of testosterone and high levels of corticosterone in adult males with a BW below 3.8 kg compared to adequate BW. Moreover, a decrease in male BW can lead to cessation of sperm production and mating activity (Casanovas, 2000). Good BW control management in males is therefore very important and separate sex feeding, special male diets, weekly weighing males and weekly feed amount adjustments are recommended by the breeding companies (Aviagen, 2018, Cobb, 2018).

9 Feather cover

In general, feather cover quality of breeders has decreased during the last 20 years for unknown reasons (van Emous and de Jong, 2013). A good feather cover quality during the production period is of the utmost importance to prevent skin damage and assist thermoregulation (de Jong and van Emous, 2017). Skin damage can be caused by mutual female competition around feeding time (van Emous and de Jong, 2013) and from rough male mating behaviour (de Jong et al., 2009). It is postulated that poor feather cover quality negatively affects the fertility of the flock due to females avoiding males (van Emous and de Jong, 2013). An on-farm study with four breeder houses showed that a poor feather cover at 30 weeks of age decreased hatchability from 40 weeks onwards (Fig. 2). Feather cover quality can be affected by a wide range of management and nutritional interventions (de Jong and van Emous, 2017; van Emous, 2015; van Emous and van Krimpen, 2019).

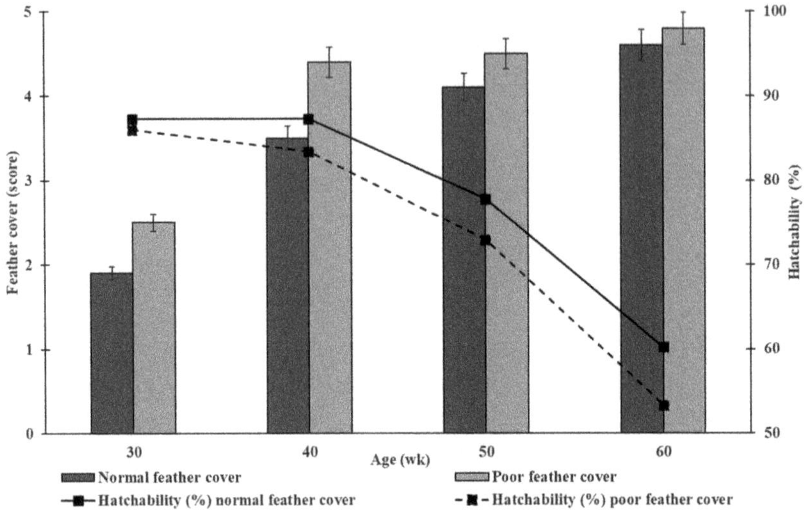

Figure 2 Effect of feather cover score (means and ± SEM) on hatchability (%). Feather cover was scored from 0 (intact feathers) to 5 (completely bald) (van Emous, unpublished data). The two groups were scored as normal or poor at the onset of 30 weeks of age.

10 Where to look for further information

A standard introduction to social and sexual behaviour of chickens is J. P. Kruijt, *Ontogeny of Social Behaviour in Burmese Red Junglefowl (Gallus gallus spadiceus)* (see Kruijt, 1964 in the References section for full details).

Global centres of broiler breeder expertise include the following:

- Wageningen Livestock Research (WLR): https://www.wur.nl/en/research-results/research-institutes/livestock-research.htm.
- University of Georgia, Department of Poultry Science: https://poultry.caes.uga.edu/.
- University of Alberta, Poultry Research Centre: https://poultry.ualberta.ca/.

11 References

Atencio, A., Edwards, H. M., Jr. and Pesti, G. M. (2005a). Effects of vitamin D3 dietary supplementation of broiler breeder hens on the performance and bone abnormalities of the offspring, *Poult. Sci.* 84(7): 1058–1068.

Atencio, A., Pesti, G. M. and Edwards, H. M., Jr. (2005b). Twenty-five hydroxycholecalciferol as a cholecalciferol substitute in broiler breeder hen diets and its effect on the performance and general health of the offspring, *Poult. Sci.* 84(8): 1277–1285.

Aviagen. (2007). *Parent Stock Nutrition Specification*, Huntsville: Aviagen.

Aviagen. (2015). *Best Practice in the Breeder House: Preventing Floor Eggs*, Aviagen.

Aviagen. (2018). *Ross Parent Stock Management Handbook*, Available at: https://en.aviagen.com/assets/Tech_Center/Ross_PS/RossPSHandBook2018.pdf.

Aviagen. (2021). *European Ross 308 Parent Stock: Nutrition Specifications*, Available at: www.aviagen.com.

Bakst, M. R., Wishart, G. and Brillard, J. P. (1994). Oviductal sperm selection, transport, and storage in poultry, *Poult. Sci. Rev.* 5: 117–143.

Barbato, G. F. (1999). Genetic relationships between selection for growth and reproductive effectiveness, *Poult. Sci.* 78(3): 444–452.

Barna, J., Végi, B., Varadi, E., Szoke, Z. and Liptói, K. (2007). Analysis of fertility in broiler breeder flocks - Male side approaches. In: *5th Viet.-Hung International Conference on "Animal Production and Aquaculture for Sustainable Farming,"* Can Tho, Vietnam.

Berrang, M. E., Frank, J. F., Buhr, R. J., Bailey, J. S., Cox, N. A. and Mauldin, J. M. (1997). Microbiology of sanitized broiler hatching eggs through the egg production period, *J. Appl. Poult. Res.* 6(3): 298–305.

Bilcik, B., Estevez, I. and Russek-Cohen, E. (2005). Reproductive success of broiler breeders in natural mating systems: The effect of male-male competition, sperm quality, and morphological characteristics, *Poult. Sci.* 84(9): 1453–1462.

Boa-Amponsem, K., Price, S. E. H., Dunnington, E. A. and Siegel, P. B. (2001). Effect of route of inoculation on humoral immune response of White Leghorn chickens selected for high or low antibody response to sheep red blood cells, *Poult. Sci.* 80(8): 1073–1078.

Bonduriansky, R. (2001). The evolution of male mate choice in insects: A synthesis of ideas and evidence, *Biol. Rev. Camb. Philos. Soc.* 76(3): 305–339.

Borges, C. A. Q., Rostagno, H. S., da Silva, J. H. V., Albino, L. F. T., Torres, C. A. A., Filho, J. J. and Ribeiro, M. L. G. (2006a). Protein requirements and carcass composition of male broiler breeders from 27 to 61 weeks of age (in Portuguese), *R. Bras. Zootec.* 35: 1971–1977.

Borges, C. A. Q., Rostagno, H. S., da Silva, J. H. V., Albino, L. F. T., Torres, C. A. A., Jordão Filho, J. and Ribeiro, M. L. G. (2006b). Exigências de energia e composição da carcaça de galos reprodutores pesados em função do consumo energético na fase de reprodução (in Portuguese), *R. Bras. Zootec.* 35(5): 1978–1984.

Bornstein, S., Plavnik, I. and Lev, Y. (1984). Body weight and/or fatness as potential determinants of the onset of egg production in broiler breeder hens, *Br. Poult. Sci.* 25(3): 323–341.

Bowling, M., Forder, R., Hughes, R. J., Weaver, S. and Hynd, P. I. (2018). Effect of restricted feed intake in broiler breeder hens on their stress levels and the growth and immunology of their offspring, *Transl. Anim. Sci.* 2(3): 263–271.

Brake, J. (1998). Equipment design for breeding flocks, *Poult. Sci.* 77(12): 1833–1841.

Brake, J. and Peak, S. D. (1999). Feeding programs for broiler breeders. In: *Proceedings of the N.C Broiler Breed Hatchery Management Conference*, Statesville, NC, North Carolina State University, Raleigh, pp. 48–56.

Bramwell, R. K., McDaniel, C. D., Wilson, J. L. and Howarth, B. (1996). Age effect of male and female broiler breeders on sperm penetration of the perivitelline layer overlying the germinal disc, *Poult. Sci.* 75(6): 755–762.

Brougher, S., Estevez, I. and Ottinger, M. A. (2005). Can testosterone and corticosterone predict the rate of display of male sexual behaviour, development of secondary sexual characters and fertility potential in primary broiler breeders?, *Br. Poult. Sci.* 46(5): 621–625.

Byrne, P. G. and Rice, W. R. (2006). Evidence for adaptive male mate choice in the fruit fly Drosophila melanogaster, *Proc. Biol. Sci.* 273(1589): 917-922.

Cahanar, A., Nitzan, Z. and Nir, I. (1986). Reproductive performance of broiler lines divergently selected on abdominal fat, *Poult. Sci.* 65(7): 1236-1243.

Calini, F. and Sirri, F. (2007). Breeder nutrition and offspring performance, *Rev. Bras. Cienc Avic* 9(2): 77-83.

Casanovas, P. (2000). *Management Techniques to Improve Male Mating Activity and Compensate for the Age-Related Decline in Broiler Breeder Fertility: Intra-Spiking, Technical News*, Quarterly Publication of Cobb-Vantress, Incorporated.

Casanovas, P. (2002). Management and management techniques to improve male mating activity and compensate for the age-related decline in broiler breeder fertility; intraspiking, *Poult. Inform. Prof.* 63: 1-13.

Chang, A., Halley, J. and Silva, M. (2016). Can feeding the broiler breeder improve chick quality and offspring performance?, *Anim. Prod. Sci.* 56(8): 1254-1262.

Chotesangasa, R. (2001). Effects of mating ratio, cock number in the flock and breeder age on fertility in Thai native chicken flock, *Kasetsart J. Nat. Sci.* 35: 122-131.

Chung, K. M. Y. (2010). *Effect of double interspiking on fertility, behavior, and blood parameters in broiler breeder males reared under heat stress conditions*, Master Thesis, University of Tennessee, Knoxville.

Cobb. (2018). *Cobb Breeder Management Guide*, Cobb-Vantress, Siloam Springs, Arkansas, US.

Collias, N. E. and Collias, E. C. (1967). A field study of the Red jungle fowl in north-central India, *Condor* 69(4): 360-386.

Collias, N. E. and Collias, E. C. (1996). Social organization of a red junglefowl, Gallus gallus, population related to evolution theory, *Anim. Behav.* 51(6): 1337-1354.

Collias, N. E., Collias, E. C., Hunsaker, D. and Minning, L. (1966). Locality fixation, mobility and social organization within an unconfined population of Red jungle fowl, *Anim. Behav.* 14(4): 550-559.

Cooper, J. J. and Appleby, M. C. (1996). Individual variation in prelaying behaviour and the incidence of floor eggs, *Br. Poult. Sci.* 37(2): 245-253.

Cornwallis, C. K. and Birkhead, T. R. (2007). Experimental evidence that female ornamentation increases the acquisition of sperm and signals fecundity, *Proc. Biol. Sci.* 274(1609): 583-590.

Couloigner, F., Rouffineau, F., Geraert, P. A. and Mercier, Y. (2015). Dietary selenium supplementation of breeders influences growth and selenium status of offspring. In: *Australian Poultry Science Symposium* 26: 80-83.

Cox, N. A., Berrang, M. E. and Cason, J. A. (2000). Salmonella penetration of egg shells and proliferation in broiler hatching eggs - A review, *Poult. Sci.* 79(11): 1571-1574.

de Beer, M. (2009). Current approaches to feeding broiler breeders. In: *Proceedings of the 17th European Symposium Poultry and Nutrition*, Edinburgh, Scotland, Attleborough, UK: World's Poultry Science Association, pp. 104-114.

de Beer, M. and Coon, C. N. (2007). The effect of different feed restriction programs on reproductive performance, efficiency, frame size, and uniformity in broiler breeder hens, *Poult. Sci.* 86(9): 1927-1939.

de Jong, I. C. and Guemene, D. (2011). Major welfare issues in broiler breeders, *Worlds Poult. Sci. J.* 67(1): 73-82.

de Jong, I. C., Lourens, A., Gunnink, H. and van Emous, R. A. (2011). Effect of stocking density on (the development of) sexual behaviour and technical performance in broiler breeders (In Dutch). Wageningen UR Livestock Research, Lelystad, Report 457, 61 p.

de Jong, I. C. and Swalander, M. (2012). Housing and management of broiler breeders and turkey breeders. In: Sandilands, V. and Hocking, P. M. (Eds.), *Alternative Systems for Poultry: Health*, Wallingford, UK: Welfare and Productivity, C.A.B. International Publishing, pp. 225–249.

de Jong, I. C. and van Emous, R. A. (2017). Broiler breeding flocks: Management and animal welfare. In: Applegate, T. (Ed.), *Achieving Sustainable Production of Poultry Meat* (vol. 3), London: Burleigh Dodds Science Publishing Limited, pp. 211–230.

de Jong, I. C., van Voorst, A. S. and Blokhuis, H. J. (2003). Parameters for quantification of hunger in broiler breeders, *Physiol. Behav.* 78(4–5): 773–783.

de Jong, I. C., Wolthuis-Fillerup, M. and van Emous, R. A. (2009). Development of sexual behaviour in commercially-housed broiler breeders after mixing, *Br. Poult. Sci.* 50(2): 151–160.

de Reu, K. (2006). *Bacteriological contamination and infection of shell eggs in the production chain*, PhD Thesis, University of Ghent, Belgium.

de Reu, K., Grijspeerdt, K., Heyndrickx, M., Uyttendaele, M., Debevere, J. and Herman, L. (2011). Bacterial shell contamination in the egg collection chains of different housing systems for laying hens, *Br. Poult. Sci.* 47(2): 163–172.

de Reviers, M. and Seigneurin, F. (1990). Interactions between light regimes and feed restrictions on semen output in two meat-type strains of cockerels. In: Brillard, J. P. (Ed.), *Control of Fertility in Domestic Birds. Les Colloques de l'INRA* (vol. 54), Paris, France: INRA, pp. 220–231.

D'Eath, R. B., Tolkamp, B. J., Kyriazakis, I. and Lawrence, A. B. (2009). 'Freedom from hunger' and preventing obesity: The animal welfare implications of reducing food quantity or quality, *Anim. Behav.* 77(2): 275–288.

Decuypere, E. and Bruggeman, V. (2007). The endocrine interface of environmental and egg factors affecting chick quality, *Poult. Sci.* 86(5): 1037–1042.

Deeming, D. C., Clyburn, V., Williams, K. and Dixon, R. A. (2002). In ovo microbial contamination of the yolk sac of unhatched broiler and pheasant embryos, *Avian Poult. Biol. Rev.* 13: 240–241.

Duncan, I. J. H. (2009). Mating behaviour and fertility. In: Hocking, P. M. (Ed.), *Biology of Breeding Poultry*, Wallingford, UK: CABI Publishing, pp. 111–132.

Duncan, I. J. H., Hocking, P. M. and Seawright, E. (1990). Sexual behavior and fertility in broiler breeder domestic fowl, *Appl. Anim. Behav. Sci.* 26: 1–13.

Dunnington, E. A., Siegel, P. B. and Anthony, N. B. (1990). Reproductive fitness in selected lines of chickens and their crosses, *J. Hered.* 81(3): 217–219.

EFSA. (2010). Scientific opinion on welfare aspects of the management and housing of the grandparent and parent stocks raised and kept for breeding purposes, *EFSA J.* 8: 81.

Eitan, Y., Lipkin, E. and Soller, M. (2014). Body composition and reproductive performance at entry into lay of anno 1980 versus anno 2000 broiler breeder females under fast and slow release from feed restriction, *Poult. Sci.* 93(5): 1227–1235.

Ekmay, R. D., de Beer, M., Mei, S. J., Manangi, M. and Coon, C. N. (2013). Amino acid requirements of broiler breeders at peak production for egg mass, body weight, and fertility, *Poult. Sci.* 92(4): 992–1006.

Ekmay, R. D., de Beer, M., Rosebrough, R. W., Richards, M. P., McMurtry, J. P. and Coon, C. N. (2010). The role of feeding regimens in regulating metabolism of sexually mature broiler breeders, *Poult. Sci.* 89(6): 1171–1181.

Etches, R. J. (1996). *Reproduction in Poultry*, Cambridge, UK: CAB International, University Press.

Fischer, G. J. (1975). The behavior of chickens. In: Hafez, E. S. E. (Ed.), *The Behavior of Domestic Animals* (3rd edn.), London, UK: Bailliere Tindall, pp. 454–489.

Fisinin, V. I., Papazyan, T. T. and Surai, P. F. (2008). Selenium in poultry nutrition. In: Surai, P. F. and Taylor-Pickard, J. (Eds.), *Current Advances in Se Research and Applications* (vol. 1), Wageningen, the Netherlands: Wageningen Academic Publishers, pp. 221–261.

Fontana, E. A., Weaver, W. D., Jr. and van Krey, H. P. (1990). Effects of various feeding regimens on reproduction in broiler-breeder males, *Poult. Sci.* 69(2): 209–216.

Fontana, E. A., Weaver, W. D., Jr. and van Krey, H. P. (1992). Intermittent periods of infertility identified in naturally mated broiler breeder hens, *J. Appl. Poult. Res.* 1(2): 190–193.

Graaf, B., Ciacciariello, M. and Tyler, N. C. (2018). The effect of crude protein intake on sperm quality of young and old male broiler breeders, *S. Afr. J. Anim. Sci.* 48(6): 1066–1072.

Haq, A. U., Bailey, C. A. and Chinnah, A. (1996). Effect of β-carotene, canthaxanthin, lutein, and vitamin E on neonatal immunity of chicks when supplemented in the broiler breeder diets, *Poult. Sci.* 75(9): 1092–1097.

Heck, A., Onagbesan, O., Tona, K., Metayer, S., Puterflam, J., Jego, Y., Trevidy, J. J., Decuypere, E., Williams, J., Picard, M. and Bruggeman, V. (2004). Effects of ad libitum feeding on performance of different strains of broiler breeders, *Br. Poult. Sci.* 45(5): 695–703.

Heier, B. T. and Jarp, J. (2001). An epidemiological study of the hatchability in broiler breeder flocks, *Poult. Sci.* 80(8): 1132–1138.

Hocking, P. M. (1989). Effect of dietary crude protein concentration on semen yield and quality in male broiler breeder fowls, *Br. Poult. Sci.* 30(4): 935–945.

Hocking, P. M. (1990). The relationships between dietary crude protein, body weight, and fertility in naturally mated broiler breeder males, *Br. Poult. Sci.* 31(4): 743–757.

Hocking, P. M. (2004). Measuring and auditing the welfare of broiler breeders. In: Weeks, C. A. and Butterworth, A. (Eds.), *Measuring and Auditing Broiler Welfare*, Wallingford, UK: CABI Publishing, pp. 19–35.

Hocking, P. M. (2007). Optimum feed composition of broiler breeder diets to maximise offspring performance. In: *16th European Symposium of Poultry Nutrition*, Strasbourg, France: World Poultry Science Association, pp. 101–108.

Hocking, P. M. and Bernard, R. (1997a). Effects of dietary crude protein content and food intake on the production of semen in two lines of broiler breeder males, *Br. Poult. Sci.* 38(2): 199–202.

Hocking, P. M. and Bernard, R. (1997b). Effects of male body weight, strain and dietary protein content on fertility and musculoskeletal disease in naturally mated broiler breeder males, *Br. Poult. Sci.* 38(1): 29–37.

Hocking, P. M. and Bernard, R. (2000). Effects of age of male and female broiler breeders on sexual behavior, fertility and hatchability of eggs, *Br. Poult. Sci.* 41(3): 370–376.

Hocking, P. M., Bernard, R. and Robertson, G. W. (2002). Effects of low dietary protein and different allocations of food during rearing and restricted feeding after peak rate of lay on egg production, fertility and hatchability in female broiler breeders, *Br. Poult. Sci.* 43(1): 94–103.

Hocking, P. M. and Duff, S. R. I. (1989). The incidence of musculo-skeletal lesions in adult male broiler breeder fowls and their relationships with body weight and fertility at 60 wk of age, *Br. Poult. Sci.* 30(4): 777–784.

Hocking, P. M., Maxwell, M. H. and Mitchell, M. A. (1993). Welfare assessment of broiler breeder and layer females subjected to food restriction and limited access to water during rearing, *Br. Poult. Sci.* 34(3): 443–458.

Hocking, P. M., Maxwell, M. H. and Mitchell, M. A. (1996). Relationships between the degree of food restriction and welfare indices in broiler breeder females, *Br. Poult. Sci.* 37(2): 263–278.

Hossain, S. M., Barreto, S. L., Bertechini, A. G., Rios, A. M. and Silva, C. G. (1998). Influence of dietary vitamin E level on egg production of broiler breeders, and on the growth and immune response of offspring in comparison with the offspring from eggs injected with vitamin E, *Anim. Feed Sci. Technol.* 73(3–4): 307–317.

Jlali, M., Briens, M., Rouffineau, F., Mercerand, F., Geraert, P. A. and Mercier, Y. (2013). Effect of 2-hydroxy-4-methylselenobutanoic acid as a dietary selenium supplement to improve the selenium concentration of table eggs, *J. Anim. Sci.* 91(4): 1745–1752.

Johnsen, T. S., Zuk, M. and Fessler, E. A. (2001). Social dominance, male behaviour and mating in mixed-sex flocks of red jungle fowl, *Behavior* 138: 1–18.

Johnson, M. L. (2013). *The effect of maternal antioxidant nutrient supplementation and age on chick post-hatch innate immune function*, PhD Thesis, University of Alberta, Canada.

Jones, E. K. M. and Prescott, N. B. (2000). Visual cues used in the choice of mate by fowl and their potential importance for the breeder industry, *Worlds Poult. Sci. J.* 56(2): 127–138.

Jones, E. K. M., Prescott, N. B., Cook, P., White, R. P. and Wathes, C. M. (2001). Ultraviolet light and mating behaviour in domestic broiler breeders, *Br. Poult. Sci.* 42(1): 23–32.

Jones, M. E. and Mench, J. A. (1991). Behavioral correlates of male mating success in a multisire flock as determined by DNA fingerprinting, *Poult. Sci.* 70(7): 1493–1498.

Katanbaf, M. N., Dunnington, E. A. and Siegel, P. B. (1989). Restricted feeding in early and late-feathering chickens. 2. reproductive responses, *Poult. Sci.* 68(3): 352–358.

Kaukonen, E., Norring, M. and Valros, A. (2016). Effect of litter quality on foot pad dermatitis, hock burns and breast blisters in broiler breeders during the production period, *Avian Pathol.* 45(6): 667–673.

Kemp, C., Wylie, L. and Fisher, C. (2001). Broiler breeder nutrition, nutrient transfer and broiler performance. In: *13th European Symposium on Poultry Nutrition*, World Poultry Science Association, Blankenberghe, Belgium, pp. 61–67.

Khabisi, M. M., Salahi, A. and Mousavi, S. N. (2012). The influence of egg shell crack types on hatchability and chick quality, *Turk. J. Vet. Anim. Sci.* 36: 289–295.

Kidd, M. T. (2003). A treatise on chicken dam nutrition that impacts on offspring, *Worlds Poult. Sci. J.* 59(4): 475–494.

Kruijt, J. P. (1964). *Ontogeny of social behaviour in Burmese Red junglefowl (Gallus gallus spadiceus)*, PhD Thesis, University of Groningen, the Netherlands.

Latshaw, J. D. (1991). Nutrition - Mechanisms of immunosuppression, *Vet. Immunol. Immunopathol.* 30(1): 111–120.

Leclercq, B., Kovassi-Kouakou, J. and Simon, J. (1985). Laying performance, egg composition and glucose tolerance of genetically lean or fat meat-type breeders, *Poult. Sci.* 64(9): 1609–1616.

Leeson, S. and Summers, J. D. (2005). *Commercial Poultry Nutrition* (3rd edn.), Guelph, Ontario: University Books.

Leeson, S. and Summers, J. D. (2009). *Broiler Breeder Production*, Guelph, Canada: University Books.

Leone, E. H. and Estévez, I. (2008). Economic and welfare benefits of environmental enrichment for broiler breeders, *Poult. Sci.* 87(1): 14–21.
Ligon, J. D., Thornhill, R., Zuk, M. and Johnsen, K. (1990). Male-male competition in red jungle fowl and the multiple roles of testosterone in sexual selection, *Anim. Behav.* 40(2): 367–373.
Ligon, J. D. and Zwartjes, P. W. (1995). Female red junglefowl choose to mate with multiple males, *Anim. Behav.* 49(1): 127–135.
Lopez, G. and Leeson, S. (1995). Response of broiler breeders to low-protein diets: 1. Adult breeder performance, *Poult. Sci.* 74(4): 685–695.
Lourens, A. and Steentjes, A. (2008). Zootechnical and veterinary factors at the broiler breeder farm that affect broiler mortality (in Dutch), Report 194, Animal Sciences Group of Wageningen University and Research Center, Lelystad, The Netherlands.
Løvlie, H. and Pizzari, T. (2007). Sex in the morning or in the evening? Females adjust daily mating patterns to the intensity of sexual harassment, *Am. Nat.* 170(1): E1–E13.
Mba, E., Renema, R. A., Pishnamazi, A. and Zuidhof, M. J. (2010). Do dietary protein:energy ratios modify growth and frame size of young broiler breeder females?, *Poult. Sci.* 89(Suppl. 1): 122.
McBride, G., Parer, I. P. and Foenander, F. (1969). The social organisation and behaviour of the feral domestic fowl, *Anim. Behav. Monogr.* 2: 125–181.
McDaniel, G. R., Brake, J. and Bushong, R. D. (1981). Factors affecting broiler breeder performance. 1, *Poult. Sci.* 60(2): 307–312.
McDaniel, G. R. and Craig, J. V. (1959). Behavior traits, semen measurements and fertility of White Leghorn males, *Poult. Sci.* 38(5): 1005–1014.
McGary, S., Estevez, I., Bakst, M. R. and Pollock, D. L. (2002). Phenotypic traits as reliable indicators of fertility in male broiler breeders, *Poult. Sci.* 81(1): 102–111.
McGovern, R. H. (2002). *Reproduction in male broiler breeders*, PhD Thesis, University of Georgia, Athens, GA.
Mejia, L., McDaniel, C. D., Lopez, K., Parker, H. M. and Corzo, A. (2012). Effects of digestible lysine intake level on Cobb 500 broiler breeder hen reproductive performance, *J. Appl. Poult. Res.* 21(4): 868–873.
Mench, J. A. (2002). Broiler breeders: Feed restriction and welfare, *Worlds Poult. Sci. J.* 58(1): 23–29.
Millman, S. T. and Duncan, I. J. H. (2000a). Effect of male-to-male aggressiveness and feed-restriction during rearing on sexual behaviour and aggressiveness towards females by male domestic fowl, *Appl. Anim. Behav. Sci.* 70(1): 63–82.
Millman, S. T. and Duncan, I. J. H. (2000b). Strain differences in aggressiveness of male domestic fowl in response to a male model, *Appl. Anim. Behav. Sci.* 66: 217–233.
Millman, S. T., Duncan, I. J. H. and Widowski, T. M. (2000). Male broiler breeder fowl display high levels of aggression toward females, *Poult. Sci.* 79(9): 1233–1241.
Moe, R. O., Guémené, D., Bakken, M., Larsen, H. J. S., Shini, S., Lervik, S., Skjerve, E., Michel, V. and Tauson, R. (2010). Effects of housing conditions during the rearing and laying period on adrenal reactivity, immune response and heterophil to lymphocyte (H/L) ratios in laying hens, *Animal* 4(10): 1709–1715.
Molenaar, R., Hulet, R., Meijerhof, R., Maatjens, C. M., Kemp, B. and van den Brand, H. (2011). High eggshell temperatures during incubation decrease growth performance and increase the incidence of ascites in broiler chickens, *Poult. Sci.* 90(3): 624–632.
Mols, M. C. P. (1978). Mating activity of males under different housing conditions (In Dutch). Internal Report 1178P. *C.O.V.P*, Het Spelderholt, Beekbergen, the Netherlands.

Montiel, E. R. (2016). *Influence of feeding programs on innate and adaptive immunity in broiler breeders*, PhD Dissertation, University of Georgia, Athens.

Moyle, J. R., Wideman, R. F., Whipple, S. M., Yoho, D. E. and Bramwell, R. K. (2011). Urolithiasis in male broiler breeders, *Int. J. Poult. Sci.* 10: 839–841.

Moyle, J. R., Yoho, D. E., Harper, R. S. and Bramwell, R. K. (2010). Mating behaviour in commercial broiler breeders: Female effects, *J. Appl. Poult. Res.* 19: 24–29.

Mphepya, L. S., van Rensburg, W. J., Mpofu, T. J., Mtileni, B. J. and Nephawe, K. A. (2019). Influence of male-male competition on reproductive performance and mortality of broiler breeders following intra-spiking, *Poul. Sci.* 98: 4549–4554.

Nazarpak, H. H. and Fallah-Khair, A. (2011). Effect of spiking of young males in different percentage on economic performance of broiler breeder flock, *J. Food Agric. Environ.* 9: 564–569.

Ordas, B., Vahedi, S., Seidavi, A., Rahati, M., Laudadio, V. and Tufarelli, V. (2015). Effect of testosterone administration and spiking on reproductive success of broiler breeder flocks, *Reprod. Domest. Anim.* 50(5): 820–825.

Pappas, A. C., Acamovic, T., Surai, P. F. and McDevitt, R. M. (2006). Maternal organo-selenium compounds and polyunsaturated fatty acids affect offspring performance and levels of selenium and docosahexaenoic acid in the chick tissues, *Poult. Sci.* 85(9): 1610–1620.

Pappas, A. C., Karadas, F., Surai, P. F. and Speake, B. K. (2005). The selenium intake of the female chicken influences the selenium status of her offspring, *Comp. Biochem. Physiol., B Biochem. Mol. Biol.* 142(4): 465–474.

Pearson, R. A. and Herron, K. M. (1982). Effects of maternal energy and protein intakes on the incidence of malformations and malpositions of the embryo and time of death during incubation, *Br. Poult. Sci.* 23(1): 71–77.

Pizzari, T., Cornwallis, C. K., Løvlie, H., Jakobsson, S. and Birkhead, T. R. (2003). Sophisticated sperm allocation in male fowl, *Nature* 426(6962): 70–74.

Pizzari, T., Løvlie, H. and Cornwallis, C. K. (2004). Sex-specific, counteracting responses to inbreeding in a bird, *Proc. Biol. Sci.* 271(1553): 2115–2121.

Powley, J. (2008). *Testes Development and Fertility. Ross Tech Note - Testes Development and Fertility - June 2008*, Huntsville: Aviagen Inc.

Prabakar, G., Gopi, M., Kolluri, G., Rokade, J. J., Khillare, G., Pearlin, B. V., Jadhav, S. E., Tyagi, J. S. and Mohan, J. (2021). Effect of supplementation of zinc-methionine on egg production, semen quality, reproductive hormones, and hatchability in broiler breeders, *Biol. Trace Elem. Res.* 199(12): 4721–4730.

Rajashree, K., Muthukumar, T. and Karthikeyan, N. (2014). Comparative study of the effects of organic selenium on hen performance and productivity of broiler breeders, *Br. Poult. Sci.* 55(3): 367–374.

Reijrink, I. A. M., Berghmans, D., Meijerhof, R., Kemp, B. and van den Brand, H. (2010). Influence of egg storage time and preincubation warming profile on embryonic development, hatchability, and chick quality, *Poult. Sci.* 89(6): 1225–1238.

Revington, W. H., Moran, E. T., Jr. and McDaniel, G. R. (1991). Performance of broiler breeder males given low protein feed, *Poult. Sci.* 70(1): 139–145.

Robinson, F. E., Wilson, J. L., Yu, M. W., Fasenko, G. M. and Hardin, R. T. (1993). The relationship between body weight and reproductive efficiency in meat-type chickens, *Poult. Sci.* 72(5): 912–922.

Romero-Sanchez, H., Plumstead, P. W. and Brake, J. (2004). Feeding broiler breeder males. 1. effect of feeding program and dietary crude protein during rearing on body weight and fertility of broiler breeder males, *Poult. Sci.* 86(1): 168–174.

Romero-Sanchez, H., Plumstead, P. W., Leksrisompong, N. and Brake, J. (2007). Feeding broiler breeder males. 2. effect of cumulative rearing nutrition on body weight, shank length, comb height, and fertility, *Poult. Sci.* 86(1): 175–181.

Rushen, J. (1983). The development of sexual relationships in the domestic chicken, *Appl. Anim. Ethol.* 11(1): 55–66.

Sarabia Fragoso, J., Pizarro Diaz, M., Abad Moreno, J. C., Casanovas Infesta, P., Rodriguez Bertos, A. and Barger, K. (2013). Relationships between fertility and some parameters in male broiler breeders (body and testicular weight, histology and immunohistochemistry of testes, spermatogenesis and hormonal levels), *Reprod. Domest. Anim.* 48(2): 345–352.

Savory, C. J., Hocking, P. M., Mann, J. S. and Maxwell, M. H. (1996). Is broiler breeder welfare improved by using qualitative rather than quantitative food restriction to limit growth rate?, *Anim. Welf.* 5: 105–127.

Savory, C. J. and Mann, J. S. (1997). Is there a role for corticosterone in expression of abnormal behaviour in restricted-fed fowls?, *Physiol. Behav.* 62(1): 7–13.

Savory, C. J., Maros, K. and Rutter, S. M. (1993). Assessment of hunger in growing broiler breeders in relation to a commercial restricted feeding programme, *Anim. Welf.* 2: 131–152.

Sexton, K. J., Renden, J. A., Marple, D. N. and Kemppainen, R. J. (1989). Effects of dietary energy on semen production, fertility, plasma testosterone, and carcass composition of broiler-breeder males in cages, *Poult. Sci.* 68(12): 1688–1694.

Sheppard, K. C. and Duncan, I. J. H. (2011). Feeding motivation on the incidence of floor eggs and extraneously calcified eggs laid by broiler breeder hens, *Br. Poult. Sci.* 52(1): 20–29.

Shini, S. and Kaiser, P. (2009). Effects of stress, mimicked by administration of corticosterone in drinking water, on the expression of chicken cytokine and chemokine genes in lymphocytes, *Stress* 12(5): 388–399.

Silveira, M. M., Freitas, Ad, Moraes, C. A., Gomes, F. S., Litz, F. H., Martins, J. M. S., Fagundes, N. S. and Fernandes, E. A. (2014). Feeding management strategy for male broiler breeders and its effects on body weight, hatchability and fertility, *Rev. Bras. Cienc Avic* 16(4): 397–402.

Smeltzer, T. I., Orange, K., Peel, B. and Runge, G. (1979). Bacterial penetration in floor and nest box eggs from mean and layer birds, *Aust. Vet. J.* 55(12): 592–593.

Sun, J. and Coon, C. N. (2005). The effects of body weight, dietary fat, and feed withdrawal rate on the performance of broiler breeders, *J. Appl. Poult. Res.* 14(4): 728–739.

Sunde, M., Turk, C. M. and DeLuca, H. F. (1978). The essentiality of vitamin D metabolites for chick development, *Science* 200: 1067–1069.

Surai, P. F. (2000). Effect of selenium and vitamin E content of the maternal diet on the antioxidant system of the yolk and the developing chick, *Br. Poult. Sci.* 41(2): 235–243.

Surai, P. F. (2006). *Selenium in Nutrition and Health*, Nottingham, UK: Nottingham University Press.

Surai, P. F. and Fisinin, V. I. (2014). Selenium in poultry breeder nutrition: An update, *Anim. Feed Sci. Technol.* 191: 1–15.

Surai, P. F., Kochish, I. I., Romanov, M. N. and Griffin, D. K. (2019). Nutritional modulation of the antioxidant capacities in poultry: The case of vitamin E, *Poult. Sci.* 98(9): 4030–4041.

Tona, K., Bruggeman, V., Onagbesan, O., Bamelis, F., Gbeassor, M., Mertens, K. and Decuypere, E. (2005). Day-old chick quality: Relationship to hatching egg quality, adequate incubation practice and prediction of broiler performance, *Avian Poult. Biol. Rev.* 16(2): 109–119.

Tyler, N. C. and Bekker, H. A. (2012). The effect of dietary crude protein on the fertility of male broiler breeders, *S. Afr. J. Anim. Sci.* 42(3): 304–309.

Tyler, N. C., Namntu, M. and Ciacciariello, M. (2021). Research note: The effect of crude protein and calcium intake on fertility of male broiler breeders, *Poult. Sci.* 100(9): 101284.

Urso, U. R. A., Dahlke, F., Mairoka, A., Bueno, I. J. M., Schneider, A. F., Surek, D. and Rocha, C. (2015). Vitamin E and selenium in broiler breeder diets: Effect on live performance, hatching process, and chick quality, *Poult. Sci.* 94(5): 976–983.

van den Brand, H., Sosef., M. P., Lourens, A. and van Harn, J. (2016). Effects of floor eggs on hatchability and later life performance in broiler chickens, *Poult. Sci.* 95(5): 1025–1032.

van Emous, R. A. (2007). Inventory of hatchability problems broiler breeder sector (in Dutch), *Pluimveehouderij*: 14–15.

van Emous, R. A. (2010). Innovative housing in broiler breeder production. In: *2nd International Symposium: Highlights in Nutrition and Welfare in Poultry Production*, Wageningen, The Netherlands, Animal Science Group, Lelystad, The Netherlands, pp. 37–44.

van Emous, R. A. (2015). *Body composition and reproduction in broiler breeders: Impact of feeding strategies*, PhD Thesis, Wageningen University, Wageningen, The Netherlands.

van Emous, R. A. and de Jong, I. C. (2013). Promising management measures to solve the major welfare problems in broiler breeders. In: *2nd International Poultry Meat Congress*, Antalya, Turkey.

van Emous, R. A. and Gunnink, H. (2011). Innovative broiler breeder housing system: 'Quality Time' house (in Dutch). Report 470. Wageningen UR Livestock Research, Lelystad, the Netherlands.

van Emous, R. A., Holleman, J. and van Schie, T. (2020). *Breeder Signals: A Practical Guide to Optimal Management of Broiler Breeders,* Zutphen, The Netherlands: Roodbont Publishers B.V.

van Emous, R. A., Kwakkel, R. P., van Krimpen, M. M. and Hendriks, W. H. (2013). Effects of growth patterns and dietary crude protein levels during the rearing on body composition and performance in broiler breeder females during the rearing and laying period, *Poult. Sci.* 92(8): 2091–2100.

van Emous, R. A., Kwakkel, R. P., van Krimpen, M. M. and Hendriks, W. H. (2015). Effects of dietary protein levels during rearing and dietary energy levels during lay on body composition and reproduction in broiler breeder females, *Poult. Sci.* 94(5): 1030–1042.

van Emous, R. A. and Mens, A. J. W. (2020). Skip-a-day male-female contact hardly has any effects (in Dutch), *Pluimveehouderij*: 14–15.

van Emous, R. A. and van Krimpen, M. M. (2019). Effects of nutritional interventions on feathering of poultry – A review. In: Olukosi, O. A. (Ed.), *Poultry Feathers and Skin:*

The Poultry Integument in Health and Welfare, Oxforshire, UK, Wallingford: CABI Publishing, pp. 133-151.
van Grunsven, P. (1978). *Sexual Behaviour of Males in the Pen or Observation Arena's (In Dutch)*. Beekbergen, the Netherlands: C.O.V.P, Het Spelderholt.
Végi, B., Váradi, E., Szöke, Z. and Barna, J. (2013). Effect of sex ratios, spiking and extra artificial insemination on the breeding efficiency of broiler breeders, *Acta Vet. Hung.* 61(3): 393-404.
Verrell, P. A. (1985). Male mate choice for large, fecund females in the red-spotted newt, Notophthalmus viridescens, how is size assessed, *Herpetology* 41: 382-386.
Virden, W. S., Yeatman, J. B., Barber, S. J., Willeford, K. O., Ward, T. L., Fakle, T. M., Wideman, R. F., Jr. and Kidd, M. T. (2004). Immune system and cardiac functions of offspring chicks from dams fed diets differing in zinc and manganese level and source, *Poult. Sci.* 83(3): 344-351.
Virden, W. S., Yeatman, J. B., Barber, S. J., Zumwalt, C. D., Ward, T. L., Johnson, A. B. and Kidd, M. T. (2003). Hen mineral nutrition impacts offspring livability, *J. Appl. Poult. Res.* 12(4): 411-416.
Vizcarra, J. A., Kirby, J. D. and Kreider, D. L. (2010). Testis development and gonadotropin secretion in male broiler breeders, *Poult. Sci.* 89(2): 328-334.
Walsh, T. J. and Brake, J. (1997). Management for optimum fertility of broiler breeders. 7. Cumulative protein consumption to 20 weeks of age influences fertility of broiler breeder females, *Poult. Sci.* 76(2): 297-305.
Walsh, T. J. and Brake, J. (1999). Effects of feeding program and crude protein intake during rearing on fertility of broiler breeder females, *Poult. Sci.* 78(6): 827-832.
Warren, D. C. and Kilpatrick, L. (1929). Fertilization in the domestic fowl, *Poult. Sci.* 8(5): 237-256.
Weil, S., Rozenboim, I., Degen, A. A., Dawson, A., Friedländer, M. and Rosenstrauch, A. (1999). Fertility decline in aging roosters is related to increased testicular and plasma levels of estradiol, *Gen. Comp. Endocrinol.* 115(1): 23-28.
Whitehead, C. C., Pearson, R. A. and Herron, K. M. (1985). Biotin requirements of broiler breeders fed diets of different protein content and effect of insufficient biotin on the viability of progeny, *Br. Poult. Sci.* 26(1): 73-82.
Wilson, H. R., Piesco, N. P., Miller, E. R. and Nesbeth, W. G. (1979). Prediction of the fertility potential of broiler breeder males, *Worlds Poult. Sci. J.* 35(2): 95-118.
Wilson, J. (2002). *Hatchery/Breeder Tip... Understanding the Factors That Influence Broiler Breeder Flock Fertility*, The University of Georgia Cooperative Extension Service.
Wilson, J. L., McDaniel, G. R., Sutton, C. D. and Renden, J. A. (1987). Semen and carcass evaluation of broiler breeder males fed low protein diets, *Poult. Sci.* 66(9): 1535-1540.
Wolanski, N. J., Renema, R. A., Robinson, F. E. and Wilson, J. L. (2004). End-of-season carcass and reproductive traits in original and replacement male broiler breeders, *J. Appl. Poult. Res.* 13(3): 451-460.
Wolc, A., White, I. M. S., Olori, V. E. and Hill, W. G. (2009). Inheritance of fertility in broiler chickens, *Genet. Sel. Evol.* 41: 47.
Wood-Gush, D. G. M. (1958). The effect of experience on the mating behaviour of the domestic cock, *Anim. Behav.* 6(1-2): 68-71.
Yassin, H., Velthuis, A. G. J., Boerjan, M. and van Riel, J. W. (2009). Field study on broilers' first-week mortality, *Poult. Sci.* 88(4): 798-804.
Yassin, H., Velthuis, A. G. J., Boerjan, M., van Riel, J. W. and Huirne, R. B. M. (2008). Field study on broiler eggs hatchability, *Poult. Sci.* 87(11): 2408-2417.

Zanussi, H. P., Shariatmadari, F., Sharafi, M. and Ahmadi, H. (2019). Dietary supplementation with flaxseed oil as source of omega-3 fatty acids improves seminal quality and reproductive performance in aged broiler breeder roosters, *Theriogenology* 130: 41–48.

Zhang, X., Berry, W. D., McDaniel, G. R., Roland, D. A., Liu, P., Calvert, C. and Wilhite, R. (1999). Body weight and semen production of broiler breeder males as influenced by CP levels and feeding regimens during rearing, *Poult. Sci.* 78(2): 190–196.

Zuidhof, M. J., Holm, D. E., Renema, R. A., Jalal, M. A. and Robinson, F. E. (2015). Effects of broiler breeder management on pullet body weight and carcass uniformity, *Poult. Sci.* 94(6): 1389–1397.

Zuk, M., Johnsen, T. S. and Maclarty, T. (1995). Endocrine-immune interactions, ornaments and mate choice in red jungle fowl, *Proc. R. Soc. Lond. B* 260(1358): 205–210.

Zuk, M., Thornhill, R., Ligon, J. D., Johnson, K., Austad, S., Ligon, S. H., Thornhill, N. W. and Costin, C. (1990). The role of male ornaments and courtship behaviour in female mate choice of red jungle fowl, *Am. Nat.* 136(4): 459–473.

Chapter 3

Advances in understanding the development of defences against pathogens in the chicken egg

Maureen M. Bain, University of Glasgow, UK

1 Introduction

Microbial contamination of hatching eggs with pathogenic bacterial species is a major concern for the poultry industry as it results in poor hatchability and chick performance (Baxter-Jones, 1991). Several bacterial infections of economic and zoonotic significance are also transmitted via the egg including *Escherichia coli* (Poulsen et al., 2017) several *Salmonella* spp. (Liljebjelke et al., 2005) and several important *Mycoplasma* spp. (Barrow, 1994). Contamination of eggs can occur in two ways: Direct or *vertical transmission* (maternal to offspring) occurs when the reproductive tissues of breeding females become colonised by the organism which is then periodically shed and becomes incorporated into the egg as it is forming. Horizontal transmission (or trans-shell penetration) occurs following oviposition. For commercial hatcheries, the risk of cross-contamination (*horizontal transmission*) is the major concern

http://dx.doi.org/10.19103/AS.2022.0118.09

(Bailey et al., 1996) as during collection and transport, eggs which are already heavily contaminated come into direct contact with clean eggs. For this reason, hatching eggs are systematically disinfected, usually by chlorine/formaldehyde fumigation but disinfection is rarely complete, and it can also compromise embryo viability (Cadirci, 2009). The effect of fumigation on the eggs natural defences is also unknown.

The bacteria found on the egg surface are mostly Gram-positive, those that contaminate the egg contents are usually Gram-negative bacteria (Legros et al., 2021; Sellier et al., 2007). *Salmonella enterica* serovar Enteritidis (*S. enteritidis*) is especially problematic, as it can be transmitted both vertically and horizontally. This bacterium can also resist the antimicrobial effects of the albumen and once they reach the nutrient-rich yolk, they divide rapidly. The egg is naturally equipped with a wide range of natural defences aimed at protecting the embryo from physical damage and microbial contamination (Baron and Jan, 2011; D'Alba and Shawkey, 2015). Physical barriers include the cuticle, eggshell, the eggshell membranes, and the vitelline membrane; physicochemical barriers associated with the albumen (alkaline pH, viscosity) and antimicrobial proteins (e.g. lysozyme and ovotransferrin). Until relatively recently, only a few proteins were considered to participate in the antimicrobial defence of the egg. However, advances in transcriptomics and proteomic approaches have led to the discovery of many more potential antimicrobial proteins and peptides in every egg compartment (Réhault-Godbert et al., 2011; Réhault-Godbert et al., 2022). The key issue and challenge here are proving the biological significance of these proteins which fundamentally depends on the availability of pure and active protein (either purified or produced by recombinant protein), which can be both difficult and time-consuming (Réhault-Godbert et al., 2011). Nevertheless, progress is being made in identifying new candidate proteins that contribute to its antimicrobial defence in each of the different egg compartments.

2 Basic components of the egg

The egg is composed of a central yolk surrounded by the albumen (or egg white), the eggshell membranes (paired), and the eggshell (Fig. 1). The cuticle forms a thin film over the surface of the calcified eggshell. In the freshly laid egg, the albumen is structured and consists of distinct physical areas that are of liquid or viscous consistency. The vitelline membrane separates the albumen from the yolk. The yolk is held in the centre of the albumen by the chalazae and is the main source of nutrients for the developing embryo (Romanoff and Romanoff, 1949). At the blunt pole of the egg, the inner and outer

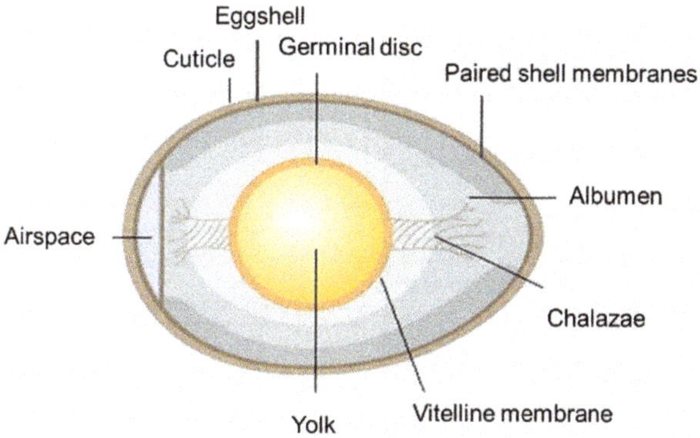

Figure 1 Egg components.

shell membranes separate to form the airspace. The inner shell membranes surround the albumen and the outer shell membranes are tightly adherent to the eggshell.

3 Formation of the egg: key events

At sexual maturity, over 12 000 oocytes are present in the ovary but only a small number of these will develop and reach maturity (Nys and Guyot, 2011) (Fig. 2). Follicular development occurs in a hierarchical fashion between five and seven follicles accumulating yolk by the process known as vitellogenesis at any one time (Réhault-Godbert and Guyot, 2018). Most yolk components are derived from the liver and transported to the ovary in the form of very low-density lipoproteins (Noble and Cocchi, 1990). Maternal immunoglobulins (IgY, IgM, and IgA) in the blood are also transferred to the yolk (Kramer and Cho, 1970). These maternal immunoglobulins are important to the chick at hatching (passive immunity) but provide minimal protection to the embryo during incubation (Réhault-Godbert et al., 2022).

At ovulation, the yolk mass from the largest follicle enters the funnel-shaped infundibulum (Fig. 2). This is where fertilisation occurs and where the formation of the vitelline membrane is completed (Nys and Guyot, 2011). The albumen is then secreted by the highly glandular region of the oviduct known as the magnum. The albumen consists of an aqueous solution of water (88%), protein (90% dry matter), minerals (6% dry matter), and free glucose (3.5% dry matter). The protein complement includes over 400 different proteins including many with known or predicted antimicrobial activity (Mann, 2007; Mann and Mann, 2011).

Eggshell membrane fibres are secreted in the isthmus and are assembled into an interwoven network which totally covers the albumen. These fibres are

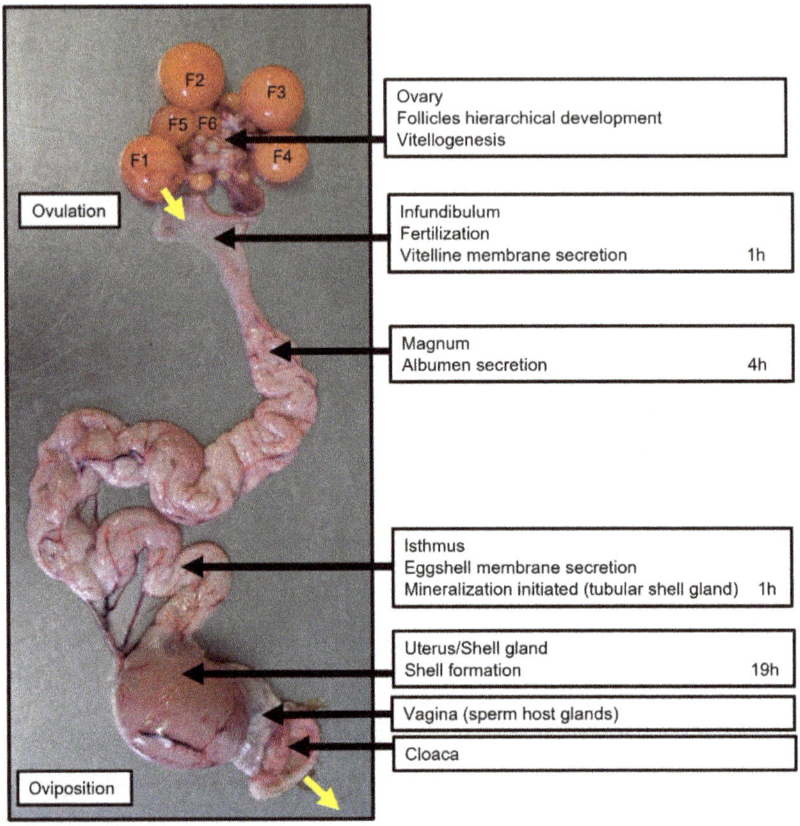

Figure 2 Only the left ovary and oviduct develop in breeding females. The main function and time spent in each region of the oviduct is indicated in this image of a chicken's reproductive system.

rich in glycoproteins, cysteine-rich eggshell membrane proteins, and collagens (types I, V, and X) (Arias et al., 1991; Kodali et al., 2011). The first stages of mineralisation occur at discrete sites on the outer shell membrane fibres just before the egg mass enters the uterus or shell gland (Gautron et al., 2021; Nys et al., 2004). The egg spends about 19 h in the uterus acquiring its calcified shell. Shell formation occurs in three distinct phases: a slow phase (~5 h), a rapid phase (~12 h), and a terminal phase (~1.5 h) (Gautron et al., 2021; Nys et al., 1999, 2004). During the slow phase, the mammillary layer is formed (~100 µm in thickness) (Fig. 3). The palisade layer (~300 µm in thickness) forms during the rapid phase. The palisade layer is interrupted by the presence of gaseous exchange pores which traverse the entire eggshell thickness and permit the exchange of gas and water. The vertical crystal layer (3–8 µm in thickness) is the final part of the mineralised shell to form. This layer is characterised by a change in crystal growth and occurs during the terminal phase of shell formation. The mechanism responsible for the arrest of mineralisation is not well understood but may involve both inorganic phosphate and phosphoproteins associated with the outer eggshell and cuticle (Gautron et al., 2021).

The eggshell is composed of calcium carbonate (95%) and an organic matrix (3.5%) which consists of proteins, glycoproteins, and proteoglycans. Several proteins associated with this organic matrix have established roles in biomineralisation. These proteins interact with the mineral phase during shell formation and modify the morphology of crystals that determine the complex architecture of the eggshell. The presence of the organic matrix also modifies the mechanical properties of the eggshell. Over 900 different matrix proteins have now been reported (Gautron et al., 2021). Among this large list are several proteins with reported antimicrobial activity that may contribute to the egg's defence. However their bioavailability to the developing embryo and mechanism of action is only just beginning to be explored (Halgrain et al., 2022; Hincke et al., 2019; Réhault-Godbert et al., 2022).

Just prior to oviposition, the cuticle (Fig. 3) is deposited onto the surface of the egg (0.5–12 μm in thickness). The cuticle contains some but not all of the eggshell pigment and consists of glycoproteins, polysaccharides, lipids, and a number of antimicrobial proteins (Bain et al., 2013). After oviposition, it takes about 3 min for the cuticle to dry (Sparks and Board, 1985). Because of its strategic positioning, the cuticle forms the eggs first line of defence and because of this, its formation, regulation and synthesis, functional properties,

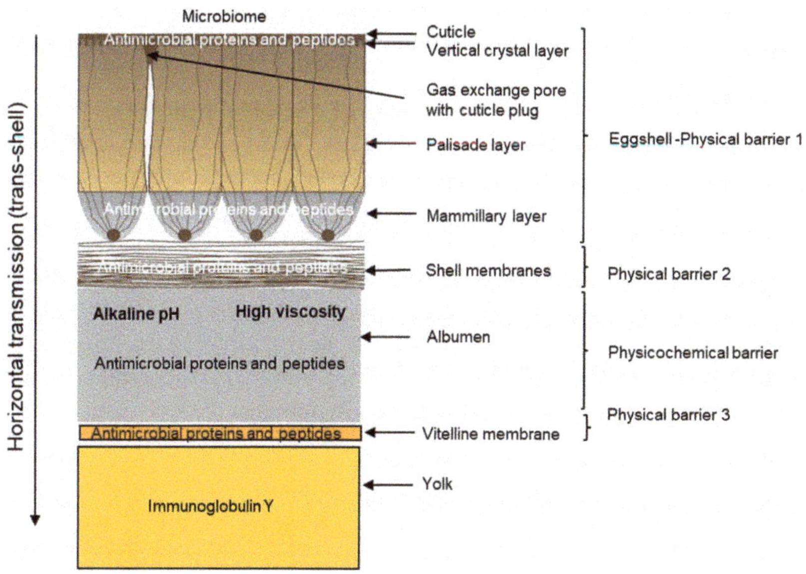

Figure 3 Physical and chemical barriers to trans-shell penetration of potentially pathogenic bacteria. (Based on a schematic presented in Réhault-Godbert et al., 2022).

composition, and measurement have attracted a lot of scientific interest (Kulshreshtha et al., 2022)–see Section 5.1 Case study.

4 Antimicrobial molecules associated with the egg and their mechanisms of action

As indicated previously, over the past 20 years, major developments in high-throughput methods (proteomics, transcriptomic approaches) have resulted in the identification of hundreds of proteins with or predicted antimicrobial properties in every egg compartment. It remains to be determined however if these molecules are secreted during egg formation to protect the reproductive system from being colonised and become incorporated into the egg or if they are actively secreted to protect the embryo (Hincke et al., 2019; Réhault-Godbert et al., 2022). Irrespective of their origin, the functional properties of many of these proteins are still to be determined.

Antimicrobial proteins in the egg display bactericidal or bacteriostatic activities via several different but complementary mechanisms as follows:

(1) **Direct interaction with the bacterial wall:** Hydrolases (e.g. lysozyme); antimicrobial peptides (e.g. avian β-defensins [AvBD] and gallin); lipopolysaccharide-binding (LBP) bactericidal/permeability increasing (BPI) proteins (e.g. TENP); C-type lectin-like proteins (e.g. Ovocleidin-17)

(2) **Chelating compounds which decrease bioavailability of nutrients required for microbial growth:** Iron-binding proteins (e.g. ovotransferrin) and proteins that bind to essential vitamins (e.g. avidin and riboflavin-binding protein)

(3) **Molecules that inhibit the activity of microbial proteases**: Inhibitors of serine proteases (e.g. ovoinhibitor, ovomucoid, and ovomucin); inhibitors of cysteine proteases (e.g. cystatin); metalloprotease inhibitors; broad spectrum inhibitors (e.g. ovostatin)

The nature and distribution of putative antimicrobial molecules in the different egg compartments have been regularly reviewed (Hincke et al., 2019; Réhault-Godbert et al., 2011, 2022; Rose-Martel and Hincke, 2017). In the next section, the key protein candidates in each egg compartment are discussed in relation to their possible function and role in the egg's defence.

5 Contribution made by each component of the egg to the egg's defence

The main physical barriers to bacterial ingress are provided by the cuticle, the eggshell, the shell membranes and the vitelline membrane. The albumen provides the main physicochemical barrier. Proteins with confirmed or predicted antimicrobial activity are found in all egg compartments and structures but are mainly concentrated in the albumen.

5.1 Case study: the cuticle

The cuticle forms as a protective barrier on the surface of the eggshell and plugs the entrance to the gaseous exchange pores. Despite being invisible, the cuticle has attracted a lot of attention from the scientific community (Fig. 4). In addition to its protective function, the cuticle also limits water ingress and at the same time it allows gaseous exchange (Sparks, 1994).

The cuticle tends to be much thicker in birds that naturally lay their eggs in humid dirty habitats (D'Alba et al., 2017) suggesting there has been some evolutionary pressure to enhance this trait. In modern poultry breeding programmes, with strict on-farm biosecurity and the widespread use of artificial incubation, emphasis has not been placed on the artificial selection for cuticle quality (Bain et al., 2019) (Fig. 5). A good coverage of cuticle, however, is critical to preventing the trans-shell penetration of bacteria: eggs with good cuticle coverage are rarely penetrated by microbes (Bain et al., 2013, 2019). Cuticle deposition is a moderately heritable trait. Thus, genetic progress to improve cuticle deposition should be possible (Dunn et al., 2019). This however will require the development of high-throughput methods that don't require eggs to be stained (Dunn et al., 2019; Kulshreshtha et al., 2022).

The cuticle also contributes to the egg's chemical defence. The main antimicrobial proteins and peptides identified in the cuticle include: Ovocalyxin-36 (OCX-36), Kunitz-like protease inhibitor, Ovocleidin-116, Ovocleidin-17, Ovocalyxin-25 (OCX-25), Clusterin and Ovocalyxin-32 (OCX-32) (Bain et al., 2013). Lower levels of other proteins have also been reported depending on the method used to recover the cuticle (Kulshreshtha et al., 2022). OCX-32 (also found in the eggshell matrix) is a potential peptidase/protease inhibitor which possesses a latexin-like domain (Gautron et al., 2001) that confers an inhibitory activity against carboxypeptidase A. This protein has been shown to affect *Bacillus subtilis* growth (Xing et al., 2007). OCX-36 is unique to birds and is related to LBPs, BPI proteins, and to palate, lung, and nasal epithelium clone (Plunc) family proteins (Gautron et al., 2007). It possesses lipopolysaccharide and lipoteichoic acid binding activity and has been shown to inhibit the growth of Gram-positive *Staphylococcus aureus* (Cordeiro et al., 2013).

The presence of antimicrobials in the cuticle is likely to influence the bacterial communities that can survive on the surface of the eggshell (the microbiome). It has been suggested that the eggshells microbiome could also provide additional protection to the egg, e.g. through competitive exclusion of potential pathogens (Kulshreshtha et al., 2022). Establishing the

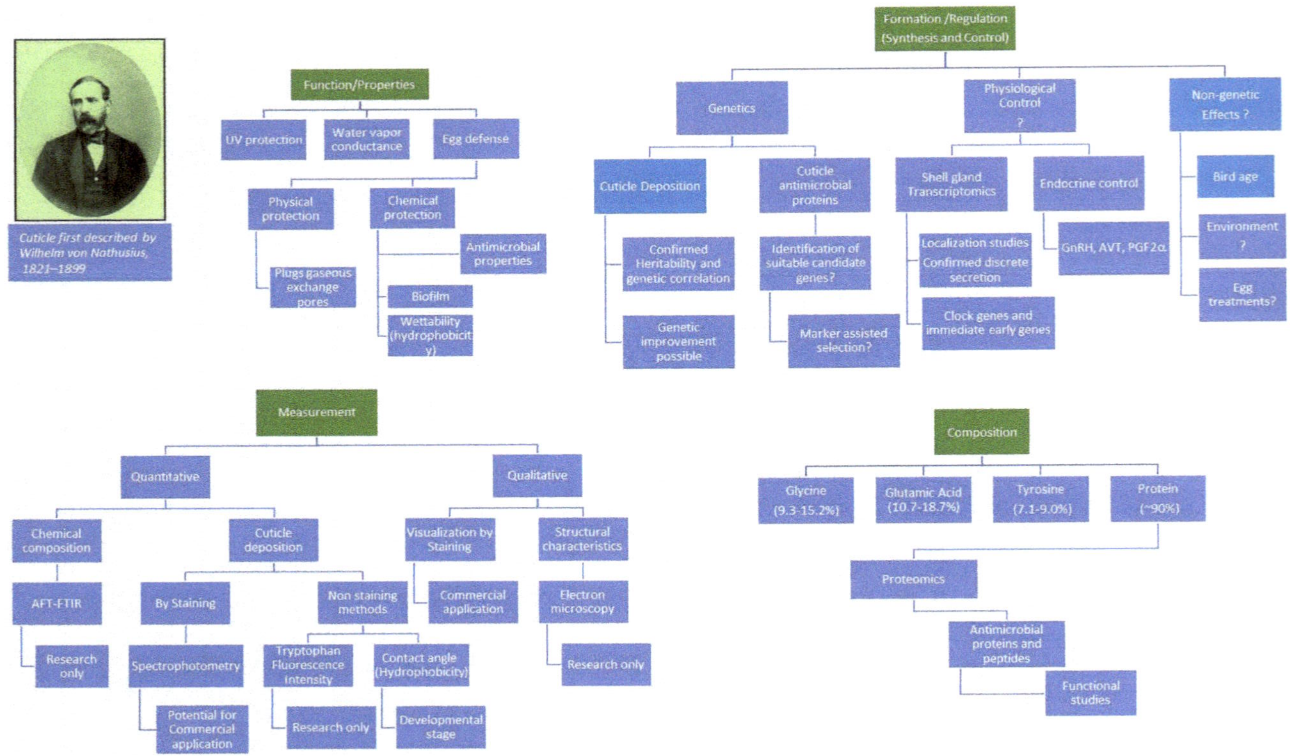

Figure 4 The evidence base confirming the importance of the cuticle as the egg's first line of defence has benefited from advances in technology including high-throughput molecular methods.

Figure 5 Broiler breeder eggs stained with MS Technologies cuticle blue to demonstrate the natural variation in the amount of cuticle deposition in one flock.

origins, the contributing factors and transmission processes that influence the eggshell microbiome is interesting as there is increasing evidence that eggs may function as transgenerational carriers of the material microbiome to the embryo (Réhault-Godbert et al., 2022).

Most of the proteins associated with the cuticle are also heavily glycosylated (Hincke et al., 2011). This could be important to the stability of the cuticle, its adhesiveness to the underlying calcified substrata, and its antimicrobial properties (Bain et al., 2019). Disinfection procedures used in the hatchery are effective at reducing bacterial loads on the surface of the eggshell but some of the new treatments available may remove the cuticle and therefore increase the risk of cross-contamination occurring at a later stage of incubation. Further work is needed to determine the impact of such treatments on the cuticle and eggshell microbiome (Olsen et al., 2017).

5.2 The eggshell and eggshell matrix

The eggshell provides the main physical protective barrier between the egg contents and the external environment. The integrity of the eggshell and its mechanical properties along with its complex architecture are critical to hatching success. Interestingly commercial hatcheries do not use crack detection devices that would allow the rapid detection of cracked and damaged eggshells, presumably because of the added risk of trauma to the egg and cross-contamination.

The eggshell's organic matrix may also contribute to the egg's chemical defence. Extracting and solubilising the eggshell matrix has presented some challenges, but the latest proteomic investigation suggests there are as many as 900 proteins in this component of the egg (Réhault-Godbert et al., 2022; Yang et al., 2020). Many of these proteins are likely to be involved in biomineralisation of the eggshell, but ~35 have been identified as having antimicrobial activity. These include well-known proteins such as lysozyme, ovotransferrin and cystatin. OCX-32 and OCX-36 which are particularly enriched in the eggshell (Hincke et al., 2019; Réhault-Godbert et al., 2022; Rose-Martel and Hincke, 2017). The bioavailability of OCX-32, OCX-36, and other putative antimicrobial proteins associated with the eggshell's organic matrix has come into question not least because these proteins form an integral part of the eggshell fabric. They could however provide local protection during the dissolution of the eggshell at the interface between the eggshell membranes and the chorioallantoic membranes (CAMs) (Halgrain et al., 2022; Hincke et al., 2019). This is discussed further in Section 6.

5.3 The eggshell membranes

The eggshell membranes form a physical semipermeable barrier between the eggshell and the albumen that allows exchange of gases and water and at the same time limit bacterial ingress. The fibres consist of collagen and proteins (~300). The inner membrane fibres closest to the albumen are lined by a lysozyme-rich limiting membrane (Hincke et al., 2000). Other potential candidates contributing to the eggs antimicrobial defence include OCX-36 (Cordeiro et al., 2013), several AvBDs (Ahmed et al., 2017), and gallin, a relatively new discovery also found in albumen (Gong et al., 2010).

5.4 The albumen

The albumen's physicochemical properties (pH and viscosity) and its high complement of antimicrobial proteins are extremely important to the egg's defence. At oviposition, the pH of the albumen is about 7.6, but within a few days, it rises to 9.0 (Guyot et al., 2016). Alkaline environments are generally unfavourable for bacterial growth. The gel-like nature of the thick albumen in the newly laid egg also limits the motility of flagellated bacteria (Schneider and Doetsch, 1974). During storage, the rise in albumen pH causes liquefaction of the thick albumen leading to a loss of viscosity (Nasri et al., 2020). At the same time, other antimicrobial proteins may be released into the milieu.

The albumen contains more than 400 proteins and has the highest complement of active antimicrobial proteins found in the egg. The major proteins involved in the egg's defence identified so far include lysozyme (3.4%), ovotransferrin (12%), ovomucin (3.5%), ovoinhibitor (1.5%), ovomucoid

(11%) plus two relatively new discoveries AvBD-11 and gallin. The antimicrobial activities and roles of these proteins in the egg's defence have been reviewed by others (Hincke et al., 2019; Réhault-Godbert et al., 2022; Rose-Martel and Hincke, 2017). The following is provided in summary.

Lysozyme exhibits catalytic muramidase activity that targets and degrades peptidoglycan, a constituent of the bacterial cell wall exposed at the surface of Gram-positive bacteria. Alternatively, antimicrobial peptides that are active against Gram-negative bacteria can be produced by proteolytic digestion of lysozyme (Mine et al., 2004). Ovotransferrin binds to iron thereby limiting the growth of Gram-negative bacteria including *Salmonella enteritis* (Legros et al., 2021). The chelating activity of ovotransferrin is reinforced at alkaline pH (Halbrooks et al., 2005). Ovomucin contributes to the gel-like properties of the thick albumen. It contains a cysteine-rich trypsin inhibitor like domain that could account for its ability to act as an antibacterial agent (Kobayashi et al., 2004). Ovoinhibitor displays potent inhibitory activity toward trypsin and subtilisin as well as a weak effect against chymotrypsin; its antimicrobial activity was demonstrated against *Bacillus thuringiensis* (Bourin et al., 2011). Ovomucoid is a trypsin inhibitor, although uncharacterised in chickens, the third domain of turkey ovomucoid inhibits *Streptomyces griseus* proteinase B (Read et al., 1983).

AvBDs are cationic peptides with biocidal activity against a wide arrange of pathogenic microorganisms (van Dijk et al., 2008). AvBD-11 isolated from the albumen (and vitelline membrane) by heparin affinity has been shown to exhibit a potent antibacterial activity against several Gram-positive and Gram-negative bacteria (Hervé-Grépinet et al., 2010). Gallin is a cationic peptide related to the AvBD family. The production of recombinant gallin peptide revealed antimicrobial activity against *E. coli* (Gong et al., 2010) making it a good candidate in terms of the egg's defence.

5.5 The vitelline membrane

The vitelline membrane limits the diffusion of nutrients from the yolk into the albumen and forms the final physical barrier to microbes gaining access to the yolk. Proteomics has revealed that the vitelline membrane contains as many as 35 proteins with confirmed or predicted antimicrobial activity (Mann, 2008). The two most significant proteins are lysozyme (21%) and AvBD-11 (1%).

5.6 The yolk

Proteomic studies have identified over 200 proteins in the yolk, but antimicrobial proteins are not very abundant in contrast to the high levels of nutrients found in this compartment. IgY is one of the most abundant proteins found in the yolk, but it does not directly contribute to the egg's natural defence. An account of how IgY is transferred to the embryo in preparedness for hatching can be found in Hincke et al. (2019).

6 Embryo-derived structures involved in the egg's defence

The natural defence of the egg at oviposition diminishes throughout incubation (Guyot et al., 2016). During the second half of incubation, the eggshell is progressively weakened by the CAM which solubilises the eggshell in order to release calcium for skeletal development (Chien et al., 2009; Gabrielli and Accili, 2010; Halgrain et al., 2022). This shell thinning process is also important in terms of facilitating the emergence of the chick at hatch. It does however mean that the egg contents and the embryo could be more vulnerable to microbial contamination during the later stages of incubation. The dissolution of the eggshell however could result in the local release of antimicrobial proteins from within the eggshell fabric (part of the eggshell organic matrix) which could provide some added protection. The strategic position of the CAM directly under the inner shell membranes may also allow for other embryonic mediators of innate immunity to act locally (Halgrain et al., 2022; Réhault-Godbert et al., 2022). Indeed, it has been suggested that the CAM may act as an early warning system to the embryo in the later stages of development in readiness for entering a challenging environment after hatching. The CAM has the capacity to contribute to the natural defences of the egg through physical, molecular, and cellular mechanisms (Halgrain et al., 2022). This makes it a particularly interesting candidate for further investigation.

7 Conclusion

Several protective mechanisms are incorporated into the design of the egg to ensure the embryo can successfully develop out with the female reproductive tract. While the role of the cuticle as the egg's first line of defence is undisputable, the discovery of putative antimicrobial proteins and peptides in all egg compartments and associated structures has raised many more questions. For example, are these molecules secreted to protect the reproductive tissues and/or are they actively incorporated into the different components of the egg to provide tailored protection to the embryo (Hincke et al., 2019). If the proteins are actively secreted in the oviduct, how are they regulated and how are these molecules then activated/mobilised to coincide with specific developmental events. What is clear is that these and other research questions will continue to attract the attention of the scientific community for years to come.

8 Where to look for further information

The following articles provide a good overview of the subject:

1. Hincke, M. T., da Silva, M., Guyot, N., Gautron, J., McKee, M. D., Guabirababa-Brito, R. and Réhault-Godbert, S. (2019). Dynamics of structural barriers and innate immune components during incubation of the Avian egg: Critical interplay between autonomous embryonic development and maternal anticipation. *Journal of Innate Immunity*. S. Karger AG. DOI: 10.1159/000493719.
2. Gautron, J., Stapane, L., Le Roy, N., Nys, Y., Rodriguez-Navarro, A. B. and Hincke, M. T. (2021). Avian eggshell biomineralization: An update on its structure, mineralogy and protein tool kit. *BMC Molecular and Cell Biology* 22(1): 11. DOI: 10.1186/s12860-021-00350-0.
3. Réhault-Godbert, S., Hincke, M., Guabiraba, R., Guyot, N. and Gautron, J. (2022). Innate defenses of the Avian egg. *Avian Immunology*. Academic Press: 365–386. DOI: 10.1016/B978-0-12-818708-1.00032-4.

Key research in this area can be found at the following organisations:

- Hincke, Maxwell | Department of Cellular and Molecular Medicine | University of Ottawa (uottawa.ca), Canada.
- Joel Gautron (researchgate.net).
- Sophie Réhault-Godbert (researchgate.net), UK.
- Prof Maureen M. Bain, University of Glasgow, UK.
- Prof Ian Dunn | The University of Edinburgh, UK.

9 References

Ahmed, T. A. E., Suso, H. P. and Hincke, M. T. (2017). In-depth comparative analysis of the chicken eggshell membrane proteome. *Journal of Proteomics* 155: 49–62. DOI: 10.1016/J.JPROT.2017.01.002.

Arias, J. L., Fernandez, M. S., Dennis, J. E. and Caplan, A. I. (1991). Collagens of the chicken eggshell membranes. *Connective Tissue Research* 26(1-2): 37–45. DOI: 10.3109/03008209109152162.

Bailey, J. S., Buhr, R. J., Cox, N. A. and Berrang, M. E. (1996). Effect of hatching cabinet sanitation treatments on salmonella cross-contamination and hatchability of broiler Eggs1. *Poultry Science* 75(2): 191–196. DOI: 10.3382/ps.0750191.

Bain, M. M., McDade, K., Burchmore, R., Law, A., Wilson, P. W., Schmutz, M., Preisinger, R. and Dunn, I. C. (2013). Enhancing the egg's natural defence against bacterial penetration by increasing cuticle deposition. *Animal Genetics* 44(6): 661–668. DOI: 10.1111/age.12071.

Bain, M. M., Zheng, J., Zigler, M., Whenham, N., Quinlan-Pluck, F., Jones, A. C., Roberts, M., Icken, W., Olori, V. E. and Dunn, I. C. (2019). Cuticle deposition improves the biosecurity of eggs through the laying cycle and can be measured on hatching eggs without compromising embryonic development. *Poultry Science* 98(4): 1775–1784. DOI: 10.3382/ps/pey528.

Baron, F. and Jan, S. (2011). Egg and egg product microbiology. In: *Improving the Safety and Quality of Eggs and Egg Products*. Elsevier, Woodhead Publishing, Cambridge UK. 330–350. DOI: 10.1533/9780857093912.3.330.

Barrow, P. A. (1994). The microflora of the alimentary tract and avian pathogens: Translocation and vertical transmission. In: *Microbiology of the Avian Egg*. Boston, MA: Springer US, 117–138. DOI: 10.1007/978-1-4615-3060-2_6.

Baxter-Jones, C. (1991). Egg hygiene: Microbial contamination, significance and control. Available at: http://europepmc.org/abstract/AGR/IND92009436. Poultry Science Symposium (Vol. 22), 269–276.

Bourin, M., Gautron, J., Berges, M., Attucci, S., Le Blay, G., Labas, V., Nys, Y. and Rehault-Godbert, S. (2011). Antimicrobial potential of egg yolk ovoinhibitor, a multidomain kazal-like inhibitor of chicken egg. *Journal of Agricultural and Food Chemistry* 59(23): 12368–12374. DOI: 10.1021/jf203339t.

Cadirci, S. (2009). Disinfection of hatching eggs by formaldehyde fumigation - A review. *Archiv fur Geflugelkunde* 73(2): 116–123. Available at: https://www.scopus.com/inward/record.uri?eid=2-s2.0-65949115948&partnerID=40&md5=bdf1ea02d784468a89cdb16055a1826d.

Chien, Y. C., Hincke, M. T. and McKee, M. D. (2009). Ultrastructure of avian eggshell during resorption following egg fertilization. *Journal of Structural Biology* 168(3): 527–538. DOI: 10.1016/j.jsb.2009.07.005.

Cordeiro, C. M. M., Esmaili, H., Ansah, G. and Hincke, M. T. (2013). Ovocalyxin-36 is a pattern recognition protein in chicken eggshell membranes. *PLoS ONE* 8(12): e84112. DOI: 10.1371/journal.pone.0084112.

D'Alba, L. and Shawkey, M. D. (2015). Mechanisms of antimicrobial defense in avian eggs. *Journal of Ornithology* 156(S1): 399–408. DOI: 10.1007/s10336-015-1226-1.

D'Alba, L., Torres, R., Waterhouse, G. I. N., Eliason, C., Hauber, M. E. and Shawkey, M. D. (2017). What does the eggshell cuticle do? A functional comparison of avian eggshell cuticles. *Physiological and Biochemical Zoology: PBZ* 90(5): 588–599. DOI: 10.1086/693434.

Dunn, I. C., Woolliams, J. A., Wilson, P. W., Icken, W., Cavero, D., Jones, A. C., Quinlan-Pluck, F., Williams, G. O. S., Olori, V. and Bain, M. M. (2019). Genetic variation and potential for genetic improvement of cuticle deposition on chicken eggs. *Genetics, Selection, Evolution: GSE* 51(1): 25. DOI: 10.1186/s12711-019-0467-5.

Gabrielli, M. G. and Accili, D. (2010). The chick chorioallantoic membrane: A model of molecular, structural, and functional adaptation to transepithelial ion transport and barrier function during embryonic development. *Journal of Biomedicine and Biotechnology* 2010: 940741. DOI: 10.1155/2010/940741.

Gautron, J., Hincke, M. T., Mann, K., Panheleux, M., Bain, M., McKee, M. D., Solomon, S. E. and Nys, Y. (2001). Ovocalyxin-32, a novel chicken eggshell matrix protein. isolation, amino acid sequencing, cloning, and immunocytochemical localization. *Journal of Biological Chemistry* 276(42): 39243–39252. DOI: 10.1074/jbc.M104543200.

Gautron, J., Murayama, E., Vignal, A., Morisson, M., McKee, M. D., Réhault, S., Labas, V., Belghazi, M., Vidal, M. L., Nys, Y. and Hincke, M. T. (2007). Cloning of ovocalyxin-36, a novel chicken eggshell protein related to lipopolysaccharide-binding proteins, bactericidal permeability-increasing proteins, and plunc family proteins. *Journal of Biological Chemistry* 282(8): 5273–5286. DOI: 10.1074/jbc.M610294200.

Gautron, J., Stapane, L., le Roy, N., Nys, Y., Rodriguez-Navarro, A. B. and Hincke, M. T. (2021). Avian eggshell biomineralization: An update on its structure, mineralogy and protein tool kit. *BMC Molecular and Cell Biology* 22(1): 11. DOI: 10.1186/s12860-021-00350-0.

Gong, D., Wilson, P. W., Bain, M. M., McDade, K., Kalina, J., Hervé-Grépinet, V., Nys, Y. and Dunn, I. C. (2010). Gallin; an antimicrobial peptide member of a new avian defensin family, the ovodefensins, has been subject to recent gene duplication. *BMC Immunology* 11(1): 12. DOI: 10.1186/1471-2172-11-12.

Guyot, N., Rehault-Godbert, S., Slugocki, C., Harichaux, G., Labas, V., Helloin, E. and Nys, Y. (2016). Characterization of egg white antibacterial properties during the first half of incubation: A comparative study between embryonated and unfertilized eggs. *Poultry Science* 95(12): 2956–2970. DOI: 10.3382/ps/pew271.

Halbrooks, P. J., Giannetti, A. M., Klein, J. S., Björkman, P. J., Larouche, J. R., Smith, V. C., MacGillivray, R. T., Everse, S. J. and Mason, A. B. (2005). Composition of pH-sensitive triad in C-lobe of human serum transferrin. Comparison to sequences of ovotransferrin and lactoferrin provides insight into functional differences in iron release. *Biochemistry* 44(47): 15451–15460. DOI: 10.1021/bi0518693.

Halgrain, M., Georgeault, S., Bernardet, N., Hincke, M. T. and Réhault-Godbert, S. (2022). Concomitant morphological modifications of the avian eggshell, eggshell membranes and the chorioallantoic membrane during embryonic development. *Frontiers in Physiology* 13: 838013. DOI: 10.3389/fphys.2022.838013.

Hervé-Grépinet, V., Réhault-Godbert, S., Labas, V., Magallon, T., Derache, C., Lavergne, M., Gautron, J., Lalmanach, A. C. and Nys, Y. (2010). Purification and characterization of avian β-defensin 11, an antimicrobial peptide of the hen egg. *Antimicrobial Agents and Chemotherapy* 54(10): 4401–4409. DOI: 10.1128/AAC.00204-10.

Hincke, M. T., da Silva, M., Guyot, N., Gautron, J., McKee, M. D., Guabiraba-Brito, R. and Réhault-Godbert, S. (2019). Dynamics of structural barriers and innate immune components during incubation of the avian egg: Critical interplay between autonomous embryonic development and maternal anticipation. *Journal of Innate Immunity* 11: 111–124. DOI: 10.1159/000493719.

Hincke, M. T., Gautron, J., Panheleux, M., Garcia-Ruiz, J., McKee, M. D. and Nys, Y. (2000). Identification and localization of lysozyme as a component of eggshell membranes and eggshell matrix. *Matrix Biology: Journal of the International Society for Matrix Biology* 19(5): 443–453. DOI: 10.1016/S0945-053X(00)00095-0.

Hincke, M. T., Gautron, J., Rodríguez-Navarro, A. B. and McKee, M. D. (2011). The eggshell: Structure and protective function.

Kobayashi, K., Hattori, M., Hara-Kudo, Y., Okubo, T., Yamamoto, S., Takita, T. and Sugita-Konishi, Y. (2004). Glycopeptide derived from hen egg ovomucin has the ability to bind enterohemorrhagic *Escherichia coli* O157:H7. *Journal of Agricultural and Food Chemistry* 52(18): 5740–5746. DOI: 10.1021/jf0353335.

Kodali, V. K., Gannon, S. A., Paramasivam, S., Raje, S., Polenova, T. and Thorpe, C. (2011). A novel disulfide-rich protein motif from avian eggshell membranes. *PLoS ONE* 6(3): e18187. DOI: 10.1371/journal.pone.0018187.

Kramer, T. T. and Cho, H. C. (1970). Transfer of immunoglobulins and antibodies in the hen's egg. *Immunology* 19(1): 157–167.

Kulshreshtha, G., D'Alba, L., Dunn, I. C., Rehault-Godbert, S., Rodriguez-Navarro, A. B. and Hincke, M. T. (2022). Properties, genetics and innate immune function of the cuticle in egg-laying species. *Frontiers in Immunology* 13: 838525. DOI: 10.3389/fimmu.2022.838525.

Legros, J., Jan, S., Bonnassie, S., Gautier, M., Croguennec, T., Pezennec, S., Cochet, M. F., Nau, F., Andrews, S. C. and Baron, F. (2021). The role of ovotransferrin in egg-white antimicrobial activity: A review. *Foods* 10(4): 823. DOI: 10.3390/foods10040823.

Liljebjelke, K. A., Hofacre, C. L., Liu, T., White, D. G., Ayers, S., Young, S. and Maurer, J. J. (2005). Vertical and horizontal transmission of *Salmonella* within integrated broiler production system. *Foodborne Pathogens and Disease* 2(1): 90–102. DOI: 10.1089/fpd.2005.2.90.

Mann, K. (2007). The chicken egg white proteome. *Proteomics* 7(19): 3558–3568. DOI: 10.1002/pmic.200700397.

Mann, K. (2008). Proteomic analysis of the chicken egg vitelline membrane. *Proteomics* 8(11): 2322–2332. DOI: 10.1002/pmic.200800032.

Mann, K. and Mann, M. (2011). In-depth analysis of the chicken egg white proteome using an LTQ Orbitrap Velos. *Proteome Science* 9(1): 7. DOI: 10.1186/1477-5956-9-7.

Mine, Y., Ma, F. and Lauriau, S. (2004). Antimicrobial peptides released by enzymatic hydrolysis of hen egg white lysozyme. *Journal of Agricultural and Food Chemistry* 52(5): 1088–1094. DOI: 10.1021/jf0345752.

Nasri, H., van den Brand, H., Najjar, T. and Bouzouaia, M. (2020). Egg storage and breeder age impact on egg quality and embryo development. *Journal of Animal Physiology and Animal Nutrition* 104(1): 257–268. DOI: 10.1111/jpn.13240.

Noble, R. C. and Cocchi, M. (1990). Lipid metabolism and the neonatal chicken. *Progress in Lipid Research* 29(2): 107–140. DOI: 10.1016/0163-7827(90)90014-C.

Nys, Y., Gautron, J., Garcia-Ruiz, J. M. and Hincke, M. T. (2004). Avian eggshell mineralization: Biochemical and functional characterization of matrix proteins. *Comptes Rendus Palevol* 3(6–7): 549–562. DOI: 10.1016/j.crpv.2004.08.002.

Nys, Y. and Guyot, N. (2011). Egg formation and chemistry. In *Improving the Safety and Quality of Eggs and Egg Products: Egg Chemistry, Production and Consumption*. Woodhead Publishing, Cambridge, UK 83–132. DOI: 10.1533/9780857093912.2.83.

Nys, Y., Hincke, M. T., Arias, J. L., Garcia-Ruiz, J. M. and Solomon, S. E. (1999). Avian eggshell mineralization. *Poultry and Avian Biology Reviews* 10(3): 143–166.

Olsen, R., Kudirkiene, E., Thøfner, I., Pors, S., Karlskov-Mortensen, P., Li, L., Papasolomontos, S., Angastiniotou, C. and Christensen, J. (2017). Impact of egg disinfection of hatching eggs on the eggshell microbiome and bacterial load. *Poultry Science* 96(11): 3901–3911. DOI: 10.3382/ps/pex182.

Poulsen, L. L., Thøfner, I., Bisgaard, M., Christensen, J. P., Olsen, R. H. and Christensen, H. (2017). Longitudinal study of transmission of Escherichia coli from broiler breeders to broilers. *Veterinary Microbiology* 207: 13–18. DOI: 10.1016/j.vetmic.2017.05.029.

Read, R. J., Fujinaga, M., Sielecki, A. R. and James, M. N. (1983). Structure of the complex of Streptomyces griseus protease B and the third domain of the turkey ovomucoid inhibitor at 1.8-A resolution. *Biochemistry* 22(19): 4420–4433. DOI: 10.1021/bi00288a012.

Réhault-Godbert, S. and Guyot, N. (2018). Vitellogenesis and yolk proteins, birds. In: Skinner, M. K. (Ed.), *Encyclopedia of Reproduction* (2nd edn.). Oxford: Academic Press, 278-284. DOI: 10.1016/B978-0-12-809633-8.20568-2.

Réhault-Godbert, S., Hervé-Grépinet, V., Gautron, J., Cabau, C., Nys, Y. and Hincke, M. (2011). Molecules involved in chemical defence of the chicken egg. In *Improving the Safety and Quality of Eggs and Egg Products: Egg Chemistry, Production and Consumption*. Woodhead Publishing, Cambridge, UK, 183–208. DOI: 10.1533/9780857093912.2.183.

Réhault-Godbert, S., Hincke, M., Guabiraba, R., Guyot, N. and Gautron, J. (2022). Innate defenses of the avian egg. *Avian Immunology* 365–386. DOI: 10.1016/B978-0-12-818708-1.00032-4.

Romanoff, A. L. and Romanoff, A. J. (1949). *The Avian Egg*. New York: Wiley Co.

Rose-Martel, M. and Hincke, M. T. (2017). The eggshell proteome yields insight into its antimicrobial protection. In: *Egg Innovations and Strategies for Improvements*. Oxford: Elsevier Inc., 157–163. DOI: 10.1016/B978-0-12-800879-9.00015-9.

Schneider, W. R. and Doetsch, R. N. (1974). Velocity measurements of motile bacteria by use of a videotape recording technique. *Applied Microbiology* 27(1): 283–284. DOI: 10.1128/am.27.1.283-284.1974.

Sellier, N., Vidal, M. L., Baron, F., Michel, J., Gautron, J., Protais, M., Beaumont, C., Gautier, M. and Nys, Y. (2007). Estimations of repeatability and heritability of egg albumen antimicrobial activity and of lysozyme and ovotransferrin concentrations. *British Poultry Science* 48(5): 559–566. DOI: 10.1080/00071660701592367.

Sparks, N. H. C. (1994). Shell accessory materials: Structure and function. In: Board, R. G. and Fuller, R. (Eds.), *Microbiology of the Avian Egg*. Boston, MA: Springer US, 25–42. DOI: 10.1007/978-1-4615-3060-2_2.

Sparks, N. H. C. and Board, R. G. (1985). Bacterial penetration of the recently oviposited shell of hens' eggs. *Australian Veterinary Journal* 62(5): 169–170. DOI: 10.1111/j.1751-0813.1985.tb07281.x.

van Dijk, A., Veldhuizen, E. J. A. and Haagsman, H. P. (2008). Avian defensins. *Veterinary Immunology and Immunopathology* 124(1–2): 1–18. DOI: 10.1016/j.vetimm.2007.12.006.

Xing, J., Wellman-Labadie, O., Gautron, J. and Hincke, M. T. (2007). Recombinant eggshell ovocalyxin-32: Expression, purification and biological activity of the glutathione S-transferase fusion protein. *Comparative Biochemistry and Physiology. Part B, Biochemistry and Molecular Biology* 147(2): 172–177. DOI: 10.1016/j.cbpb.2007.01.015.

Yang, R., Geng, F., Huang, X., Qiu, N., Li, S., Teng, H., Chen, L., Song, H. and Huang, Q. (2020). Integrated proteomic, phosphoproteomic and N-glycoproteomic analyses of chicken eggshell matrix. *Food Chemistry* 330: 127167. DOI: 10.1016/j.foodchem.2020.127167.

Chapter 4

Assessing poultry semen quality

Murray R. Bakst, formerly USDA-ARS, USA*

1 Introduction

Poultry breeder and hatchery operations seek to maximize the production of healthy chicks and poults using the most efficient and cost-effective approaches. This aspiration would be enhanced if the technology existed that would allow the selection of future breeder stock from young chicks and poults rather than mature stock. Advances in genomic and associated technologies in identifying biomarkers to predict fertility in domestic animals have been reported (Long et al., 2020) and have been incorporated into some breeding programs at the primary breeder level. Notwithstanding possible future advances in the selection and breeding for high fertility, there will always be a need for the evaluation of semen to both evaluate advances in breeder male selection and to check semen quality prior to artificial insemination (AI).

Currently, most semen samples are routinely being evaluated on commercial breeder farms using AI to produce fertilized eggs. The vast majority of commercial turkey breeder and multiplier farms rely on AI for the production of fertile eggs. In contrast, broiler breeder farms generally rely on natural mating, although AI has been used in some breeding programs. More wide-scale use of AI for broiler breeders has been applied in those countries where labor is relatively cheap. Attempts to manually collect semen for an evaluation from

* Author email: murray_bakst@hotmail.com

http://dx.doi.org/10.19103/AS.2022.0118.05

broiler males that only experienced natural mating can be problematic and lead to samples with semen/sperm traits that are not reflective of the quality at the time of mating. AI is a two-step process whereby the semen is collected from the male and then inseminated into the female. How the semen is handled (diluted, evaluated, transported, and stored) between collection and insemination is dependent on breeder company management goals and the capabilities and willingness of personnel to adhere to company guidelines. These guidelines should include instructions detailing the methods and steps used to evaluate semen collection, its quality and storage, and fertility determination. For the majority of breeder farms using AI, a simple visual examination of the semen is not a sufficient evaluation of its quality. Most often a combination of semen quality traits is evaluated (primarily sperm concentration, motility, and viability) and if one or more of these traits are below a minimum value, the semen is not used for AI (Table 1).

This chapter will discuss how to optimize various aspects of selected semen evaluation procedures currently used on most breeder farms. As most semen evaluation and fertility determination techniques have

Table 1 A summary of the advantages and disadvantages of semen evaluation techniques used on breeder farms

Sperm concentration
Hemocytometer: subjective, time consuming if doing multiple samples, best used to establish the calibration curves (see chapter in Bakst and Long, 2014) used with other semen evaluation techniques
Photometers (table top or portable): objective, must be kept clean and dust free to obtain reproducible optical density measurements
Packed cell volume: objective, relative fast, multiple samples simultaneously prepared, and reproducible when directions precisely followed
Sperm viability
Nigrosin/eosin stain: subjective, time consuming if doing multiple samples, could be coupled with the evaluation of sperm morphology
Fluorescent stains: objective if sperm sorted by flow cytometry, time consuming, not used on breeder farms due to cost and need for dedicated personnel
Sperm motility and mobility
Hanging-drop: subjective, quick
Computer-assisted semen analysis (CASA): objective, detailed information only on sperm motility characteristics, not used on breeder farms due to cost and need for dedicated personnel
Sperm mobility assay: objective, time consuming, identifies most fecund males, portable commercial version has been used on breeder farms (see Appendix 1, Bakst and Long, 2014)
Semen quality analyzers
Objective, integrates sperm motility, concentration, and morphology into a Sperm Quality Index, strong indicator of semen quality

remained essentially unchanged for the past decade, the step-by-step procedures will not be repeated in this chapter. However, these techniques are accessible to the reader by connecting to the link https://midwestpoultry.com/wp-content/uploads/2023/01/Semen-Techniques-Manual-2nd-edition-2014-2022.pdf provided by the Midwest Poultry Federation (MPF). This link provides free access to the manual *Techniques for Semen Evaluation, Semen Storage, and Fertility Determination* (Bakst and Long, 2014) for viewing and downloading. Bakst and Long (2014) includes detailed step-by-step directions and illustrations of the most commonly used semen evaluation and fertility determination techniques applicable to both breeder farm and poultry research laboratories. Additional information regarding AI, particularly the history of AI in poultry, can be found in Mohan et al. (2018).

Before delving directly into semen evaluation, a brief overview of the biological events leading to fertilization is presented. Such knowledge will foster a better understanding of the rationale and biological basis of some of these techniques and their application.

2 Biology of fertilization

2.1 Fertilization versus fertility

For the purpose of this study, fertilization is the result of complex interactions between sperm and an ovum, fusion of their genetic material, and the ensuing formation of the earliest stage of an embryo, a zygote (blastodisc in birds). Alternatively, fertility is a quantitative term used to describe the ratio of the number of fertilized eggs observed relative to the total number of eggs examined. Two other related terms commonly used on the breeder farm and hatchery are 'candling fertility' and 'true fertility'. Candling fertility (generally determined 8–12 days after setting the eggs in an incubator) is the ratio of number of fertilized eggs visualized (evidence of a viable embryo) to the number of eggs examined at candling. For true fertility, personnel must break open (perform break-outs) the candled clear (no evidence of a viable embryo, hence clear) eggs and determine if the eggs were fertilized but failed to develop (nonviable embryos) (Fig. 1). Evidence of fertilization includes an intact or remnants of a blastoderm or embryo, extra-embryonic membrane formation, blood spots, or blood vessel formation. If the above indicators of a nonviable embryo are not readily observed, examine the texture of the yolk. The yolk of a fertilized ovum has a mottled-like, irregular appearance, whereas the unfertilized ovum has the appearance one would expect from a table egg. Once the number of nonviable embryos is determined, true fertility is derived by calculating the ratio of the sum of nonviable and viable embryos to the number of eggs examined at candling.

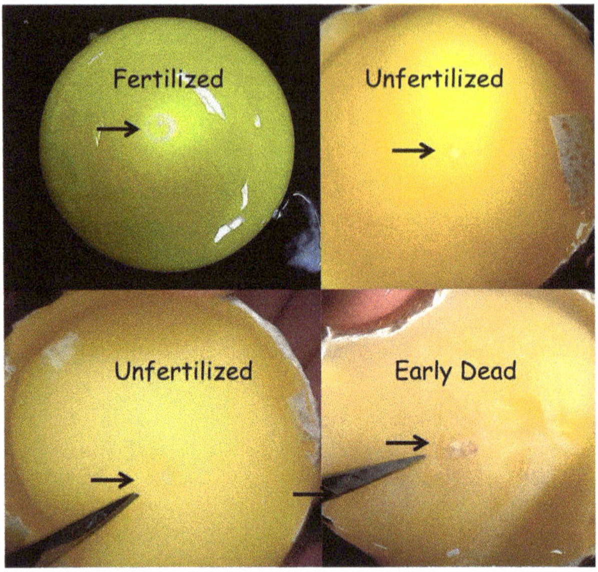

Figure 1 Images in the top row show a fertilized (left) and an unfertilized (right) egg during a fresh egg breakout. Images in the bottom row show an unfertilized (left) and an early dead embryo (right) during a clear egg breakout conducted after 10 days of incubation. Note the mottled-like appearance of the yolk in the early dead egg but not the unfertilized egg after 10 days of incubation. (All the images were turkey eggs.)

2.2 Overview of egg formation and fertilization

Egg formation begins with the freshly ovulated ovum (egg yolk) being guided into the infundibulum, the most anterior segment of the oviduct adjacent to the ovary. At ovulation, the ovum is enveloped by a single fibrous layer - the inner perivitelline layer (IPL). Sperm must be present in the anterior end of the infundibulum at this time in order to interact with the outer surface of the IPL as it is a partial barrier to sperm penetration into the ovum. While sperm can digest a path (a hole) anywhere in the IPL, fertilization can only take place at the germinal disc (GD) region. The GD is a 3-4 mm diameter whitish disc composed of white yolk spheres, lipids, and cell organelles including the haploid nucleus (a single set of unpaired chromosomes, referred to as a pronucleus). Another unique property of the GD is that sperm preferentially bind to its overlying IPL in far greater numbers (per square mm) than the IPL overlying the yellow yolk. Several hundreds of sperm can penetrate the IPL overlying the GD and enter this region (polyspermy). However, only one haploid male pronucleus derived from a single sperm will fuse with the single haploid female pronucleus in the GD. This fusion, referred to as syngamy, results in the paired, diploid number of chromosomes. What then follows is a cascade of events that give rise to the first visual evidence of fertilization – the initial cleavage divisions of the fertilized GD. This occurs shortly after the egg mass (term describing an oviductal egg) enters the uterus, about 5.5 h after fertilization. Further differentiation of the fertilized GD over the next several hours gives rise to a disc-shaped sheet of cells of uniform thickness, the blastoderm (Eyal-Giladi and Kochav, 1976).

Whether sperm are present or not, the ovulated ovum continues through the infundibulum and acquires another investment consisting of albumen-like proteins, the outer perivitelline layer (OPL). The OPL adds strength to the IPL and also serves as an impenetrable barrier to sperm thereby preventing pathological polyspermy. An hour or so after ovulation, the ovum enters the magnum where over the next 3 h the proteins and fluids that form the albumen layer (egg white) are deposited around the ovum. The egg mass is transported to the isthmus where the shell membrane is formed and then transported to the uterus where the shell is formed over the next 20 h. The hard-shelled egg mass then passes through the utero-vagina junction (UVJ) and makes a brief transit through the vagina before oviposition. The embryo of a fresh laid egg (blastoderm) consists of 60 000 to 100 000 cells.

Sustained fertility can only be attained if there are sufficient numbers of sperm entering the vagina at the time of semen transfer. With turkeys, the number of sperm inseminated is a management decision and can vary between 200 million and 400 million sperm. This decision is based on the ages of both the tom and hen flocks, the true fertility status of the hens, and the frequency of AI. Regardless if 200 million or 400 million sperm are inseminated, only 4–5 million sperm reside in the sperm storage tubules (SSTs) located in the UVJ with the vast majority of sperm presumably expelled from the vagina. Those sperm that reach the SSTs were subjected to an intense selection process orchestrated by the vagina. The competition between sperm to reach and enter the SSTs favors those sperm possessing the highest mobility (ability to swim through a viscous medium at 41°C; Froman et al., 1997). But high mobility is not the only criterion for sperm selection. Additional factors that may influence sperm selection and transport through the vagina could include sperm surface antigens and glucoconjugates; neuroendocrine-like cells in the vaginal epithelium (Bakst and Akuffo, 2008); and other yet to be defined factors in the vaginal luminal milieu at the time of semen transfer. Relatively new evidence suggests that the vaginal microbiome (Rowe et al., 2020) as well as extracellular vesicles in semen (Cordeiro et al., 2021) may influence the fate of sperm in the oviduct for a variety of species. For a more comprehensive understanding of the sperm-oviduct interactions addressed above, see Stepinska and Bakst (2007), Birkhead and Brillard (2007), Bakst and Dymond (2013), Bakst (2018), Assersohn et al. (2021), and Brady et al. (2022).

3 Evaluation of semen

3.1 Purpose of semen evaluation

The weekly evaluation of semen provides data on male reproductive fitness that can also serve as an indicator of male health/nutritional status. Breeder company management can use such data to monitor and compare the week-to-week status of a single flock, or multiple flocks, at a single or multiple geographical locations. While it is expected that there will be some variation in weekly semen quality estimates between similarly aged flocks at various locations, sudden, highly variable semen quality estimates in and between flocks suggest a potential problem. More than likely, one will find that the problem is not due to the biology of the male but some change in the management of one or more flocks. Approaches addressing variable semen quality within and between flocks should include an assessment of the personnel performing semen collection, dilution, storage, evaluation, and insemination. For this assessment to be meaningful, baseline instructions and guidelines that define each task are required to be provided to relevant personnel by the breeder company. These company-wide standard operating procedures (SOPs) are not only an important educational tool but a management tool as well. Their implementation will lead to increased efficiency and uniformity in personnel performance, while reducing variability in the results. Given widespread accessibility to the internet, electronic (recorded) and virtual (real-time and interactive) learning is best suited for the introduction, instruction, and follow-up training of the SOPs.

3.2 Categories of semen evaluation techniques

These can be divided into two broad categories. The first category comprises the relatively simple qualitative procedures, e.g. determination of semen volume and appearance (is the semen contaminated or off color), sperm motility and morphology, sperm concentration using a hemocytometer, and slides with stained sperm (live/dead stains) to microscopically determine the percentage of live sperm. The second category includes those techniques requiring technical, often computer-based equipment that produce more precise quantitative data, e.g. the ratio of sperm viability (fluorescent cell sorting), sperm motility/mobility characteristics by computer-assisted-semen-analysis (CASA), and various techniques for determining an index of semen quality such as a sperm quality analyzer (SQA). Earlier versions of the SQA [see Wishart and Wilson, Chapt. IV, Sect. 2, *in* Bakst and Long (2014)] measured sperm motility characteristics and gave rise to the sperm motility index (SMI), whereas later versions of the SQA integrated sperm concentration, morphology, and motility to arrive at a more comprehensive SQI (Dumpala et al., 2006). The distinction between qualitative estimates and quantitative data is that qualitative data is subjective, that is, derived by an individual who subjectively decides how the test sample compares to learned guidelines. In contrast, quantitative data derived from calibrated precision instruments is consistent and does not reflect biases that may be expressed by the evaluator (Table 1).

The hanging droplet method used to rate sperm motility is a prime example of a subjective test (see Appendix for methodology and scoring guide). Reviewing the guidelines used for scoring a hanging-drop preparation cited in the Appendix, one can see how it could be interpreted differently by different personnel. When viewing swirls consisting of hundreds of thousands of sperm, discerning whether most or a few sperm are either vigorously or slowly moving in a forward motion is clearly a decision that can be subject to personal biases. A revised method of the hanging-drop procedure that includes an improved scoring system is available in Bakst and Long (2014). As noted previously, objective data describing sperm motility and associated kinetics (sperm velocity, tail beat amplitude and frequency, path linearity) can be obtained with a CASA system and can be used to improve fertility (Tesfay et al., 2020). However, due to the initial and subsequent operational costs and the need for trained personnel for its operation it is unlikely that a CASA system will be used on a breeder farm for routine motility evaluation of poultry semen. Alternatively, a semen quality analyzer (SQA) and the sperm motility assay are both objective and applicable to breeder farm use. Of all the semen quality techniques currently available, only the sperm mobility assay is capable of identifying the most fecund males (Froman et al., 1997; Froman and Rhoads, 2013).

Whether a subjective or objective method for evaluating semen is used, it is critical that the instructions and guidelines in the SOPs are precisely followed by all personnel to attain reproducible results. For example, when estimating sperm concentration using the packed cell volume procedure (PCV, also referred to as a spermatocrit) a 3 min centrifugation of the semen will not give the same result as a 6 min centrifugation of the same sample. Other common examples where omissions or inconsistencies in laboratory techniques will influence the outcome of test results include pooled semen samples not thoroughly mixed before evaluation; staining or incubation times are not consistent; not routinely wiping the outside of the tip on a transfer pipette and not examining the tip contents for bubbles and debris before transfer; not checking that all the semen has been discharged from a pipette tip; not checking cuvettes for dust and fingerprints immediately before use; or not using the same magnification each time PL sperm holes are counted or in the assessment of sperm motility. If these simple and quick procedures and checks are not performed, data will be unreliable and comparisons useless.

3.3 From collection to breakout: an encompassing perspective of semen quality

Sustained high fertility is only possible when high-quality semen is used for AI. The initial screening of the semen, and arguably the most critical, begins with the semen collector (milker) during semen collection. The milker, having sufficient light directed on the cloaca, must be able to quickly discern uncontaminated semen seconds after it appears on the phallus and collect only that semen into a collection vial.

Whether housed in cage or a floor pen, how breeder males are handled before and during the semen collection process will impact semen quality. Stressed males may be unwilling to cooperate during semen collection and may require extra cloacal stokes (the manual squeezing of the cloacal region to express semen) to complete the semen collection process. Generally, two cloaca strokes will express the maximum volume of good quality semen. A third cloacal stroke is up to the discretion of the collector. Additional cloacal strokes may stress the male causing the collector to squeeze harder increasing the risk of semen contamination with fecal debris, urates, blood, and possible bruising of the cloaca.

Excessive handling of the hens by other members of the AI crew, such as the chasers and catchers, or by the inseminator during both the cloacal breaking (exteriorizing the vagina) and hen release after insemination may stress the hens and possibly impact egg production and shell quality. With AI, the successful transfer of semen into the hen is dependent on proper coordination between the inseminator shooting the semen into the hen's exteriorized vagina and the breaker releasing the pressure around the cloaca, thereby facilitating semen retention by the vagina. If semen is observed on the cloacal lips after the breaker releases pressure, the hen should be inseminated again. One approach to checking the success of semen transfer at insemination is to examine the discarded semen straws a few minutes after they have been used and look for residual semen. If more than 10% of the original semen volume remains in the used straws most likely the inseminating gun is not functioning in an optimal manner. The insemination gun should be checked for debris obstructing the semen expulsion mechanism. If the problem persists, examine the seals and gaskets in the insemination gun and replace if necessary.

With turkeys, to maximize the sperm population in the SSTs, the first two inseminations must be performed about 7–10 days before the first eggs are laid (Christensen and Johnston, 1977), or when 50% or more of the hens show squatting behavior. These pre-lay inseminations result in higher initial and sustained fertility than if the hens were inseminated after the first egg is laid. Males should be photo-stimulated early enough so that they can be collected and the semen evaluated prior to or around the onset of hen squatting behavior. This practice assures males are producing good quality semen for the first two inseminations prior to the onset of lay. It also presents an opportunity to both cull those with poor semen quality traits and identify those males with superior semen quality traits.

4 Fertility determination

Techniques for determining and predicting fertility were presented in Bakst and Long (2014) and are considered measures of semen quality following AI. There are two types of assessments. One relies on the visual examination of the GD determine if the GD was fertilized or not, and if fertilized, whether the embryo was viable or nonviable. The other types of procedures require a microscope to both determine the numbers of sperm trapped in the OPL or the number of sperm holes in the IPL. The holes observed in the IPL overlying the GD are actual sites where sperm penetrated the IPL to gain access to the GD at the time of fertilization. These data can be applied to both estimate true fertility and predict the duration of fertility.

Determining true fertility during egg-breakouts is a very useful management tool. If the hatchery reports low candling fertility, the question that must be immediately addressed is if this is due to poor hen fertility (a breeder farm issue) or high embryo mortality (most likely due to poor egg handling during storage at the breeder farm, transportation to the hatchery, or storage at the hatchery). The longer this information is unavailable, the lower the production of hatching eggs. For example, if eggs are set 6-14 days after lay and candled after 8-12 days of incubation, then 14–28 days of egg production elapsed before low true fertility was detected and corrective measures initiated. On the breeder farm using AI, assuming that the flocks are in good health, and there are no housing or AI personnel issues, low true fertility may be remedied by immediately increasing the number of sperm per insemination dose and/or decreasing the number of days between successive inseminations. With natural mating, the issue of low true fertility is more complex. From a behavioral perspective, males may be losing interest in mating. Likewise, the hen may be unreceptive to the male or if initially receptive the hen may abandon the male in the course of mating. More rarely, males may fail to successfully or adequately transfer semen into the hen's vagina during copulation either due size differences or to musculoskeletal conformation issues (McGary et al., 2003). (For more detailed information on fertility issues with natural mating broilers, see Chapter 2 of this book).

As mentioned above, while egg breakouts provide the basis for true fertility determination, IPL-hole and OPL-sperm counts could provide estimates of not only true fertility but also the duration of fertility. Low sperm hole counts also indicate that sperm numbers in the SSTs may be nearly depleted (Brillard and Bakst, 1990; Wishart, 1995; Wishart and Staines, 1999). Both assays are best performed with unincubated eggs within 72 h after lay. With eggs stored longer or incubated the IPL and OPL become progressively more fragile and difficult to isolate and prepare properly for the assays. An IPL sperm-hole number below 6 indicates that relatively few sperm are reaching the site of fertilization in the infundibulum and the probability of that ovum and subsequent ova being fertilized is less that 5%. Conversely, if the IPL sperm holes are so dense that they appear to have merged, forming gaps in the PL, excessive numbers sperm were in the infundibulum at ovulation and had penetrated the PL. Lorenz and Ogasawara (1968) observed high rates of early embryonic mortality in eggs laid by hens inseminated in the anterior end of the oviduct. By introducing sperm into the oviduct in this manner, the vaginal and UVJ sperm selection mechanisms are by-passed thereby allowing nearly all sperm transferred to reach the infundibulum.

In the course of inseminating a flock of hens, it is likely that a small percentage of the hens are inseminated within a 20–30 min window following oviposition. During this window, the breaker (the individual that manually everts the vagina at the time of insemination) may inadvertently cause a partial prolapsed of the vagina (this is sometimes referred to as a 'rose' by AI crewmembers). Without much difficulty, the inseminator can insert the inseminating straw up to or beyond the UVJ thus bypassing the normal sperm selection mechanisms in the vagina, thus leading to possible pathological polyspermy.

5 Conclusion

The majority of the semen evaluation procedures discussed above should not be used to predict the fertilizing capacity of semen samples. However, they may be considered as a basis for culling breeder males (J.P. Brillard, personal communication). Semen evaluation techniques that indicate more fecund sperm are restricted to two procedures. Of the single-trait semen quality tests currently available, only the sperm mobility assay developed by D. Froman and colleagues [see Bakst and Long (2014) for both the laboratory method and its equivalent for commercial farm use] actually reveals individual males producing sperm with high mobility rates. When semen for these males is inseminated into hens, the probability of fertilization is higher than when semen from males producing low mobility sperm was inseminated. Semen quality analyzers integrate multiple sperm traits (motility, concentration, and viability) into a SQI that objectively differentiates good quality semen from lesser quality semen samples (McDaniel et al., 1998; Dumpala et al., 2006). Finally, as noted in the beginning of this chapter, regardless of the advances in the genetic selection of breeder males for high fertility, on-farm evaluation of semen will continue to remain an integral component in poultry breeding management.

6 Acknowledgements

I want to thank the Midwest Poultry Federation for generously posting the link to *Techniques for Semen Evaluation, Semen Storage, and Fertility Determination* on their website and specifically thank Lara Durben for making that happen. I also thank J. P. Brillard and John Sharp for most informative and helpful discussions regarding the contents of this text.

7 Appendix

The hanging-drop method was described by Parker, et al. (1942) as follows: 'A drop of undiluted semen was examined microscopically using the hanging-drop technique for the amount and type of movement.' The hanging-drop method used in some laboratories today involves placing a drop of undiluted semen on a standard slide cover-glass and placing the cover-glass, semen side down, over the well in a 1.5 cm diameter well-slide. Sperm motility is evaluated using a low power objective (10x) and scored by the method described by Parker et al. (1942) as 5 (vigorous and progressive), 4 (few sperm vigorously motile and few sperm inactive), 3 (many sperm moderately active and a number of inactive sperm), 2 (many sperm showing slight motility and a number of inactive sperm), 1 (few sperm showing slight motility), and 0 (no motility).

The hanging-drop method used in some laboratories today can be done by one of the two methods. The first is described in detail in Bakst and Long (2014, see Chapter 4, Section 1, p. 45). The second is performed by placing a drop of undiluted semen on a standard slide cover-glass and placing the cover-glass, semen side down, over the well in a 1.5 cm diameter well-slide. Both are scored using the method in Bakst and Long (2014).

8 References

Assersohn, K., Brekke, P. and Hemmings, N. (2021), Physiological factors influencing female fertility in birds, *R. Soc. Open Sci.*, 8(7), 202274, http://doi.org/10.1098/rsos.202274.

Bakst, M. R. (2018), Reproduction in poultry: an overview, in Scanes, C. G. and Hill, R. A. (Eds.), *Biology of Domestic Animals*, pp. 213–240, Boca Raton, FL: CRC Press, Taylor & Francis Group.

Bakst, M. R. and Akuffo, V. (2008), Serotonin localization in the turkey vaginal but not sperm storage tubule epithelia, *J. Avian Biol.*, 39, 348–354.

Bakst, M. R. and Dymond, J. S. (2013), Artificial insemination in poultry, in Lemma, A. (Ed.), *Success in Artificial Insemination – Quality of Semen and Diagnostics Employed*, https://doi.org/10.5772/54918.

Bakst, M. and Long, J. (2014), Techniques for semen evaluation, semen storage, and fertility determination, midwest poultry federation, Available at: https://midwestpoultry.com/speakers/additional-resources.

Birkhead, T. T. and Brillard, J. P. (2007), Reproductive isolation in birds: postcopulatory prezygotic barrier, *Trends Evol.*, 22, 266–272.

Brady, K., Krasnec, K. and Long, J. A. (2022), Transcriptome analysis of inseminated sperm storage tubules throughout the duration of fertility in the domestic turkey, *Meleagris gallopavo*, *Poult. Sci.*, 101(4), 101704, https://doi.org/10.1016/j.psj.2022.101704.

Brillard, J. P. and Bakst, M. R. (1990), Quantification of spermatozoa in the sperm-storage tubules of turkey hens and the relation to sperm numbers in the perivitelline layer of eggs, *Biol. Reprod.*, 43(2), 271–275.

Christensen, V. L. and Johnston, N. P. (1977), Effect of time of day of insemination and the position of the egg in the oviduct on fertility of turkeys, *Poult. Sci.*, 56(2), 458–462.

Cordeiro, L., Lin, H. H., Carvalho, A. V., Grasseau, I., Uzbekov, R. and Blesbois, E. (2021), First insights on seminal extracelluar vesicles in chickens of contrasted fertility, *Reproduction*, 161(5), 489–498.

Dumpala, H. M., Parker, H. M. and McDaniel, C. D. (2006), The sperm quality index from fresh semen predicts semen quality after storage, *Int. J. Poult. Sci.*, 5, 850–855.

Eyal-Giladi, H. and Kochav, S. (1976), From cleavage to primitive streak formation: complementary normal table and a new look at first stages of the development of the chick. I. general morphology, *Dev. Biol.*, 49(2), 321–337.

Froman, D. P., Feltmann, A. J. and McLean, D. J. (1997), Increased fecundity resulting from semen donor selection based in vitro sperm motility, *Poult. Sci.*, 76(1), 73–77.

Froman, D. P. and Rhoads, D. D. (2013), Breeding and genetics symposium: a systems biology definition for chicken semen quality, *J. Anim. Sci.*, 91(2), 523–529.

Long, X., Wang, Y., Wu, F., Li, R., Chen, L., Qian, W. and Qiao, J. (2020), Pregnancy outcomes of single/double blastocysts and cleavage embryo transfers: a retrospective cohort study of 24,422 frozen-thawed cycles, *Reprod. Sci.*, 27(12), 2271–2278.

Lorenz, F. W. and Ogasawara, F. X. (1968), Distribution of spermatozoa in the oviduct and fertility in domestic birds. VI. The relations of fertility and embryo normality with the site of experimental insemination, *J. Reprod. Fertil.*, 16(3), 445–455.

McDaniel, C. D., Hannah, J. L., Parker, H. M., Smith, T. W., Schultz, C. D. and Zumwalt, C. D. (1998), Use of sperm analyzer for evaluating breeder males. 1. Effects of altering semen quality and quantity on the sperm motility index, *Poult. Sci.*, 77(6), 888–893.

McGary, S., Estevez, I. and Bakst, M. R. (2003), Potential relationships between physical traits and male broiler breeder fertility, *Poult. Sci.*, 82(2), 328–337.

Mohan, J., Sharma, S. K., Kolluri, G. and Dhama, K. (2018), History of artificial insemination in poultry, it's components and significance, *Wld's, Poult. Sci. J.*, 74, 475–488. https://doi.org/10.1017/S0043933918000430.

Parker, J. E., McKenzie, F. F. and Kempster, H. L. (1942), Fertility in the male domestic fowl, *Mo. Ag. Exp. Sta. Res. Bull.*, 347.

Rowe, M., Veerus, L., Trosvik, P., Buckling, A. and Pizzari, T. (2020), The reproductive microbiome: an emerging driver of sexual selection, sexual conflict, mating systems, and reproductive isolation, *Trends Ecol. Evol.*, 35(3), 220–234.

Stepinska, U. and Bakst, M. R. (2007), Fertilization, in Jamieson, B. G. M. (Ed.), *Reproductive Biology and Phylogeny of Birds*, pp. 553–587, Enfield, NH: Science Publishers.

Tesfay, H. H., Sun, Y., Li, Y., Shi, L., Fan, J., Wang, P., Zong, Y., Ni, A., Ma, H., Mani, A. I. and Chen, J. (2020), Comparative studies of semen quality traits and sperm kinematic parameters in relation to fertility rate between 2 genetic groups of breed lines, *Poult. Sci.*, 99(11), 6139–6146.

Wishart, G. J. (1995), New approaches to evaluating male and female fertility, in Bakst, M. R. and Wishart, G. J. (Eds.). *Proceedings of the First International Symposium on The. Artificial Insemination of Poultry*, pp. 207–223, Savoy, IL: The Poultry Science Association, Inc.

Wishart, G. J. and Staines, H. J. (1999), Measuring sperm:egg interaction to assess breeding efficiency in chickens and turkeys, *Poult. Sci.*, 78(3), 428–436.

Chapter 5

Key stages of embryo development in poultry

Marleen Boerjan, formerly R&D Royal PasReform, The Netherlands

1 Introduction

In this chapter we follow the key stages in the development of the chicken embryo: from the fertilized oocyte, the zygote, through to the hatched chicken. Extra-embryonic tissues like sub-embryonic fluid, yolk sac, amnion and chorion-allantois membrane (CAM) develop in synchrony with the embryo and play essential roles in the reallocation of albumen water, nutrient uptake from the yolk, removal of metabolic waste products and protection of the embryo.

In this chapter the main focus is the basic morphological and physiological aspects of the development of the chicken in general. This means that we ignore possible differences in (timing) of embryo development between

http://dx.doi.org/10.19103/AS.2022.0118.03

commercial poultry breeds selected for layer or meat production. The focus on the development of the chicken embryo serves as an example for other commercial poultry species such as turkey, duck, goose and ostrich (Sellier, et al., 2006; Li, et al., 2019).

2 Nomenclature describing the different phases of embryonic development

The morphological description of the earliest embryo stages uses the roman numbering system defined by Eyal-Giladi and Kochav (1976). Stages EGK I-EGK X describe the morphological development in-utero before egg-laying/oviposition. Stages EGK-XI, EGK-XII and EGK-XIII describe stages during the diapause period, egg storage in hatcheries and early incubation (Fig. 1). Hamburger and Hamilton (1951) produced a standard table based on easily recognized morphological criteria for each embryonic stage of development during incubation. They used Arabic numbers to classify stages e.g. stage EGK-XIII corresponds with HH stage 1. Comparative classifications of embryonic stages for other commercial poultry species can be found in Sellier et al. (2006; turkey, duck, goose, guinea fowl, Japanese quail and chicken) and Li et al. (2019; duck, goose and chicken). A detailed chronology of chicken embryo

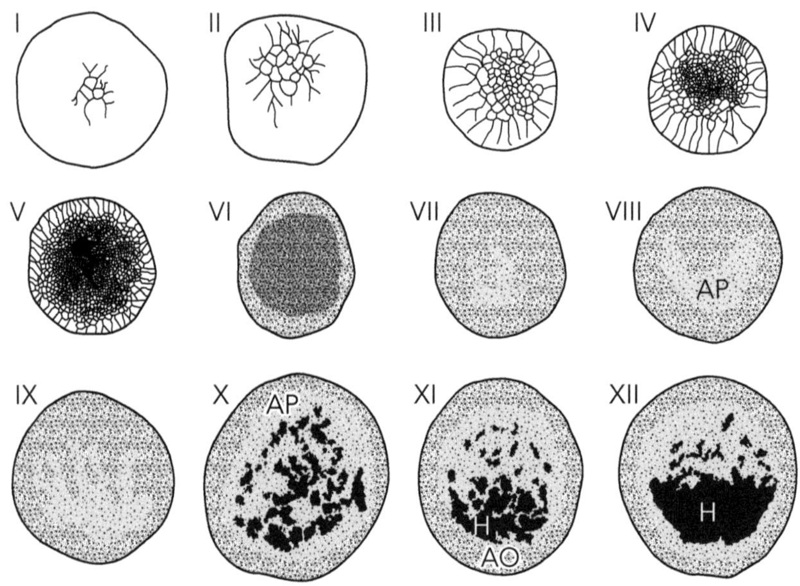

Figure 1 Normal stages of chick development based on Eyal-Giladi and Kochav (1976) which includes stages I-XIII from cleavage to primitive streak formation. This figure only shows stages I-XII Source: Khaner (2017) based on Eyal-Giladi and Kochav (1976). AP =area pellucida; AO = Area Opaca; H=hypobast.

development is provided in Table 1 (which, given its length, is positioned as an Appendix after the References section).

3 Fertilization: polyspermy and formation of the one-cell embryo/zygote

Embryonic development starts with fertilization: the fusion of the female oocyte and sperm cells from the male to form a one-cell embryo/zygote. The bird oocyte is typically telolecithal. It is characterized by a small disc of cytoplasm including the germinal disk on top of a large yolk: called the animal pole. The yolk forms the vegetal pole of the oocyte and provides nutrition for the growing embryo during the incubation of the egg and first days of the chicken after hatching.

The completed preovulatory oocyte is covered by the inner-perivitelline-layer (IPVL) before being released (ovulated) from the dominant ovarian follicle into the infundibulum of the oviduct. Fertilization should occur within 15 min after the oocyte is released from the dominant follicle and before the outer perivitelline membrane (OPVL) is formed by the epithelium cells of the infundibulum (Etches, 1996; Krawczyk and Jaworska-Adamu, 2018).

Like the mammalian sperm the avian sperm has a flagellum/tail, a midpiece loaded with mitochondria to provide energy for sperm motility, a head containing (haploid) nuclear material and a vesicle known as 'the acrosome'. The latter is loaded with proteolytic enzymes which become activated after sperm binding to the ZP1 and ZP3 glycoproteins present in the bird's inner peri-vitelline layer (IPVL). The chicken ZP1 and ZP3 glycoproteins are homologues to the mammalian zona pellucida glycoproteins ZP1 and ZP3 (Ichikawa et al., 2016).

Sperm binding to the ZP1 and ZP3 proteins initiates the release of local sperm protease from the acrosome followed by the entry of the sperm nuclear material which leaves small holes in the inner perivitelline membrane (IPVL) of the avian oocyte/ovum (Wishart and Staines, 1999). Sperm binding to ZP1 and ZP3 glycoproteins also initiates the formation of the female pronucleus.

In contrast to mammals, where polyspermy is pathogenic, physiological polyspermy is the rule in birds and is needed for optimum development (Etches, 1996). The fact that the avian oocyte needs to fuse with more than one sperm cell for optimum embryonic development was described for pigeons by Harper in 1902. More recent studies of polyspermy in birds include Blesbois et al. (2012), Hemmings and Birkhead (2015), Ichikawa, et al. (2016) and Krawczyk and Jaworska-Adamu (2018). Krawczyk and Jaworska-Adamu (2018) describe the interaction of multiple sperm cells binding to ZP1 and ZP3 proteins of the inner perivitelline membrane and the formation of multiple sperm pronuclei near and in the germinal disc. Finally, only one sperm-derived pronucleus fuses with the pronucleus from the oocyte and forms a one-cell embryo or zygote.

Hemmings and Birkhead (2015) have shown that more than a few sperm cells are needed to fuse with the avian oocyte in order to stimulate optimum early development and survival of the early embryo during incubation (i.e. reducing early embryonic mortality).

The entry of multiple sperm cells activates the single (female) nucleus in the germinal disk to complete meiotic cell division and the development of the haploid female pronucleus. Meiotic cell divisions occur only during the formation of haploid male and female germ cells, respectively the sperm and oocytes. The process of meiosis consists of a sequence of two different cell division cycles and only one cycle of DNA synthesis (Fig. 2). During the first meiotic cell division the homologous chromosomes, in a diploid cell, connect and exchange genetic information (cross-over of chromatids). Daughter nuclei receive one of the homologous chromosomes. In the second meiotic division the sister chromatids/chromosome are divided between the next two daughter cells (Fig. 2). Thereafter the chromosomes decondense and a round of DNA synthesis follows to form a sister chromatid/chromosome: the final haploid germ cell. In conclusion: meiosis results in germs cells which contain half the number of chromosomes such that, after fusion (fertilization) of the female and the male haploid germ cells, a diploid one-cell embryo, the zygote, is formed.

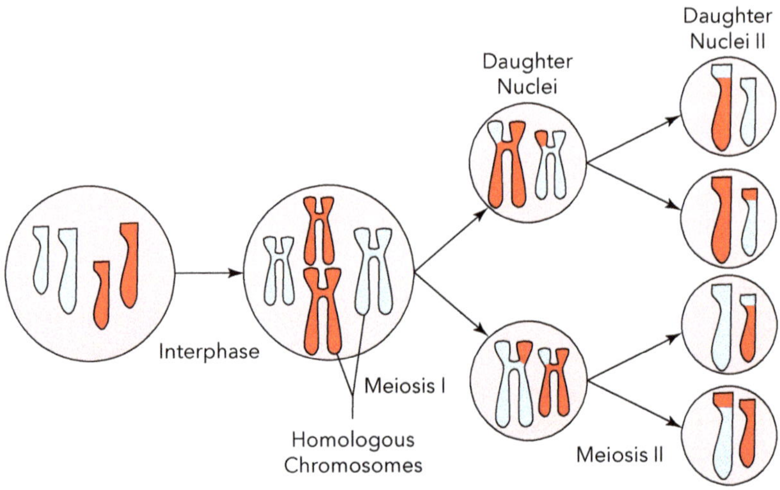

Figure 2 Meiotic cell divisions occur only during the formation of haploid male and female germ cells respectively sperm and oocytes. The process of meiosis consists of a sequence of two different cell divisions and only one cycle of DNA synthesis. The first cell division is called *meiosis I* with the main event: exchange of genetic information between the paired homologous chromosomes (cross-over of chromatids) and daughter nuclei receive one of the homologous chromosomes. During *meiosis II* the sister chromatids/chromosomes are divided over the next two daughter cells. Finally, one meiotic cell cycle results in four haploid germ cells.

The multiplication of the diploid zygote is governed by mitotic cell divisions. The first step of mitotic cell division results in two cells each with a total number of species-specific chromatids. The total amount of DNA is restored by a process of DNA replication during the interphase of the cell. In interphase cells the decondensed chromatids are organized in the nucleus which facilitates DNA replication and formation of chromosomes. In the chicken the total number of chromosomes is 78: 38 pairs autosome chromosomes + 1 pair sex chromosomes: female ZW and male ZZ.

In synchrony with zygote formation in the germinal disk, the multiple sperm heads from the extra male pronuclei degenerate. Yolk-related DNases are involved in the degeneration of these supernumerary male pronuclei (Blesbois, 2012).

4 Intra-uterine development: from zygote to Eyal-Giladi stage X at oviposition

4.1 Zygotic Gene Activation and formation of primordial germ cells

In birds the first stages of embryonic development occur in-utero during egg formation (Fig. 3). In the chicken, stages EGK-I to EGK-X develop during the first 25 h of egg formation in the oviduct and shell gland also known as the uterus (Eyal-Giladi and Kochav, 1976; Khaner, 2017; Hwang et al., 2018).

The one-cell embryo/zygote is totipotent which means that all embryonic cell types, including primordial germ cells (PGCs) and cells forming the extra-embryonic tissues, originate from the totipotent zygote. In birds the first cleavage divisions are incomplete i.e. the first cleavage divisions result in cells lacking a cell membrane at the yolk side. These membranes are first formed in the central cells after a few cleavage divisions. By that time these cells are loaded with yolk granules which are metabolized during early development.

Most of the cells in the Eyal-Giladi stages I to IV are still undifferentiated, pluripotent and have the potential to develop into different structures, tissue or organs of the final embryo. The PGCs are an exception. The PGCs are programmed to form the future gametes i.e. oocytes and sperm. Recently germ cell-specific gene expression was found in EGK stage I-IV embryos (Rengaraj et al., 2020).

After the fusion of the maternal and paternal pronuclei, only maternal mRNA genes and proteins are available to activate embryonic genes in a process called Zygotic Gene Activation (ZGA). In chicken the first cleavage divisions and development to stage EGK-V are governed solely by genes activated from the maternal genome. Hwang et al. (2018) showed that, in the Eyal-Giladi cleavage stages I to IV, only maternal White Leghorn (fWL) genes

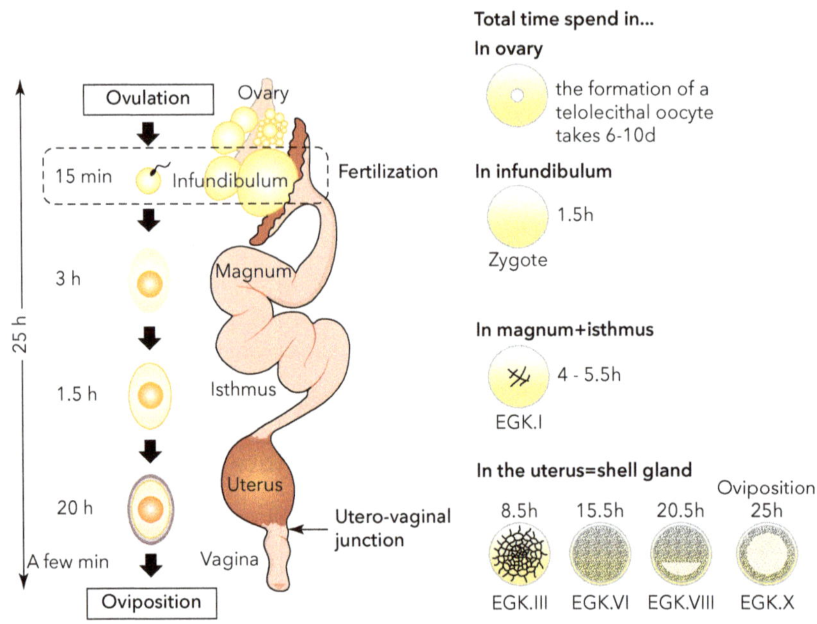

Figure 3 Overview of fertilization and first stages of embryonic development in-utero during egg formation.

are activated i.e. before the embryo reaches the shell gland (uterus). Transcripts from paternal Korean Oge (mKO) genes were observed from EGK stage V and were fully active from EGK stage X (see Fig. 3 in Hwang et al., 2018). Hwang et al. (2018) hypothesized that delayed activation of paternal genome in the embryonic cells prevents activation of supernumerary sperm pronuclei present in the egg after polyspermic fertilization.

4.2 Primordial germ cells and their role in reproduction

PGCs can be defined as precursor cells for future oocytes and sperm cells and thus carry cellular and genetic information from parents. In day-old chicks the PGCs are stored in gonads as precursor cells of oocytes and sperm cells. These cells are unique in forming a continuum between generations.

For decades scientists have been trying to find answers to the following question:

- what is the origin of PGCs?
- are they set aside after fertilization and first cleavage divisions or do they originate from embryonic cells induced to form primordial germ cell?

In 1885 Weismann put forward his 'continuity of the germ plasm' theory. His idea was that in the maternal oocyte a certain amount of future 'cellular germ' plasma is separated from ordinary (somatic) cellular plasma. In the early embryo as a consequence, after a few cell divisions, germ cells as well as somatic cells arise from different/distinct populations. The presence of 'germ plasm' has recently been demonstrated by a Japanese/Korean research group. Rengaraj et al. (2020) found germ cell-specific gene expression in the cleavage stage of chicken embryos (EGK stage I–IV).

Tracking the movement of PGCs in early chicken blastoderms (EG&K stages X–XII) became possible with specific staining techniques and in-vitro culture techniques developed by Meyer (1960). Using these methods Meyer (1964) was able to study the migration of PGCs via the blood vessels in the vasculosa area to the bilateral gonadal region in stage HH15–HH20 (E3) (Fig. 4). Eyal-Giladi et al. (1981) and Ginsburg and Eyal-Giladi (1986) showed that PGCs were randomly present in the epiblast and had migrated to form the germinal crescent in the HH stage 6–10 embryo (Fig. 4). These findings have been confirmed by immune-histochemistry using PGC-specific antigens (SSEA-1) (Karagenç et al., 1996).

As embryonic development proceeds, the PGCs migrate via the blood vessels of what is known as the 'area pellucida' into the developing but still indifferent (undifferentiated) gonadal ridge of HH stage 20 (Fig. 4). The gonadal ridge loaded with PGCs develops into a testis in males carrying ZZ chromosomes. If the embryo has the ZW genotype, the loaded gonadal ridge including PGCs develops into the ovary. Recent research has enabled (Kong et al., 2018; Szczerba et al., 2020):

- easy collection/purification of chicken PGCs from embryonic blood
- in-vitro techniques to keep PGCs undifferentiated.

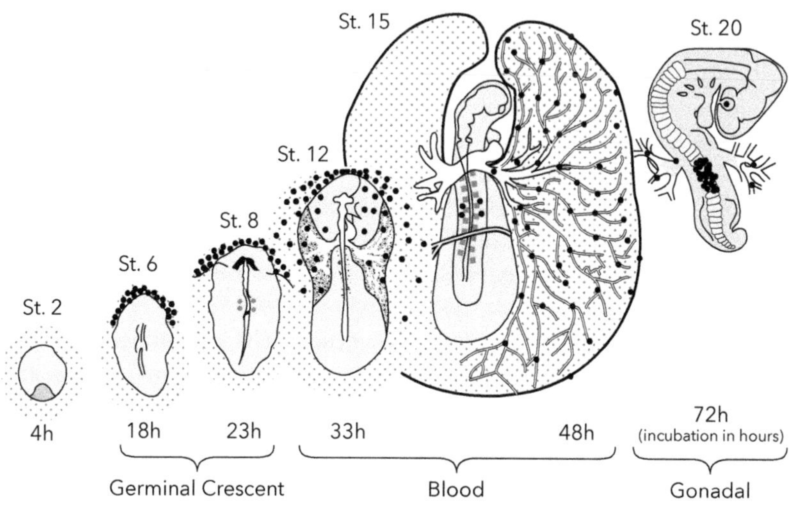

Figure 4 Origin and migration of primordial germ cells (black dots). Drawn after Nieuwkoop and Sutasurya, 1979.

This has made it possible to produce germlines of chimeric/transgenic birds commercially (as reviewed by Choi, et al., 2010).

4.3 Morphological development of Eyal-Giladi and Kochav stage V to Eyal-Giladi and Kochav stage XII

The embryo stages EGK-V and EGK-VI are characterized by central areas consisting of several layers of cells as a result of mitotic activities in the previous stages. During the later VII–X stages the central area transforms to the one-cell thick area pellucida (epiblast) by shedding most 'deep' cells into the sub-embryonic cavity between the area pellucida and a multi-nuclear cellular layer called the yolk syncytial layer (Nagai et al., 2015). The area pellucida is surrounded by what is known as the 'area opaca', a multicellular layer of endodermal cells filled with yolk granula and covered with one layer of ectodermal cells which is a continuation of pellucida ectoderm cells. These ectodermal cells have intimate contact with the vitelline membrane in order to stretch the blastoderm (Kunz, et al., 2021; Bellairs and Osmond, 1998; New, 1959). The shedding of deep cells is not random but starts at a specific area called Koller's Sickle, a ridge formed by a cluster of cells in the marginal zone (MZ) of the area pellucida which touches the area opaca.

Koller's Sickle marks the future posterior side of the embryo and shedding of deep cells progresses to the anterior side. Regular egg rotation 10–15 times per hour during shell formation in the shell gland plays an important role in induction of this posterior-anterior axis (Khaner, 2017). In the EGK-X stage the future posterior-anterior axis is determined next to the ventral-dorsal axis which is already present in the fertilized egg since the ventral-dorsal axis runs along the epiblast-yolk line (Kochav and Eyal-Gyladi, 1971; Callebaut and Nueten van, 1994).

Intra-uterine development occurs at the body temperature of the hen (40–41°C). After oviposition the egg (and thus the EGK-X embryo) cools at a certain rate to 'nest' or room temperature. Embryonic activity slows down and, at a controlled rate of cooling, the embryo reaches the EGK-XII stage which can be defined as a diapause stage (Etches, 1996; Fasenko, 2007 and Pokhrel et al., 2018) until the egg temperature increases to incubation temperature (37.8°C) and gastrulation is initiated.

At oviposition, and after cooling down of eggs, chicken embryos are mainly at the EGK-XII stage, although embryos from older flocks might have been developed further to stage EGK-XIII (Pokhrel et al., 2017; Pokhrel et al., 2021). Nest type and egg handling in the farm might influence cooling-down profiles and thus the embryonic stage at the start of incubation. Most commercial chicken-hatching eggs are assumed to be at the EGK-X to EGK-XIII stage at the start of incubation. Sellier et al. (2006) studied the embryos of different commercial breeds and found that embryos from quail were in stage EGK XI while duck embryos were in stage VIII at the start of incubation.

5 Embryo development during incubation

From the viewpoint of successful hatchery management, the most vulnerable period is the interval between egg laying/oviposition and starting incubation. The embryos in freshly laid eggs are in a state of diapause. However, temperature fluctuations can activate embryonic cells while cooling down of these activated embryos to storage temperature could increase early mortality (Fasenko, 2007; Pokhrel et al., 2021).

Embryonic and extra-embryonic tissue development continues when eggs are warmed to an incubation temperature of 37.8°C (100°F) either in the nest or in a commercial incubator. For successful development and hatching of a healthy (good quality) chicken, the micro-climate including temperature, moisture content and CO_2 need to be optimum for every single egg during incubation (Lundy, 1969), taking into account the breed and age of the parents and type of incubator (Pokhrel et al., 2018; Bergoug et al., 2013; Oviedo-Rondón et al., 2020).

5.1 Embryo and extra-embryonic tissue development (E1-E9): first four (E0-E4) days of embryo development (including the physics of gastrulation)

At the onset of incubation most chicken embryos are in a stage EGK-X to XII. This means that the primary hypoblast consists of single cells or small clusters of cells (Stern and Downs, 2012; Nagai et al., 2015; Lee HC et al., 2020). During the first hours of incubation (E0: 0–24 h) cell clusters merge and form the hypoblast in HH1 (EGK-XIII/XIV) (Lee et al., 2020). Hypoblast formation starts at the posterior side (future tail) and progresses to anterior (future head) (Fig. 5).

The EGK-XIII-EGK-XIV (HH1) embryonic stages are often referred to as the blastoderm. The blastoderm includes the area pellucida (epiblast) plus the area opaca. The area opaca consists of an ectoderm which is a continuation of the epiblast ectodermal layer and a multi-layered endoderm loaded with yolk granules. The area pellucida and area opaca are separated by a thin layer of epithelial cells called the MZ (Khaner and Eyal-Giladi, 1989). The posterior side of the MZ (PMZ) is the Koller's Sickle, a cluster of cells initiating the formation of the hypoblast and later of the primitive streak and gastrulation processes (Callebaut, 2005; Callebaut et al., 2017; Lee et al., 2020; Nájera and Weijer, 2020).

The increasing interest in the physics of cellular interactions during embryogenesis helps to explain the development of different structures and

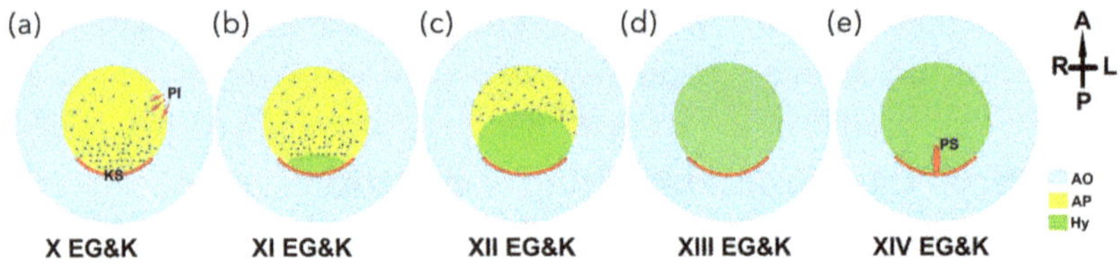

Figure 5 At oviposition the chick embryo is in stage X: the Koller's Sickle (KS) marks the future posterior side of the embryo. After cooling down the embryo is in diapause. Development continues when temperature increases with the formation of the hypoblast (Hy) followed by primitive streak (PS) formation in stage XIV EG&K. Redrawn from Pokhrel, et al. (2021).

the induction of differential gene expression during the formation of functional tissues (Forgacs and Newman, 2005). External physical factors (like temperature, pH, ion-gradients and tension between cells) might activate enzymes which, for example, degrade protein filaments in the cytoskeleton and allow cells to migrate. The chicken embryo is an example. It was already known in 1959 that the area opaca exerts tension on the area pellucida (New, 1959). New (1959) published his data from in-vitro experiments with chicken blastoderms attached to vitelline membranes and concluded: 'It seems reasonable to conclude that normal expansion of the blastoderm is dependent on an adhesion between the edge of the blastoderm and the vitelline membrane'. Bellairs et al. (1969) confirmed New's results and provided new evidence supporting the hypothesis that the area opaca influenced the development of the embryo by exerting tension upon the area pellucida.

Recently Kunz et al. (2021) concluded that the outer ectoderm cells of the area opaca are physically connected to the vitelline membrane, with the moving vitelline membrane stretching the blastoderm via physical forces. They showed that the stiffness of the vitelline membrane hampers the formation of the primitive streak, closure of the neural tube and further development. They also found out that the spontaneous increase in the pH of the albumen during egg storage reduced the stiffness of the vitelline membrane through a reduction of the number of glycoprotein fibres in the vitelline membrane. The spontaneous increase of the albumen pH is the result of CO_2 release from the albumen after oviposition, during cooling down and storage. In fresh eggs the partial pressure of CO_2 in the fresh albumen is high because the albumen becomes saturated with CO_2 during egg formation in the oviduct (Sharp and Powell, 1930; Banerjee and Keener, 2012).

5.2 Day one: first day of incubation (E0)

The reduced tension on the epiblast combined with increased incubation temperature supports, firstly, a transition of the round blastoderm to an elongated embryo at about 5–9 h of incubation. Secondly, the migration of posterior Koller sickle endoblast cells along the elongated axis forms a central posterior-anterior furrow in the embryo (epiblast). This furrow is the first sign of the primitive streak and also marks the initiation of gastrulation (E0: 5–6 h; HH 2) (Lillie, 1952).

Gastrulation is the process that forms/defines the future body plan of the embryo which is based on the germ layers ectoderm, endoderm and mesoderm. The initiation of gastrulation is recognized by massive, controlled cell movements towards the centre of the epiblast and formation of the final primitive streak including Hensen's node (a group of cells that constitutes the organizer of the embryo) at the anterior tip of the fully formed primitive streak (18–19 h incubation). The centre of Hensen's node serves as a pit through which cells from the primitive streak stream into the blastocoel, a compartment between the epiblast and hypoblast (Lillie, 1952; Gilbert, 2014). These migrating cells are precursors of the embryonic as well as extra-embryonic mesoderm. The primitive streak transforms into the neural tube and mesoderm cells.

Mesoderm cells cluster in paired blocks forming the somites: first somite forms at 23–26 h (HH 7) (Lillie, 1952). These mesodermal-paired somites are formed along the neural tube from anterior to posterior. The somites form the embryo and develop into the basic structures of the body e.g. ribs, vertebrae, and all skeletal muscles (Gilbert, 2014).

However, mesodermal cells also migrate laterally from the primitive streak into the expanding area pellucida and area opaca. These mesodermal cells are responsible for the vascularization of the area pellucida and yolk sac. The migration laterally of mesodermal cells is facilitated through intimate contact with the vitelline membrane. These migrating mesoderm cells split the area opaca into two different cell layers: the outer mesoderm cells merge with the external ectodermal chorion membrane forming the somatopleure. The internal mesoderm merges with endodermal cells to form the splanchnopleure which finally shall develop into the yolk sac. The space between the somatopleure and splanchnopleure is called the extra-embryonic body cavity (EEBC).

The controlled gastrulation-related movements of epiblast cells initiate controlled cell–cell interaction followed by cell-differentiation along the primitive streak.

5.3 Development of germ layers: endoderm, ectoderm and mesoderm during the differentiation phase E1 (HH 7) to E4(HH 24)

In synchrony with the embryo, both the lateral area pellucida and the area opaca differentiate into the extra-embryonic membranes and compartments: the future yolk sac, amnion, chorion, allantois and chorio-allantois. These extra-embryonic membranes and compartments outside the embryo are essential for the embryo to obtain nutrients, oxygen and to hold waste products from metabolism.

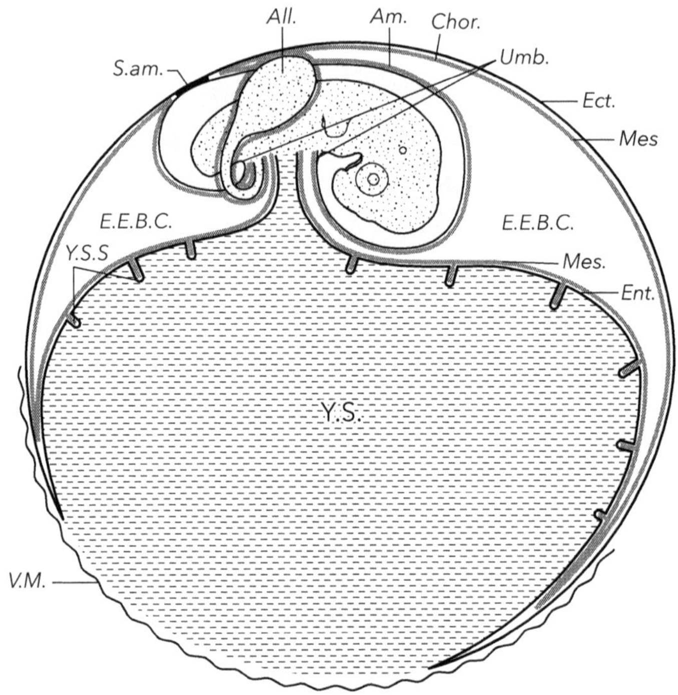

Figure 6 E4 embryo and compartments. Alb., albumen; Alb. S., albumen Sac; All. I., inner wall of the allantois; All. C., cavity of allantois; All. S., stalk of allantois; All.+ Am., fusion of allantois and amnion; Am., amnion; Am. C., amniotic cavity; Chor. chorion; C.T.R., connective tissue ring; Ect. ectoderm; EEBC., extra-embryonic body cavity; Ent., endoderm; Mes., Mesoderm; S-am., sero-amniotic connection; S.Y.S.U., sac of the yolk sac umbilicus; Umb., umbilicus V.M., vitelline membrane; Y.S., yolk sac; Y.S.S.,s of the yolk sac. Source: Lillie (1952, Fig 131 and 132).

Both somatopleure and splanchnopleure membranes grow rapidly during the first 3-4 days of incubation. Between incubation days 4-6, the mesoderm of splanchnopleure becomes highly vascularized and, from this point on, is called the yolk sac membrane. This replaces the vitelline membrane almost completely leaving a small piece of vitelline membrane above a clump of left-over concentrated albumen in the sharp end of the egg (Fig. 8).

In the yolk sac membrane the endodermal epithelial cells (EEC) are oriented towards the yolk and the vascularized mesoderm is part of the EEBC. Gilbert (2014) concludes that, by day 4 (HH stage 23), the formation of the yolk sac ectodermal-mesoderm layer marks the end of gastrulation and the three distinct germ layers have formed.

The single germ layers give rise to distinct organ systems (Gilbert, 2014):

- The ectoderm forms the skin, brain and nervous system.
- The endoderm forms the epithelium of digestive tract and lungs.

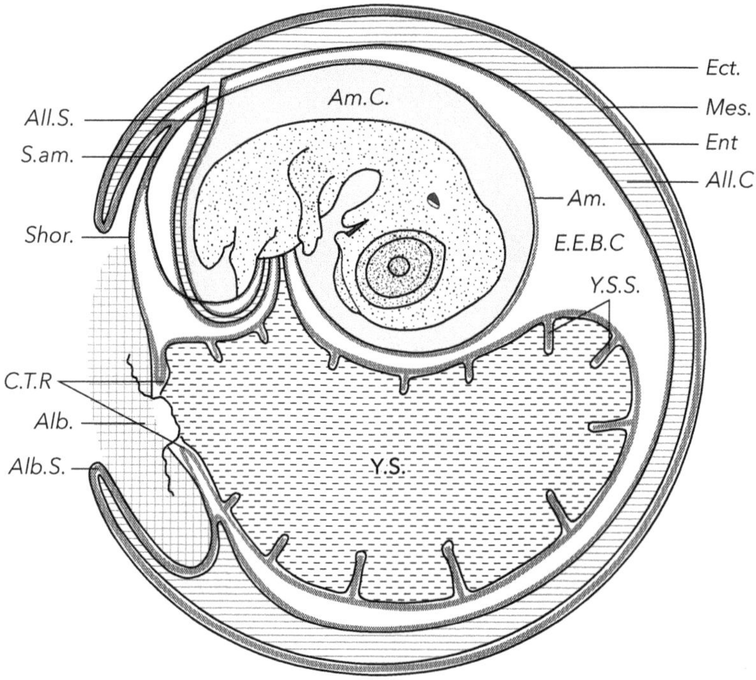

Figure 7 E9 embryo and extra-embryonic tissues. Alb., albumen; Alb. S., albumen Sac; All. I., inner wall of the allantois; All. C., cavity of allantois; All. S., stalk of allantois; All.+ Am., fusion of allantois and amnion; Am., amnion; Am. C., amniotic cavity; Chor. chorion; C.T.R., connective tissue ring; Ect. ectoderm; EEBC., extra-embryonic body cavity; Ent., endoderm; Mes., Mesoderm; S-am., sero-amniotic connection; S.Y.S.U., sac of the yolk sac umbilicus; Umb., umbilicus V.M., vitelline membrane; Y.S., yolk sac; Y.S.S.,s of the yolk sac. Source: Lillie (1952, Fig 131 and 132).

- The mesoderm gives rise to the heart, blood, kidney, gonads, bones, muscles and connective tissues.

In hatchery management these first days are crucial to achieve optimum results (hatchability and chick quality). Incubation should not be disturbed: avoiding opening the incubator and/or removing trolleys. Turning of eggs is crucial during this phase of incubation.

5.4 Morphological differentiation of the embryo E1-E9 (HH 7-HH 35)

At the end of the first incubation day the E0 blastoderm is no longer flat and round but 'pear-shaped' with the primitive streak. The latter forms a central groove and a head fold anteriorly which is part of the embryo.

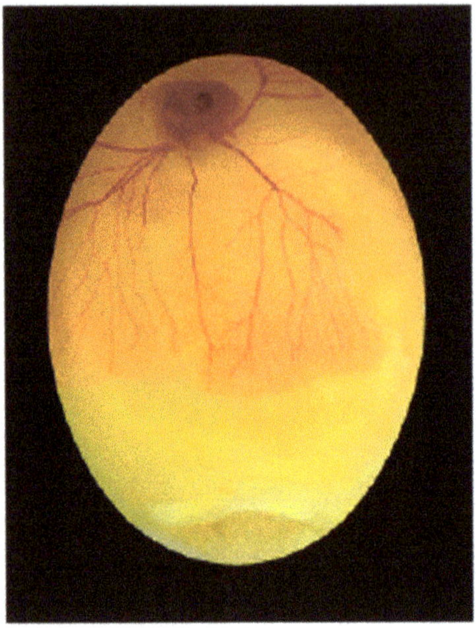

Figure 8 Formation of the yolk sac membrane after 6 days of incubation (E6).

In the first four days (E0–E4) of the differentiation phase, the embryonic structures develop very fast: from a flat disk, blastoderm, on the first day (E0) to a 3D structure of the E4 (HH24) embryo.

During the second day (E1) of incubation, the head of the embryo has turned to the right and four pairs of somites are visible. At the posterior side faint clusters of blood cells can be recognized in the area vasculosa, while endothelial precursor cells have developed from mesoderm precursors in posterior primitive streak of HH4 (E0) embryos (Nagai, et al., 2018). The EEBC filled with sub-embryonic fluid (SEF) shows a margin called sinus terminalis (Fig. 7). This is the margin of overgrowth formed from ectodermal cells and is responsible for maintaining epiblast tension (Bellairs et al., 1967).

The E3 and E4 embryos have turned onto their left side and, in the E4 embryo, the head flexure is complete (axes of forebrain and hind-brain form a right angle). In the E3 embryo the beating heart is clearly visible: heart beats/min depend on temperature. If the eggs are cooling down during inspection the beats/min decrease significantly. In the E4 embryo the forebrain vesicle is clearly visible and the first signs of eye pigmentation can also be observed. The area vasculosa has developed and a clear sinus terminalis forms the demarcation of the area vasculosa and the area vitellina.

At day 4 the embryo is protected by the fluid-filled amnion cavity and, in the embryo (E4), we can recognize small wings and limb buds. The ventral body wall has been closed (Martinsen, 2005) and the beating of the heart is still visible.

After 5/6 days of incubation, both active and passive embryonic movements can be clearly observed because of the transparency of the amnion membrane and clear amnion fluid. The smooth muscle in the mesodermal layer of the amnion responsible for passive movements which start as soon as the amnion has closed. These passive movements occur in waves starting at the anterior/head of the embryo and spread over to the posterior/tail area. The frequency of these passive movements increases with the age of the embryo and is highest at day 9. Frequency of these passive movements stabilizes between days 13 and 15 (Wu et al., 2001).

Active movements are made by the embryo: short periods of movements are alternated by longer periods of inactivity (Freeman and Vince, 1974; Wu et al., 2001). These active movements start with simple trunk movements on about day 3. As the embryo grows movements of limbs and wings are also involved. From day 7 onwards, different parts of the body or the whole body are involved in movements. Wu et al. (2001) describe four different types of movements based on time of onset, part of body and frequency.

Prenatal movements of the chicken embryo and vertebrate embryos are generally essential for optimum skeletal, muscle and neuromuscular development and postnatal maturation (Bradley et al., 2005).

The E6 (HH29) embryo shows toes and bent wings at the elbow. The day 6 and day 7 embryo 'floats' on a maximum volume of SEF (Fig. 9) which can be noticed during the breakout of unhatched incubated eggs: yolk and yolk membrane can no longer be recognized and are hidden in the milky SEF fluid, the embryo has been transformed in a grey clump of dead tissue in which we can recognize the dark eye and a beak with a clear egg tooth if the embryo died around E7.

In the E7 (HH30-HH31) embryo almost all yolk sac primitive circulating erythrocytes are replaced by the definitive erythrocytes loaded with adult type of haemoglobin (Sheng, 2010).

In the E8 (HH32-HH34) embryo three rows of tail feather primordia are visible dorsally.

The E9 (HH 35) embryo has reached a certain size and most of embryonic structures/organs and tissues have reached their final form. Species-specific body contours can also be recognized: a day 9 chicken embryo cannot be mixed up with a day 9 duck embryo (Li et al., 2019). The beak is either a chicken beak with a clear egg tooth or a flat duck beak with a prominent central egg tooth. In most poultry species, the webbing between toes disappears between E8 and E10, the exception being waterfowl (Li et al., 2019).

In male day 9 embryos, carrying ZZ sex chromosomes, the indifferent gonads have differentiated into a testis (Etches, 1996). In females carrying ZW sex

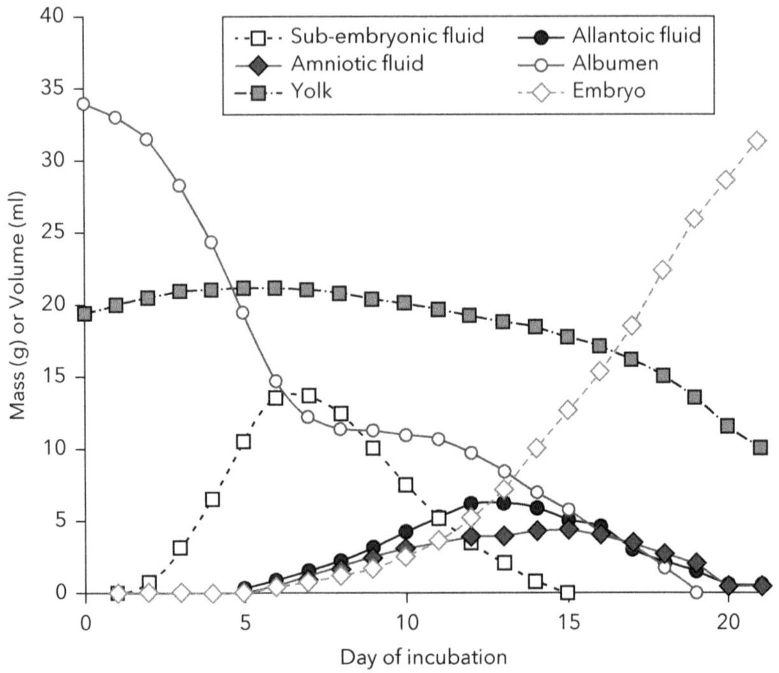

Figure 9 The pattern of changes in the mass of the embryo, yolk and albumen, and volume of the fluid compartments of the developing fowl egg. Data from Romanoff (1967).

chromosomes, the indifferent left gonad differentiates to an ovary carrying about 28 000 oocytes, and the right gonad has degenerated (Bellairs and Osmond, 1998).

In incubation practice, the day 9/10 male and female embryos can be separated through the analysis of the concentration of estradiol and estrone-sulphate in the allantoic fluid. The allantoic fluid collected from female embryos contains significantly more estrogens when compared with allantoic fluid collected from male embryos (Weissmann et al., 2013).

In birds, in contrast to mammals, secondary sexual characteristics development is only partly under the control of sex hormones. Birds have acquired Cell-Autonomous Sex Identity (CASI): which means that sexual phenotype depends on the sex chromosomes present in the individual cells and is independent of circulating sex hormones (Clinton et al., 2012; Ioannidis et al., 2021).

6 Endocrine control systems and maternal thyroids

Hormones are molecules involved in the regulation of many body functions including development, growth and reproduction. Specific endocrine organs/glands are involved in the production of hormones. After being released into the bloodstream, hormones bind to specific receptors on target tissues.

The maturation of endocrine control systems follows specific morphological as well as functional developmental pathways (Freeman and Vince, 1974; Bellairs and Osmond, 1998). The development of the chicken endocrine system has been reviewed by De Groef et al. (2008 and 2013) who provide a more detailed description of the process.

The hypothalamus-hypophyseal complex can be defined as the 'master-endocrine-regulator'. This complex releases a whole complex of stimulatory and inhibitory hormones to control specific endocrine organs, for example, gonads, pancreas and thyroid (Scanes, 2015). The hypothalamus-hypophyseal complex is fully matured and active shortly before internal pipping to control the last phases of hatching (Decuypere et al., 1991).

Thyroid hormones (TH) are essential controllers for the differentiation of tissues and embryonic structures throughout incubation (Decuypere et al., 1991). TH need to be available during the initiation of gastrulation (E0–E1) until the hatching phase (Decuypere et al., 1991). The embryonic hypothalamic-hypophysis-thyroid axis does not develop until after E10, when thyroid stimulating hormone is first detected and E12 when T3 (3,5,3'-triiodothyronine) hormone can first be detected (Yahav and Brake, 2014). Before the final development of the embryonic hypothalamic-hypophysis-thyroid axis, maternal TH deposited in the yolk play essential roles from day one of incubation (McNabb and Wilson, 1997; Flamant and Samarut, 1998; Too et al., 2017; Darras, 2019).

7 Growth and development of the embryo E10-E15

At E10 (HH36) toes are completely separated. Feather germs are visible over most of the body while eyelids are clearly visible and begin to close. The first feathers are visible at E11 (HH37).

The E10 (HH36) embryo moves actively in the amnion while the amnionic volume reaches its maximum at day 11 (Romanoff, 1967; Silva Da et al., 2017). Movements are random and jerky (Freeman and Vince, 1974).

Between day 11 and day 12, the sero-amniotic connection becomes perforated and albumen concentrated in the albumen sac mixes with the amniotic fluid. Albumen is swallowed by the embryo with albumen proteins entering the digestive tract and finally the yolk sac of the embryo.

In this second week of development, the mineralization of bones continues with calcium recruited from the shell via the chorio-allantois. Mineralization of bones started on day 8/9 with calcium originating from the yolk (Freeman and Vince, 1974).

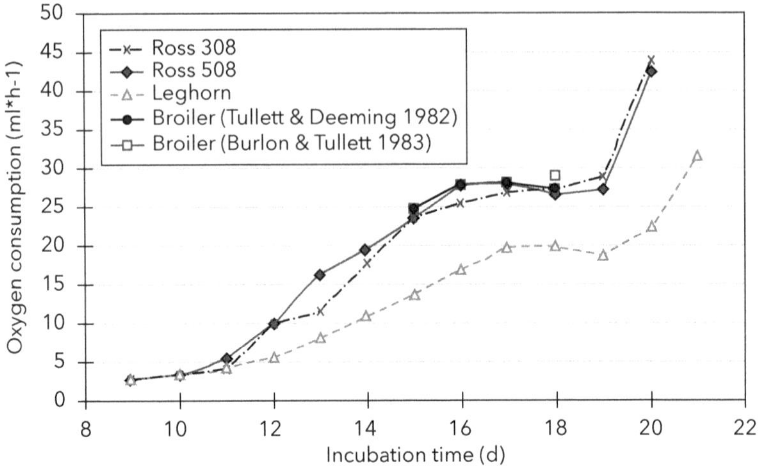

Figure 10 Janke et al. (2004): Courses of oxygen consumption (directly proportional to heat production) during incubation of broiler embryos of Ross 308, Ross 508, White Leghorn and two examples of the same broiler breeder line (n = 23, Tullet and Deeming, 1982; and N = 39, Burton and Tullet, 1983).

As described earlier, the bone marrow becomes erythropoietic around day 12 and the majority of erythrocytes are produced by the bone marrow in E13-E15 embryos (Sheng, 2010).

The second week of embryonic development is characterized by an increase in the size of the embryo. Romanoff (1967) showed an exponential increase in embryonic weight after day 10 as discussed in detail by Freeman and Vince (1974). In this period oxygen consumption by embryos increases as a result of increased metabolic activity to support embryonic growth (Romijn and Lokhorst, 1960). An example is broiler embryos from breeds Ross 308 and 508, where oxygen consumption has been shown to increase from day 10 to day 15 when oxygen consumption reaches a plateau (Fig. 10) (Janke et al., 2004). In these experiments oxygen consumption by White Leghorn layer embryos was lower and reached a plateau 2 days later on day 17.

8 E16-E19 plateau phase: the embryo shows endothermic responses

The plateau phase in oxygen consumption is explained by the fact that the influx of oxygen via CAM is limited by the diffusive capacity of the egg shell (Rahn et al., 1979). The plateau in oxygen consumption and as a consequence the metabolic heat production lasts about 3 days. Metabolic heat production increases again when oxygen becomes available after internal pipping followed by lung activation and later external pipping: the perinatal period (Decuypere et al., 1991). The end of the plateau phase, marked by increased metabolic heat production, could also be a trigger for the development of endothermic responses as hypothesized and discussed by Price and Dzialowski (2018): an endodermic response upon cooling is characterized by a metabolic response to control body temperature (Tzschentke, 2007; Yahav, 2015). Typically metabolic heat production increases at day 20 for broiler breeds and at day 21 for the White Leghorn layer breed (Janke et al., 2004).

9 Extra-embryonic fluids and compartment development and function

For successful incubation, along with the development of the embryo itself, a system of compartments and membranes within the egg needs to develop to provide the embryo with the correct environment, including the supply of nutrients and oxygen. These extra-embryonic fluids, compartments and membranes develop outside of the body of the embryo but are an extension of the embryo itself (Sheng, 2010). The functions of these compartments and membranes are:

- to allow the embryo to develop within an aqueous environment;
- to allow the embryo to take up nutrients from the yolk and albumen;
- to transport oxygen that has diffused through the eggshell to the embryo;
- to release carbon dioxide through the eggshell;
- to support the redistribution of albumen water to the yolk sac, amnion and chorio-allantois (Fig. 9); and
- to hold metabolic waste products.

The extra-embryonic compartments change during the course of incubation. The first to develop is the EEBC containing the SEF as noted earlier.

9.1 Sub-embryonic fluid formation (E1-E4) and other compartments: yolk sac, amnion, chorion and allantois

SEF formation is an active process of intracellular water transport from the albumen to the EEBC. This process is facilitated by active intercellular Na^+-ion transport facilitated by Na^+/K^+ ATPase and membrane-associated carbon anhydrase (Babiker and Baggott, 1992; Latter and Baggott, 2001). The activation of SEF secretion by the blastoderm to the sub-embryonic cavity during the first day of incubation becomes clearly visible as a ring of light-yellow yolk around the E1 embryo after 24 h of incubation (see E1 in table 1 (appendix)).

The area vasculosa membrane and the EEBC develop during days 3-7 and both play an important role in SEF formation and fluid storage (Romanoff, 1967; Deeming, 2002). During days 6–7 the SEF volume reaches its maximum, after which the SEF is redistributed over the yolk sac, amnion/chorion-allantois compartments as shown by Romanoff (1967) (Fig. 9).

Little is known about the composition of the SEF. However, Baggott et al. (2002) have suggested that turning during incubation stimulates the growth of the area vasculosa, the volume and quality of SEF and, finally, the volume of the allantoic fluid and yolk sac. The most critical period for turning is days 3-7 of incubation (Tazawa, 1980; Deeming, 1989a, 1989b and 2002). Tona et al. (2005) confirmed the importance of turning during the incubation of broiler eggs for the transfer of albumen to embryos.

Embryonic death during the first day of incubation can easily be discriminated from an infertile egg, even after 21 days of incubation. If the clear egg is opened very carefully, the infertile egg shows a normal yolk with a (very) thin/weak membrane. If the clear egg is fertile, the yolk changes in colour and, when poured in a dish, the yolk membrane breaks immediately and the whitish SEF becomes visible.

9.2 Yolk sac and extra-embryonic coelom: feed the embryo and more

Transfer of the nutrients from the yolk to the embryo is effectively a three-phase process (Williams, 1967):

- *Phase 1.* The cleavage cells in the area pellucida are laden with yolk inclusions to provide the necessary nutrients.
- *Phase 2.* On day 2 of incubation, the primitive streak becomes visible and then the yolk stores in the area opaca are utilized.
- *Phase 3.* At the end of day 2 the first blood vessels become visible on the posterior side of the epiblast and this is the start of the area vasculosa development and in fact the initiation of the yolk sac membrane formation.

The yolk sac membrane will be the main mechanism for transporting nutrients from the yolk to the embryo for the rest of the incubation.

More recent studies have further refined our understanding of the process of nutrient transfer to the embryo and the three phases have been redefined as follows:

- *Phase 1.* E0 (HH stage 1).
- *Phase 2.* E2 (HH13)-E4 (HH 23-24).
- *Phase 3.* E7-E21.

These phases are discussed in more detail in the following sections.

9.2.1 Phase 1

In the E0 (HH stage 1) blastoderm, several morphogenetic processes are activated. These include: expansion (cell division) of the blastoderm; formation of the hypoblast; initiation of primitive streak formation; and active transport of albumen water to the sub-embryonic cavity. The metabolic energy needed to support these processes is delivered from glucose present in thin albumen adjacent to the yolk and from the yolk granules present in endodermal cells of the area opaca. The oxygen needed for aerobic metabolism of glucose is delivered by diffusion (Kucera et al., 1984; Moran, 2007). The maternal thyroids needed to support the above-mentioned processes are present in the yolk and can be taken up by the embryo from the first day of incubation (Darras, 2019).

9.2.2 Phase 2. E2 (HH13)-E4 (HH 23-24)

The synchronous development of embryo and yolk sac becomes visible. During the second day of incubation, embryonic mesoderm as well as lateral extra-embryonic mesoderm develop in the area pellucida. The lateral extra-embryonic mesoderm follows the EEC of the splanchnopleure and ectoderm of the somatopleure (Fig. 6). The migration of mesoderm activates EEC processes which (1) activates the uptake of yolk nutrients (mainly lipids) through the endodermal epithelial cells and (2) stimulates the vascularization of the mesoderm cells so that yolk nutrients become available and feed the early embryo (Bauer et al., 2013).

At the posterior side of the area pellucida, the first blood islands become visible. Circulation starts on the second day (Bauman and Meuer, 1992; Sheng, 2010). The blood islands, also called angioblasts (Sabin, 1917; 1920), consist of precursor primitive blood cells (primitive erythrocytes) as well as precursors for blood vessels (endothelial) cells (Sheng, 2010). After 3 days of incubation (E3, HH18), the embryonic heart and embryonic blood vessels form a continuum with developed arteria, veins and network of fine blood vessels in the extra-embryonic area vasculosa (blood ring). The sinus terminalis is formed at the outer limit of the extra-embryonic mesoderm ingression of the splanchnopleure. The sinus terminalis thus separates the area vasculosa and the area vitellina in the developing yolk sac.

Primitive embryonic versus definitive adult erythrocytes: In contrast to mammalian erythrocytes, avian erythrocytes are nucleated. This is true for the primitive as well as for the definitive (adult) erythrocytes. The transition from primitive to adult erythrocytes starts in E4–E5 embryos and can be recognized by a change in the morphology of the erythrocytes: the primitive erythrocytes are somewhat larger compared to the definitive erythrocytes. The primitive erythrocytes are loaded with an embryonic type of haemoglobin. The embryonic haemoglobin has a greater oxygen affinity and a lower Bohr effect (Scanes, 2015). The latter means that, at higher tissue CO_2 concentrations (= lower tissue pH), oxygen is released more easily, thereby activating oxidative glycolysis and preventing lactate accumulation through anaerobic glycolysis.

By day E7 (HH30–HH31), almost all yolk sac primitive circulating erythrocytes are replaced by definitive erythrocytes loaded with an adult type of haemoglobin. The source of these yolk sac adult erythropoietic cells is not yet resolved, as reviewed by Sheng (2010).

9.2.3 Phase 3. E7-E21

The origin of yolk sac tissue lies in the area vasculosa (Sheng, 2010). By day 3–4 of incubation, the yolk sac splanchnopleure connects to the embryonic midgut. The yolk stalk has been formed so that, later on, the yolk content can be transferred directly into the gut which occurs shortly before withdrawal of the yolk sac into the body cavity.

In synchrony with the development of the embryo, the total surface of the yolk sac membrane (YSM) increases and the EEC form ridges and folds which hang into the yolk (Freeman and Vince, 1974).

As the YSM continues to develop it becomes a multifunctional tissue (Uni et al., 2012; Wong and Uni, 2021):

1. The YSM acts as a gut, absorbing nutrients from the yolk and expresses genes that are normally found in the intestine (Uni et al., 2012).
2. The YSM is an active tissue for glycogenolysis and glycogenesis and is a major store of glycogen prior to hatch, storing 12 times more than the liver. Glycogen is an essential energy source for the hatching chick (Wong and Uni, 2021).
3. In addition to its role in glycogen metabolism and storage, the yolk sac can function as the liver in the production of plasma proteins as shown from transcriptome analysis (Yadgary et al., 2014).
4. From E1 to E15 yolk sac tissue is involved in erythropoiesis (Nagai et al., 2018).

9.3 Formation and function of amnion and chorion

After 24 h (E1) of incubation, a head fold (pre-amnion membrane) is visible on the anterior side. Both the amnion and chorion develop from the somatopleure, the extra-embryonic ectodermal-mesodermal layer after ingression of mesoderm in the blastoderm (area opaca). During the first days of incubation, there is no distinct separation of embryonic and extra-embryonic tissue which both originate from the blastoderm. The epiblast and area opaca forming the blastoderm are still united.

On the first day of incubation, pre-amnion folds of ectodermal and avascular mesodermal tissue from the somatopleure are formed at the anterior as well as at posterior sides of the embryo. These membranes grow from all sides and meet in the region where the limbs should form. At this meeting-point the anterior and posterior ectoderm as well as the endodermal layers fuse to form the so-called sero-amniotic connection at about 60 h (HH 17) of incubation. The amnion becomes filled with fluid, with the ectoderm layer covering the inner surface as a result of the typical folding of the anterior and posterior ectodermal layers. The outer mesodermal layer of the amnion differentiates to a smooth muscle-rich layer and supports embryonic movements from day 5. The amnion is recognizable after 4 or 5 days of incubation if the hatchery manager opens an incubated egg at the position of the air cell and carefully removed the membrane.

9.4 Amnion functions

The amnion protects the embryo from desiccation, external shocks, extreme temperature and pH fluctuations and stimulates passive embryonic movements. Between day 11 (E11) and E(12) the sero-amniotic connection opens and albumen proteins are released in the amniotic cavity. These proteins are swallowed by the embryo into the digestive tract and finally end up in the yolk sac and partly appear in the allantoic fluid as metabolic waste products.

The change in the amnion-protein profiles has been analysed by Silva Da (2017). Several albumen proteins have been detected in the amnion fluid such as immunoglobulin Y (IgY), ovotransferrin and serum albumin (ALB). Other proteins were found originating from the yolk such as Transthyretin (TTR) a transporter TH (Too et al., 2017). Silva Da et al. (2017) also found several proteolytic proteins in the allantois and the amnion fluid. This observation raises the hypothesis that both the amnion fluid and especially allantoic fluids could have a digestive function and might play roles in morphogenesis.

9.5 The chorion and allantois membranes develop separately and fuse to the chorio-allantois membrane

The chorion also arises from the anterior extra-embryonic blastoderm but continues to grow as the somatopleure with ectodermal membrane adjacent to the egg shell. Consequently, the avascular mesoderm layers from the amnion and chorion lie face-to-face, a typical result of the folding and growth of the pre-amnion folds of ectodermal and avascular mesodermal layers (Etches, 1996).

At the end of day 3 (HH stage 18), the allantois becomes visible: this is a thick-walled endodermal-mesodermal pocket which is a continuation of the primitive hindgut. The allantois grows rapidly and, by day 4 (HH 24) of incubation, the outer layer is transformed into the vascular mesoderm as part of the extra-embryonic coelum. At about days 6-7 (HH 30), the allantois starts to fuse with the chorion to form the chorio-allantois membrane (CAM) which is in close contact with the shell. The CAM consists of three layers:

- chorion ectoderm close to the shell;
- fused somatic mesoderm and vascularized mesoderm from the allantois; and
- allantois endoderm.

By day 16 the CAM covers most of the yolk sac and fuses in some places with the YSMs (Bellairs and Osmond, 1998). The CAM is in close contact with the egg shell. After day 12 the membrane is being pressed against the shell as the embryo grows and oxygen is needed for lipid metabolism.

The CAM and allantois have four functions:

- Supporting metabolic gas exchange: oxygen uptake and release of carbon dioxide via the pores of the shell.
- Calcium resorption from the shell is needed to ossify the bones from day 11. Both these functions can only be realized because the membranes of the chorio-allantois are pressed against the egg shell.
- From day 5 urine (consisting of urea, uric acid and ammonia produced by the premature kidney) flows continuously into the allantois, reaching a maximum of about 7 ml at E11.
- The allantois acts as a water reservoir.

10 Hatching

10.1 Synchronous hatching

Freeman and Vince (1974) presented an overview of adapted behaviour related to synchronizing the hatching process. In domestic fowl and bobwhite quail, breathing as well as sounds from beak clicking are involved in synchronizing hatching. In bobwhite quail clutches, clicking mainly acts as an accelerating stimulus to synchronize hatching.

In natural as well as commercial incubation, rates of embryonic development vary mainly for several reasons: most commonly due to variations in pre-incubation and incubation temperature, and the stage of embryo development at the start of incubation. In the case of commercial incubation, the variation in hatching time, also called 'hatch window', is also determined by mixing eggs from different breeds, breeder flock age, storage periods and storage temperatures (Molenaar et al, 2010; Tong et al., 2013.)

Synchronous hatching is also essential in commercial incubation in order to minimize variation in day-old chick quality: at the moment of hatching all chicks should be dry, have a closed navel and hydrated legs. A consequence of an asynchronous hatch (with a large hatch window) is that some chicks could have dehydrated legs while others are very 'fresh', showing an open navel with umbilicus muscle still swollen and feathers still damp.

10.2 Physiology including endocrinology of hatching

In an individual incubated egg, the hatching process starts between day E16 and E17 at the end of the plateau phase with integrated jerky embryonic movements alternated with a period of 15 sec of rest (Freeman and Vince, 1974). Integrated or coordinated movements finally result in a pre-hatching position of the embryo at day 18: the body is oriented lengthwise in the shell with the head under the right wing and the beak pointing to the air cell. The joints of the legs rest at the small end of the egg. Prior to these movements on day 18 the embryo has its head between its legs pointing down to the pointed end of the egg.

The process of retraction of the yolk sac also starts on day 17 and is characterized by the withdrawal of the small intestine into the body. The withdrawal of the small intestine should be finished before the yolk sac is retracted into the body via the umbilicus. The yolk sac withdrawal is completed on day 20, leaving swollen sphincter muscles which flatten during the drying of the chicken and closes the umbilicus.

During day 19 the absorption of allantoic and amniotic fluids is completed (Freeman and Vince, 1974). The embryo is oriented with the beak directed to the air cell ready to pip internally at about day 20, a process which involves piercing the air cell membrane; the embryo also starts breathing while gas exchange via the chorio-allantois gradually decreases.

The CO_2 concentration in the air cell reaches 6–8% which could stimulate the production of surfactants: a mixture of lipids and proteins which reduces the surface tension in the lungs (Blacker et al., 2004). The lungs need surfactants to prevent adherence of lung surfaces i.e. to keep lungs open and free from exudation of fluids into the lungs.

After internal pipping and lung inflation, the embryo occupies the whole egg with the beak close to the inner shell membrane. At about 10–12 h after internal pipping, external pipping occurs, where the beak with egg tooth pierces the shell and external oxygen becomes available.

After about 10–12 h, intensive movement of legs and wings turn the chicken in the shell while the head and beak, with help of the hatching muscle, break the shell in a circular pattern so the chick can push the shell open and leave the shell during day 21. Metabolic energy for hatching is delivered from glycogen stores in the yolk sac and liver (Wong and Uni, 2021).

Towards hatching, the overall endocrinal master controller hypothalamus-hypophyseal complex is matured and prepared to stimulate the thyroid gland to support the hatching process: thyroid hormone levels are optimized to support the hatching process. (Decuypere et al., 1991; Decuypere and Bruggeman, 2007; Givisiez et al., 2020).

11 Conclusion

This chapter was written with the intention to produce a complete and comprehensive chapter on embryo development in poultry for biologists and veterinarians working in the poultry incubation business. With this goal in mind, a few important decisions had to be made when I started writing and searching for new information. An easy decision was to focus on the chicken because the chicken embryo has been the basic subject of developmental biology since Aristoteles (384–322 BC) and still is. A second decision was not to go into details of gene expression but to focus on those morphological aspects easy to follow during egg break-outs in order to analyse poor hatching results in practice.

Searching for new information directed me to focus, firstly, on the embryonic development during egg formation and stage of diapause after laying and cooling down. Secondly, the development of extra-embryonic tissues and compartments received attention with the focus on synchronous development of embryo and extra-embryonic tissues and compartments.

Especially the development of the yolk sac is intriguing. The yolk sac develops, in synchrony with the embryo, from 'a ball filled with nutrients' to a complex metabolic organ. The yolk sac endoderm and mesoderm layers become involved in the process of blood cell and veins development. This whole process seems to be supported by maternal thyroid hormones present in the yolk before the maturation of the embryonic hypothalamic-hypophysis-thyroid axis during the second week of incubation. In the latest phases, before hatching, the yolk sac has become a multifunctional tissue/organ also involved in glycogenesis and as such becomes a major store of glycogen needed to support hatching.

12 Acknowledgements

The author would like to acknowledge Professor Oded Khaner (Department of Health and Life Sciences, Hadassah Academic College, Israel) and Dr Yuval Cinnamon (Institute of Animal Science, Poultry and Aquaculture Science Dept, Agricultural Research Organization, The Volcani Institute, Israel) for their permission to include published figures in this chapter. Figure 1 was prepared by Oded Khaner and Figure 5 was made by Yuval Cinnamon.

13 Where to look for further information

Incubation Fertility Research Group (www.ifrg.be) offers an informal meeting of colleagues and scientists from all over the world interested in incubation and embryonic development in poultry. The IFRG is a working group (no. 6) of the World's Poultry Science Association (www.WPSA.com).

13.1 Books on the morphological development of embryonic structures in poultry

- Bellairs, R. and Osmond, M. (1998). *The Atlas of Chick Development*, London, UK: Academic Press Limited.
- Lillie, H. R. (1952). *Lillie's Development of the Chick (revised by H. L. Hamilton)*. Edited by Willier, B.H. New York: Henry Holt and Company.

13.2 General Books on fundamentals of embryonic development: tissue development and gene expression and physics of embryonic development in birds

- Gilbert, S. F. (2014). *Developmental Biology*, (10th edn). Sunderland, MA: Sinauer Associates, Inc. Publishers.
- Forgacs, G. and Newman, S. A. (2005). *Biological Physics of the Developing Embryo. Cambridge: University Press.*
- Deeming, D. C. (2002). *Avian incubation, Behaviour, Environment and Evolution*. Oxford Ornithology Series. Oxford: Oxford University Press.

13.3 Books and papers related to physiological topics of chicken development

- Freeman, B. M. and Vince, M. A. (1974). Development of the avian embryo. In: *A Behavioural and Physiological Study*. London: Chapman and Hall.
- Decuypere, E., Dewil, E. and Kühn, E. R. (1991). Chapter 15: The hatching process and the role of hormones. In: *Avian Incubation, Poultry Science symposium Number Twenty-two*. Heinemann, Butterworth.

- De Groef, B., Grommen, S. V. H. and Darras, V. M. (2008). The chicken embryo as a model for developmental endocrinology: Development of the thyrotropic, corticotropic, and somatotropic axis, *Molecular and Cellular Endocrinology,* 293 (1-2), 17–24. doi.org/10.1016/j.mce.2008.06.002.
- Scanes, C. G. (Ed.). (2015). *Sturkie's Avian Physiology* (6th edn. or 7 th edn.). Amsterdam: Academic Press, Elsevier.

14 References

Babiker, E. M. and Baggott, G. K. (1992). The effect of turning upon the sub-embryonic fluid and albumen of the Japanese quail, *British Poultry Science* 33(5), 973-991.

Baggott, G. K., Deeming, D. C. and Latter, G. V. (2002). Electrolyte and water balance of the early avian embryo: Effects of egg turning, *Avian and Poultry Biology Reviews* 13(2), 105–119.

Bakst, M. R., Wishart, G. and Brillard, J.-P. (1994). Oviducal sperm selection, transport and storage in poultry, *Poultry Science Review* 5, 114–143.

Banerjee, P. and Keener, K. M. (2012). Maximizing carbon dioxide content of shell eggs by rapid cooling treatment and its effect on shell quality, *Poultry Science* 91(6), 1444–1453.

Bauer, R., Plieschnig, J. A., Finkes, T., Riegler, B., Hermann, M. and Schneider, W. J. (2013). The developing chicken yolk sac acquires nutrient transport competence by an orchestrated differentiation process of its endodermal epithelial cells, *The Journal of Biological Chemistry* 288(2), 1088–1098.

Bauman, R. and Meuer, H. J. (1992). Blood oxygen transport in the early avian embryo, *Physiological Reviews* 72(4), 941-965.

Bellairs, R., Boyde, A. and Heaysman, J. E. M. (1969). The relationship between the edge of the chick blastoderm and the vitelline membrane, *Wilhelm Roux' Archiv Fur Entwicklungsmechanik Der Organismen* 163(2), 113–121.

Bellairs, R., Bromham, D. R. and Wylie, C. C. (1967). The influence of the area opaca on the development of the young chick embryo, *Journal of Embryology and Experimental Morphology* 17(1), 195–212.

Bellairs, R. and Osmond, M. (1998). *The Atlas of Chick Development*. London, UK: Academic Press Limited.

Bergoug, H., Burel, C., Guinebretière, M., Tong, Q., Roulston, N., Romanini, C. E. B., Exadaktylos, V., McGonnell, I. M., Demmers, T. G. M., Verhelst, R., Bahr, C., Berkmans, D. and Eterradossi, N. (2013). Effect of pre-incubation conditions on hatchability, hatch time and hatch window, and effect of post-hatch handling on chick quality at placement, *World's Poultry Science Journal* 69(2), 313–334.

Blacker, H. A., Orgeig, S. and Daniels, C. B. (2004). Hypoxic control of the development of the surfactant system in the chicken: Evidence for physiological heterokairy, *American Journal of Physiology. Regulatory, Integrative and Comparative Physiology* 287(2), R403–R410.

Blesbois, E. (2012). Biological features of the avian male gamete and their application to biotechnology of conservation, *Journal of Poultry Science* 49(3), 141–149.

Bradley, N. S., Solanki, D. and Zhao, D. (2005). Limb movements during embryonic development in the chick: Evidence for a continuum in limb motor control antecedent to locomotion, *Journal of Neurophysiology* 94(6), 4401-4411. DOI: 10.1152/jn.00804.2005.

Burton, F. G. and Tullet, S. G. (1983). A comparison of the effects of eggshell porosity on the respiration and growth of domestic fowl, duck and turkey embryos, *Comp, Biochemistry and Physiology* 75A, 167–174.

Callebaut, M. (2005). Origin, fate, and function of the components of the avian germ disc region and early blastoderm: Role of Ooplasmic determinants, *Developmental Dynamics* 233(4), 1194–1216.

Callebaut, M. and Neuten van, E. (1994). Rauber's (Koller's) sickle: The early gastrulation organizer of the avian blastoderm, *European Journal of Morphology* 32(1), 35–48.

Callebaut, M., Van Neuten, E. and Hubens, G. (2017). Uptake and interaction of ooplasms in avian blastoderms: Impact on neurogastrulation and coelomo-cardiovascular system formation. A review, *Histology, Cytology and Embryology* 1(3), 3–12. DOI: 1015761/HCE.1000116.

Choi, J. W., Kim, S., Kim, T. M., Kim, Y. M., Seo, H. W., Park, T. S., Jeong, J. W., Song, G. and Han, J. Y. (2010). Basic fibroblast growth factor activates MEK/ERK cell signaling pathway and stimulates the proliferation of chicken primordial germ cells, *PLoS ONE* 5(9), e12968. DOI: 10.1371/journal.pone.

Clinton, M., Zhao, D., Nandi, S. and McBride, D. (2012). Evidence for avian cell autonomous sex identity (CASI) and implication for the sex-determination process?, *Chromosome Research* 20(1), 177–190.

Darras, V. M. (2019). The role of maternal thyroid hormones in avian embryonic development, *Frontiers in Endocrinology* 10, 66. DOI: 10.3389/fendo.2019.00066.

De Groef, B., Grommen, S. V. H. and Darras, V. M. (2008). The chicken embryo as a model for developmental endocrinology: Development of the thyrotropic, corticotropic, and somatotropic axis, *Molecular and Cellular Endocrinology* 293(1–2), 17–24. DOI: 10.1016/j.mce.2008.06.002.

De Groef, B., Grommen, S. V. H. and Darras, V. M. (2013). Hatching the cleidoic egg: The role of thyroid hormones, *Frontiers in Endocrinology* 4, 63. DOI: 10.3389/fendo.2013.00063.

Decuypere, E. and Bruggeman, V. (2007). The endocrine interface of environmental and egg factors affecting chick quality, *Poultry Science* 86(5), 1037–1042.

Decuypere, E., Dewil, E. and Kühn, E. R. (1991). Chapter 15: The hatching process and the role of hormones. In: *Avian Incubation, Poultry Science Symposium Number Twenty-Two*. Heinemann, Butterworth.

Deeming, D. C. (1989a). Characteristics of unturned eggs: A critical period, retarded embryonic growth and poor albumen utilisation, *British Poultry Science* 30(2), 239–249.

Deeming, D. C. (1989b). Failure to turn eggs during incubation: Development of the area vasculosa and embryo growth, *Journal of Morphology* 201(2), 179–186.

Deeming, D. C. (2002). *Avian Incubation, Behaviour, Environment and Evolution*. Oxford Ornithology Series. Oxford: Oxford University Press.

Etches, R. J. (1996). Reproduction. In: *Poultry*. Wallingford, UK: CAB International.

Eyal-Giladi, H., Ginsburg, M. and Farbarov, A. (1981). Avian primordial germ cells are of epiblastic origin, *Journal of Embryology and Experimental Morphology* 65, 139–147.

Eyal-Giladi, H. and Kochav, S. (1976). From cleavage to primitive streak formation: A complementary normal table and a new look at the first stages of the development of the chick. I. General morphology, *Developmental Biology* 49(2), 321-337.

Fasenko, G. M. (2007). Egg storage and the embryo, *Poultry Science* 86(5), 1020–1024.

Flamant, F. and Samarut, J. (1998). Involvement of thyroid hormones and its alpha receptor in avian neurulation, *Developmental Biology* 197(1), 1-11.

Forgacs, G. and Newman, S. A. (2005). *Biological Physics of the Developing Embryo.* Cambridge: Cambridge University Press.

Freeman, B. M. and Vince, M. A. (1974). Development of the avian embryo. In: *A Behavioural and Physiological Study*. London: Chapman and Hall.

Gilbert, S. F. (2014). *Developmental Biology* (10th edn.). Sunderland, MA: Sinauer Associates, Inc. Publishers.

Ginsburg, M. and Eyal-Giladi, H. (1986). Temporal and spatial aspects of the gradual migration of primordial germ cells from the epiblast into the germinal crescent in the avian embryo, *Journal of Embryology and Experimental Morphology* 95, 53-71.

Givisiez, P. E. N., Moreira Filho, A. L. B., Santos, M. R. B., Oliviera, H. B., Ferket, P. R., Oliveira, C. J. B. and Malheiros, R. D. (2020). Chicken embryo development: Metabolic and morphological basis for *in ovo* feeding technology, *Poultry Science* 99(12), 6774-6782. DOI: 1016/j.psj.2020.09.074.

Hamburger, V. and Hamilton, H. L. (1992). A series of normal stages in the development of the chick embryo. 1951, *Developmental Dynamics* 195(4), 231-272.

Harper, E. H. (1904). The fertilization and early development of the pigeon's egg. *American Journal of Anatomy* 3(4), 349-386.

Hemmings, N. and Birkhead, T. R. (2015). Polysperm in birds: Sperm numbers and embryo survival, *Proceedings of the Royal Society of London, Series B* 282, 2015168. DOI: 10.1098/rspb.2015.1682.

Hwang, Y. S., Seo, M., Kim, S. K., Bang, S., Kim, H. H. and Jae, Y. (2018). Zygotic gene activation in the chicken occurs in two waves, the first involving only maternally derived genes. *eLife* 7, e39381.

Ichikawa, Y., Matsuzaki, M., Hiyama, G., Mizushima, S. and Sasanami, T. (2016). Sperm-egg interaction during fertilization in birds, *Journal of Poultry Science* 53(3), 173-180.

Ioannidis, J., Taylor, G., Zhao, D., Liu, L., Idoko-Akoh, A., Gong, D., Lovell-Badge, R., Guioli, S., McGrew, M. J. and Clinton, M. (2021). Primary sex determination in birds depends on DMRT1 dosage, but gonadal sex does not determine adult secondary sex characteristics, *Proceedings of the National Academy of Sciences of the United States of America* 118(10), e2020909118. DOI: 10.1073/pnas.20209118.i.

Janke, O., Tzschentke, B. and Boerjan, M. (2004). Comparative Investigations of heat production and body temperature in embryos of modern chicken breeds, *Avian and Poultry Biology Reviews* 15(3), 191-196. DOI: 10.3184/147020604783637868.

Karagenç, L., Cinnamon, Y., Ginsburg, M. and Petitte, J. N. (1996). Origin of primordial germ cells in the prestreak chick embryo, *Developmental Genetics* 19(4), 290-301.

Khaner, O. (2017). Hefzibah Eyal-Giladi (1925-2017): Over fifty years of embryological research in Israel, *International Journal of Developmental Biology* 61(3-4), 121-126.

Khaner, O. and Eyal-Gyladi, H. (1989). The chick's marginal zone and primitive streak formation. I. Coordinative effect of induction and inhibition, *Developmental Biology* 115, 275-281.

Kochav, S. and Eyal-Gyladi, H. (1971). Bilateral symmetry in chick embryo, determination by gravity, *Science* 171(3975), 1027-1029.

Kong, L., Qiu, L., Guo, Q., Zhang, X., Chen, B., Zhang, Y. and Chang, G. (2018). Long-term in vitro culture and preliminary establishment of chicken primordial germ cell lines, *PLoS ONE* 13(4), e0196459. DOI: 10.1371/journal.pone.0196459.

Krawczyk, A. and Jaworska-Adamu, J. (2018). Oocyte-sperm interactions in the fertilization of birds, *Medycyna Weterynaryjna* 75(1), 6181-2019. DOI: 10.21521/mw.6181.

Kucera, P., Raddatz, E. and Baroffo, A. (1984). Oxygen and glucose uptake in the early chick embryo. In: Seymour, S. and Junk, W. (Eds.), *Respiration and Metabolism of Embryonic Vertebrates*. The Netherlands: Dordrecht.

Kunz, D., Wang, A., Chan, C. U., Pritchard, R. H., Wang, W., Gallo, F., Bradshaw, C. R., Terenzani, E., Müller, K. H., Huang, Y. and Xiong, F. (2021). Downregulation of extraembryonic tension controls body axis formation in avian embryos. *bioRxiv Preprint*. DOI: 10.1101/2021.02.24.432525.

Latter, G. V. and Baggott, G. K. (2002). Role of carbon dioxide and ion transport in the formation of sub-embryonic fluid by the blastoderm of the Japanese quail, *British Poultry Science* 43(1), 104-116.

Lee, H. C., Lu, H. C., Turmaine, M., Oliveira, N. M. M., Yang, Y., Almeide De, I.. and Stern, C. D. (2020). Molecular anatomy of the pre-primitive streak chick embryo, *Open Biology* 10(2), 190299.

Li, S., Bai, S., Qin, X., Zhang, J., Irwin, D. M., Zhang, S. and Wang, Z. (2019). Comparison of whole embryonic development in the duck (*Anas platyrhynchos*) and goose (*Anser cygnoides*) with the chicken (*Gallus gallus*), *Poultry Science* 98(8), 3278-3291.

Lillie, H. R. (1952). *Lillie's Development of the Chick (Revised by H. L. Hamilton)*. Edited by Willier, B. H. New York: Henry Holt and Company.

Lundy, H. (1969). A review of the effects of temperature, humidity, turning and gaseous environment in the incubator on the hatchability of the hen's egg. In: Carter, T. C. and Freeman, B. M. (Eds.), *The Fertility and Hatchability of the Hen's Egg*. Edinburgh: Oliver and Boyd, 143-176.

Martinsen, B. J. (2005). Reference guide to the stages of chick heart embryology, *Developmental Dynamics* 233(4), 1217-1237.

McNabb, F. M. and Wilson, C. M. (1997). Thyroid hormone deposition in avian aggs and effects on embryonic development, *American Zoologist* 37(6), 553-560.

Meyer, D. B. (1960). Application of the periodic acid-Schiff technique to whole chick embryos, *Stain Technology* 35(2), 83-89.

Meyer, D. B. (1964). Migration of primordial germ cells in chick embryo, *Developments in Biologicals* 10, 154-190.

Molenaar, R., Reijrink, I. A. M., Meijerhof, R. and Van den Brand, H. (2010). Meeting embryonic requirements of broilers throughout incubation: A review, *Revista Brasileira de Ciência Avícola* 12(3), 137-148.

Moran, E. T. (2007). Nutrition of the developing embryo and hatchling, *Poultry Science* 86(5), 1043-1049.

Nagai, H., Sezaki, M., Kakiguchi, K., Nakaya, Y., Lee, H. C., Ladher, R., Sasanami, T., Han, J. Y., Yonemura, S. and Sheng, G. (2015). Cellular analysis of cleavage-stage chick embryos reveals hidden conservation in vertebrate early development, *Development* 142(7), 1279-1286. DOI: 10.1242/dev.118604.

Nagai, H., Shin, M., Weng, W., Nakazawa, F., Jakt, L. M., Alev, C. and Sheng, G. (2018). Early hematopoietic and vascular development in the chick, *International Journal of Developmental Biology* 62(1-2), 137-144. DOI: 10.1387/ijdb.170291gs.

Nájera, G. S. and Weijer, C. J. (2020). Cellular processes driving gastrulation in the avian embryo, *Mechanisms of Development* 163, 103-624.

New, D. A. T. (1959). The adhesive properties and expansion of the chick blastoderm, *Journal of Embryology and Experimental Morphology* 7(2), 146-164.

Nieuwkoop, P. D. and Sutasurya, L. A. (1979). *Primordial Germ Cells in the Chordates*. Cambridge: Cambridge University Press.

Oviedo-Rondón, E. O., Velleman, S. G. and Wineland, M. J. (2020). The role of incubation conditions in the onset of avian myopathies, *Frontiers in Physiology* 11, 545045. DOI: 10.3389/fphys.2020.545045.

Pokhrel, N., Cohen, E. B.-T., Genin, O., Ruzal, M., Sela-Donenfeld, D. and Cinnamon, Y. (2018). Effects of storage conditions on hatchability, embryonic survival and cytoarchitectural properies in broiler from young and old flocks, *Poultry Science* 97(4), 1429–1440.

Pokhrel, N., Cohen, E. B.-T., Genin, O., Sela-Donenfeld, D. and Cinnamon, Y. (2017). Cellular and morphological characterization of blastoderms from freshly laid broiler eggs, *Poultry Science* 96(12), 4399–4408.

Pokhrel, N., Sela-Donenfeld, D. and Cinnamon, Y. (2021). The chick blastoderm during diapause, a landmark for optimization of preincubation storage conditions, *Poultry Science* 100(8), 101227.

Price, E. P. and Dzialowski, E. M. (2018). Development of endothermy in birds: Pattern and mechanisms, *Journal of Comparative Physiology. Part B* 188, 373–391.

Rahn, H., Ar, A. and Paganelli, C. V. (1979). How bird eggs breathe, *Scientific American* 240(2), 46–55.

Rengaraj, D., Won, S., Han, J. W., Yoo, D. K. H. and Han, J. Y. (2020). *DAZL* and its interacting Gene during germ cells specification and zygotic genome activation in chickens, *International Journal of Molecular Sciences* 21, 8170. DOI: 10.3390/ijms21218170.

Romanoff, A. L. (1967). *Biochemistry of the Avian Embryo*. New York: Macmillan.

Romijn, C. and Lokhorst, W. (1960). Foetal heat production in the fowl, *Journal of Physiology* 150(1), 239–249.

Sabin, F. R. (1917). Preliminary note on the differentiation of angioblasts and the method by which they produce blood-vessels, blood-plasma and red blood-cells as seen in the living chick, *Anatomical Record* 13(4), 199–204.

Sabin, F. R. (1920). Studies on the origin of blood-vessels and of red blood corpuscles as seen in the living blastoderm of chicks during the second day of incubation, *Carnegie Institution of Washington* 9, 213–262.

Scanes, C. G. (Ed.). (2015). *Sturkie's Avian Physiology* (6th edn.). Amsterdam: Academic Press, Elsevier.

Sellier, N., Brillard, J.-P., Dupuy, V. and Bakst, M. R. (2006). Comparative staging of embryo development in Chicken, Turkey, Duck, Goose, guinea fowl and Japanese Quail assessed from five hours after fertilization through seventy-two hours of incubation, *Journal of Applied Poultry Research* 15(2), 219–228.

Sharp, P. F. and Powell, C. K. (1930). Decrease in internal quality of hen's eggs during storage as indicated by the yolk, *Industrial and Engineering Chemistry* 22, 909–910.

Sheng, G. (2010). Primitive and definitive erythropoiesis in the yolk sac: A bird's eye view, *International Journal of Developmental Biology* 54(6–7), 1033–1043.

Silva Da, M., Labas, V., Nys, Y. and Réhault-Godbert, S. (2017). Investigating proteins and proteases composing amniotic and allantoic fluids during chicken embryonic development, *Poultry Science* 96(8), 2931–2941. DOI: 10.3382/ps/pex058.

Stern, C. D. and Downs, K. M. (2012). The hypoblast visceral endoderm: An evo-devo perspective, *Development* 139(6), 1059-1069.

Szczerba, A., Kuwana, T., Paradowska, M. and Bednarczyk, M. (2020). In vitro culture of chicken circulating and gonadal primordial germ cells on a somatic feeder layer of avian origin, *Animals (Basel)* 10(10), 1769. DOI: 10.3390/ani10101769.

Tazawa, H. (1980). Adverse effect of failure to turn the avian egg on the embryo oxygen exchange, *Respiration Physiology* 41(2), 137-142.

Tona, K., Onagbesan, O., Bruggeman, V., Mertens, K. and Decuypere, E. (2005). Effects of turning duration during incubation on embryo growth, utilization of albumen and stress regulation, *Poultry Science* 84(2), 315-320.

Tong, Q., Romanini, C. E., Exadaktylos, V., Bahr, C., Berckmans, D., Bergoug, H., Eterradossi, N., Roulston, N., Verhelst, R., McGonnell, I. M. and Demmers, T. (2013). Embryonic development and the physiological factors that coordinate hatching in domestic chickens, *Poultry Science* 92(3), 620-628. DOI: 10.3382/ps.2012-02509.

Too, H. C., Shibata, M., Yayota, M., Darras, V. M. and Iwasawa, A. (2017). Expression of thyroid hormone regulator genes in the yolk sac membrane of the developing chicken embryo, *Journal of Reproduction and Development* 63(5), 463-472.

Tullet, S. G. and Deeming, D. C. (1982). The relationship between eggshell porosity and oxygen consumption of the embryo in the domestic fowl, *Comparative Biochemistry and Physiology* 72A, 529-533.

Tzschentke, B. (2007). Attainment of thermoregulation as affected by environmental factors, *Poultry Science* 86(5), 1025-1036.

Uni, Z., Yadgary, L. and Yair, R. (2012). Nutritional limitations during poultry embryonic development, *Journal of Applied Poultry Research* 21(1), 175-184.

Weismann, A. (1885). *Die Continuität des Keimplasma's als Grunslage Einer Theorie des Vererbung*. Jena: Fisher Book Company Verlag.

Weissmann, A., Reitemeier, S., Hahn, A., Gottschalk, J. and Einspanier, A. (2013). Sexing domestic chicken before hatch: A new method for *in ovo* gender identification, *Theriogenology* 80(3), 199-205. DOI: 10.1016/j.theriogenology.2013.04.014.

Williams, J. (1967). Yolk utilization. In: Edit Weber, R. (Ed.), *The Biochemistry of Animal Development* (Vol. II). New York: Academic Press, 341-382.

Wishart, G. J. and Staines, H. J. (1999). Measuring sperm:egg interaction to assess breeding efficiency in chickens and turkeys, *Poultry Science* 78(3), 428-436.

Wong, E. A. and Uni, Z. (2021). Centennial review: The chicken yolk sac is a multifunctional organ, *Poultry Science* 100(3), 100821.

Wu, K. C., Streicher, J., Lee, M. L., Hall, B. K. and Müller, G. B. (2001). Role of motility in embryonic development I: Embryo movements and amnion contractions in the chick and the influence of Illumination, *Journal of Experimental Zoology* 291(2), 186-194.

Yadgary, L., Wong, E. A. and Uni, Z. (2014). Temporal transcriptome analysis of the chicken embryo yolk sac, *BMC Genomics* 15(1), 690. DOI: 10.1186/1471-2164-15-690.

Yahav, S. (2015). Chapter 37: Regulation of body temperature: Strategies and mechanisms. In: *Sturkie's Avian Physiology* (6th edn.).

Yahav, S. and Brake, J. (2014). Chick: A unique platform to study the effects of environmental factors on embryo development, *Journal of Stem Cells* 9(1), 17-37.

15 Appendix: Table 1 Chronology of the development of the chicken embryo

Table 1 Chronology of the development of the chicken embryo (Photography by Dr M.L. Boerjan, former Director, R&D Pas Reform; edited for publication by H. Arts, Marketing Director, Pas Reform)

Embryonic stage	Description of embryonic stage	Notes: what do you see in unhatched eggs after 21 days of incubation = a Breakout analysis in the hatchery
E0 (EGK XI-XIII) 	Fresh eggs (storage 0-3 days) - Albumen: thick and thin - Yolk: ball-shaped i.e. strong membrane - Embryo 3.5-4 mm: clear center in a white ring (donut-like form)	No initiation of embryonic growth: - Albumin watery thin - Yolk membrane thin Consequently yolk forms a flat disk but does not break immediately
E1 (0-24 h) (HH 7) 	- Elongated embryo - Primitive streak visible - One somite stage - Embryo surrounded by a light-yellow ring of sub-embryonic fluid (SEF)	- Albumin watery thin - Yolk thin and breaks easily - Whitish membranes become visible
E2 (25-48 h) (HH 8-12) 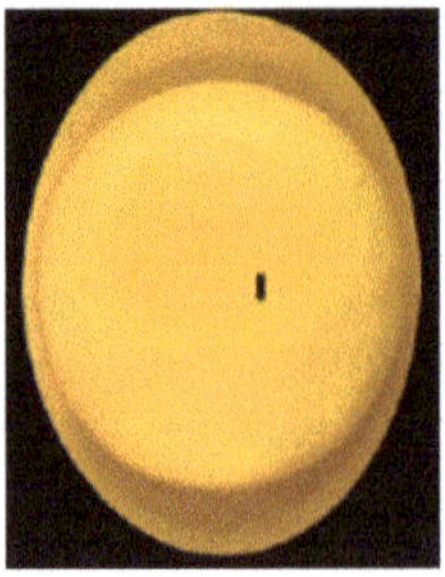	- Faint blood islands visible - Primitive blood cells in blood islands - Head and tail folds of amnion visible - Heart is S-shaped - Allantoic bud at hind gut is formed - 16 somites - Surface/volume of subembryonic fluid increased (=extra-embryonic compartments)	- Yolk breaks immediately - Watery yolk together with whitish membranes becomes visible - Small trails of blood might be seen

(*Continued*)

Table 1 (*Continued*)

Embryonic stage	Description of embryonic stage	Notes: what do you see in unhatched eggs after 21 days of incubation = a Breakout analysis in the hatchery
E3 (49–72 h) (HH 13–20) 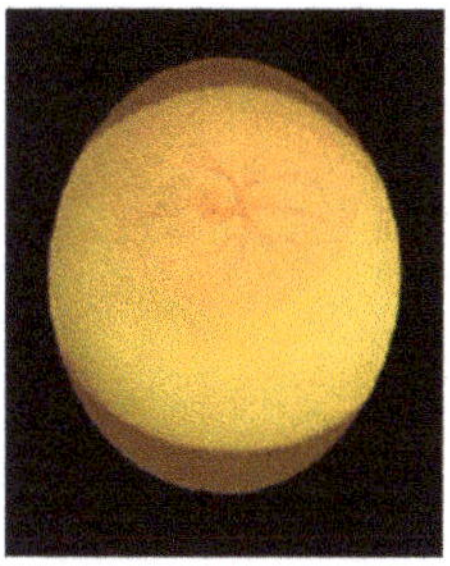	- Head turned left - Allantois now distinct - First passive movements - 30-36 somites - Embryo started to turn on left side - Area vasculosa prominent - Sinus terminalis clear - Vitelline membrane more than 50% over yolk	- Yolk thin and hardly recognized - SEF mixed with yolk, membrane - Blood ring clear - Embryo small clump
E4 (73–96 h) (HH 21–24) 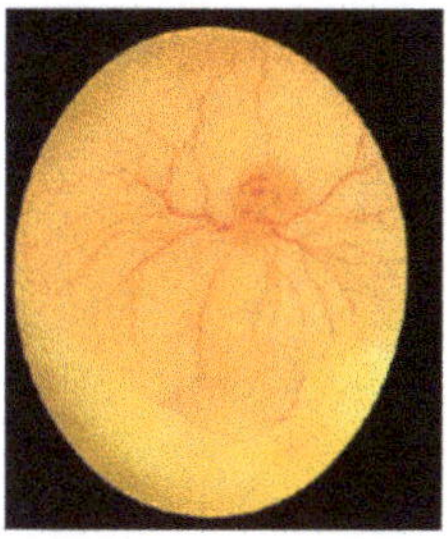	- Active movements of head and neck - Embryo fully turned on left side - Area vasculosa covers 50% of yolk - Eye pigmentation visible	- Embryo easy to find because of black contours of the eyes
E5 (97–120 h) (HH26) 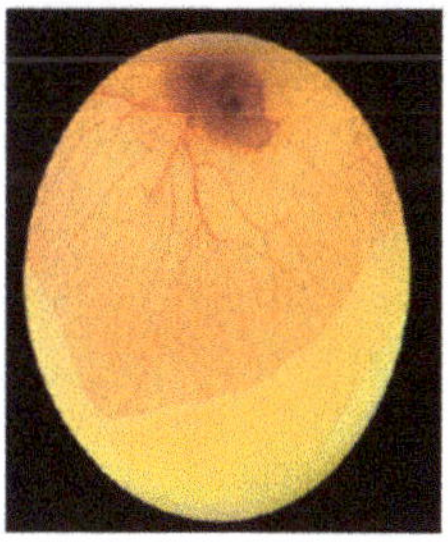	- Embryo turned to its left side - Wing and leg buds develop - Eye pigmentation: distinct - Elbows and knees develop	- Large embryo - Eye of the embryo very clear - Almost all yolk is mixed with SEF

(*Continued*)

Table 1 (*Continued*)

Embryonic stage	Description of embryonic stage	Notes: what do you see in unhatched eggs after 21 days of incubation = a Breakout analysis in the hatchery
E6 (121–144 h) (HH 27–29) 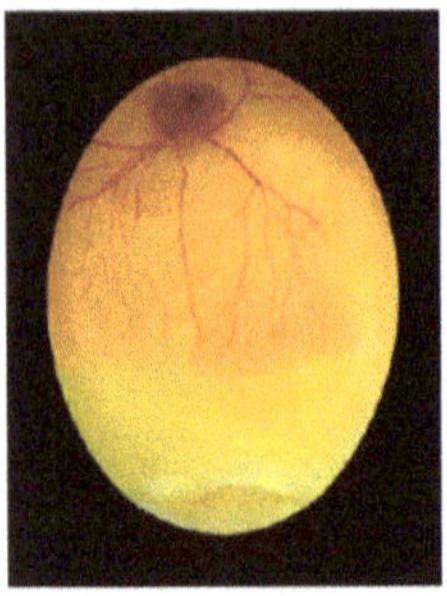	- Embryo visible in amnion - Beak becomes visible - Area vasculosa covers 75% of yolk sac - Albumen proteins are concentrated in the sharp end of the egg - Chorion and allantois fuse to chorio-allantois	- Egg content mixed with SEF watery, no yolk - Embryo a grey clump with a black eye - Wings might be visible - Jelly albumen in small end of egg
E7 (145–168 h) (HH 30–31) 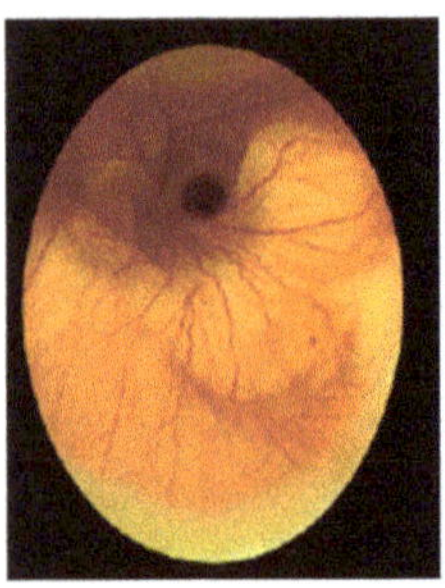	- Egg tooth and comb appear - Digits in wings and toes are visible - Webs between digits and toes developed - Legs move	Like day 6 but embryo is larger and a clear egg tooth is visible on beak
E8 (169–192 h) (HH 32–34) 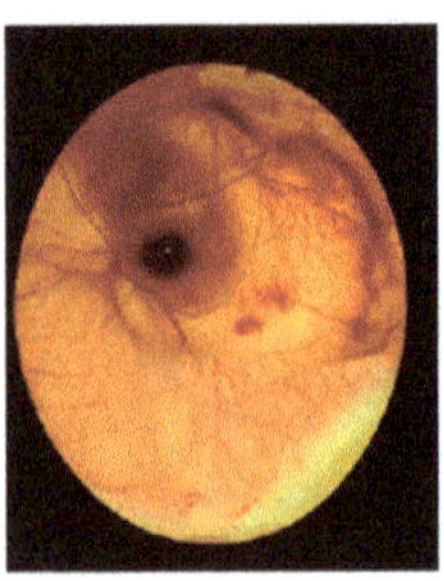	- Three rows of feather buds are visible - Calcification of bones is initiated - Upper and lower beak equal in length	- Three rows of feather on tail - Early small feather germs visible on head, neck and in the eye area - Third toe present - Webs between wing digits and toes clearly visible

(*Continued*)

Table 1 (*Continued*)

Embryonic stage	Description of embryonic stage	Notes: what do you see in unhatched eggs after 21 days of incubation = a Breakout analysis in the hatchery
E9 (193–216 h) (HH 35) 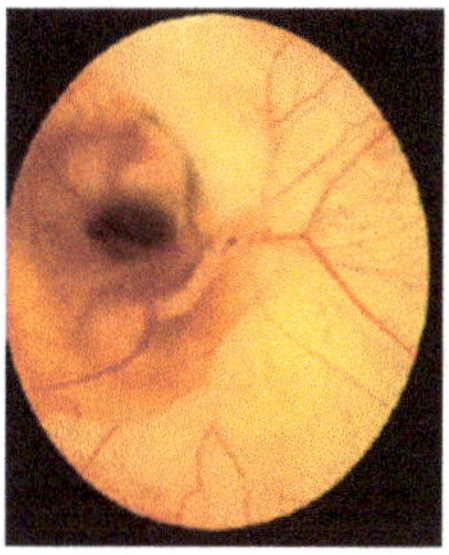	- Chorion-allantois covers 80% of embryo and yolk sac - Mouth opening appears	- Limb webs between digits start to disappear - Beak lengthened - Eye lids visible - Feathers on sides of the eyes conspicuous - New feathers on ventral side
E10 (217–240 h) (HH 36) 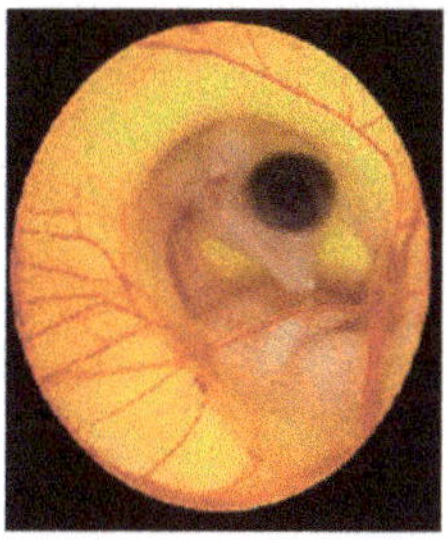	- The rhythmic contractions of amniotic muscles rock the embryo in the amniotic fluid - Growth of eyelids visible	- Toes are completely separated
E11 (241–264 h) (HH 37) 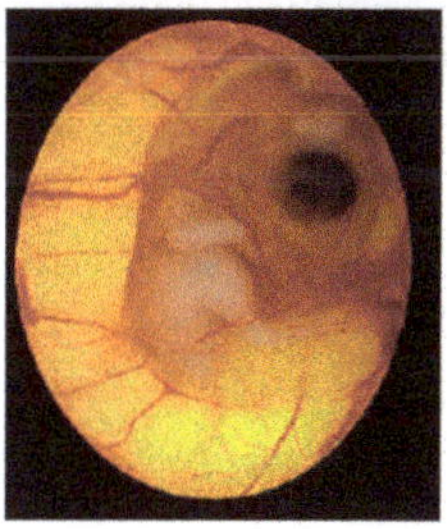	- Volume of amniotic cavity at maximum - Albumen is released in the amniotic cavity via sero-amniotic connection - comb is prominent visible	- Scales on legs can be recognized

(*Continued*)

Table 1 (*Continued*)

Embryonic stage	Description of embryonic stage	Notes: what do you see in unhatched eggs after 21 days of incubation = a Breakout analysis in the hatchery
E12 (265–288 h) (HH 38) 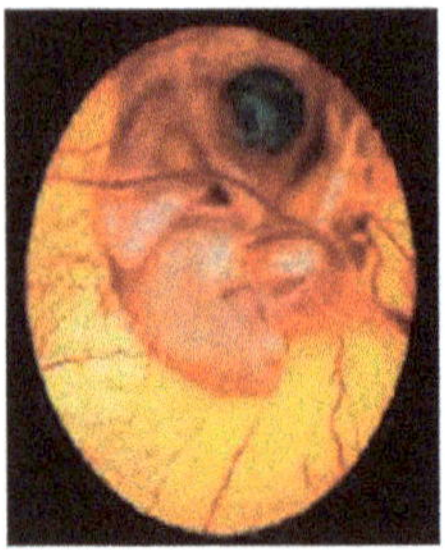	- Chorion-allantois complete and closed - Albumen proteins transported in yolk sac - Premature scales on legs	- Feather germs on eyelids and wings
E13 (289–312 h) (HH 39) 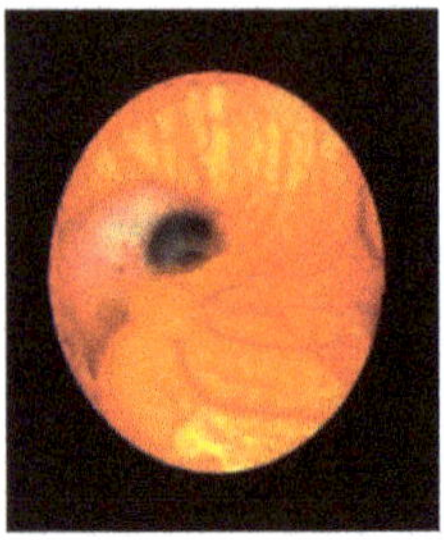	- Head in yolk sac - Metabolic heat production and oxygen consumption increase exponentially - Periods of passive embryonic activity increase - Calcium is mobilized from the shell	- Scales on legs overlapping - Feathers have increased in length
E14 (313–336 h) (HH 40) 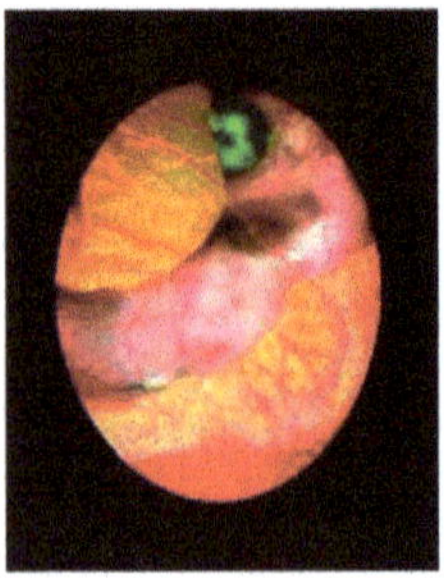	- Turning of embryo towards air cell starts - Eye lids closed - In the yolk sac ridges of endoderm epithelium can be observed	- Yolk sac developed and connected to the body via umbilical cords - Small intestine is still outside the body

(*Continued*)

Table 1 (*Continued*)

Embryonic stage	Description of embryonic stage	Notes: what do you see in unhatched eggs after 21 days of incubation = a Breakout analysis in the hatchery
E15 (337-360 h) (HH 41) 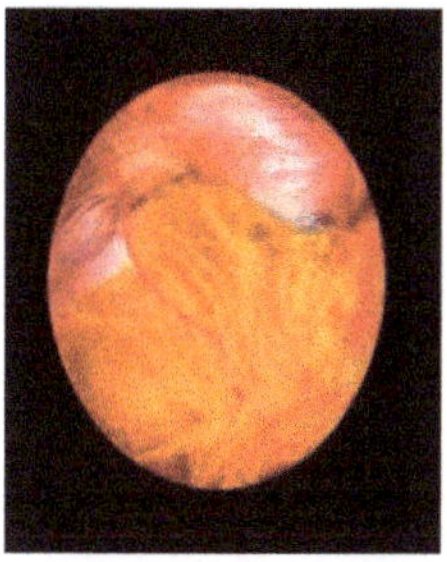	- The embryo continues to grow - Activity becomes less - Body covered with feathers	- Head between legs
E16 (361-384 h) (HH 42) 	- Metabolic heat production at maximum - Beak is open	- Small intestine is still visible
E17 (385-408 h) (HH 43) 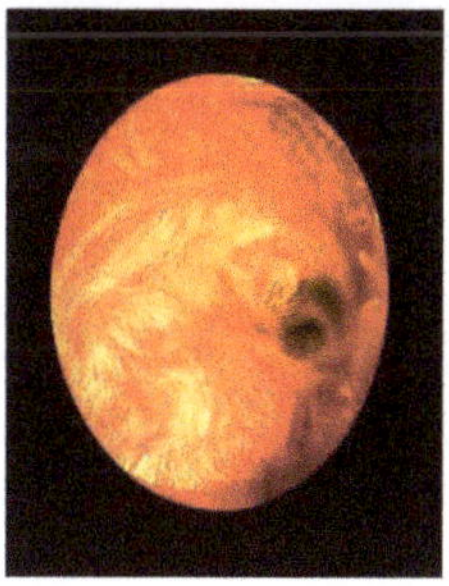	- Head has turned towards the air cell - Volume of amniotic cavity decreases - Growth rate decreases	- Small intestine withdrawal complete - Uptake of yolk sac starts

(*Continued*)

Table 1 (*Continued*)

Embryonic stage	Description of embryonic stage	Notes: what do you see in unhatched eggs after 21 days of incubation = a Breakout analysis in the hatchery
E18 (409-432 h) (HH 44) 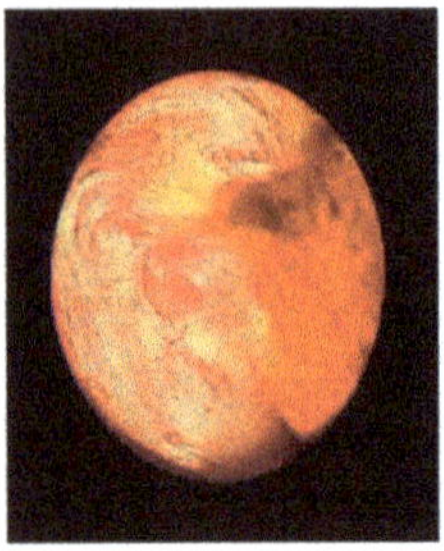	- First behaviour response to light: discrimination to blue and red from spectrum - Head under right wing - Beak towards air cell - Oxygen consumption in plateau phase	
E19 (433-456 h) (HH 45) 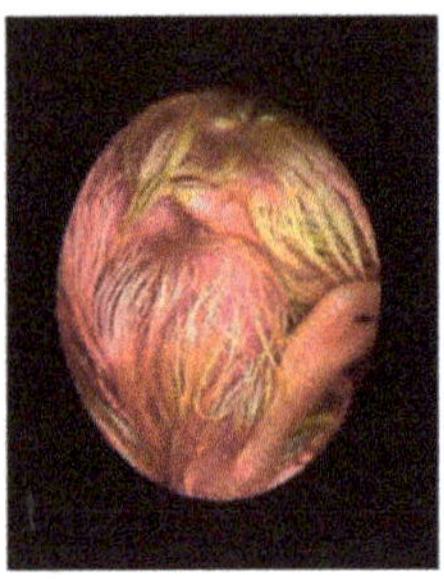	- Hatching muscle matures - Head under right wing - Absorption of allantois fluid completed - Glycogen stores from liver and yolk mobilized - Yolk sac withdrawal begins	- Head over right wing might be a sign of overheating during last days in setter

E20 (HH 45) at 457–480 h of incubation

- Chorio-allantois start to degenerate
- Lungs are inflated/activated and lined with surfactant to reduce surface tension
- Inner shell membrane pierces
- Yolk sac completely withdrawn
- Navel is closed but umbilical sphincter (ring of muscles) swollen (hydrated)
- Shell is pipped

E21 (HH 46) at 481-205 h of incubation

- Navel is closed with shrunken/flat umbilical sphincter
- Hatched chick ready after drying of feathers (damp neck feathers are OK)

Hatching time chicken depends on:

- Storage time: the longer eggs are stored, the later chicks hatch
- Incubation program (temperature; humidity, ventilation
- Species and Breed within the species

Chapter 6

Chicken egg storage and transport

Dinah Nicholson, Aviagen Ltd, UK

1 Introduction

For a chicken breeding operation, one of the major production challenges is to manage the rolling stock of hatching eggs so every egg is set while still at peak quality. While hatchability holds up well for the first week or so after the egg is laid, it will start to fall in the second week, with the rate of decline increasing rapidly into the third and fourth week. A typical profile is shown in Fig. 1.

The profile shows a slight increase in hatchability over the first 4 days of storage, an uneven plateau to 10 days followed by a steady fall to 21 days. If storage conditions are inappropriate - too warm or with a fluctuating storage temperature - then hatchability will start to fall sooner, and the rate of decline will be more extreme.

A hatchery will usually be supplied with hatching eggs from multiple farms, similar in size and with flock ages managed so that in total they deliver a similar number of eggs every week. In the past, the egg utilisation versus egg age conundrum was often handled by combining eggs from

http://dx.doi.org/10.19103/AS.2022.0118.12

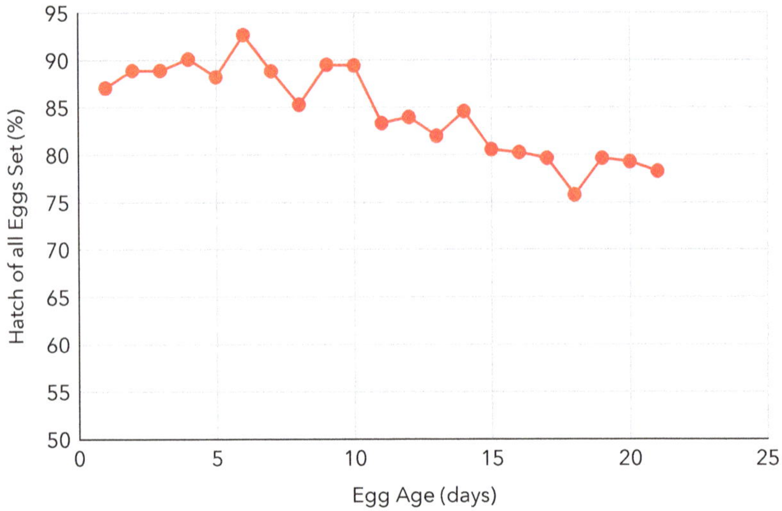

Figure 1 Profile of hatchability changes as egg age increases – Ross 308 broiler hatching eggs. Inappropriate storage conditions accelerate the rate of decline. Source: Aviagen data (2008).

different breeder flocks (and even breeds at the broiler generation) so that eggs could be set together to meet the needs of a day's hatch with minimal egg age. However, modern production has become steadily more demanding of uniformity in broiler flocks. In addition, food retailers often expect reliable traceability of meat deliveries back to the breeder source farm. These constraints, added to those of variable demand through the year and destination flock sizes getting steadily larger, make it more and more likely that eggs from one source flock will be collected over several days in order to fill a single broiler or breeder house. As a consequence, egg age at set in commercial hatcheries can sometimes be significantly higher than is optimal for good hatchability.

The effects of egg storage on hatchability, chick quality and progeny performance and the techniques that might improve storability have been of interest to researchers for many years, with reviews published by Mayes and Takeballi (1984), Butler (1991), Meijerhof (1992), Brake et al. (1997), Fasenko (2007), Reijerink et al. (2008) and most recently by Nasri et al. (2020). Rather than attempt to improve on an already well-covered topic, this chapter will focus on the techniques that are in practical use in commercial poultry operations, followed by a section describing the experiments and field trials implemented by specialists employed by Aviagen to develop a technique using short periods of incubation during egg storage (SPIDES) to improve the hatchability of stored eggs.

The most important factors that make achieving good hatch and chick quality from stored eggs so difficult are caused by

- changes to the blastoderm during egg storage causing an increase in the percentage of abnormal and dead embryos identified by Mather and Laughlin (1977), Fasenko et al. (1992b), Bakst and Akuffo (1999) and Bakst et al. (2012);
- deterioration in internal egg quality, especially albumen and the vitelline membrane - Walsh et al. (1995), Lapao et al. (1999) and Fasenko et al. (2009); and
- bacterial growth in eggs contaminated immediately after lay, leading to an increase in rots and exploding eggs during incubation - Gesche and Schuler (1979) and Humphry and Whitehead (1993).

2 Changes to the blastoderm during egg storage

Eyal-Giladi and Kochav (1976) published a system for staging chicken embryos in the very early stages of development - from fertilisation to the period immediately after the eggs are laid. They described the 14 stages of development that happen before Hamburger and Hamilton (1951) stage 2, which is usually achieved after 6-7 h of incubation.

At oviposition, a fertile hatching egg will hold a blastoderm on the yolk surface which contains around 60000 cells and will have developed to on average EGK stage 10 (Eyal-Giladi and Kochav, 1976). Various factors will affect the mean and distribution of the embryo stage at and immediately after oviposition. Fasenko et al. (1991), using eggs from a naturally mated layer breed, showed that eggs left in the nest for 6 h after oviposition had blastoderms which were more advanced than those in eggs collected and cooled within an hour of lay (12.33 vs 11.11, $P = 0.0001$). She did not report hatchability but did suggest that there might be an optimal stage of development to permit the best hatchability. She also reported (Fasenko et al., 1992a), this time using artificially inseminated broiler breeders, that the first egg in a sequence tends to have a slightly more advanced embryo at oviposition than all the other eggs (10.36 vs 10.05, $P = 0.0001$). In this trial, hatchability was reported and shown to be significantly higher in eggs from within the sequence, i.e. those with a lower embryo stage.

If cooling is impeded in parts of the egg pack, the eggs which remain at or close to incubation temperature longer will continue to develop past EKG stage 10. Figure 2 shows data collected on behalf of Aviagen (Bakst, 2012, personal communication).

The embryos were staged on the day after the eggs were laid (using eggs from a flock aged 36 weeks). Rather than a tight distribution around stage EGK

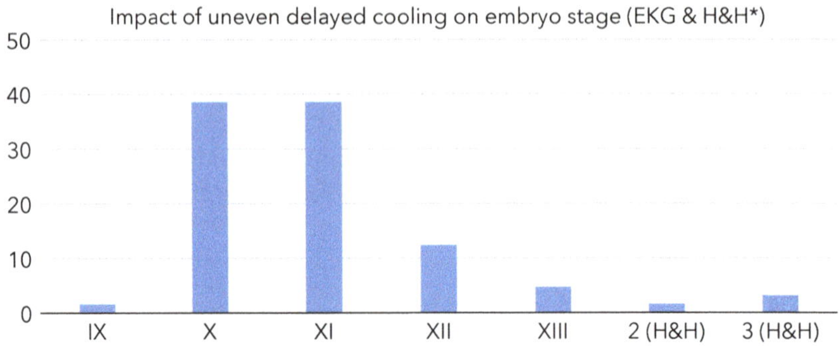

Figure 2 The effect of uneven cooling of hatching eggs packed in close-stacked plastic setter trays on the stage of embryo development 24 h after egg collection on farm. *EGK = Eyal-Giladi and Kochav (1976), H&H = Hamburger and Hamilton (1951).

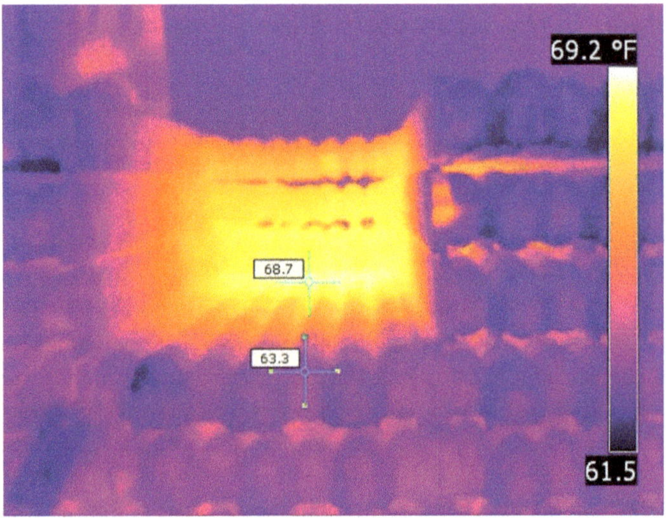

Figure 3 Thermal image of eggs the morning after collection when packed close and unable to cool evenly. Eggs in the centre of the stack were 5°F warmer than those at the periphery.

X, as reported by Selliers et al. (2006), there was a skewed distribution (Fig. 2), with a small number of eggs developing as far as Hamburger and Hamilton stages 2 and 3.

Rather than the more usual system for collecting hatching eggs onto setter trays, which are then placed on well-spaced racks in buggies, this operation collects eggs onto close-stacked plastic trays which delay cooling if eggs are packed when still warm. Thermal images of the system, taken the day after eggs were laid, showed that eggs in the centre of the pack were 5°F warmer than those at the edges (Fig. 3).

Published by Burleigh Dodds Science Publishing Limited, 2023.

The hatchery involved was not reporting any worries about hatchability using this system, but the likelihood that the wide range of embryo stages in the egg pack would also widen the hatch window and the possibility of raised embryo mortality in the embryos which had reached the two Hamburger and Hamilton (H&H) stages remain of concern.

When fertile hatching eggs are stored, cells in the blastoderm start to die and the percentage of viable cells drops from 27.3% at 3-4 days of age to 17.4% and 10.5% at 10-12 and 17 days, respectively (Bakst et al., 2012). Mather and Laughlin (1976), looking at delays in early development, showed that eggs incubated for 42 h were delayed by 5.3 h after 7 days storage and 12.2 h after 14 days. Similarly, Fasenko (2009) also showed that embryos stored for 14 days can be slower to start growing; after 96 h at incubation temperature, they were some 10 h behind embryos from eggs stored only 4 days. Bakst (personal communication) showed that eggs stored for 17 days and then incubated for 7 days had an embryo stage of H&H 27.6, two stages, behind eggs stored for 3-4 days before set. It is standard practice in hatcheries where egg age is variable to advance the set time of stored eggs to take account of the delay and allow the chicks to emerge at roughly the same time after incubation.

By contrast, the EGK stage does not necessarily change much when eggs are stored for up to 21 days at a low enough temperature; stored at 14°C, the EGK stage remained the same but the number of dead blastoderms was significantly ($P < 0.00001$) increased after 14 days and 21 days of storage (Fasenko et al., 1992b). This ties in with observations by Mather and Laughlin (1976) and Bakst and Akuffo (1999) that long storage leads to a proportion of the blastoderms developing abnormally. In practical hatchery terms, these will be the embryos which never start to grow again after storage or which start to grow and then very soon die in the early embryonic development stage.

Brady et al. (2022) analysed the transcriptome activity of blastoderms exposed to prolonged egg storage. Eggs were stored for 21 days at 17°C, and their gene expression was compared with blastoderms from eggs stored for 4 days. They concluded that prolonged storage was associated with the enrichment of pathways associated with cell stress and death.

3 Changes to egg quality during storage

For hatching eggs, the most significant components of the egg in terms of egg storage are the shell cuticle, the vitelline membrane and the albumen. All show signs of change with egg storage, with changes generally being detrimental to hatchability.

3.1 The cuticle

The cuticle, the final layer of the egg shell to be formed before oviposition, forms the main barrier to bacterial penetration of the egg. Sparks (1985) showed that the cuticle will only prevent bacterial penetration of the egg shell once it has cured - this broadly corresponds to the newly laid egg losing the appearance of wetness which is associated with the uncured cuticle, typically within 15 min after oviposition. It follows that any contaminated material that the egg encounters immediately after it is laid, particularly if that material is wet, may allow bacteria to enter the egg shell.

Egg storage does not seem to have a commercially significant effect on cuticle quality. Ball et al. (1975) used a green stain (Edicol Supra Pea Green) to show the cuticle cover of eggs. The dye is protein specific, and the intensity of the stain reflects the cuticle quality of the egg. Eggs stored at 24°C showed a worsening of cuticle cover after 18 days and again after 49 days. For eggs stored at 5°C, there was no change up to 36 days. Given that eggs are unlikely to be stored as long, or as warm, as these were, storage *per se* is unlikely to be an issue.

3.2 Albumen

Albumen is deposited around the egg yolk in the magnum section of the oviduct. The absolute weight increases with egg weight as flocks age, but in general albumen makes up 57-65% of the total weight of the hen's egg (Nys and Guyot, 2011), with the higher values generally seen in eggs laid by commercial layer lines and by relatively immature hens (Peebles et al., 2000). The albumen can be divided into four parts: an inner liquid layer in contact with the vitelline membrane; a layer of thick albumen; the chalazae, which help to hold the yolk in position; and an outer liquid layer. The thick albumen makes up 57% of the total albumen by weight (Nys and Guyot, 2011). With increasing egg age, the thick albumen deteriorates. Walsh et al. (1995), working with broiler lines, showed that long storage (14 days vs 7 days) and high storage temperature (23.9°C vs 12.8°C) were detrimental to albumen quality (as albumen height, mm). The observed effects of both factors were significantly different ($P < 0.001$), and there was also a significant interaction ($P < 0.001$) between storage time and storage temperature where the impact on early mortality + infertility due to storage temperature was much greater in eggs stored for 14 days. Plotting mean albumen height against the percentage of infertile and early dead embryos recorded in the trial showed a steady increase in embryo mortality as albumen height deteriorated (Fig. 4).

Lapao et al. (1999), storing eggs at 16°C, showed that albumen height was reduced as eggs were stored and was able to show a similar

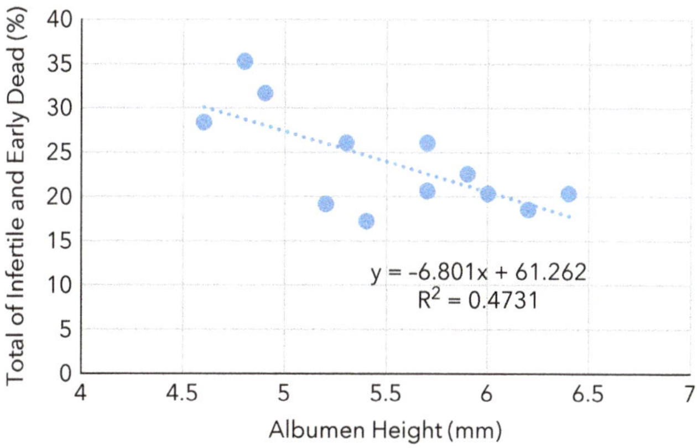

Figure 4 The relationship between albumen height and losses due to early embryo mortality in broiler hatching eggs. Source: data from Walsh et al. (1995).

deterioration due to hen age. Differences due to egg and flock age were both significant ($P < 0.0001$) but there was no interaction between egg age and flock age.

3.3 The vitelline membrane and mottling

The vitelline membrane is formed in the infundibulum, immediately after ovulation, and is complete within 20 min of ovulation (Nys and Guyot, 2011). It is composed of an inner and outer fibrous layer, sandwiching an amorphous middle layer. The membrane tends to deteriorate with egg age; at the most basic level, any attempt to isolate the blastoderm for staging becomes increasingly difficult in stored eggs (Fasenko et al., 2009). Bakst (personal communication), isolating blastoderms for staging from broiler hatching eggs, showed an increase in the number of broken yolks from 2% after 3 days storage up to 12% after 23 days. When the vitelline membrane is not functioning properly, water moves from the albumen into the yolk, causing a phenomenon called mottling, where the egg surface is marked with darker areas, as shown in the photograph in Fig. 5. It is known that mottling gets worse with egg age (Lyon et al., 1972).

Fasenko et al. (2009) suggested that in long-stored eggs, the embryo does not cling to the inner peri-vitelline layer as closely as it does in fresh eggs and that any alteration to the membrane may influence early embryo development, reducing survival rates. But the fragility of the membrane under stress also means that in some eggs the yolk will rupture early in incubation, and in others it will rupture as the residual yolk is drawn into the body cavity just before hatch (personal observation).

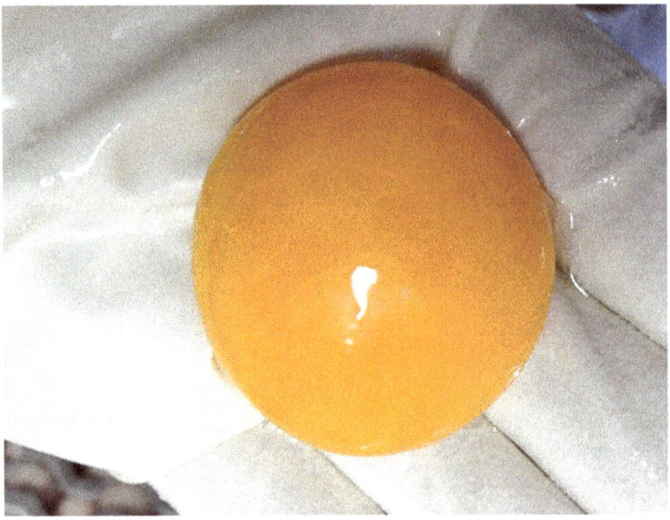

Figure 5 Egg yolk with extensive mottling after 5 days of storage.

4 Improving the survival of embryos in stored eggs

4.1 Storage temperature and humidity

Breeding companies and integrators tend to define acceptable storage temperatures depending on their usual egg age at set. Eggs stored for a short time can cope with a higher temperature than eggs stored for longer (Meijerhof, 1994). The ideal will be a compromise between the temperature needed to maintain egg contents in good condition, protecting the embryo and avoiding condensation at any point in the egg collection - storage-transport-storage-setting chain. For a commercial operation, there will also be some pressure to keep the costs of building the eggs store down and to minimise the running costs by not overcooling.

Edwards (1902) suggested a value of 20.5°C for physiological zero - the temperature below which all embryonic development stops. By contrast, Funk and Biellier (1944) who simply measured the diameter of the blastoderm, showed no change in the diameter at storage temperatures between -1°C and 24.4°C (30°F and 76°F). Possibly, the most satisfactory measure is that of Fasenko et al. (1992b), who used the EGK staging system to show that at a storage temperature of 14°C, the embryo shows no sign of further development during egg storage of 21 days (although the number of dead embryos still rose after 14 days and 21 days). More recently, Özlü et al. (2022), using EGK staging over a range of egg storage temperatures, were able to show significant increases in embryo stage over storage periods of 3, 7 and

Published by Burleigh Dodds Science Publishing Limited, 2023.

14 days at temperatures of 18°C, 21°C and 24°C but no development in eggs stored at 15°C. The possibility of embryo development, particularly over a long storage period, explains why fluctuating temperatures, even if below physiological zero, can still be detrimental to hatchability. Özlü et al. (2018) compared the EGK score and hatchability of broiler embryos and hatching eggs from a young and an old flock which were either kept at a constant 17.9°C or allowed to rise from that baseline to 21°C 3 times a day. Set at 7 days, the eggs on the fluctuating treatment had a significantly ($P < 0.05$) raised EGK score. However, it was only in the eggs laid by older hens that the fluctuating temperature affected hatchability - the temperature fluctuation was associated with a hatch drop of 3.3% points, due to an increase in early dead embryos. The young flock, despite an increase in the EGK score, did not lose any hatchability.

Meijerhof et al. (1994) stored hatching eggs laid by hens of 37 weeks or 59 weeks of age at 10°C, 15°C or 20°C for 2, 6, 9 or 12 days. The eggs laid by the older birds were much less able to cope with adverse treatment. High storage temperatures and storage over 9 days both reduced hatchability. The 10°C treatment gave numerically better hatch than the 15°C treatment for all egg storage periods, although the difference was only statistically significant ($P < 0.05$) at 12 days storage. For the younger birds, the 10°C treatment still performed well, but only the 20°C temperature held for 12 days significantly reduced hatchability.

Relative humidity during storage is not particularly important (Brake et al., 1997), provided it is not too low so as to cause eggs to dehydrate (Proudfoot and Hulan, 1976) or too high so as to permit mould growth on walls and ceiling of the egg store (personal observation). In most egg stores, humidity will build naturally due to evaporation from the stored eggs and after the routine washing of floors and walls. Provided the weight loss during storage is included in any assessment of incubation weight loss to 18 days, this is probably the best way to manage egg store humidity.

The temperature response data cited above tend to suggest that keeping hatching eggs consistently below 15°C will improve the overall hatch results. This is probably true, but only if the operation can manage the risk of condensation when the eggs are moved from the cold store to a warmer, more humid environment. Surface moisture seems to be necessary for bacterial penetration of the egg shell pores, which is why it is so important to avoid conditions likely to cause condensation on the shell surface (Bruce and Drysdale, 1994). For this reason, a dew point chart is a useful tool, helping to identify which moves are safe, and which might allow condensation and thus cause problems. Aviagen (2018) publish such a table, focused on changes often observed in hatching egg storage and incubation. If eggs are held at 15°C, and moved to a warmer room, then condensation will occur at 26°C if the humidity is held at 50% relative humidity (RH). If the storage temperature is dropped to 12°C, then the dew point will be at 23°C at the same humidity level. When the eggs are stored at 15°C, they can safely be moved straight to the setter room corridor (around 23°C and 50% RH) to make up the set. Storing the eggs at 12°C can only be done safely if the humidity is kept below 50%. Otherwise, an additional store held at an intermediate temperature will be needed. The risk of condensation on the egg shell is the main reason that many integrations are reluctant to reduce egg store temperatures to a more appropriate level when egg age is high.

5 Turning eggs during storage

Landauer (1961) referred to four publications in the first half of the twentieth century, three of which showed no benefit to turning during storage and the fourth (Funk and Forward, 1951) showed a benefit only if eggs were stored for more than 7 days.

Proudfoot (1966) stored White Leghorn eggs from a 60-week-old flock for between 14 and 28 days. Half the eggs were held at 50° from the vertical and turned through 180° once a day. The other half were held with the air cell facing upwards. Storage periods of up to 14 days showed no benefit to turning eggs during storage, but by 21 days a 10% point gap had opened up which persisted to the 28th day of storage. A later, more complex series of experiments (Proudfoot, 1969), this time using meat type lines of unspecified age, looked at the effect of storing eggs with the air cell downwards (small end up, SEU), as well as that of the same turning treatment as in the earlier experiments. He showed that for a storage of 7 days, hatch was better if eggs were stored with the air cell upwards (small end down, SED), compared with eggs stored SEU. Hatchability was 78.3% for the SED and 70.4% for the SEU treatments, $P < 0.05$). Longer storage periods, of 2, 3 or 4 weeks, showed numeric advantage to storage SEU, although the difference was statistically significant ($P < 0.05$) in only one of the three trials. There was no significant advantage to turning nor was there any treatment interaction. Again, the treatment effects were only apparent in eggs of 21 and 28 days, with an average gap of 10.5% points at 21 days widening to 16.55% points at 28 days. Although not statistically significant, turning the SEU and SED eggs daily narrowed the gap between them from 10–16% points down to 2–3% points.

Becker et al. (1969) looked at storage periods of up to 5 weeks, turning boxed eggs either through 180° (as did Proudfoot) or through 90° (equivalent to turning in a commercial setter), twice a day. They showed that the 90° turn gave better results than either no turning or a 180° turn.

Bowman (1969) and Insko and Martin (1933) both showed that lines with different selection histories may respond to in-storage turning differently. Given the considerable amount of time that has elapsed since both publications, it is unlikely that either paper is of any predictive value, but they do raise the possibility that lines with different selection histories may differ in their response to storage treatment.

Elibol et al. (2002) reported trials investigating the interaction of broiler breeder flock age, egg age and turning frequency during storage. They compared eggs from young (30–31 weeks) and old (52–53 weeks) flocks over two experiments. The experiment used more eggs per replicate and more replicates per treatments than most earlier reports. Eggs were stored for 3, 7 or 14 days and either not turned or turned through 90°, 4 or 24 times a day. Trays were distributed throughout both setter and hatcher to avoid any systematic interaction between treatment and local conditions in the incubators. The two experiments showed a significant benefit to turning four times a day, which was especially apparent in eggs from the older flocks. Turning every hour showed a small numeric improvement in experiment 1 and a small decrease in performance in experiment 2. Again, the older flock was more sensitive. However, this 2002 trial showed turning to have an effect at 14 days storage, while Proudfoot (1966, 1969) only showed improvements from 21 days storage.

Elibol and Brake (2008) looked at eggs from 60- to 62-week-old breeders, stored for 3 days or 14 days at 18°C. The eggs were either stored SED (controls) or SEU. The rate of fall of hatchability with increasing egg age was quite high in this trial (82.9% vs 53.6% hatch of fertile at 3 days and 14 days respectively,

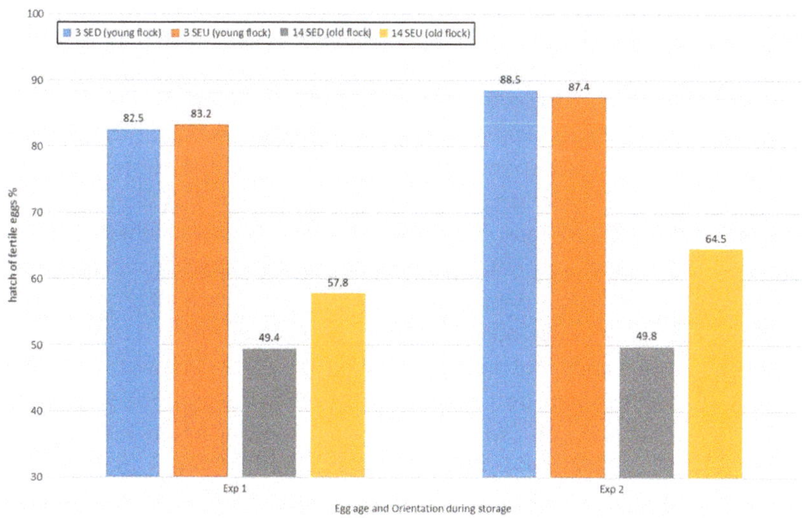

Figure 6 Elibol and Brake (2008) hatchability of eggs laid by young or old flocks, stored for 3 days or 14 days small end down or up.

$P < 0.05$); this was probably because of the relatively high storage temperature. Storing eggs SEU improved hatchability from 49.4% to 57.8% in eggs stored for 14 days. Fresh eggs were not significantly improved by the SEU orientation during storage but were no worse (Fig. 6).

Turning during storage can make a useful contribution to improved hatchability, especially if eggs are being stored for longer than a week. For most hatcheries, the most practical method for turning the eggs will be to use setter trollies which can be connected up to the appropriate turning mechanism installed in a part of the egg store. This will turn the eggs through 90° and can be programmed to turn as often as desired - although it is probably better to stick to frequencies on or below 4 times per day. When planning which eggs to turn, eggs from older (>50 weeks) flocks which will be stored for more than 7 days should be prioritised.

Holding eggs SEU is also very effective in improving hatchability of stored eggs, but in the majority of commercial hatcheries, it is not particularly practical to implement this on a large scale. It is likely to be most useful in production systems which store cooled eggs in boxes, which can then be inverted as required. Turning the eggs as well as inverting them may give additional benefits provided the boxes can be turned daily without damaging the contents. It is probably safer to turn eggs through 180° than 90° if they are packed in boxes, to make the most of the structural integrity of the fibre trays.

6 Transport

Strategies for handling hatching egg holding temperatures from farm to hatchery egg store tend to vary depending on local climate. In temperate regions, with good roads, eggs will usually be packed straight on to setter trays and allowed to cool to the agreed storage temperature on farm. They will be collected from the farm one or more times per week, in a temperature-controlled truck with air suspension and taken to the hatchery store where they will be stored until they are set. The main vulnerabilities in this scenario are the transfers from farm store to truck in periods of hot, humid weather when cold eggs are briefly taken into hot, humid air and condensation occurs, bringing with it the risk of bacterial contamination. In tropical climates, the condensation risk of moving cooled eggs outside becomes constant, and as a consequence many operations do not cool the eggs on the farm at all but move them daily to the hatchery to be cooled. Both systems work well in the appropriate climate.

Poor road conditions can cause problems, especially if the eggs are transported on plastic rather than fibre trays. Jarring due to uneven road surfaces or potholes will increase the number of cracked eggs but may also cause the air cell membrane to detach from the shell membrane (Knox and

Olsen, 1936). The affected eggs had a hatchability some 20% points worse than eggs where the air cells appeared normal. A subsidiary study reported in the same paper showed that jarring only caused the membranes to detach if the eggs were packed SEU. Torma and Kovácsné (2012), comparing outcomes of hatching eggs produced in Hungary and either hatched locally or in Russia, observed that the transported eggs hatched less well, with a distinctive pattern of malformed embryos showing duplications of body parts. Torma and Kovácsné (2019) reported that vibration of between 10 Hz and 30Hz lasting for 10 min was enough to depress hatchability and that the effect was worse in eggs packed on a plastic tray than a fibre one. (From a control value of 91.06%, hatch of fertile was reduced by 4.16% in fibre trays and 8.24% in plastic trays.) No solution other than avoidance of transporting eggs over poor road conditions was identified, although the results do suggest that regular maintenance of farm tracks and driveways under a production company's control could be worthwhile in cases where road conditions are otherwise good.

Finally, Landauer (1961) cited Dareste writing between 1883 and 1891, who reported that eggs allowed to rest for 2 days after transport hatched normally, while if set immediately the same batch of eggs produced largely abnormal embryos. More than a century later, most hatchery managers still aim for at least 24 h rest between the eggs' arrival at the hatchery and starting incubation.

7 Egg shell contamination and disinfection

The hatching egg is well protected from bacterial contamination under most conditions (as discussed by Bain elsewhere in this book). This section covers some of the main reasons why eggs become contaminated under commercial conditions and some methods to limit contamination.

The first line of defence against bacterial contamination is the cuticle, which once cured covers the openings of the egg shell pores and limits the ability of bacteria to penetrate the shell. Haines and Moran (1940) showed that eggs immersed in a suspension of pseudomonas while still warm all went rotten. Similar exposure once the egg was fully cooled substantially reduced the incidence of rots (from 100% to just under 20%). This observation is behind still-current advice to only wash eggs (if unavoidable) in water warmer than the egg (Sparks, 1994). When eggs were placed on straw which had recently been sprayed with pseudomonas, none of the eggs rotted; the authors concluded that moisture is also important. Sparks (1985) showed how in a freshly laid egg, with the cuticle still wet, pushing the egg shell against a fresh dropping allowed bacteria to penetrate the shell. Once the cuticle was dry, similar exposure did not allow any bacteria into the egg. In practical terms, this means that nest boxes should be checked daily, and any droppings or other source of bacteria removed. Floor eggs should be discarded.

It is important to avoid any cleaning or disinfection treatment that might damage the cuticle. Scraping or sanding the egg surface to remove patches of dirt come into this category, as do disinfectants such as peracetic acid which can strip the protein coat from the egg surface (Burgess, 1994) or spray application methods which are excessively vigorous (Sparks and Burgess, 1993).

The risk of condensation during egg storage and transport has already been discussed and is an extremely frequent cause of contamination problems.

The first possibility of bacterial contamination, before the cuticle has cured, occurs in nest boxes before the majority of eggs are collected. The only way to prevent contamination at this point is by keeping next boxes and belts free of droppings and other sources of bacteria.

It is still advisable (and often a legal requirement) to disinfect the egg shell surface as soon as possible after the eggs have been collected. The disinfectant used should be checked to make sure it does not damage the cuticle (either due to formulation or application method), is effective on the main organisms of concern (bacteria and fungi), does not wet the eggs in the process of application and does not damage the embryo. Safety of the people applying the chemical is an important consideration.

Formaldehyde gas remains the most effective disinfectant for hatching eggs, effective on bacteria, viruses and fungal spores, with a good residual effect (Cardirci, 2009). Unfortunately, formaldehyde is also a carcinogen and as such has been removed from use in some countries. Ultraviolet (UV) light (Scott, 1993; Burgess 1994; Coufal et al., 2003) can be a good alternative for on-farm use - it is not effective on as many organisms as formaldehyde and has no residual activity, but it is effective on bacteria and although capable of causing damage to the operator can be effectively shielded so that workers are safe. Disinfection with UV light on farm should usually be followed by formaldehyde fumigation on arrival at the hatchery to control bacterial and fungal spores.

8 Short periods of incubation during egg storage

When a hen lays a series of eggs in a nest, the eggs laid early in the sequence will be partially re-warmed each time she returns to the nest to lay the next egg. Trials exploring whether the partial rewarming makes any difference to the hatchability of the early eggs in the clutch have been reported since Jackson (1912), who warmed the eggs by placing them under a broody hen and found that stored eggs hatched better as a result. Kosin (1956), Kan et al. (1962), Meir and Ar (1998) and Ar and Meir (2002) all reported trials where short periods

Published by Burleigh Dodds Science Publishing Limited, 2023.

of heating repeated once or twice a day for 1-5 h gave some hatch benefit. However, Reijrink et al. (2009) who heated eggs for half an hour every other day in a 37.8°C water bath did not report any difference in hatchability or hatch of fertile after 15 days storage.

Despite the positive reports in the literature, heat treatment of eggs during storage was not being widely used - occasional parent stock and commercial layer hatcheries were using some sort of heat treatment during storage, but they were very much in the minority. It seemed likely that some aspect of how treatments were implemented in a research setting was not transferring adequately to commercial practice. Given the perennial problems seen in keeping egg age at set within reasonable limits, hatchery specialists employed by Aviagen started a programme of trials designed to develop robust advice for operations needing to preserve the hatchability of stored eggs.

8.1 Replicated experiments

The first three experiments were carried out at the Aviagen Product Development Centre in Albertville, AL, USA. The trials hatchery is equipped with fixed-rack multistage Chickmaster setters.

For the first trial, carried out in 2010, the heat treatment followed that of Meir and Ar (1998) and was defined in terms of time spent in the incubator. The treated eggs were placed in the setter corridor for 4 h on 8 days and 15 days of

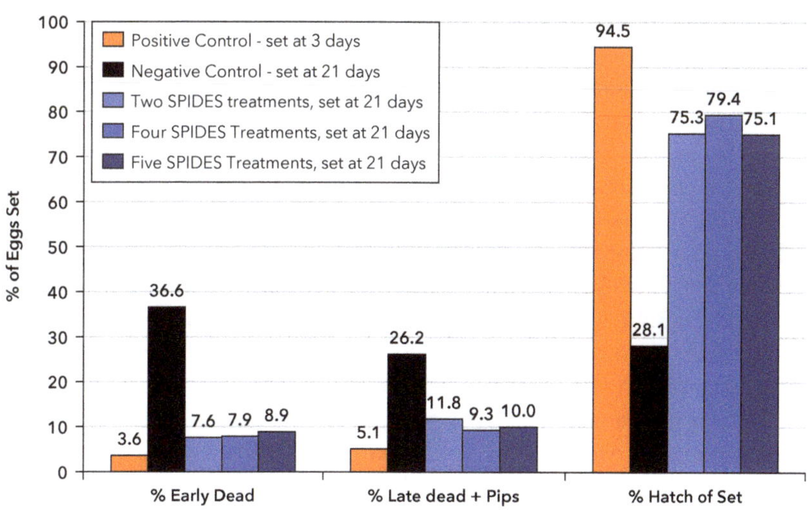

Figure 7 Experiment 1 - hatchability and late and early embryo mortality after 3, 4 or 5 heat treatments.

storage (treatment 1), on 5, 10, 15 and 18 days of storage (treatment 2) or for 4 h on days 3, 8, 12, 15 and 19 of storage (treatment 4). They were then returned to the cooled egg store. There was a positive control where the eggs were set 3 days after they were laid and a negative control where the eggs were not heat treated but were set with all the heat-treated groups after 21 days of storage. Twelve replicate setter trays of 162 eggs per treatment were used (1944 eggs per treatment). The broiler hatching eggs used were from a 33-week-old Ross 308 flock, and the positive control treatment hatched at 94.5%. The negative control eggs hatched at 28.1% - far worse than expected. This was probably because the first trial was carried out during a very hot August and the egg store could not keep the temperature down to the targeted 15.5°C. The three SPIDES treatments hatched at 75.3%, 79.4% and 75.1%, with treatment 2 being significantly ($P < 0.05$) better than the other two. Figure 7 shows hatchability and embryo mortality levels for all five treatments, showing that the SPIDES treatments improved both early and late mortality compared to the untreated stored control eggs.

The outcome of the heat treatment was very much better than expected, so a second trial was set up to make sure that they were repeatable. In this trial, sample eggs at the top, middle and bottom of the stack were fitted with thermistors attached to mini data loggers (Gemini Tinytag Talk 2 - TK 4023) to monitor internal temperature. The trial was expanded to look at the impact of heat treatment on eggs stored for 7 and 14 days, as well as the original 21 days. Eggs stored for 7 days were treated once on day 4 of storage. Eggs stored for 14 days were treated on days 4, 7 and 11 and eggs stored for 21 days were treated on days 5, 10, 15 and 19, repeating the best treatment in experiment 1.

The internal egg temperatures recorded in experiment 2 are shown in Fig. 8. It can be seen that the eggs at the top heated up the fastest and to the highest peak temperature. The eggs at the bottom of the trolley were next, and the ones in the middle were the slowest, only reaching just over 32.5°C.

In this second trial, the untreated eggs which were stored for 21 days hatched much better than before, achieving 62.6% rather than 28.1% hatchability. The trial was carried out in December, and the egg store was better able to hold temperature. Hatchability and early and late embryo mortality are shown in Fig. 9. SPIDES treatment improved hatchability by 2.3% after 7 days ($P < 0.05$), by 3.4% after 14 days ($P < 0.005$) and by 14.3% after 21 days ($P < 0.001$). For the 21-day storage period, SPIDES treatment reduced early embryo mortality by half and late embryo mortality by a third.

Despite the variability in peak temperature within the eggs due to their position on the trolley as they were heat treated, there was no correlation between peak temperature and hatchability of individual replicates, which suggested that all the trays had been able to exceed some undefined threshold temperature.

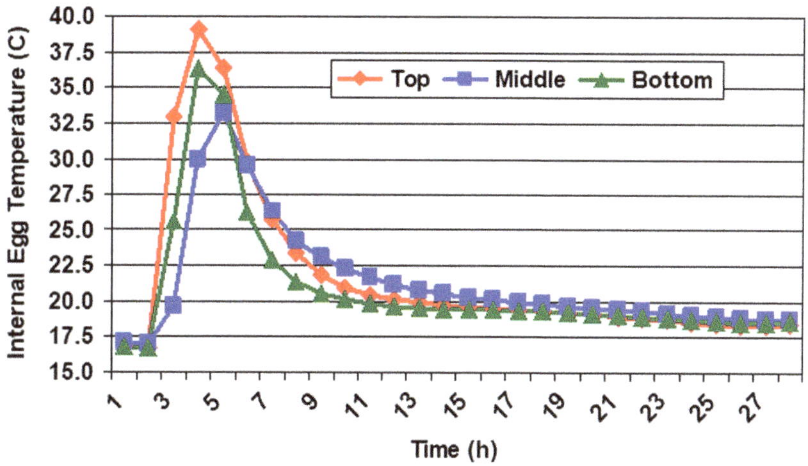

Figure 8 Internal temperatures of eggs in different positions – experiment 2.

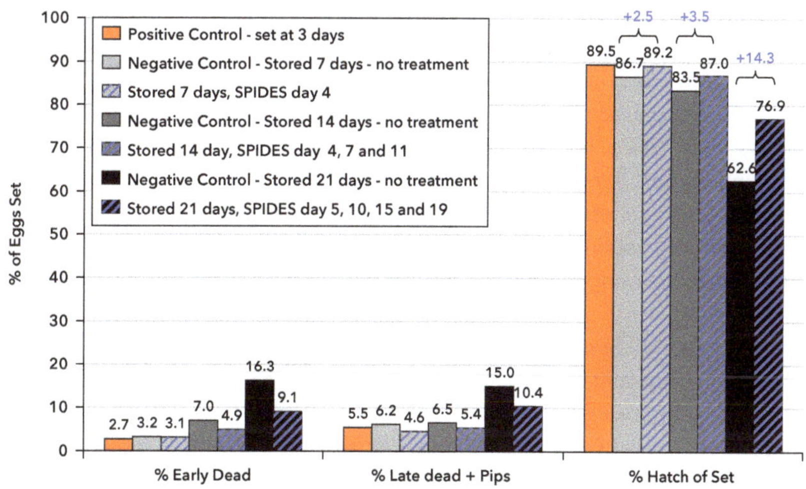

Figure 9 Experiment 2 – hatchability and early and late embryo mortality of eggs stored for 7, 14 or 21 days with or without heat treatment.

Having shown that the results of the first experiment were repeatable, and that heat treatment was effective in eggs as fresh as 7 days, the third experiment explored the duration and frequency of heat treatment on Ross 308 eggs stored for 21 days. As with the previous trial, there was a positive control set at 3 days storage and a negative control set after 21 days with no heat treatment. The SPIDES treatments explored exposure times of 2, 4 or 6 h, each of which was repeated 3, 4 or 6 times.

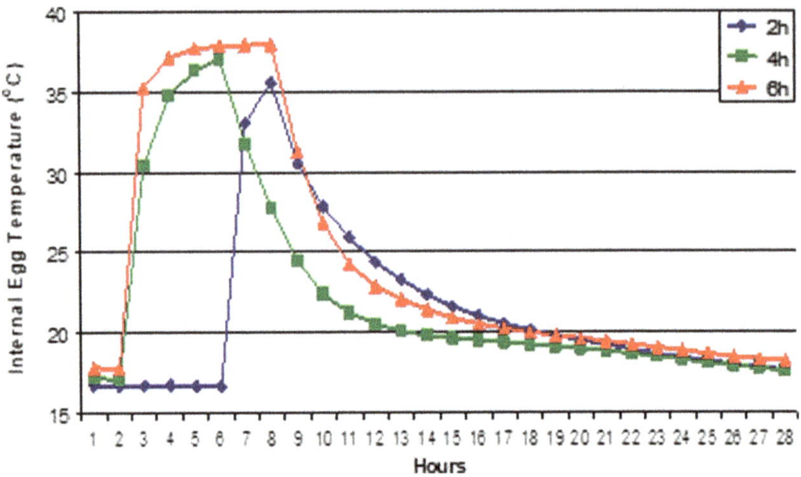

Figure 10 Internal egg temperature of eggs held for 2, 4 or 6 h in the corridor of a Chickmaster fixed-rack multistage setter.

Figure 10 gives the internal egg temperatures of eggs heated in the corridor of a Chickmaster fixed-rack multistage setter for 2, 4 or 6 h. It shows that the longest treatment allowed the eggs to reach incubation temperature and be held there for 3 h. Four-hour exposure just allowed the eggs to reach incubation temperature, but they were over 35°C for 2 h. The shortest exposure just reached 35°C before the eggs were moved back to the egg store. Temperature variability between top, bottom and middle trays was greatest in the 2-h treatment and least in the 6-h one.

Once again, the hatchability of the untreated, stored eggs was poor (Fig. 11). Of the SPIDES-treated eggs, the longest treatments gave worse hatchability – catastrophically so for the six treatments of 6 h, which hatched at 1.1%. Six repetitions gave the worst hatch for all three exposure times, but the losses were less with 4- and 2-h exposures.

Analysing the results further, the hatchability for each treatment was recalculated as the percentage of lost hatchability due to storage that was recovered after SPIDES treatment, which in this experiment had a range between 60% recovery at the upper end and 60% detriment for the worst treatment (i.e. the treated eggs were substantially worse than the negative control). Logging the internal temperature made it possible to calculate the cumulative time above key temperatures for all nine treatments and examine its relationship with hatch performance. The amount of time the eggs were above 30°C and 32°C was identified, including the time accumulated as the eggs were warming up and as they were (more slowly) cooling down. Plotting the time spent above 32°C against the percentage recovery of the hatch lost due to storage as a second-order polynomial curve gave an R^2 value of 0.967,

Published by Burleigh Dodds Science Publishing Limited, 2023.

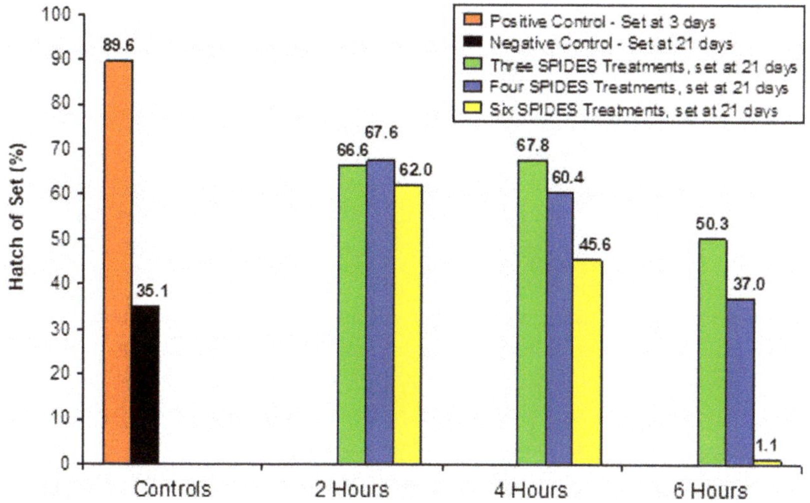

Figure 11 Experiment 3 – hatchability of fresh and stored eggs given heat treatments of 2, 4 or 6 h 3, 4 or 6 times.

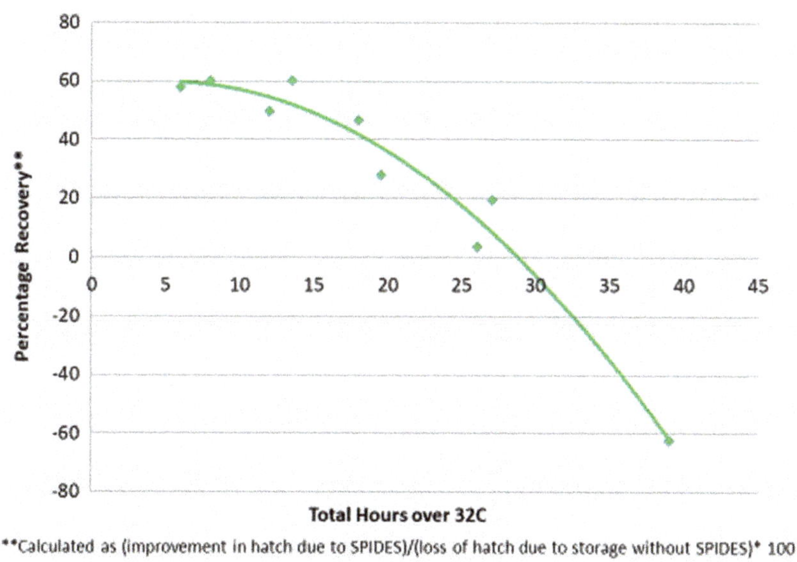

Figure 12 The relationship between cumulative exposure to temperatures above 32°C and the percentage of the hatchability lost due to storage recovered after SPIDES treatments.

marginally higher than the R^2 of the fitted curve when the baseline was set at 30°C. The results of this trial showed that for the best outcome, the total amount of time that the eggs spent above 32°C should not exceed 15 h (Fig. 12).

When Funk and Biellier (1944) explored rates of growth in early-stage embryos at different temperatures, they not only looked at the low temperatures that allow little or no development, but they also looked at the temperatures approaching incubation temperature (37.8°C). They showed that while at 26.7°C the blastoderms increased in size from 4.9 mm to 6.3 mm in 7 days, at 29.4°C, they grew from 4.8 mm to 32.1 mm. Higher temperatures further increased the growth rate, to a point where normal incubation temperature caused the embryo to reach 27.2 mm after 3 days. It would seem that for SPIDES, this upper physiological zero is a key indicator.

One final experiment was carried out in the UK. Having relied so far on an incubator which could heat the eggs relatively fast, it was necessary to test a slower warm up, to make sure that SPIDES would work in large single-stage setters. Experiment 4 was performed in two 12-tray Bristol trial setters, which could be programmed to heat up at different speeds. For this experiment, they were programmed to reach 37.8°C in either 4 h or 8 h, with heat treatments given at 9, 12, 16 and 19 days of age. The eggs were all set on the same day, when individual trays were between 22 days old and 25 days old. The number of trays of each egg age was the same in each treatment, and data were collected to setter tray. The two heating speeds gave identical hatchability, 52.9%, having recovered 70% of the hatchability lost due to storage.

Concurrent with the experiments carried out in Aviagen facilities, Dymond et al. (2013) compared various heat treatment regimens with eggs set fresh or after 21 days. The SPIDES programme involved treating eggs for 4 h on days 5, 9, 15 and 19 prior to set on day 21 post oviposition. Additional treatments gave 6-h or 12-h heat treatment on day 5 only, with the eggs again set on day 21. The egg shell temperature during the 4-h heat treatment showed that the eggs heated up rather faster than in the Chickmaster setters, and the total time above 37.5°C was about 100 min per treatment, rather than only just reaching 37.5°C in the larger incubators. Starting from a base of EGK 10.5 in the untreated eggs, each 4-h SPIDES treatment increased the embryo stage by 1.5 to 2 stages, such that after four treatments the embryos had reached H&H stage 3. The single, longer heat treatments of 6 or 12 h delivered 5.2 h and 10.1 h above 37.5°C respectively, with embryo stage reaching EGK 14 and H&H 3.5. As with the internal trials, the eggs set fresh hatched very well, at 92%. Storage for 21 days without heat treatment caused hatch to drop to 71%, more than 60% of which was recovered after SPIDES treatment. The single, longer treatments, on the other hand, were both associated with hatchability worse than that of untreated eggs. It would seem that a single, early treatment is not sufficient to restore hatchability in eggs stored for 21 days, although it is also possible that the 12-h treatment, especially, was sufficiently long to exceed the optimal time above 32°C suggested in experiment 3. The results of the third internal experiment made it very clear why hatcheries had found SPIDES such a hit or miss technique to implement. Working from small-scale experiments where exposures were defined in terms of time meant very different things depending on the type and capacity of the setters being used to deliver the treatments. Access to Tinytag data loggers allowed us to measure the actual egg temperatures achieved. Help from Murray Bakst and James Wade allowed us to see what those temperatures were doing to embryo development. The combination was a very powerful one.

8.2 Field trials

After the second experiment, SPIDES was discussed widely within Aviagen, and managers of individual hatcheries encouraged to try it. Most of the hatcheries saw some improvement, and plotting hatch lift versus egg age at set showed a very similar relationship to that seen in experiment two. Generally, the poorer the storage conditions, the more the hatchability was improved, to the extent that recoveries of over 60% can probably be assumed to indicate that storage temperatures were either too high or were fluctuating. Results improved once it was possible to give guidance in terms of internal egg temperatures and cumulative time above 32°C.

This stage of the investigation made it very obvious that giving only a treatment duration, without considering its impact on egg temperature, was one of the reasons that heat treatment had taken so long to be adopted. If the first Aviagen experiment had used a large, single-stage setter to heat the eggs, the heat treatment would probably not have improved hatchability.

Feedback from the Aviagen hatcheries told us as follows:

- SPIDES can only work if the eggs are well separated in setter trays and on racks. Bulk packed eggs will not heat or cool evenly.
- Putting warm eggs straight back into the egg store causes the temperature of neighbouring eggs to fluctuate and they hatch worse than they should. Eggs needed to be cooled to a maximum of 24°C before they were returned to the egg store.
- Most single-stage setters delivered SPIDES treatments more effectively if only half full, with alternate tray racks empty.
- SPIDES consistently gave a small benefit even in eggs of only 7 days old.
- The hatchery manager needed to see the benefit in order to tolerate the extra planning required to keep up regular treatments in a full egg store.

8.3 Dedicated short periods of incubation during egg storage machines

The information gathered in the experiments reported here was sufficient to work with incubator manufacturers to design a dedicated SPIDES machine, which could be used in egg stores remote from the hatchery which had no incubators on site. Most incubator companies now offer dedicated SPIDES machines; the most effective use egg surface temperature to control the process and have significantly increased heating, cooling and fan capacity to ensure that the heating and cooling are rapid and evenly distributed.

8.4 Future trends in research

There has been a steadily increasing body of publications investigating the aspects of how SPIDES affects embryos since the early publications by French et al. (2011) and Nicholson et al. (2013), with an increasing number of trials reported which were not initiated by Aviagen.

Collaborative trials with the University of Ankara have shown, among other things, that the effect of SPIDES and egg turning are additive (Özlü et al., 2021), suggesting that the two techniques target different processes which both cause hatchability to fall with egg age. Most recently, Brady et al. (2022) compared the activation of transcriptome processes in blastoderms in fresh eggs, in eggs which had been stored for 21 days and others which had been both stored and SPIDES treated. Compared to the fresh eggs, the blastoderms in stored eggs showed enrichment of pathways associated with cell stress and death. SPIDES treatment, despite the eggs being the same age as the stored ones, was associated with enrichment of pathways associated with basic cell and anti-apoptotic functions. This is probably the start of an understanding of why heating to incubation temperature makes such a big difference to hatchability and also why it seems to work particularly well when storage conditions are suboptimal. Further work should be both fascinating and valuable.

9 Acknowledgements

The SPIDES project was a success due to the knowledge and hard work of a lot of people. Thanks are especially due to Steve Tullett, Nick French, Murray Bakst, Vanessa Kretzschmar, Andy Hogg, James Wade, Timea Torma, Serap Yasun, Daniel Chapman, Okan Elibol, Serdar Özlü and Tolga Erkus, all of whom contributed an enormous amount of knowledge, practical experience, enthusiasm and tough feedback from the replicated experiments and field trials. The combination generated a body of advice that has allowed SPIDES to be implemented safely and effectively in a variety of incubator makes and types and provided the confidence to do so over different lines and generations.

10 Conclusion

Chicken hatching eggs will not hatch as well if they are stored for more than a week than they do if set fresh. Poultry researchers have investigated the causes of hatchability deteriorating and many methods that might affect the rate of change, and there are excellent and exhaustive reviews of the literature. Techniques which are helpful on a small scale in the laboratory are helpful but only a starting point. Some may be impossible to multiply up at current levels of technology and others may need more work to explore the limits. At a practical commercial level, egg age probably changes faster than an organisation is able to change its egg storage protocols. Hatcheries which are able to manage egg storage temperatures, turn eggs from older flocks, and deliver SPIDES treatments will usually cope best with fluctuating egg temperatures.

11 Where to look for further information

- Birkhead, T. (2016). The Most Perfect Thing - Inside (and Outside) a bird's egg. Bloomsbury, London, UK.
- Board, R. G. and Fuller, R. (1984). (eds) Microbiology of the Avian Egg Chapman and Hall, London, UK.
- Meijerhof, R. (1994). Theoretical and empirical studies on temperature and moisture loss of hatching eggs during the pre-incubation period. PhD thesis Wageningen University, The Netherlands.
- Reijrink, I. A. M. (2010). Storage of Hatching Eggs-Effects of storage and early incubation conditions on egg characteristics, embryonic development, hatchability and chick quality. PhD thesis Wageningen University, The Netherlands.
- Incubation and Fertility Research Group, WPSA working group 6, https://www.ifrg.be/.

12 References

Ar, A. and Meir, M. (2002). Principles of preservation of embryonic viability during pre-incubation egg storage. In: *11th European Poultry Conference. Proceedings, Bremen*. World's Poultry Science Association, Beekbergen, The Netherlands, 1-14.

Aviagen. (2018). *Ross PS Management Handbook*, 179.

Bakst, M. R. and Akuffo, V. (2002). Impact of egg storage on embryo development. *Avian and Poultry Biology Reviews* 13(3):125-131.

Bakst, M. R., Akuffo, V., Nicholson, D. and French, N. (2012). Comparison of blastoderm traits from 2 lines of broilers before and after egg storage and incubation. *Poultry Science* 91(10):2645-2648.

Ball, R. F., Logan, V. and Hill, J. F. (1975). Factors affecting the cuticle of the egg as measured by intensity of staining. *Poultry Science* 54(5):1479-1484.

Becker, W. A., Spencer, J. V. and Hawkes, B. W. (1969). Angle of turning chicken eggs during storage. *Poultry Science* 48:1748.

Bowman, J. C. (1969). Storage of hatching eggs. 2. Evidence of genotype-environment interactions. *British Poultry Science* 10(1):19–28.

Brady, K., Talbot, Jr. C. C., Long, J. A., Welch, G., French, N., Nicholson, D. and Bakst, M. R. (2022). Transcriptome analysis of blastoderms exposed to prolonged egg storage and short periods of incubation during egg storage. *BMC Genomics* 23(1):262.

Brake, J., Walsh, T. J., Benton, C. E., Petitte, J. N., Meijerhof, R. and Peñalva, G. (1997). Egg handling and storage. *Poultry Science* 76(1):144–151.

Bruce, J. and Drysdale, E. M. (1994). Trans shell transmission. In: Board, R. G. and Fuller, R. (Eds.), *Microbiology of the Avian Egg*. London: Chapman and Hall, 63–92.

Burgess, D. (1994). Ultra violet light disinfection. *International Hatchery Practice* 9:21.

Butler, D. (1991). Egg handling and storage at the farm and hatchery pp 195-204. In: *Poultry Science Symposium 22, Avian Incubation Edit Tullett, S. G.* Butterworth-Heinemann.

Cardirci, S. (2009). Disinfection of hatching eggs by formaldehyde fumigation – a review. Arch.*Geflugelk* 73:116–123.

Coufal, C. D., Chavez, C., Knape, K. D. and Carey, J. B. (2003). Evaluation of a method of ultraviolet light sanitation of broiler hatching eggs. *Poultry Science* 82(5):754–759.

Dymond, J., Vinyard, B., Nicholson, A. D., French, N. A. and Bakst, M. R. (2013). Short periods of incubation during egg storage increases hatchability and chick quality in long-stored broiler eggs. *Poultry Science* 92(11):2977–2987.

Edwards, C. L. (1902). The physiological zero and the index of development for the egg of the domestic fowl, *Gallus domesticus*. *American Journal of Physiology-Legacy Content* 6(6):351–397.

Elibol, O. and Brake, J. (2008). Effect of egg position during three and fourteen days of storage and turning frequency during subsequent incubation on hatchability of broiler hatching eggs. *Poultry Science* 87(6):1237–1241.

Elibol, O., Peak, S. D. and Brake, J. (2002). Effect of flock age, length of egg storage and frequency of turning during storage on hatchability of broiler hatching eggs. *Poultry Science* 81(7):945–950.

Eyal-Giladi, H. and Kochav, S. (1976). From cleavage to primitive streak formation: Complementary normal table and a new look at first stages of the development of the chick. I. General morphology. *Developments in Biologicals* 49:321–337.

Fasenko, G. M. (2007). Egg storage and the embryo. *Poultry Science* 86(5):1020–1024.

Fasenko, G. M., Hardin, R. T., Robinson, F. E. and Wilson, J. L. (1992a). Relationship of hen age and egg sequence position with fertility, hatchability, viability and preincubation embryonic development in broiler breeders. *Poultry Science* 71(8):1374–1383.

Fasenko, G. M., Robinson, F. E., Hardin, R. T. and Wilson, J. L. (1992b). Research Note: Variability in preincubation embryonic development in domestic fowl 2. Effects of duration of egg storage period. *Poultry Science* 71(12):2129–2132.

Fasenko, G. M., Robinson, F. E., Armstrong, J. G., Church, J. S., Hardin, R. T. and Petitte, J. N. (1991). Variability in preincubation embryo development in domestic fowl. 1. Effects of nest holding time and method of egg storage. *Poultry Science* 70(9):1876–1881.

Fasenko, G. M., Robinson, F. E. and Christensen, V. L. (2009). Effects of long term storage on the egg, embryo and chick. *Avian Biology Research* 2(1–2):73–79.

French, N. A., Nicholson, D., Kretzschmar, V., Goyne, D. and Veal, J. (2011). *Effect of Applying Short Periods of Incubation Temperature During Pre-Incubation Egg Storage on the Hatchability of Broiler Eggs Poultry Science Association Meeting* (Vol. 2011).

Funk, E. M. and Biellier, H. V. (1944). The minimum temperature for embryonic development in the domestic fowl (*Gallus domesticus*). *Poultry Science* 23(6):538-540.

Funk, E. M. and Forward, J. (1951). Effect of humidity and turning of eggs before incubation on hatching results. University of Missouri College of Agriculture Agricultural Experimental Station Bulletin 554.

Gesche, E. and Schuler, A. M. (1979). Bacterial flora of fresh eggs and eggs stored for four weeks (In Spanish). *Alimentos* 4J3:11.

Haines, R. B. and Moran, T. (1940). Porosity of, and bacterial invasion through, the shell of the hens egg. *Journal of Hygiene, Cambridge* 40:453-461.

Hamburger, V. and Hamilton, H. L. (1951). A series of normal stages in the development of the chick embryo. *Journal of Morphology* 88(1):49-92.

Humphry, T. J. and Whitehead, A. (1993). Egg age and the growth of *Salmonella enteriditis* PT4 in egg contents. *Epidemiology and Infection* 111(2):209-219.

Insko, W. M., Jr. and Martin, J. H. (1933). Effect of Frequent Turning on hatchability and distribution of embryo mortality. *Poultry Science* 12(5):282-286.

Jackson, H. W. (1912). Poultry experiments. Pennsylvania State College Agriculture Experimental Station Bulletin 120:12-15.

Kan, J., McPherson, B. N. and Gyles, N. R. (1962). Pre-incubation warming of chicken eggs. *Poultry Science* 41(5):1478-1480.

Knox, C. W. and Olsen, M. W. (1936). The effect of tremulous air cells upon the hatchability of eggs. *Poultry Science* 15(4):345-348.

Kosin, I. L. and St. Pierre, E. (1956). Studies on pre-incubation warming of chicken and turkey eggs. *Poultry Science* 35(6):1384-1392.

Landauer, W. (1961). The hatchability of chicken eggs as influenced by environment and heredity. *Storrs Agricultural Experimental Station 1*. https://opencommons.uconn.edu/saes/1.

Lapao, C., Gama, L. T. and Chaveiro Soares, M. (1999). Effects of broiler breeder age and length of egg storage on albumen characteristics and hatchability. *Poultry Science* 78:840-845.

Lyon, C. E., Newell, G. W. and Morrison, R. D. (1972). Vitelline membrane strength and egg yolk mottling. *Poultry Science* 51(2):480-487.

Mather, C. M. and Laughlin, K. F. (1976). Storage of hatching eggs: The effect on total incubation period. *British Poultry Science* 17(5):471-479.

Mather, C. M. and Laughlin, K. F. (1977). Storage of hatching eggs: The effect on early embryonic development. *British Poultry Science* 18(5):597-603.

Mayes, F. J. and Takeballi, M. A. (1984). Storage of eggs of the fowl (*Gallus domesticus*) before incubation: A review. *World's Poultry Science Journal* 40(2):131-140.

Meijerhof, R. (1992). Pre-incubation holding of hatching eggs. *World's Poultry Science Journal* 48(1):57-68.

Meijerhof, R. (1994). *Theoretical and empirical studies on temperature and moisture loss of hatching eggs during the pre-incubation period*. PhD Thesis, University of Wageningen.

Meijerhof, R., Noordhuizen, J. P. T. M. and Leenstra, F. R. (1994). Influence of pre-incubation treatment on hatching results of broiler breeder eggs produced at 37 and 59 weeks of age. *British Poultry Science* 35(2):249-257.

Meir, M. and A. Ar. (1998). Pre-incubation warming as a means of lengthening storage time of fertile eggs. In: *10th European Poultry Conference, Israel, June*. Jerusalem, Israel. World's Poultry Science Association, Israel Branch, 825-829.

Nasri, H., van den Brand, H., Najjar, T. and Bouzouala, M. (2020). Egg storage and breeder age impact on egg quality and embryo development. *Journal of Animal Physiology and Animal Nutrition* 104(1):257-268.

Nicholson, D., French, N., Tullett, S., van Lierde, E. and Jun, G. (2013). Short periods of incubation during egg storage - SPIDES Lohmann. *Information* 48:51-61.

Nys, Y. and Guyot, N. (2011). *Egg Formation and Chemistry in 'Improving the Safety and Quality of Eggs and Egg Products'*. Cambridge: Woodhead Publishing, 83-132.

Özlü, S., Elibol, O. and Brake, J. (2018). Effect of storage temperature fluctuation on embryonic development and mortality, and hatchability of broiler hatching eggs. *Poultry Science* 97(11):3878-3883.

Özlü, S., Uçar, A., Erkuş, T., Yasun, S., Nicholson, A. D. and Elibol, O. (2021). Effects of flock age, storage temperature, and short period of incubation during egg storage, on the albumen quality, embryonic development and hatchability of long stored eggs. *British Poultry Science* 62(4):611-619.

Özlü, S., Uçar, A., Erkuş, T., Yasun, S., Nicholson, A. D. and Elibol, O. (2022). Physiological status of embryos from broiler breeder hens under different temperatures during storage period Proceedings Incubation and Fertility Research Group. Leiden.

Peebles, E. D., Zumwalt, C. D., Doyle, S. M., Gerard, P. D., Latour, M. A., Boyle, C. R. and Smith, T. W. (2000). Effects of breeder age and dietary fat source and level on broiler hatching egg characteristics. *Poultry Science* 79(5):698-704.

Proudfoot, F. G. (1966). Hatchability of stored chicken eggs as affected by daily turning during storage and prewarming and vacuuming eggs enclosed in plastic with nitrogen. *Canadian Journal of Animal Science* 46(1):47-50.

Proudfoot, F. G. (1969). Effect of packing orientation, daily positional change and vibration on the hatchability of chicken eggs stored up to four weeks. *Canadian Journal of Animal Science* 49(1):29-35.

Proudfoot, F. G. and Hulan, H. W. (1976). *Care of Hatching Eggs Before Incubation*. Research Station, Kentville, NS: Agriculture Canada Publication, 1-17.

Reijrink, I. A. M., Meijerhof, R., Kemp, B., Graat, E. A. M. and van den Brand, H. (2009). Hatchability and chick quality. *Poultry Science* 88(12):2649-2660.

Reijrink, I. A. M., Meijerhof, R., Kemp, B. and Van Den Brand, H. (2008). The chicken embryo and its micro environment during egg storage and early incubation. *World's Poultry Science Journal* 64(4):581-598.

Scott, T. A. (1993). The effect of UV-light and air filtering system on embryo viability and microorganism load on the egg shell. *Journal of Applied Poultry Research* 2(1):19-25.

Sellier, N., Brillard, J. -P., Dupuy, V. and Bakst, M. R. (2006). Comparative staging of embryo development in chicken, turkey, duck, goose, guinea fowl and Japanese quail assessed from five hours after fertilization through seventy-two hours of incubation. *Journal of Applied Poultry Research* 15(2):219-228.

Sparks, N. H. C. (1985). *The Hen's eggshell - A resistance network*. PhD Thesis, University of Bath.

Sparks, N. H. C. (1994). *Shell Accessory Materials: Structure and Function in Microbiology of the Avian Egg*. Edited by Board, R. G. and Fuller, R. London: Chapman and Hall, 25-42.

Sparks, N. H. C. and Burgess, A. D. (1993). Effect of spray sanitising on hatching egg cuticle efficacy and hatchability. *British Poultry Science* 34(4):655-662.
Tímea, T. and Katalin, K. G. (2019). A MECHANIKAI HATÁSOK BEFOLYÁSOLÓ SZEREPE HÚSHIBRID TENYÉSZTOJÁSOK SÉRÜLÉSÉRE ÉS A KELÉSI EREDMÉNYEKRE KÜLÖNBÖZŐ TÍPUSÚ TOJÁSTÁLCÁKON. *AWETH* 15(2):64-72.
Torma, T. and Kovácsné, K. G. (2012). Effects of mechanical impacts on hatchability of broiler breeders. *Agricultural Science and Technology A* A4:535-540.
Walsh, T. J., Rizk, R. E. and Brake, J. (1995). Effects of temperature and carbon dioxide on albumen characteristics, weight loss and early embryonic mortality of long stored hatching eggs. *Poultry Science* 74(9):1403-1410.

Chapter 7

In ovo sexing in poultry chicks

Anke Förster, Agri Advanced Technologies GmbH, Germany; and Thomas Bartels, Institute of Animal Welfare and Animal Husbandry, Germany

1 Introduction

Since 1 January 2022, the killing of day-old chicks has been prohibited by law in Germany; on 1 January 2023, France followed suit, and Italy and many other countries in Europe are also planning to ban the culling of day-old chicks. How did this step come about? What is the background? And how will this law be implemented? These questions will be answered in this chapter - at least in broad outline.

Until the middle of the last century, poultry farming was characterised by small-scale production of eggs and meat for own consumption; surpluses were sold directly or on regional markets. Rising populations, growing cities and the shift of large populations away from subsistence food production have encouraged the development of egg and poultry meat production into a competitive industry. Research into poultry diseases and their control in the feeding sector, and in performance testing, selection and breeding planning quickly led to expanded production capacity and increased efficiency. Due to a high negative correlation between reproductive and growth traits, specialist hybrid breeding developed in the mid-twentieth century offered a great opportunity for the commercial poultry industry. Broiler lines bred for meat yield are reaching the desired slaughter weight in an increasingly shorter time, whilst laying hybrids specifically selected for high laying performance and egg

http://dx.doi.org/10.19103/AS.2022.0118.07

quality have achieved significant increases in egg production. In addition to the classic breeding progress in pure breeding, significant heterosis effects have also been realised. As specialisation progressed, male chicks from these lines lacked economically viable uses and it became common practice worldwide to kill them as day-old chicks. The differentiation between male and female chicks on the first day of life ('sexing') was initially only possible via the so-called 'Japanese method', in which male chicks are visually identified by their rudimentary sexual organ (phallus) after the cloaca is protruded (vent sexing). The high risk of vertical transmission of pathogens with this method, stress for both animals and humans with the increased mortality of day-old chicks, and high costs led to attempts to determine the sex of the chicken embryo whilst still in the incubation process. However, these efforts receded after phenotypic sex differences became clearly recognisable through breeding of sex-linked crosses for day-old sex identification. In brown-egg-laying lines, the sex-linked segregation of the silver (S) and gold (s+) alleles led to sex-specific down colouration in day-old chicks (Smyth, 1990), whilst in white-egg-laying lines, the feather development gene K, which is located on the Z chromosome, is used for sexing of day-old chicks (Somes, 1990). Crosses between fast feathering (k+/k+) males and slow-feathering (K/-) females result in fast-feathering female chicks (k+) and slow-feathering male chicks (K/k). With these crosses, precise sexing of large numbers of animals at high speed became possible at low cost in commercial hatcheries.

This was the situation at the end of the 1990s. Increasing ethical and legal concerns regarding the common practice of killing day-old chicks (practised not only by conventional egg producers but also for the most part in laying hen production under organic guidelines) renewed research focus on sex determination in the egg (Klein et al., 2003a). Sperm-sexing - which is used as a tool to avoid offspring of the undesired sex in farm animals such as cattle - cannot be used in poultry, since unlike in mammals, the sex of the bird offspring is determined by the female germ cell.

While automated sex determination was still in its infancy in the early 2000s (Preisinger, 2003; Phelps et al., 2003), intensive research and development work led to at least three proven methods coming to practical maturity and application as of 2020 (Richard, 2020).

The development of these and other methods, as well as the challenges still to be overcome from idea to practical application, are presented as follows, complete with case studies and a future outlook.

2 Key issues and challenges

The requirements for in ovo sexing have been described by Phelps et al. (2003): 'Commercial acceptance of automated gender sorting of avian embryos will require a sex specific marker that can be detected with 98.5% accuracy, at a rate of 20 000 to 30 000 eggs per hour, and at a cost equivalent to manual sexing.'

Kaleta and Redmann (2008) specified these requirements with their frequently cited list of criteria for an ideal method for in ovo sexing of hybrid lines of layer-type chickens:

i. Methods must be applicable to eggs prior to setting for incubation.
ii. Methods must not affect the integrity of eggshells and the interior of eggs.
iii. Eggs containing male blastoderm cells need to be used for reasonable purposes.
iv. Applied methods should have no negative effects on embryonic development.
v. Rate of hatchability of chicks must not be affected.
vi. Necessary technical equipment must be available and its use economically feasible.
vii. Methods must be applicable to large numbers of eggs.
viii. Sorting by gender must be possible within short periods of time.
ix. Methods must be free of any negative effect on female reproduction.
x. Post-embryonic physical and behavioural development of chicks remains unchanged.
xi. Methods must be acceptable from a humanitarian and ethical point of view.

All development approaches have been and are still measured according to these criteria, with many challenges to be overcome from idea to practical application. The scheme on technology maturity levels (Federal Agency for Agriculture and Food Fig. 1: technology readiness level) can be used to illustrate this path and examples will be used to illustrate impasses, promising approaches and possible compromises.

In basic research, the first task is to find a 'source of information' that provides reliable information about the sex of the embryo as early as possible. A biologically reliable and comprehensible identification of sex-specific distinguishing features (such as phenotypic characteristics, levels of male or female sex hormones, different DNA of the sex chromosomes, sex-typical biomarkers or spectral differences) provides a starting point for further developments.

In a further step, the application scenarios are described, i.e. in which developmental stage can information regarding sex differences be obtained (before hatching or during incubation? In which developmental stage?); which prerequisites are necessary (e.g. opening of the eggshell, taking tissue samples or embryonic fluids, genetic modification); is damage to the embryo and

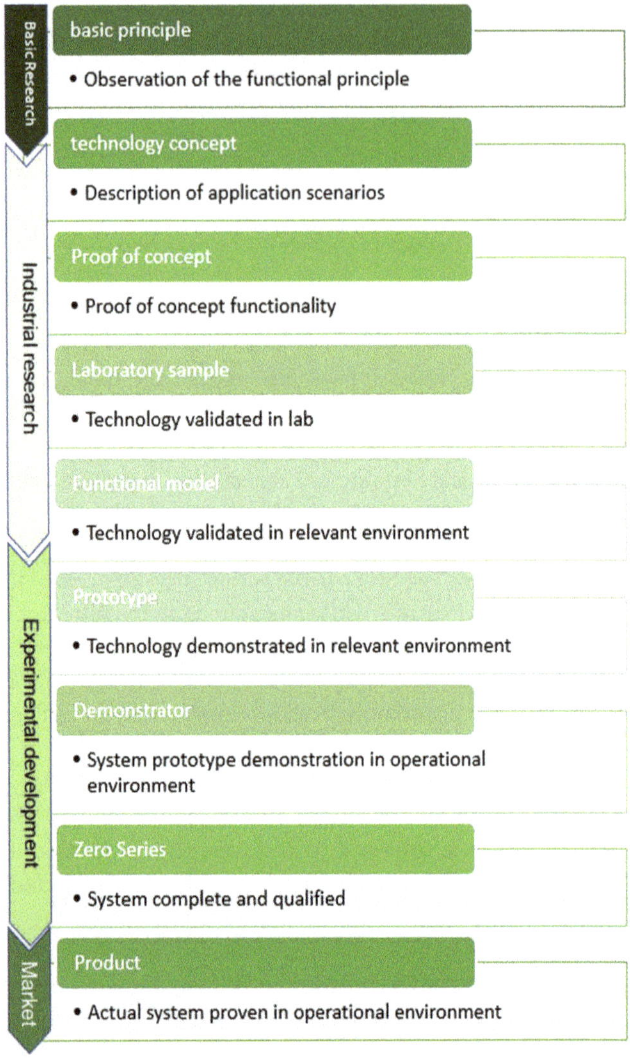

Figure 1 Technology readiness level (adapted from Wikipedia) 'Technology Readiness level' and BLE 'DIP Merkblatt TRLs'.

therefore a reduced hatching result possible? And what diagnostic certainty (or alternatively accuracy) can be expected with the respective procedure? Is there scope for interpretation of the results?

Initial trials are performed to test the effectiveness of the principle and allow a statement if the proposed method might be considered promising (in most cases based on very small numbers). Ideas are abandoned in this first step if results cannot be reproduced and/or no differences between sexes are visible (e.g. the use of a pendulum (Kallenberg, 1913), egg shape as an indicator for the sex of the embryo (Imholt, 2010; Yilmaz-Dikmen and Dikmen, 2013), egg contents in the unincubated egg and hen condition (Aslam et al., 2013)). Once a so-called 'proof of concept' has been established, a laboratory-scale test will be set up - often with the participation of the industry. This is characterised by rather small quantities and predominantly manual activities (e.g. egg openings, collection of blood, allantoic fluid or similar, closure of the egg openings, etc.), and due to a lack of infrastructure the decision is often made to stop incubating the eggs before the chicks hatch.

This is followed by optimisation steps, experiments to verify the results, improvements to the method and transition from experimental development at the laboratory to a larger scale in the hatchery. This is one of the most important steps in process development, in which many initially promising ideas find their 'end'. As one example, a method that identified the IR spectroscopy of unincubated eggs had been favoured in a large research project, but could not be pursued further since hatching tests revealed hatchability losses of more than 90% of the eggs treated in this way (Steiner et al., 2011; Bartels et al., 2014).

The transition from a proof of concept of the process to the development of a prototype suitable for daily use is fluid and, from experience, takes a long time. In this process step, it is important to check various aspects for later practical application. Are the individual steps required suitable for automation? Can damage to the embryo and/or a negative impact on hatchability be ruled out? Can the information and measurement data obtained be easily interpreted and is it reproducible? Is the respective methodology socially acceptable? Automation and verification in extensive hatchery trials over the entire incubation period are the most important steps in this phase - the hatchery trials must show whether (and to what extent) the intervention has a negative impact on the hatching rate. The accuracy and variability of a sexing method can only be established with certainty based on large amounts of data. Ideally, the parameters relevant to the successful development of a demonstrator, i.e. the testing of the prototype in the hatchery under practical conditions, should be evaluated in several trials. In parallel, automation should be developed for the necessary measures and 'ancillary work' required for high throughput use in the hatchery. These include, for example, the handling of eggs before and after treatment, the collection or sampling of liquid and analysis of ingredients (liquid-based methods such as hormone or DNA determination), egg opening and closure, detection of blood vessels, automated image evaluation (spectroscopic methods) and so on.

The functionality of the demonstrator developed in this way is then proven under real conditions in high throughput. This is where the capacity of the process becomes apparent, as well as its accuracy, which ultimately determines its use in practice. An unreliable reproducibility of results at this late stage can still lead to the decision not to implement the method into practice. For example, the market launch of an automated system for Raman spectroscopy on day 5 was stopped due to variable accuracy results, and further investigations were necessary and undertaken to achieve reliable results for practical application (Richard, 2020).

These rather technocratic challenges are not the only factors impeding the development of a practical and operational method for sex determination in the egg. A successful system also has to fit into the political and societal requirements, and its costs have to be compared to rearing the roosters. In Germany, for example, the time point of sex determination and the potential for the embryo to experience pain at that developmental stage has become a particular focus of attention in recent years. The political response to this discussion has been an amendment to the Animal Welfare Act, which led to an additional constraint - from 1 January 2024, only methods applicable before day 7 of incubation will be permitted in Germany (TierSchG, 2006). Methods that allow the identification of female and male embryos due to genetic engineering in the parent stock or attempt to inhibit male embryos early in their development through genetic engineering integration of a sex-linked lethal factor are under pressure and will likely not be used due to a lack of consumer acceptance, at least in Western European countries.

3 Case studies

At the time of writing this review, three methods for in ovo sexing have successfully followed the above developmental parameters to practical application in Europe. Some have been used in commercial hatcheries since 2019. To date, the highest amount of 'in ovo chicks' - i.e. female layer chicks without killing the male - have been produced using hyperspectral analysis (AAT, Germany, 'CHEGGY'), followed by endocrinological sexing (Seleggt, Germany) and DNA analysis (PlantEgg, Germany). Less information is available regarding the results of mass spectrometry (In Ovo, The Netherlands, 'Ella').

These practically applied methods, their development and their special characteristics are described below.

3.1 Hyperspectral analysis

Hyperspectral analysis is a spectral technique - in this case vibrational micro spectroscopy is used to generate three-dimensional images (two spatial and one spectral dimension) that map identification features of inhomogeneous samples (Bellon-Maurel and Gorretta, 2014). Hyperspectral imaging in the near-infrared (NIR) range has been used for several years in numerous applications, especially in the agricultural and food sectors. Bellon-Maurel and

Gorretta (2014) describe, for example, quality monitoring, analysis of product composition and product properties and detection of foreign objects.

Sexual dimorphism in plumage colour in commercial brown laying hybrids can be determined by the hyperspectral analysis method for sex-specific sorting of embryos. The expression of the first feather system begins as early as day 6 of incubation and, from day 11, feathers are increasingly evident, especially on the back and tail. From day 12, the covering feathers of the wings develop (Hamburger and Hamilton, 1951). In brown laying hybrids with sex-specific feather colour, sex differences in pigmentation levels at this stage are already visible. Göhler et al. (2017) investigated the use of hyperspectral analysis between day 11 and day 14 of incubation and achieved an accuracy of up to 97%. The major advantage of this method is that - unlike other spectroscopy methods - analysis can be performed non-invasively through the intact eggshell because the eggshell does not form a barrier for the light spectra captured by the hyperspectral camera. Cross-contamination and negative influences on the development of the embryos, the hatching rate and later performance and health of the laying hens are not an issue.

The process steps of this method (illumination of the eggs in a measuring chamber from below; detection of the air chamber with a thermal imaging camera; recording of the spectra by the hyperspectral camera from above; analysis of the recorded spectra with a specially developed algorithm; classification of the hatching eggs into 'female', 'male', 'infertile/not developed') can be relatively easily automated and, together with the necessary upstream and downstream transport steps, integrated into the routine process in a hatchery. The equipment of the hatchery is considered by adapting the system to the dimensions of the setter trays. The hatching eggs are loaded into the machine on the usual trays. Automated in this way, sexing capacity can be up to 20000 hatching eggs/h. Since 2020, the 'Cheggy' system developed by AAT has been used in large-scale hatcheries in several European countries (Zootecnica, 2021; AAT, 2021).

The production data of all hatcheries using the 'Cheggy' method are centrally recorded and analysed. This allows conclusions to be drawn regarding potential sources of error and factors influencing hatching results and accuracy, with the aim of continuously improving sorting accuracy. Some of this data was used in a field study to analyse the production parameters and results of this system in a French hatchery (Förster et al., 2022). In this study, the accuracy of sexing was > 95% with an average sexing error of 4.1%, and the impact of the sorting technology on hatchability was very low. Average female hatchability was 38% or - converted to the economical figure used in commercial hatcheries - 2.66 hatching eggs per female chick. Compared to an expected hatchability rate without in ovo sex determination, a surplus of less than 10% has to be targeted (Table 1 and 2).

Table 1 Number of hatches per strain, average hatchability and sexing error

Strain	No. hatches	No. eggs set	% females/ eggs set*	% hatch/ transferred eggs**	% sexing error
Lohmann Brown	22	657 939	37.4%	92.6%	4.4%
Hy-Line Brown	18	726 472	39.1%	91.4%	3.8%
Brown Nick	4	158 400	41.3%	95.1%	4.3%
ISA Brown	4	62 541	33.8%	94.2%	3.5%
Novogen Brown	1	3300	36.8%	95.4%	4.1%
Total	49	1 608 652	38.0%	92.5%	4.1%

*Female chicks of all eggs set
**Hatched chicks of eggs incubated after in ovo sex determination, male and unfertile eggs sorted out

Table 2 Specific population parameters of two strains

	% sexing error		% females/ eggs set		% expected female hatchability*		No. hatching eggs/female chick	
	Hy-Line	LB	Hy-Line	LB	Hy-Line	LB	Hy-Line	LB
N	18	22	18	22	18	22	18	22
Average	3.8%	4.4%	39.1%	37.5%	42.4%	41.6%	2.57	2.73
Std. dev.	2.0%	1.5%	2.8%	4.7%	2.3%	1.8%	0.19	0.45
Min.	1.2%	1.4%	34.3%	23.1%	36.1%	38.0%	2.31	2.33
Max.	8.5%	7.4%	43.2%	43.0%	44.2%	43.5%	2.91	4.33
Median	3.0%	3.9%	39.2%	38.4%	43.4%	42.0%	2.55	2.61

*Expected from experience with the same flock under similar conditions

3.2 Endocrinology

As early as the late 1990s, work was being carried out on an automated method to analyse certain components in the allantoic fluid that would make it possible to reliably distinguish between male and female chicken embryos. Studies showed that the allantoic fluid of male embryos contained little estradiol between the 13th and 18th day of incubation, whilst the concentration in female embryos was more than three times higher (Phelps et al., 2003; Tran et al., 2010).

From an animal welfare point of view, sex determination and sorting towards the end of embryonic development is too late and not an acceptable alternative to killing day-old male chicks directly after hatching. Efforts have therefore been made since about 2010 to detect sex-specific hormonal concentration differences in samples from allantoic fluid much earlier (Weissmann et al., 2013).

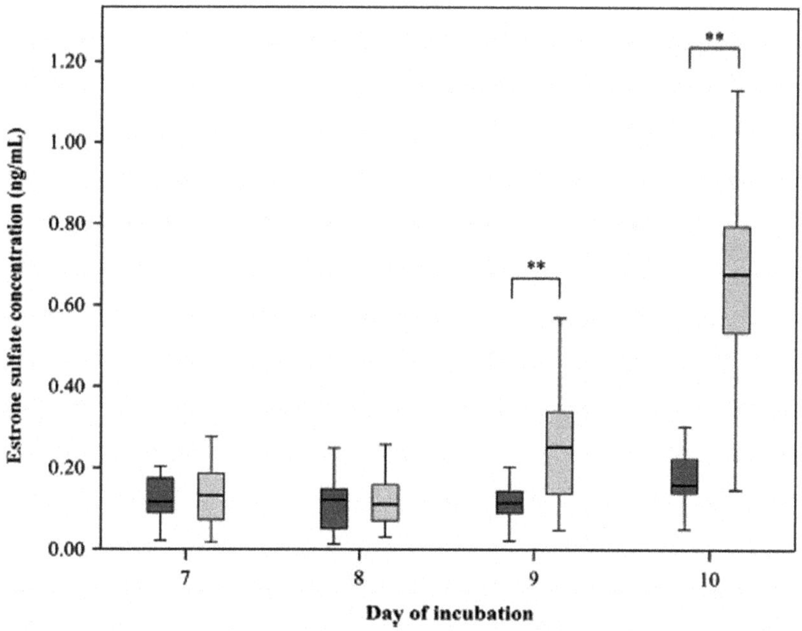

Figure 2 Box plot diagram of estrone sulphate concentrations (ng/mL) of male (dark grey bar) and female (light grey bar) embryos in the allantoic fluid on incubation days 7–10. Highly significant distinctions are indicated by double asterisks (Weissmann et al., 2013).

For this purpose, allantoic fluid was manually collected from eggs incubated for 7–10 days, closely following the method described by Phelps et al. (2003). The fluid was analysed with an ELISA assay for the levels of estradiol (E2), estrone sulfate (E1S) and testosterone (T), hormones for which sex-specific differences in concentrations are expected in female and male embryos. No differences were found in testosterone levels on any of the 4 developmental days, so further investigations focused on estradiol and estrone sulphate levels only. On day 7, only estrone sulphate was present, but in similar amounts in both male and female embryos. On day 8, estrone sulphate as well as estradiol were present, but also without significant differences. Significant differences were not apparent until day 9, and were even more pronounced on day 10, allowing precise sex identification at that developmental stage (see Figure 2).

In the study, fluid collection was initially performed as described by Phelps et al. (2003) (horizontally placed egg, at the equator), but another position for collection was also investigated (tilted egg, just below the air cell). The latter position proved to be advantageous in three ways:

1. Fluid could be collected more reliably (by slanting the fluid 'pools' in a more localisable region).

2. Hatching losses were significantly less, i.e. the embryo was less damaged (embryo tends to be located further away from the puncture site).
3. Better position for possible automation, since eggs do not have to be removed from the tray to be placed horizontally.

Using this method, an accuracy of 98% was achieved in further hatching trials, and the hatching rate was only reduced by about three percentage points due to manual treatment compared to control groups. In this study, it was also found that control and experimental groups did not differ in terms of performance parameters in the rearing and laying period (Weissmann et al., 2014).

The promising results of these investigations, which were part of a research project funded by the German Federal Ministry of Food and Agriculture, led to the involvement of a leading food retail company, which, together with a project partner, aimed at automating the process and commercialising it.

Since 2019, a fully automated process has been available that carries out the process steps shown in the following diagram (Figure 3: candling; hole at a defined point with laser for fertilised eggs; removal of a small amount of allantoic fluid; analysis of the estrone sulfate content) with robot support (Seleggt, 2022). According to the manufacturer, 3600 hatching eggs/h can be sorted in this way.

3.3 DNA analysis

The PLANTegg process of in ovo gender determination of chicken embryos is based on genetic differences between male and female individuals (ZW/ZZ-system). Male chicks possess homozygote sex chromosomes (ZZ), whilst females possess hemizygote sex chromosomes (ZW). A modern molecular technique - the PCR (polymerase chain reaction) method - is used to determine the differences between the two genders. Using the PCR method, tiny amounts of genetic information (DNA) present in the allantoic fluid can be withdrawn from the egg and amplified by PCR. After amplification of the DNA, a visible optical signal indicates the presence of female DNA, whilst in the case of the male, no signal is visible. A clear and reliable distinction between male and female eggs at an early stage of the incubation process is therefore feasible, yielding 'yes' or 'no' results. No intermediate results are generated, no interpretations of the results are needed and no minimum concentration of metabolite is necessary. The PLANTegg process is used commercially in hatcheries with a sexing precision of more than 99% to avoid the killing of day-old male chicks. The different steps of the PLANTegg process within the hatchery are described below (Kleine, 2022; PLANTegg, 2021).

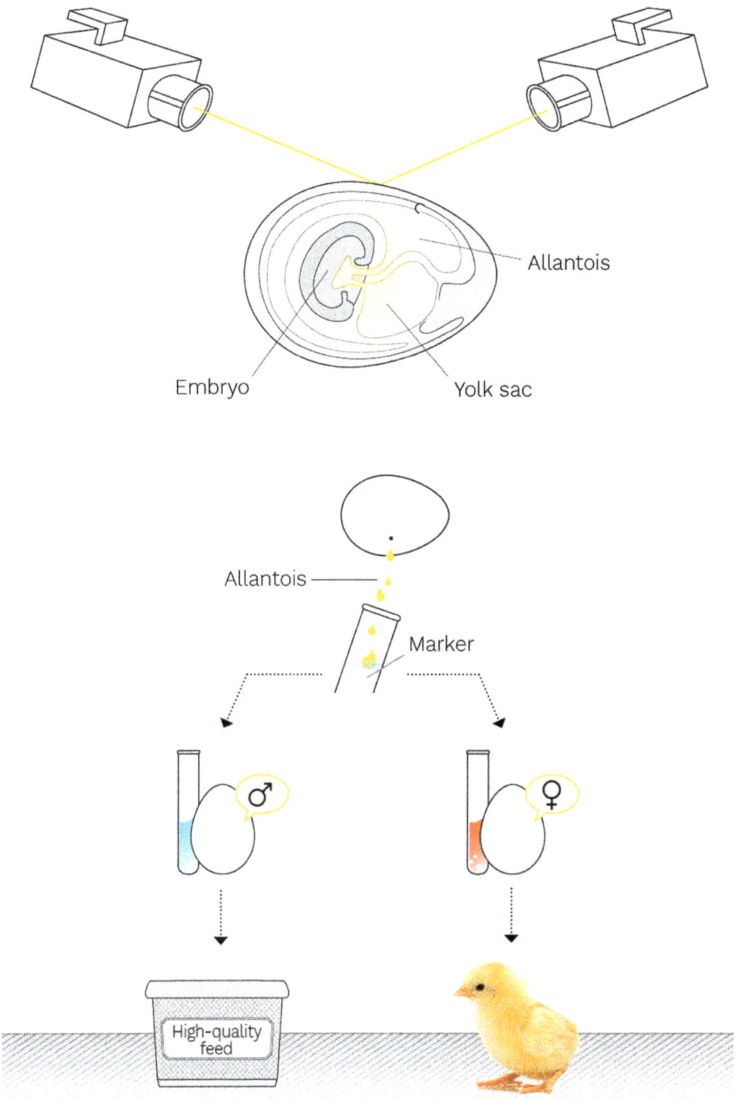

Figure 3 The Seleggt process (www.seleggt.com).

1. Processing: Hatching eggs are candled and only fertilised eggs are transferred at day nine to a roboting device, installed within the hatchery. This device receives hatching eggs from the hatchery. A laser beam creates a tiny opening (0.3 mm) in the hatching egg and a pipette

retrieves a few drops of allantoic fluid from the opening of the hatching egg. The allantoic fluid contains a few embryonic cells sufficient for gender determination. A total of 3600 eggs/h can be processed with one device. The sampling technology is similar to the procedure used by Seleggt.

2. PCR analysis: The allantoic fluid contains genetic material (DNA) of the embryonic cells. The fluid is collected in reaction tubes and transferred to an extra lab room within the hatchery or to an external container lab attached to the hatchery. There, the allantoic fluid is combined with the PCR reaction chemistry. The tubes are then transferred to the PCR machine(s) where the PCR reaction takes place and the results (male/female) are generated (after about 1 h processing time).
3. Sorting: The result (male/female) is transferred electronically to the hatchery and the selection of the male and female eggs takes place in an automated selection device. Female eggs are returned to the incubators whilst male eggs are sorted out for feed purposes.

The flow chart (Figure 4) shows the different steps performed by the hatchery and by PLANTegg. The PLANTegg system is installed within the hatchery and operated by PLANTegg. The modular organisation of the equipment - roboting device(s), PCR lab with PCR machine(s) - can be adopted according to the capacity of the hatchery.

3.4 Mass spectrometry

The allantoic fluid of a developing embryo not only contains hormones and small amounts of genetic information but also glucose, choline and valine are found to be more concentrated in allantoic fluid from females and could thus serve as biomarkers (Bruins and Stutterheim, 2014). In Ovo, a spin-off of Leiden University, identified a single biomarker that can be detected by mass spectrometry in a fast and high throughput way. Its Alpha prototype realised 1800 eggs/h with 95% accuracy. In an ongoing project, it plans to upscale this prototype and implement this technology outside the Netherlands (In Ovo, 2021; CORDIS, 2021).

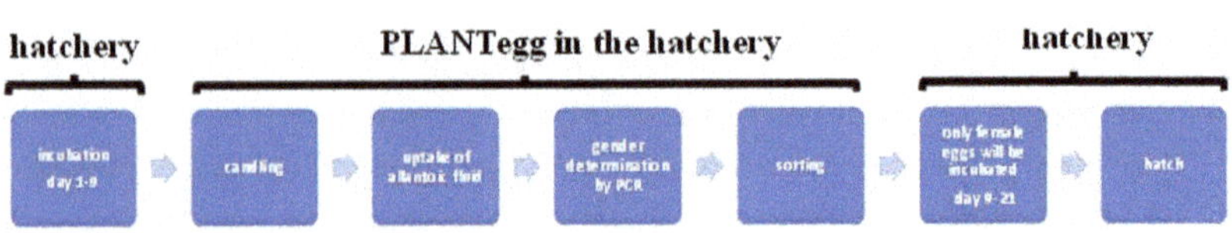

Figure 4 The PLANTegg process (Kleine, 2022, www.plantegg.de).

4 Future research

Figure 5 presents an overview of research projects worldwide - a snapshot without claiming completeness. Patents have already been disclosed or granted for numerous process inventions to solve the problem of early sex determination in eggs, whilst 'breakthroughs in early sex determination' have been announced in the press. However, it is usually not possible to determine from the available documentation to what extent the respective inventions could be successfully implemented in practice. There is also often a lack of scientifically proven information on the methodology, diagnostic accuracy and other relevant aspects (precision, speed, hatching rate, operational costs, animal health and performance parameters of the hens, etc.) (Krautwald-Junghanns et al., 2018). This applies, for example, to the 'Hypereye' method developed in Canada, which claims to not only allow fertility control but also sex determination by means of hyperspectral imaging on the unincubated egg. The US company ONCE claims to have developed a 'genetic switch' that can be used to selectively activate or deactivate light absorption centres in sex-determining genes using narrowband light-emitting diodes. In this way, it should be possible to selectively produce only offspring of the desired sex (hens in laying hen production, roosters in broiler production). SOOS, a company located in Israel, follows a similar approach of turning genetically male embryos into functional female chicks. According to its own information, this is carried out non-invasively during the incubation process by applying sound waves and other specific incubation parameters. Unfortunately, no reliable results have yet been published, and no information is available concerning trials or the practical use of these methods.

Other experimental methods have proven to be unsuitable for large-scale use. In studies by Klein et al. (2002), the germinal disc was first localised by magnetic resonance tomography for in ovo sex determination on the unincubated egg using molecular genetic methods, following which blastoderm cells were removed from the germinal disc by biopsy, and sex-specific sequences of the W chromosome were determined by PCR or in situ hybridisation (Klein and Ellendorff, 2000). No negative effects of cell removal on embryonic development, hatchability, or reproductive performance of tested individuals were observed (Klein et al., 2003b). However, the development of practicable procedures failed due to the very high technical and financial expenditure for the localisation of the germinal disc (Klein et al., 2003a, b).

In ovo sex differentiation based on the differing DNA content in cells of male and female domestic chickens was the subject of other studies. According to Tiersch (2003), the DNA content in cells of male individuals (ZZ) is approximately 2.50 pg, which is about 2% higher than in cells of female individuals (ZW; approximately 2.45 pg), but breed- or lineage-specific differences have to be taken into account (Nakamura et al., 1990). DNA content is determined

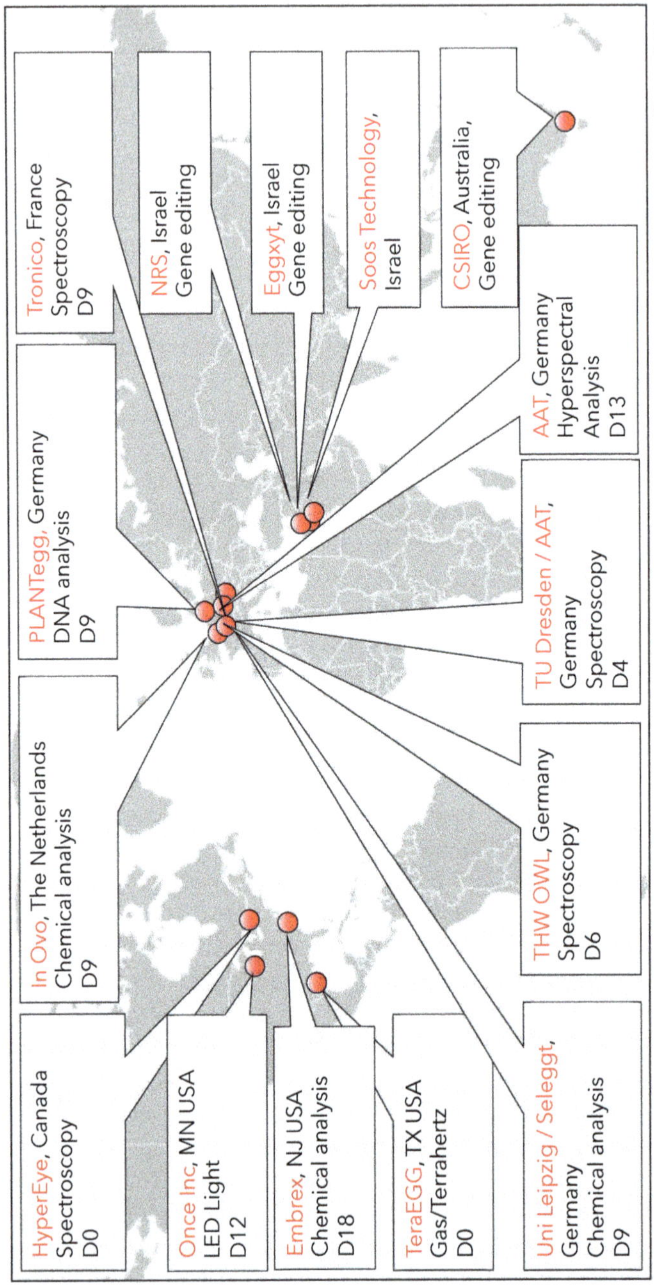

Source: AAT

Figure 5 Research projects and approaches for in ovo sex determination worldwide.

by flow cytometry (Tiersch, 2003), yet detailed information on the sampling methodology and the analysed cell material is missing. It is only mentioned that very small samples of about 1 μl blood are sufficient for an analysis. If blastoderm cells are analysed, suitable methods for germinal disc localisation are indispensable. Due to the unsatisfactory accuracy of sex determination and the high cost of the equipment, this method has not yet been developed to practical maturity, but there are other approaches with more chance of success, which are described below.

4.1 Vibrational spectroscopic methods

Compared to methods based on biochemical analysis, spectroscopic methods offer several advantages. These include, among others, contactless examination, as well as low consumption costs (e.g. sealing material for eggs identified as 'female'). Furthermore, examinations can be performed as early as 3.5 days of incubation. During spectroscopic measurement, eggs that are not capable of development can be detected and removed from the incubation process. Spectroscopic examinations are non-invasive in the sense that no biological material has to be removed. However, the eggshell remains a barrier for light spectra, so that a hole is necessary to allow access to the egg content to be examined. The actual measurement, including obtaining the sex information, takes only a few seconds, allowing sorting of eggs in near real time. Several spectroscopic methods have been investigated to date for in ovo sex determination with varying degrees of success.

4.1.1 Fourier transform infrared spectroscopy

Fourier transform infrared (FT-IR) Spectroscopy is an optical technique used to determine the structure of molecules and to characterise chemical interactions between molecules (Steiner and Koch, 2009). The basic principle of FT-IR spectroscopy is based on the selective absorption of light in the infrared wavelength range (2.5–20 μm; corresponding wavenumbers 4000-500 cm^{-1}). When infrared light of a defined wavelength hits a molecule, the atoms can be set into vibration without breaking the bonds between them. Spectral decomposition of the light yields an infrared spectrum in which the vibrations of the atoms are represented as absorption bands. Based on their pattern, the substance under investigation can be clearly identified ('fingerprint method'). FT-IR spectroscopy allows the investigation of samples without additional markers and can therefore also be used for in situ analyses. In addition, no damage to the samples occurs, since there is neither heating of the samples nor destruction of the molecules. A short measurement time and high spatial resolution are further decisive advantages of IR spectroscopy, which makes it particularly suitable for studying biological samples.

Expectations for this method to determine sex on the unincubated egg were high, and initial results were quite promising (Steiner et al., 2011). It was shown that a defined opening in the calcareous shell can be made contactless by the use of suitable lasers. Precise position determination of the germinal disc through this opening could be achieved by optical coherence tomography (OCT) (Burkhardt et al., 2010). Fourier transform infrared (FT-IR) spectroscopy can then be applied for sex diagnosis. This method has already been used to identify sex-specific spectral differences on germinal discs of unincubated chicken eggs, allowing unambiguous sex assignment (Steiner et al., 2011). However, large-scale hatchability experiments have shown that opening the calcareous shell of unincubated eggs results in a significant decrease in hatch rate, as the germinal disc appears to be highly sensitive to the slightest changes in the environment (Krautwald-Junghanns et al., 2014).

4.1.2 Raman and fluorescence spectroscopy

Raman spectroscopy (named after the Indian physicist, Chandrasekhara Venkata Raman) is the spectroscopic investigation of the inelastic scattering of light by molecules. In this process, incident light strikes a vibrating molecule and emerges as scattered light. The inelastically scattered light has a different frequency or energy from the incident light. Raman spectroscopy thus represents a subfield of vibrational spectroscopy based on the so-called 'Raman effect'. In a Raman spectroscopic measurement, light of a defined wavelength is irradiated onto the object under investigation and the spectrum of the scattered light is analysed. Shifts in the frequency of the light compared to the incident light allow conclusions regarding the vibrations of functional molecular groups. Raman spectroscopic investigations offer several advantages compared to other spectroscopic techniques such as nuclear magnetic resonance (NMR) spectroscopy or infrared (IR) spectroscopy. For example, Raman spectroscopy requires zero or minimal sample preparation in advance, which enables prompt sample characterisation (Schnepel, 1980; Rösch, 2002).

In the ultraviolet (UV) wavelength range, DNA and protein information is amplified by the so-called resonance effect, since these substances exhibit a particularly high absorption in this frequency range. Based on investigations of feather pulp using UV resonance Raman spectroscopy, it could be demonstrated that this minimally invasive method is principally suitable for sex determination. To clarify the origin of the feather pulp Raman bands, the spectra were compared with the Raman spectra of DNA and various proteins. Thus, an assignment of the Raman signals of the feather pulp spectrum could be made. Using chemometric methods, it was possible to elucidate spectral characteristics of male and female chickens. For sex determination, the Raman spectrum was compared with a database and assigned to the male or female group (cluster) by chemometric methods. Support vector machines (SVM) were used to classify the Raman spectra. Recovery rates could be calculated from the classification results. It was shown that the SVM method is well suited to distinguish male and female chickens based on tissue samples with recognition rates of about 95% (Harz et al., 2008). However, ultraviolet resonance Raman (UV-RR) spectroscopy proved to be unsuitable for in ovo studies of germinal discs, as the high energy input due to the UV radiation required resulted in cell damage to the blastoderm.

The use of Fourier transform infrared Raman spectroscopy could overcome these effects, since the energy content of the waves in the near-infrared range is very low. Incubating hatching eggs prior to examination also proved to be advantageous. If the hatching eggs are incubated for approximately 3-5 days, necessary manipulations of the egg have significantly less effect on the further development of the embryo, according to previous findings. In addition, at this time already recognisably unfertilised eggs and those with irregularly

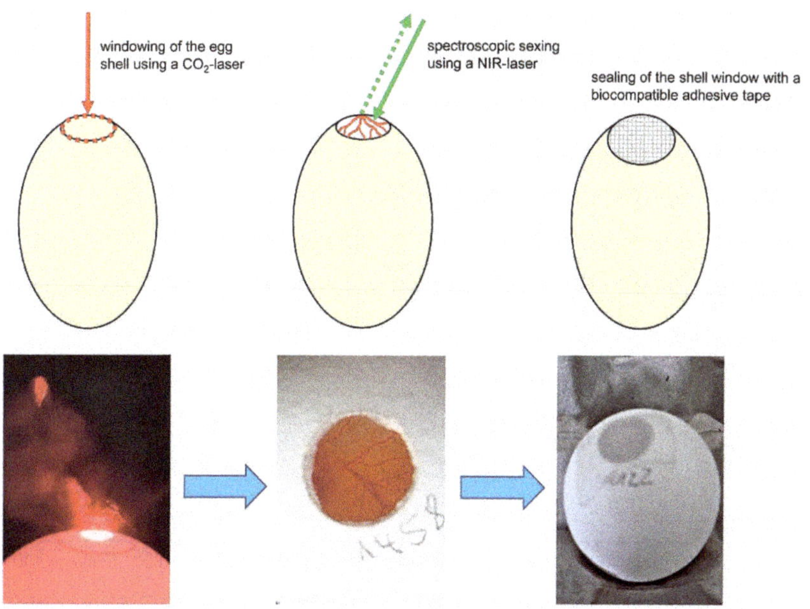

Figure 6 Sequence of the optical spectroscopic in ovo sex determination on incubation day 3.5. In the first step, a hole of approximately 12 mm is made in the egg at the blunt pole (automated process). This is followed by contactless Raman and fluorescence spectroscopic analysis of the embryonic blood (note: for better visibility of the blood vessels, the egg was opened at the pointed pole for the photo). In the third step, female eggs are resealed and further incubated (T. Bartels, FLI Celle).

developed embryos can be removed from the incubation process. In addition, sex-specific information can also be obtained from the developing embryonic vascular system. To identify the sex of the opened egg, a hole of about 12 mm must first be made in the calcareous shell and the 'lid' removed (Figure 6). The procedure is well established and can also be automated. It is advantageous to uncover the blunt egg pole, as the inner shell membrane can remain intact. On the third day of incubation, the embryonic blood vessel system has already developed, by which the egg reveals its sex. The latest generation of the method, which was primarily developed at TU Dresden, uses two laser beams of different wavelengths. The scattered light from the haemoglobin is used to identify the sex in a combination of Raman and fluorescence spectroscopy. Whilst Raman spectra provide molecular structural features, fluorescence primarily provides information on the concentration of haemoglobin (Galli et al., 2016, 2017a, 2017b, 2018). A sophisticated combination of the two information channels precisely provides sex information (Figure 7). Compared to previous developments, the enhanced method not only allows for clearer signal differences between the sexes and thus faster measurement times but also intrinsic referencing, which provides a decisive advantage, especially for rapid automation. Male eggs can be sorted out and processed further, whilst

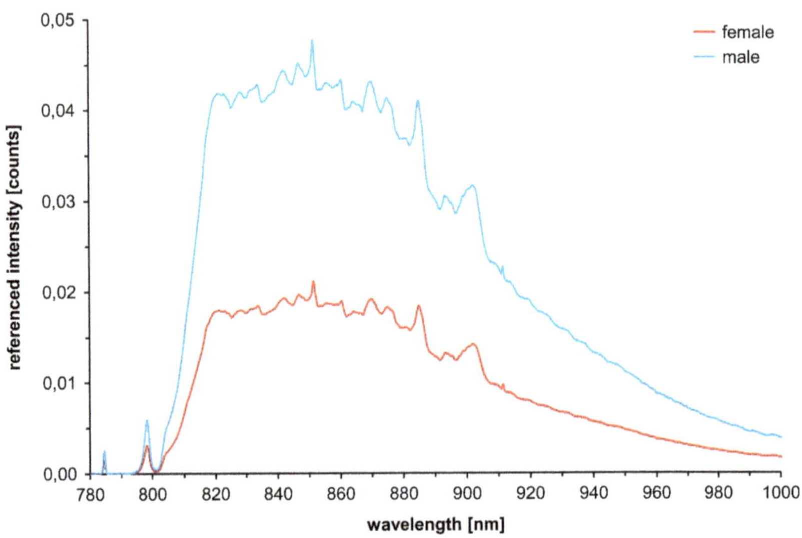

Figure 7 Selected in ovo spectra of a female and a male egg on incubation day 5.5. The large signal differences are mainly due to fluorescence. Significantly weaker, but molecularly very informative Raman signals can be seen on top of the fluorescence peaks and are also very important for differentiation. Although the figure shows clear differences between the sexes, in practice the signals scatter considerably from egg to egg, requiring the use of artificial intelligence methods for sex recognition (G. Preusse, TU Dresden).

female eggs are resealed with a medical plaster and incubated further until hatching.

This highly demanding process has not yet been scaled up beyond the laboratory. In larger and more extensive trials in hatcheries, AAT has not yet been able to fully reproduce the published results obtained on a smaller number (AAT, 2022). Furthermore, the accuracy was too variable to be sufficient for practical use. In a current follow-up project, further trials should bring this application to market maturity with a consistently high level of accuracy and speed.

Nevertheless, optical spectroscopic sex determination has turned out to be a fairly promising method for early in ovo sex determination and could - in line with the German Animal Protection Act - prevent the killing of potentially sentient embryos.

4.1.3 Time-resolved laser-induced fluorescence spectroscopy

This method is based on the well-known measurement of laser-induced fluorescence (LIF), which has an extremely high detection sensitivity and precision. In combination with the detection of the time-dependent behaviour of the fluorescent light (zLIF) and with the algorithms adapted and optimised for sex determination, the sex of an embryo can be detected reliably before the sixth day of incubation. Only a very small opening of the eggshell (< 2 mm) is necessary, with the membrane kept intact, so that the risk of contamination is low. A current research project aims to further develop the measuring system and the algorithms. A study in combination with other imaging systems is also under investigation, to find a way to keep the eggshell intact. Figure 8 describes schematically the measurement system. It consists of a pulsed excitation laser, a multi-channel detector, an ICCD camera with polychromator, a control system with delay generator, an optical fibre with sensor head and an evaluation unit.

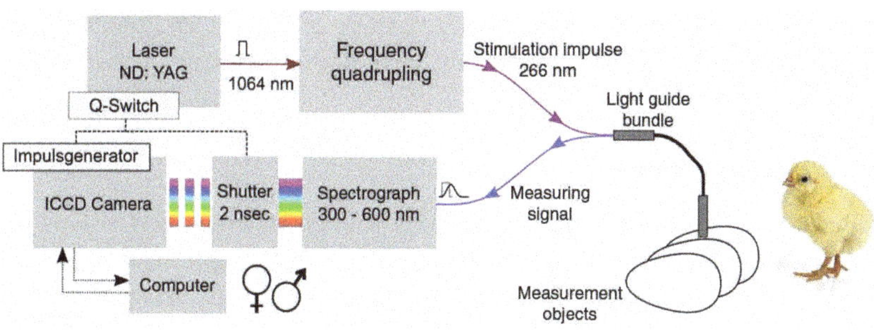

Figure 8 Schematic illustration of a system for time-resolved laser-induced fluorescence spectroscopy (Dörksen, 2022).

4.2 Magnetic resonance imaging

ORBEM is another example of a magnetic resonance imaging (MRI) process. Founded as a spin-off from the Technical University of Munich in 2019, it combines artificial intelligence (AI) and imaging technology to detect and distinguish the gonadal development of an embryo after 13 days of incubation (ORBEM, 2022). This non-invasive method is applicable to all layer chicken genetics but must be performed at a relatively late stage of embryonic development. High precision (near 100%) is claimed, but at present the throughput of eggs of 900 eggs/h is insufficient for industrial use and subject to further studies (Quentin et al., 2022).

4.3 Gene editing

The genetic marking of sex chromosomes is also being studied as a potential route for in ovo sex determination in chickens. Studies by Quansah et al. (2013) and Doran et al. (2018) focused on the production of genetically engineered hens and described the marking of the Z chromosome of breeding hens with green fluorescent protein. The so-called 'EggXYt' method is also being considered as a viable option for in ovo sex determination in chickens (cf. https://www.eggxyt.com/) and is based on sex identification after gene manipulation using nonspecific fluorescent marker genes. This can be achieved by producing genetically modified hens at the parental level. By labelling the Z chromosome of breeding hens with fluorescent biomarkers, when mated to non-genetically modified roosters at the reproduction level, male-determined eggs can be identified and sorted out based on their fluorescence behaviour prior to incubation. The laying hen chicks hatching from such mating do not show any foreign genes, according to current studies. With a comparable technique, the Israelian company NRS claims to change the parental Z chromosome in such a way that male offspring cease development at a very early stage of embryogenesis (cf. https://www.nrspoultry.com/project-sexing/). Due to negative perceptions of genetic modification of animals and the resulting lack of consumer acceptance, the use of genetic engineering for in ovo sex determination is not viable, at least in Western European countries.

4.4 Avian egg odour

Webster et al. (2015) reported that volatiles from developing eggs of Japanese quail (*Coturnix japonica*) carry information regarding egg fertility, along with the sex and developmental status of the embryo. Specifically, it was shown that egg volatiles undergo change over the course of incubation, and not only differ between fertile and infertile eggs but are also sex-predictive as early as day 1 of incubation. One volatile ketone that differs between eggs containing male and female embryos (2-undecanone) has previously been identified as a hormone-linked constituent of avian odour (Whittaker et al., 2011). Xiang et al. (2022a) used solid-phase microextraction/gas chromatograph-mass spectrometry (SPME/GC-MS) for analysis of volatile compounds of chicken eggs. They also found sex-specific differences in the composition (or content) of volatiles between (hatched) male and female chicken eggs during E0–E9 and between unhatched male and female chicken eggs of different lines.

So-called 'TeraEgg' technology is a spectroscopic method that uses terahertz radiation in the sub-millimetre wave range of the electromagnetic spectrum to study and elucidate the structure of matter, for the analysis of volatile substances. This method is designed to determine not only the fertilisation of an unincubated egg but also the sex of the 'embryo-to-be'. Thus, both unfertilised and male-determined hatching eggs can be sorted out before incubation and utilised as eggs for consumption. To date, there are no published results using this technique (Gonzalez, 2020).

4.5 Protein quantification

According to Xiang et al. (2022b), the quantification of yolk proteins is a promising future approach. The authors were able to identify 35 differentially expressed proteins using tandem mass tags (TMT) proteomics, which in principle allow in ovo sex determination on the unincubated fertilised egg.

5 Discussion

Avoiding the routine killing of day-old male chicks has now become a concern of considerable ethical, legal and sociopolitical consequence (Ort, 2010; Buhl, 2013). In Germany alone, approximately 45 million male offspring are killed annually in the context of laying hen reproduction and marketed primarily as 'frozen chicks' for feeding zoo animals and pets. Globally, the laying hen population is currently estimated at around seven billion, meeting the current world demand of around 68.3 million tons of eggs per year. On an international level, the development of applicable and more marketable procedures could avoid the need to kill a corresponding number of male layer hybrids in the future, thus making a valuable contribution to animal welfare in poultry farming.

The limited use of male layer hybrids and their resulting culling shortly after hatching affects not only conventional egg production but to a large extent layer husbandry, according to organic standards. The lack of profitability of rearing 'laying hen brothers' stems from, among other things, their longer fattening period, a reduced fattening performance with higher feed expenditure, and a lower proportion of breast muscle meat, which is particularly appreciated by consumers. The carcasses also do not correspond visually to the consumer ideal of the 'full-fleshed broiler' and also require special cooking methods, which makes marketing on a broad basis even more difficult. Therefore, the fattening of male layers and their marketing as so-called 'poussins', 'spring chickens' or 'brother cockerels' is currently considered a niche market (Koenig et al., 2010, 2012a, b). Alternatively, the breeding of so-called 'dual-purpose chickens' is being pursued by crossing fattening and laying lines as a compromise solution with a dual focus on meat production and laying performance. However, such breeding lines, which are also already commercially available, still show significantly lower economic performance (lower laying performance and egg size, higher feed consumption with longer fattening duration, little breast musculature) than specialist laying or fattening lines respectively (Icken et al., 2013). Despite consumer acceptance, only a very specialist market segment can be served with dual-purpose chicken products at this time (Brümmer et al., 2018; Gangnat et al., 2018; Lambertz et al., 2018). Marketing opportunities for brother cockerels and dual-purpose chickens are currently found in the production of high-quality meat and sausage products in the upper price segment, especially after longer fattening periods (Schütz et al., 2018; Siekmann et al., 2018).

Twenty years ago, the challenges for a functioning method of in ovo sex determination were of a biological and technical nature; today, political and ethical challenges have been added to the mix. In ovo sexing has the potential to end the culling of male chicks in the egg-producing industry. However, to meet industry requirements for precision and speed, full automation of the process is required to make this technology applicable in hatcheries (Galli et al., 2017b). In Germany and other European countries, for example, there are currently (at least) three alternatives on the market that can successfully identify the sex of an embryo during the second trimester of incubation. Organic associations prefer rearing of the males and in some cases completely reject in ovo sex determination. Consumers now have the choice to buy eggs from supply chains 'without chick culling', but often do not know if the laying hens' brothers were raised or if an in ovo sexing process was used.

6 Conclusion and future trends

Critics of in ovo sexing argue that this approach does not represent any progress for animal welfare, as the killing of the chicks is merely brought forward. Depending on the method, embryos in different stages of development are subject to in ovo sex determination. In this context, the question arises as to at what age domestic chicken embryos are capable of nociception and can therefore be considered sentient. The capacity of avian embryos for in ovo nociception is currently no longer in question (Aleksandrowicz and Herr, 2015; Bjørnstad et al., 2015). A prerequisite for sentience is the presence of functional pathways that allow conduction of stimuli to the brain. The development of the central nervous system in the chicken embryo is a dynamic process that begins on the second day of incubation and matures continuously until hatch. Currently, in the absence of reliable data, it is scientifically accepted that the chicken embryo is not yet capable of nociception before day 7 of incubation, whilst pain perception is likely to be present from day 15 onwards. At present, it is not possible to name a specific time point when nociception begins.

Since all currently available techniques to detect an embryo's sex are applied during this period, the question arises to whether the rejected male eggs can be handled in an animal welfare-friendly and consumer-accepted way. The use of electrical current flow as a method of anaesthesia of chicken embryos has been investigated, in order to avoid a possible perception of pain when the incubation process is terminated (Zumbrink et al., 2020). Based on the promising results of this study, a high throughput system for electrical anaesthesia ('Stunny') was constructed by AAT in Germany. The stunning unit, in combination with hyperspectral in ovo sex determination (or other techniques), can be seen as a bridge technology and provides an important step in discussions regarding chick culling, until earlier techniques are applicable (Zumbrink et al., 2022).

6.1 Outlook

This is currently the subject of controversial discussion, not only in Germany (Bartels et al., 2020) but also in Switzerland, Austria and France, which will have a considerable effect on political decisions and the future of hatcheries and egg layer production. Efforts in Germany to only allow in ovo procedures that can be used before the seventh day of incubation are fuelling research but will ultimately result in rearing the male chicks as the only remaining method if a proven technology in the first trimester of incubation is not available.

For technical or biological reasons, current in-ovo techniques have little practical ability to detect the sex of an embryo at an earlier stage. The sex-specific hormones and the colour of the feathers are not detectable before the 8th or 10th day of incubation. With liquid-based methods, it is essential to obtain a certain amount of allantoic fluid, which is also only possible from the second trimester of incubation.

Further development of spectroscopic techniques as well as other approaches is needed for solutions that perform reliably as early as possible in embryonic development. Much hope lies in this technical progress since the alternatives

- rearing of the males or use of dual-purpose breeds - are neither sustainable nor environmentally friendly and can only be a solution for niche markets.

7 Where to look for further information

The following books and articles provide a good overview of the subject:

- Bellairs, R. and Osmond, M. (2005), *The Atlas of Chick Development*, Elsevier, Amsterdam.
- Crawford, R. D. (1990), *Poultry Breeding and Genetics*, Elsevier, Amsterdam.
- Douglas, C. and Turner, J. M. A. (2020), 'Advances and challenges in genetic technologies to produce single-sex litters', PLoS Genet., 16, e1008898.
- Gautron, J., Réhault-Godbert, S., Van de Braak, T. G. H. and Dunn, I. C. (2021), 'Review: What are the challenges facing the table egg industry in the next decades and what can be done to address them?', *Animal*, 15, 100282.
- Gremmen, B., Bruijnis, M. R. N., Blok, V. and Stassen, E. N. (2018), 'A public survey on handling male chicks in the Dutch egg sector', *J. Agric. Environ. Ethics*, 31, 93–107.
- Lee, J., Kim, D.-H. and Lee, K. (2020), 'Current approaches and applications in avian genome editing', *Int. J. Mol. Sci.*, 21, 3937.
- Reithmayer, C. and Mußhoff, O. (2019), 'Consumer preferences for alternatives to chick culling in Germany', *Poult. Sci.*, 98, 4,539–4,548.
- Reithmayer, C., Danne, M. and Mußhoff, O. (2021), 'Societal attitudes towards in ovo gender determination as an alternative to chick culling', *Agribusiness*, 37, 306–323.
- De Haas, E. N., Oliemans, E. and van Gerwen, M. A. A. M. (2021), 'The need for an alternative to culling day-old male layer chicks: A survey on awareness, alternatives, and the willingness to pay for alternatives in a selected population of Dutch citizens', *Front. Vet. Sci.*, 8, 662197.

Key research in this area can be found on the websites of these organisations:

- Agri Advanced Technologies GmbH (https://www.agri-at.com).
- In Ovo B.V. (https://inovo.nl).
- PLANTegg GmbH (https://www.plantegg.de).
- Seleggt GmbH (https://www.seleggt.com).

8 References

AAT (Agri Advanced Technologies GmbH) (2021). In ovo sex determination of layer chicks in the egg, Retrieved on 11 November 2021. Available at: https://www.agri-at.com/en/products/in-ovo-sex-determination.

AAT (Agri Advanced Technologies GmbH) (2022). The Raman spectroscopic method, Retrieved on 11 November 2021. Available at: https://www.agri-at.com/en/products/in-ovo-sex-determination/the-raman-spectroscopic-method.

Aleksandrowicz, E. and Herr, I. (2015). Ethical euthanasia and short-term anesthesia of the chick embryo, *Altex* 32(2), 143–147.

Aslam, M. A., Hulst, M., Hoving-Bolink, R. A. H., Smits, M. A., de Vries, B., Weites, I., Groothuis, T. G. G. and Woelders, H. (2013). Yolk concentrations of hormones and glucose and egg weight and egg dimensions in unincubated chicken eggs, in relation to egg sex and hen body weight, *Gen. Comp. Endocrinol.* 187, 15–22.

Bartels, T., Steiner, G., Preusse, G., Galli, R., Förster, A., Preisinger, R., Cramer, K. and Krautwald-Junghanns, M.-E. (2014). Spektroskopische Methoden zur Geschlechtsfrühdiagnose in der Legehennenvermehrung, *Rundsch. Fleischhyg Lebensmittelüberw.* 66, 440–442.

Bartels, T., Wilk, I. and Schrader, L. (2020). *Entwicklung von Nozizeption und Schmerzempfinden bei Hühnerembryonen, [Stellungnahme des FLI].* Insel Riems, Greifswald: Friedrich-Loeffler-Institut.

Bellon-Maurel, V. and Gorretta, N. (2014). NIR hyperspectral imaging for food and agricultural products. In: Salzer, R. and Siesler, H. W. (Eds.), *Infrared and Raman Spectroscopic Imaging*. Weinheim: John Wiley & Sons, 295–338.

Bjørnstad, S., Austdal, L. P. E., Roald, B., Glover, J. C. and Paulsen, R. E. (2015). Cracking the egg: Potential of the developing chicken as a model system for nonclinical safety studies of pharmaceuticals, *J. Pharmacol. Exp. Ther.* 355(3), 386–396.

Bruins, W. S. and Stutterheim, W. M. (2014). Gender, viability and/or developmental stage determination of avian embryos in ovo (NL), *Patent WP 2014021715A2/EP 2880440 B1, 20181031 (EN).*

Brümmer, N., Luy, J., Rovers, A. and Christoph-Schulz, I. B. (2018). Mehr als eine Nische? Das Potential des Zweinutzungshuhns als Alternative zum Kükentöten, *J. Consum. Prot. Food Saf.* 13, 226–230.

Buhl, A. C. (2013). Legal aspects of the prohibition on chick shredding in the German state of North Rhine-Westphalia, *Glob. J. Anim. Law* 2, 1–8.

Burkhardt, A., Geißler, S. and Koch, E. (2010). Localization of the germinal disc in ovo for further chicken sexing using optical coherence tomography, *Biomed. Tech.* 55 (Suppl.1).

CORDIS. (2021). High-throughput solution for in-egg selection of laying hens, Retrieved on 8 April 2022. Available at: https://cordis.europa.eu/project/id/959321.

Dörksen, H. (2022). Personal Communication, March 2022.

Doran, T. J., Morris, K. R., Wise, T. G., O'Neil, T. E., Cooper, C. A., Jenkins, K. A. and Tizard, M. L. V. (2018). Sex selection in layer chickens, *Anim. Prod. Sci.* 58(3), 476–480.

Förster, A., Zumbrink, L. and Hurlin, J. (2022). *'L'imagerie hyper-spectrale comme outil d'ovo-sexage pour les embryons des souches d'œufs bruns', 14e Journées de la Recherche Avicole et Palmipèdes à Foie Gras*, 444–448.

Galli, R., Preusse, G., Uckermann, O., Bartels, T., Krautwald-Junghanns, M. E., Koch, E. and Steiner, G. (2016). In ovo sexing of domestic chicken by Raman spectroscopy, *Anal. Chem.* 88(17), 8657–8663.

Galli, R., Preusse, G., Uckermann, O., Bartels, T., Krautwald-Junghanns, M. E., Koch, E. and Steiner, G. (2017a). In-ovo sexing of chicken eggs by fluorescence spectroscopy, *Anal. Bioanal. Chem.* 409(5), 1185–1194.

Galli, R., Koch, E., Preusse, G., Schnabel, C., Bartels, T., Krautwald-Junghanns, M.-E. and Steiner, G. (2017b). Contactless in ovo sex determination of chicken eggs, *Curr. Dir. Biomed. Eng.* 3(2), 131–134.

Galli, R., Preusse, G., Schnabel, C., Bartels, T., Cramer, K., Krautwald-Junghanns, M. E., Koch, E. and Steiner, G. (2018). Sexing of chicken eggs by fluorescence and Raman spectroscopy through the shell membrane, *PLoS ONE* 13(2), e0192554.

Göhler, D., Fischer, B. and Meissner, S. (2017). In-ovo sexing of 14-day-old chicken embryos by pattern analysis in hyperspectral images (VIS/NIR spectra): A non-destructive method for layer lines with gender-specific down feather color, *Poult. Sci.* 96(1), 1-4.

Gangnat, I. D. M., Mueller, S., Kreuzer, M., Messikommer, R. E., Siegrist, M. and Visschers, V. H. M. (2018). 'Swiss consumers' willingness to pay and attitudes regarding dual-purpose poultry and eggs, *Poult. Sci.* 97(3), 1089-1098.

Gonzalez, P. F. J. (2020). *Instrumentations Radiofréquences pour l'Agriculture de Précision: Application à la Filière Aviaire pour le Sexage Précoce in et Ex Ovo.* Toulouse: Electronique, Université Paul Sabatier.

Hamburger, V. and Hamilton, H. L. (1951). A series of normal stages in the development of the chick embryo, *J. Morphol.* 88(1), 49-92.

Harz, M., Krause, M., Bartels, T., Cramer, K., Rösch, P. and Popp, J. (2008). Minimal invasive gender determination of birds by means of UV-resonance Raman spectroscopy, *Anal. Chem.* 80(4), 1080-1086.

Icken, W., Schmutz, M., Cavero, D. and Preisinger, R. (2013). Dual purpose chicken: The breeder's answer to the culling of day-old male layers. In: *Proceedings of the 9th European Symposium on Poultry Welfare*, Uppsala, Sweden.

Imholt, D. (2010). *Morphometrische Studien an Eiern von Hybrid- und Rassehühnern mit Versuchen zur Detektion einer Beziehung zwischen der Form von Eiern und dem Geschlecht der darin befindlichen Küken.* Giessen: VVB Laufersweiler Verlag.

In Ovo. (2021). In ovo hatches first 150,000 chicks without culling, Press Release https://inovo.nl/news/in-ovo-hatches-first-150000-chicks-without-culling/.

Kaleta, E. F. and Redmann, T. (2008). Approaches to determine the sex prior to and after incubation of chicken eggs and of day-old chicks, *World Poult. Sci. J.* 64(3), 391-399.

Kallenberg, F. (1913). Offenbarungen des Siderischen Pendels, Verlagsanstalt J. S. Huber, Diessen vor München.

Klein, S. and Ellendorff, F. (2000). Localization of Xho1 repetitive sequences on autosomes in addition to the W chromosome in chickens and its relevance for sex diagnosis, *Anim. Genet.* 31(2), 104-109.

Klein, S., Rokitta, M., Baulain, U., Thielebein, J., Haase, A. and Ellendorff, F. (2002). Localization of the fertilized germinal disc in the chicken egg before incubation, *Poult. Sci.* 81(4), 529-536.

Klein, S., Flock, D. and Ellendorff, F. (2003a). Management of newly hatched male layer chicks - Current knowledge on sex determination and sex diagnosis in chicken: Potential solutions, *Worlds Poult. Sci. J.* 59, 62-64.

Klein, S., Baulain, U., Rokitta, M., Marx, G., Thielebein, M. and Ellendorff, F. (2003b). Sexing of freshly laid egg: Development of embryos after manipulation; analytial approach and localization of the blastoderm in the intact egg, *Worlds Poult. Sci. J.* 59, 39-45.

Kleine, M. (2022). *A PCR-Based Method for Gender Determination of Hatching Eggs*, Personal Communication, January 2022.

Koenig, M., Hahn, G., Damme, K. and Schmutz, M. (2010). Utilization of laying type cockerels as coquelets - Growth performance and carcass quality, *Fleischwirtschaft.* 90, 92-94.

Koenig, M., Hahn, G., Damme, K. and Schmutz, M. (2012a). Utilization of laying type cockerels as "coquelets": Influence of genotype and diet characteristics on growth performance and carcass composition, Arch. Geflügelk 76, 197-202.

Koenig, M., Hahn, G., Damme, K. and Schmutz, M. (2012b). Untersuchungen zur Mastleistung und Schlachtkörperzusammensetzung von Stubenküken aus verschiedenen Legehybridherkünften, *Züchtungskunde.* 6, 511-522.

Krautwald-Junghanns, M.-E., Bartels, T., Cramer, K., Einspanier, A., Fischer, B., Förster, A., Galli, R., Koch, E., Meissner, S., Preisinger, R., Preusse, G., Steiner, G. and Weißmann, A. (2014). Lösungsansätze zu Vermeidung der Tötung männlicher Eintagsküken aus Legehennenlinien, *Dtsch. Tierärzteblatt.* 9, 1228-1232.

Krautwald-Junghanns, M. E., Cramer, K., Fischer, B., Förster, A., Galli, R., Kremer, F., Mapesa, E. U., Meissner, S., Preisinger, R., Preusse, G., Schnabel, C., Steiner, G. and Bartels, T. (2018). Current approaches to avoid the culling of day-old male chicks in the layer industry, with special reference to spectroscopic methods, *Poult. Sci.* 97(3), 749-757.

Lambertz, C., Wuthijaree, K. and Gauly, M. (2018). Performance, behavior, and health of male broilers and laying hens of 2 dual-purpose chicken genotypes, *Poult. Sci.* 97(10), 3564-3576.

Nakamura, D., Tiersch, T. R., Douglass, M. and Chandler, R. W. (1990). Rapid identification of sex in birds by flow cytometry, *Cytogenet. Cell Genet.* 53(4), 201-205.

ORBEM (2022). Early stage sexing, Retrieved on 11 April 2022. Available at: http://www.orbem.ai/solutions.

Ort, J.-D. (2010). Zur Tötung unerwünschter neonater und juveniler Tiere, *NuR* 32(12), 853-861.

Phelps, P., Bhutada, A., Bryan, S., Chalker, A., Ferrell, B., Neuman, S., Ricks, C., Tran, H. and Butt, T. (2003). Automated identification of male layer chicks prior to hatch, *World Poult. Sci. J.* 59, 32-37.

PLANTegg (2021). *The PLANTegg Process*, Retrieved on 21 January 2022. Available at: https://www.plantegg.de/en/.

Preisinger, R. (2003). Sex determination in poultry - A primary breeder's view, *World Poult. Sci. J.* 59, 54-58.

Quansah, E. S., Urwin, N. A. R., Strappe, P. and Raidal, S. (2013). Progress towards generation of transgenic lines of chicken with a green fluorescent protein gene in the female specific (w) chromosome by sperm-mediated gene transfer, *Adv. Genet. Eng.* 2, 29.

Quentin, M., Bourin, M., Gautron, M. and Réhault-Godbert, S. (2022). *'L'ovosexage en Europe: Etat des lieux et perspectives éthiques et techniques', 14ᵉ Journées de la Recherche Avicole et Palmipèdes à Foie Gras*, 434-443.

Richard, L. (2020). Vorfahrt für gesexte Legehennen, *DGS Mag.* 31, 14-19.

Rösch, P. (2002). *Raman-spektroskopische Untersuchungen an Pflanzen und Mikroorganismen*, Dissertation, Julius-Maximilians-Universität Würzburg.

Schnepel, F.-M. (1980). Physikalische Methoden in der Chemie: Raman-Spektroskopie, *Chem. Unserer Zeit* 14(5), 158-167.

Schütz, K., Mergenthaler, M. and Wittmann, M. (2018). Marktpotential für Geflügelprodukte aus Hahnenfleisch von Lege- und Zweinutzungshybriden. *Fachhochschule Südwestfalen, Forschungsberichte des Fachbereichs Agrarwirtschaft Soest Nr. 45.* ISBN (print), 978-973-940956-73-6.

Seleggt (2022). The Seleggt process, Retrieved on 26 February 2022. Available at: https://www.seleggt.com/seleggt-process/.

Siekmann, L., Meier-Dinkel, L., Janisch, S., Altmann, B., Kaltwasser, C., Sürie, C. and Krischek, C. (2018). Carcass quality, meat quality and sensory properties of the dual-purpose chicken Lohmann Dual, *Foods* 7(10), 156.

Smyth, J. R. (1990). Genetics of plumage, skin and eye pigmentation in chickens. In: Crawford, R. D. (Ed.), *Poultry Breeding and Genetics*. Amsterdam: Elsevier, 109-167.

Somes, R. G. (1990). Mutations and major variants of plumage and skin in chickens. In: Crawford, R. D. (Ed.), *Poultry Breeding and Genetics*. Amsterdam: Elsevier, 169-208.

Steiner, G. and Koch, E. (2009). Trends in Fourier transform infrared spectroscopic imaging, *Anal. Bioanal. Chem.* 394(3), 671-678.

Steiner, G., Bartels, T., Stelling, A., Krautwald-Junghanns, M. E., Fuhrmann, H., Sablinskas, V. and Koch, E. (2011). Gender determination of fertilized unincubated chicken eggs by infrared spectroscopic imaging, *Anal. Bioanal. Chem.* 400(9), 2775-2782.

Tiersch, T. R. (2003). Identification of sex in chickens by flow cytometry, *Worlds Poult. Sci. J.* 59, 25-32.

TierSchG. (2006). Tierschutzgesetz in der Fassung der Bekanntmachung vom 18. Mai 2006 (BGBl. I S. 1206, 1313), das zuletzt durch Artikel 2 Absatz 20 des Gesetzes vom 20. Dezember 2022 (BGBl. I S. 2752) geändert worden ist.

Tran, H. T., Ferrell, W. and Butt, T. R. (2010). An estrogen sensor for poultry sex sorting, *J. Anim. Sci.* 88(4), 1358-1364.

Webster, B., Hayes, W. and Pike, T. W. (2015). Avian egg odour encodes information on embryo sex, fertility and development, *PLoS ONE* 10(1), e0116345.

Weissmann, A., Reitemeier, S., Hahn, A., Gottschalk, J. and Einspanier, A. (2013). Sexing domestic chicken before hatch: A new method for in ovo gender identification, *Theriogenology* 80(3), 199-205.

Weissmann, A., Förster, A., Gottschalk, J., Reitemeier, S., Krautwald-Junghanns, M.-E., Preisinger, R. and Einspanier, A. (2014). In ovo-gender identification in laying hen hybrids: Effects on hatching and production performance, *Eur. Poult. Sci.* 78, 199-205.

Whittaker, D. J., Soini, H. A., Gerlach, N. M., Posto, A. L., Novotny, M. V. and Ketterson, E. D. (2011). Role of testosterone in stimulating seasonal changes in a potential avian chemosignal, *J. Chem. Ecol.* 37(12), 1349-1357.

Xiang, X. L., Hu, G., Jin, Y. G., Jin, G. F. and Ma, M. H. (2022a). Nondestructive characterization gender of chicken eggs by odor using SPME/GC-MS coupled with chemometrics, *Poult. Sci.* 101(3), 101619.

Xiang, X., Yu, Z., Liu, Y., Huang, Y., Wang, J., Chen, L. and Ma, M. (2022b). Differential proteomics between unhatched male and female egg yolks reveal the molecular mechanisms of sex-allocation and sex-determination in chicken, *Poult. Sci.* 101(7), 101906.

Yilmaz-Dikmen, B. and Dikmen, S. (2013). A morphometric method of sexing white layer eggs, *Rev. Bras. Cienc Avic* 15(3), 203-210.

Zootecnica. (2021). European hatcheries are introducing CHEGGY, Retrieved on 4 April 2022. Available at: https://zootecnicainternational.com/news/european-hatcheries-are-introducing-cheggy-sex-determination-in-the-egg-is-ready-for-high-volume-applica-tion/.

Zumbrink, L., Förster, A. and Hurlin, J. (2022). *Anesthésie électrique d'embryons de poussins males au deuxième tiers de l'incubation dans le respect du bien-être animal, 14e Journées de la Recherche Avicole et Palmipèdes à Foie Gras*, 449-453.

Zumbrink, L., Brenig, B., Förster, A., Hurlin, J. and von Wenzlawowicz, M. (2020). Electrical anaesthesia of male chicken embryos in the second third of the incubation period in compliance with animal welfare, *Eur. Poult. Sci.* 84. DOI: 10.1399/eps.2020.315.

Chapter 8

In ovo vaccination of chicken eggs

Christopher J. Williams, Consultant, USA

1 Introduction

The vaccination method to safely present a live antigen in an efficacious dosage to the developing chicken embryo advanced from the research laboratories of the US Department of Agriculture in the 1980s to the commercial poultry industry during the early 1990s with the development of the Embrex®[1] Inovoject® egg injection system. Following the commercial inception in 1993, the application of vaccines in ovo has grown to encompass over 35 different vaccine types (see Table 1) and an application scope including over 95% of North American broiler chickens. Additionally, commercial in ovo vaccination is in broiler production in more than 35 countries worldwide with an excess of 24 billion eggs vaccinated annually. The success of the equipment application as a veterinary medical process is due to many factors; however, two basic principles drive the technology. One, the mass application with individual dosage control during administration delivers a uniform flock vaccination in commercial production at an ergonomic and rapid rate. Two, the process serves to safely deliver to the broiler egg biologically active vaccines at the earliest possible time of embryonic development, limiting stress from the procedure and preparing the chick 'immunologically' for several disease challenges that may occur later in life, during the grow-out period. From these two basic principles, the application scope has continued to grow, expand and adapt

[1] Embrex®, Inovoject®, Bursaplex® and Magniplex® are registered trademarks of Zoetis, Inc., Parsippany, NJ, USA.

http://dx.doi.org/10.19103/AS.2022.0118.10

Table 1 USDA licensed in ovo vaccines by manufacturer as of December 2021

	Manufacturer				
Vaccine type	Boehringer Ingelheim	Ceva	Huvepharma	Merck	Zoetis
HVT	+	+			
SB1	+				
HVT SB1		+			
CVI988				++	+
HVT CVI988					+
HVT-IBD	+	+			+
HVT-ND	+	+		+	+
HVT-ILT		+		+	
HVT-IBD-ND	+	+		+	
HVT-IBD SB1	+	+		+	
HVT-ILT SB1				+	
HVT-ND-ILT				+	
HVT-IBD CVI988		+			
HVT-ND CVI988		+			
HVT-ND SB1		+		+	
HVT-IBD SB1 CVI988		+			
FP	+				
FP-ILT		+			
FP-ND	+				
IBD				+	
IBD plus AB	+				+
Coccidia			+		

to regional priorities of disease pressure and different hatchery environments across the global commercial poultry industry. Many factors continue to interact to both challenge and define the technology, and it is the goal of the author to describe and detail how these factors shaped and continue to evolve. Additionally, there are adjacent considerations that are present and will be discussed to fully understand the impact and scope of embryonic vaccination in the chicken egg.

2 History and evolution of in ovo vaccination

The inoculation of embryonated chicken eggs has been used for many decades in different evaluative and production capacities targeting the developing embryo as a 'bio-reactor' or growth medium for viral and bacterial replication.

Published by Burleigh Dodds Science Publishing Limited, 2023.

There has been a wide range of use in immunological and virology, as the embryonated egg serves as an ideal growth and support media for many viruses based upon the embryo itself as well as the extra-embryonic structures found within the developing chicken egg. For example, human influenza vaccines are produced commercially in embryonated chicken eggs (day 9-11 of incubation) via inoculation and subsequent growth/collection of vaccine virus. Many vaccines utilized in animal health are grown in embryonic tissue in vivo as well as chicken embryo fibroblasts (CEFs) in culture. Virus isolation and pathogenicity determinations utilizing viral and bacterial applications to the developing embryo have long been a standard practice in disease control and epidemiology. In addition to these 'diagnostic' uses of egg injection, embryonic gender determination has been accomplished via sampling of embryonic waste and/or tissue late in development.

Vaccination of the chicken egg as a method of disease control is a different perspective, where the integrity of the vaccine and the safety of the embryo are critical to the application technology to ensure the efficacy of the vaccine against subsequent disease challenge. We shall review the application and understandings with respect to both safety and efficacy, but also the challenges of maintaining these two important factors as part of a veterinary medical device process. For additional information on inoculation of eggs, an excellent overview and inoculation site studies via compartmental delivery have been provided by Manders et al. (2021). Another perspective of use of in ovo delivery to improve poultry health has been investigated by many in academia via injection of nutrients, competitive exclusion bacteria and other select compounds. While the injection of nutrients and other entities are not available or successful in commercial use, a good review of egg injection and in ovo applications of various products in poultry was completed by E. D. Peebles (2018), and a review by P. Ferket is included in a separate chapter of this publication. What differs from all the described in ovo applications and that of the information presented herein is the focus of this review specifically upon egg injection as an application method for vaccination of the chicken embryo.

The in ovo vaccination technology can be traced back to discoveries in the laboratories of the US Department of Agriculture in the early 1980s. The initial work encompassed in ovo application of the herpes virus of turkeys (HVT) as a live viral antigen given to late development stage chicken embryos to protect the hatched chick against subsequent Marek's disease challenge. (Sharma and Burmester, 1982; Sharma and Graham, 1982). The discovery initiated a group focus by Marek's disease researchers to understand key scientific and immunological properties of embryonic vaccination in poultry for protection against the disease (Sharma and Witter, 1983; Sharma et al., 1984; Witter and Lee, 1984). While many research groups have continued to develop the scientific 'library of knowledge' and expand the vaccine platform to other disease challenges regarding in ovo vaccination, the possibility of the transfer of this technology to the poultry industry spawned the formation of a start-up company focused solely on bringing new technologies to the poultry industry (Embrex®). The foundation and focus of the company were secured in 1985 via a long-term sole-lease agreement of the Sharma/Burmester patent (US Patent #4,458,630, 10 July 1984; Disease control in avian species by embryonal vaccination) between the company and the US government. Extensive research and collaborations between the company and poultry integrators in the late 1980s and early 1990s were completed and involved several different prototypical egg injection devices, extensive laboratory evaluations and a multitude of field-based trials before commercial success was made possible (Miles et. al., 1992; Gildersleeve, et. al., 1993).

Today successful in ovo vaccination involves many options for disease prevention with over 35 different licensed vaccine products (USDA, Table 1). Academic and pharmacological research continues to expand the family of in ovo approved vaccines as the technology represents a robust commercial methodology with accurate and safe access to each bird for individual dosage. The challenging environments and working conditions of the commercial hatchery have demanded robust equipment and supportive processes as well as a greater focus on sanitation and environmental control at the hatchery and breeder farm level. The focus of quality control has improved the overall conditions and reduced the microbial challenges at facilities successfully incorporating the technology. Additionally, more and more reliance upon process automation has become a part of standard operations of the veterinary medical devices used for in ovo vaccination as securing skilled and qualified labor continues to be a challenge globally in agriculture and food animal production.

3 Interacting factors of in ovo vaccination success

3.1 Device adaptation to commercial egg incubation: five critical points

A basic understanding of the commercial incubation process complexity is paramount to the comprehension of the challenges and associated variables of applying vaccines into the developing embryos as part of their everyday routines. A short overview of commercial incubation is provided herein; however, further details can be found in poultry management textbooks and hatchery management guidelines, including *Poultry Meat and Egg Production* by Parkhurst and Mountney (1995) or similar.

Hatching eggs, unlike the more familiar table eggs for human consumption, are produced on breeder farms and are by design mostly fertile eggs. The farms typically have 2–6 houses or barns of hens and roosters with 8000 to 12 000 birds in each house with approximately 8–10 hens per rooster. These birds have been selected over many generations to produce offspring that are grown for meat production. These birds grow and fulfill market protein needs as to size- and product-specific targets of chicken meat for consumption. Specifically, the target weight and size may define the required breed cross, and these target markets fall into three general categories: small (4.0 lb or 2.5 kg), medium (6.0 lb or 3.75 kg) and large (8.0 lb or 5.0 kg). The hens supply fertile eggs to the hatchery for approximately 40 weeks of production (age 25–65 weeks), and each house will produce approximately 1.5 million eggs over the 40-week production cycle. So, if we imagine a hatchery that incubates one million eggs per week, breeder farm support will require approximately 27–30 houses or barns of laying hens in production at any given time, each averaging 37 500–40 000 hatching eggs per week supplying the hatchery. The integrated production would also include a continuous growth cycle of young hens (pullets) and roosters to supply the ongoing needs of egg production as the older birds go out of production. The young pullets and roosters are typically placed every 2–4 weeks and are grown under strict guidelines and target weights in preparation for egg production at 25 weeks of age. This large-capacity production environment is typical to commercial broiler meat production globally with hatcheries producing 0.5–4 million chicks per week from single facilities. On the average, broiler hatcheries produce between 1.0 million and 1.8 million birds per week with the largest complexes producing close to 4 million broilers per week. In the United States alone, approximately 200 commercial hatcheries produce 9 billion broiler chickens annually. It is important to realize the multitude of variables presented by the large scope of farms, breeds, hatcheries and environmental conditions.

Understanding the production scope of commercial hatcheries is the beginning of the understanding of how the embryonic vaccination can be applied in the industry. We must consider the volume and hatchery processes on a large scale to understand how the ergonomics and production must match the science and biology of vaccination in the egg. Key to hatchery operations is the production of a consistent supply of quality hatching eggs. After the eggs are laid at the farm, they are graded as part of collection process from nests, typically three to four times daily. The acceptable hatching eggs are placed individually with pointed or small end down in plastic incubation egg trays or flats with 25–150 eggs per tray (incubator type specific) and then placed in cool storage on the farm. Hatching eggs require cool temperature storage after being laid prior to incubation. The storage period enables accumulation of eggs from many sources and therefore can provide routine scheduling of a large volume of eggs to set (or begin incubation) at one time. Eggs are typically set four times per week (Monday, Tuesday, Thursday, Friday). The hatching eggs are usually transported to the hatchery from the farms two or three times weekly. The amount of storage time prior to set can affect incubation length, and this represents just one example of the many variables that impact incubation and normal population characteristics of embryonic development in the production environment.

The incubator egg tray design conforms to the type and manufacture source of the incubation equipment found at the hatchery. Currently, there are over 25 different incubation egg tray designs in use globally, and the tray types are supported by different incubation companies with approximately 20–25 different incubator designs. Incubator designs fall into two general categories: multi-stage or single-stage incubation. Multi-stage incubators have multiple aged embryos at various stages of development found in one machine. The eggs are set in a coordinated repeating schedule to utilize embryonic heat of older embryos and therefore run more efficiently than single-stage or single-age incubators. Both types of incubators have specific beneficial traits; however, single-stage incubation provides better biosecurity and more targeted programable control for the interior environments of the machines (specific temperature and ventilation profiles are different for eggs from different hen ages, storage time, breeds, etc.).

Regardless of incubator type, all chicken eggs require 21 days of incubation. Eggs remain for 18–19 days in the incubator flat (in the incubator or setters) and then subsequently are transferred to hatching baskets where they spend the last 2–3 days in specialized hatching compartments or hatchers. The transfer process enables the division of incubation (clean) and hatch (dirty). Transfer of eggs out of the incubator flats into the hatch basket maintains a relatively clean and stable environment (intact eggs) for incubation. This is also important as transfer to the hatching cabinets provides an enclosed controlled environment for the final challenges of production including disinfection and rigorous cleaning of equipment after hatch (shell debris, hatch debris, unhatched egg waste, etc.). If we consider the calendar of events, eggs set on a Monday will hatch 3 weeks later, also on a Monday. They require removal/transfer on Friday or Saturday preceding the hatch. After 18/19 days of incubation in the egg tray, the eggs are removed from the incubator and 'transferred' into a hatching basket for final development and hatching. The incubator trays are then cleaned and returned to the breeder farm for use again. The 3-week incubation cycle allows for a repeating program of egg set, egg transfer, incubator cleaning, chick hatch, hatcher cleaning, etc. for each hatch day as well as 'recycling' of equipment such as the hatch basket, hatcher compartment, incubation flat, etc. The opportunity presented by the removal of the eggs from incubation and transfer out of the incubation tray into the hatch basket and hatch compartments is the moment that egg vaccination for disease prevention takes place.

The window of opportunity for egg vaccination exists at day 18–19 of incubation in normal production, but exact timing for optimal performance can be incubation equipment-type dependent. Specific challenges and critical factors are present that can enable or effectively destroy the vaccination process. We must consider that many factors are interactive and revolve around the device interface with the developing embryo in the egg. The factors fall into five general categories: injection device location and positioning on the egg, accurate shell penetration, site of vaccine delivery, sanitation, and vaccine specifications, including handling, safety and efficacy. Each of these five critical factors have unique hurdles to address, including different flat/incubator type as it relates to injection device design, embryonic developmental stages (injection timing) that affect vaccine delivery site, embryo size and safety, population dynamics (multi-stage vs single-stage incubation, relative 'window' of development), hatchery environmental and injection device sanitation and the differences in egg size and microbial challenges due to breeder hen age and management.

3.2 Device and eggshell interface

Due to the somewhat fragile nature of the eggshell, egg handling for injection has unique challenges. In design of a device to inject an egg, several important functions are quickly realized. Conveyance of the incubator egg tray must be gentle and uniquely adapted to the specific tray characteristics. Some incubation trays have a 'flat' bottom edge, while others are supported by internal posts that act as 'feet' to raise or suspend the egg as part of the characteristic features of the tray for that incubator type. Some incubator trays have a surrounding rigid 'frame' that suspend the grid of plastic that supports the egg off the horizontal surface below the tray, exposing the small end or point of the egg. All the different tray configurations require uniquely designed conveyors to move the eggs and allow for sequential processes of loading, injection and transfer/unloading of hatching baskets and empty trays. It is important to realize that every time the egg is handled, some will inevitably be broken. Acceptance levels for cracked eggs in any commercial production process is very low, with a general target of 0.25–0.50% as a maximum total loss. The less the start/stop motions during incubator tray conveyance, the lower the damage risk to the eggshell. It is also important to consider the differences in shell strength and size as the hen goes through her 40 weeks of production. In general, the smaller eggs produced by younger flocks have stronger thicker shells, and as the hen ages and the eggs increase in size, the shell becomes thinner and more fragile. This is important to consider as the incubation tray remains the same, and the egg will sit differently due to the changing egg size. Eggs from younger hens (smaller) tend to roll or tip off center more often, especially with abrupt start/stop motions. Eggs from older breeder hens are larger in both length and circumference, and therefore egg breakage can occur more readily on both the bottom of the shell at the tray interface and along the sides due to egg-to-egg contact.

The device that injects the egg must be able to locate correctly and repeatably upon the blunt or large end of the egg as it sits upon presentation in the incubation tray. While at first consideration, this action seems simple; however, characteristics such as of angle of needle entry, different egg height, egg orientation (leaning), tray rigidity and surface friction dynamics (shell, device) make egg location by the injection device a very critical 'first step'. Without proper egg location by the injection device, the vaccination process is compromised. Acceptable levels of 'miss-location' as a quality control point should be extremely low in design and application (0.003%). Desired angle of entry for shell piercing is perpendicular to the shell surface with the egg presented as it has been incubated and nestled in the incubation tray.

3.3 Embryo orientation and late-stage embryo development

We must consider some of the characteristics and physiology of the developing embryo to better understand the angle of entry and therefore the target delivery site for in ovo vaccination. The embryo is orienting for hatch at day 18/19 by positioning its head under the right wing with the upper body positioned toward the blunt end of the egg. The lower body is crouching in the small end of the egg with the feet positioned toward the sides of the egg (Fig. 1). The embryonic positioning for hatch begins on day 17 of incubation

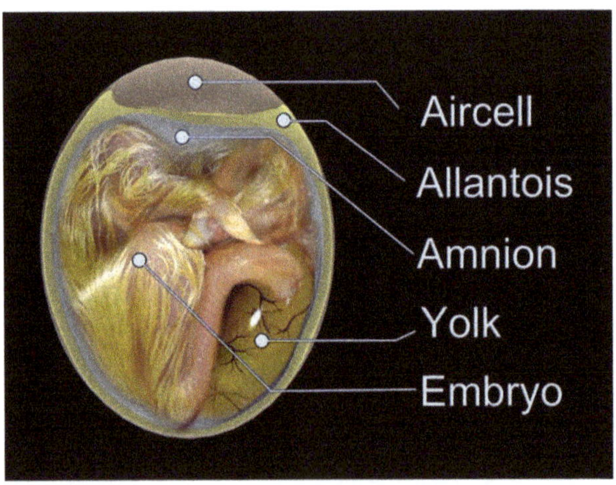

Figure 1 Embryo compartments and positioning for hatch at day 18 of incubation.

Published by Burleigh Dodds Science Publishing Limited, 2023.

and is influenced by gravity, as eggs set upside down will complete similar orientation; however, the egg itself is misaligned and positioned small end up. Thus, the head will be found in the small end of the egg (embryo malposition). These 'upside down' embryos will not have developed normally due to many reasons primarily associated with compartmental management of nutrients and waste. Many upside down 'malpositioned' embryos do not hatch or, if they hatch, are of poor quality.

It is important to understand the developmental stage and progression during the day 18/19 time frame. Chronological time of development and developmental stage are not necessarily the same between incubation type, breed type, hen age and storage time. Small differences in average heat during incubation (hot spots or cold spots) and egg storage time before incubation can significantly affect embryonic growth rate. This will be discussed further in later section regarding vaccine delivery site in ovo. It is important to consider that the earliest moment for in ovo vaccination is when the embryo is in the normal position for hatching with the head under the right wing and the stalk of yolk has begun to ascend into the abdomen. Intestinal loops should not be visible in the yolk and the yolk has a bi-lobed appearance upon examination. The latest moment of development for successful in ovo vaccination is just prior to external pip. The corresponding chronological time in incubation can be as broad as from day 17 plus 14–16 h to day 19 plus 2–4 h of the normal 21 day (504 h) of incubation.

Figure 1 shows the five major compartments in the egg at day 18 of embryonic development. It is important to understand that the internal structures are changing as the embryo develops. The amnion and allantois will be disappearing, the yolk sac will be pulled into the body cavity and the embryo's body will be growing significantly larger. The air cell chamber grows slightly larger from day 17.5 to day 18.5 as the amnion and allantois regress. Then the chamber becomes slightly smaller due to the embryo's final growth surge just prior to hatch. The 'pip-line' becomes evident as the chick breaks through the shell during hatch. It is defined by the bottom edge of the air cell chamber, where the inner shell/outer shell membrane separate. Here we find a 'weakening' of the shell. Calcium has been removed from the shell from below the base of the air cell membrane by the chorioallantoic (CAM) blood supply. The calcium from the shell is required for bone growth. The shell area above the air cell membrane retains the original calcium level deposited by the hen, thus forming a thick/thin demarcation forming the 'pip-line' around the egg. The air cell structure in the blunt end of the egg also serves as an important 'staging' area and function to the hatching chick. Prior to externally pipping the shell, the embryo's beak breaks through or 'pips' internally through the inner shell membrane early on day 19 of incubation and initiates pneumatic respiration. Once the lungs and air sacs are clear of fluids and functioning (typically 6–18 h) and the yolk sac is withdrawn into the abdomen, the embryo will externally pip through the shell and begin the final stages of hatch.

Basic knowledge of embryo positioning at injection time is important as changes in egg orientation from an exact vertical stance will affect where the embryo eventually resides inside the egg. An egg tilted one direction will result in embryo positioning with the head pointing away from the angle of tilt. To vaccinate correctly, the site of vaccine placement during injection must target the embryo body or amnion (more detail in subsequent text). The 'target' inside the egg is the area identified by the small bullet point showing 'Amnion' in Fig. 1. This is a central location in upper portion of the blunt end of the egg and can be visualized as a small pocket or cup-shaped area formed by the curled head and neck as it rests beneath the right wing. When the egg is vertically tilted, the angle of entry needs to be opposite the angle of tilt to reach the central 'pocket' or target point. The device location must therefore rotate axis (not just straight down) and create an angle entry toward the direction of the egg tilt while maintaining a 'flat' adjusted approach, directing shell punch and needle trajectory perpendicular to the shell surface. Without rotational freedom and translational movement to locate on the eggshell surface, the angle of entry may not allow the needle to permeate the shell (deflect) at all or may incorrectly direct the shell punch and injection needle toward an improper site such as the head of the embryo. So not only does the device require the freedom of movement to physically locate correctly on the egg, the material and design of the device interface (locator) should permit a 'sliding' or free movement along the shell surface. Light dense polymers of plastics (such as Hydex®[2], Rynite®[3], Ertalyte®[4], Delrin® or polyethylene terephthalate [PET]) provide superior characteristics and performance as part of material and design of egg locator fixtures of injection devices.

3.4 Shell penetration and the injection needle

Force of pressure required to cleanly permeate the shell with a needle stylus (shell punch) depend primarily upon speed (acceleration) and needle tip (punch) design. The greater the speed, the lower the required pressure. In current designs utilizing needle of 16 gauge or 18 gauge, the force to permeate an average individual shell average between 3 and 4 pounds of pressure (1.4–1.8 kg) at a speed of 8–14 in./s. The punch tip itself should be designed to repeatably produce entry (eggshell punch) without radiating cracks at entry point in the shell. It is also important to consider the possible transfer of energy that may break the shell in another area such as the small end that is supported

[2] Hydex® is a registered trademark of AL Hyde Company (Ensinger), Grenlock, NJ, USA.
[3] Rynite® and Delrin® are registered trademarks of DuPont de Nemours, Inc., Wilmington, DE, USA.
[4] Ertalyte® is a registered trademark of Quadrant EPP Europe N.V., New York, NY, USA.

by the tray or the sides that are in contact with adjacent eggs. Due to the inconsistencies in egg support as provided by the incubation tray, the injection device may require an individualized egg support mechanism that lifts the egg slightly out of the incubation tray (from below). The injection device also requires a mechanism to secure positioning for the entire injection sequence once the device is located upon the eggshell surface. It is important to note that during egg support there should be minimal change in orientation of the egg along its vertical axis. The goal during egg support for injection is to keep the original angle of entry secure and minimize 'rolling' of the egg. Robustness of the needle tip should also be considered as the process requires repeated entry in production flow with minimal need for replacement. Custom tip design of straight cut 30° bevel has shown superior performance when compared to standard B bevel (subcutaneous injection needle, 28° tip) as measured over time with minimal burring or bending. As the punch stylus must also undergo repeated sanitation (between injections), the material of choice is high-quality 316 stainless steel.

The perspective of egg injection for embryonic vaccination as a two-step process (permeate shell, insert needle to inject vaccine to the embryo/amnion) lends itself to the design of a 'needle inside a needle'. Separate needles allow for separate process priorities for each operation as well as specific needle tip design. Shell penetration exposes the entry needle to both a physical (shell) and microbiological (bacteria/fungus) challenge. Fecal organisms found within the shell can be transferred more easily and directly to the embryo or amnion if the same stylus or needle is utilized for both shell entry and injection. A 'needle inside a needle' approach to egg injection for embryo vaccination significantly reduces the possibility that the needle serves as a vector of transmission and creates a microbial challenge at vaccine delivery point (the embryo or amnion). A short stroke of the outer needle or eggshell punch is possible, and the larger-shell punch needle stops before touching the embryo or chorioallantoic membrane (CAM). The air cell chamber is present directly below entry site on the blunt end of the egg, and the shell punch entry is only into this space. The shell punch needle should enter the egg 0.635 cm (0.25 in.) as measured from the shell surface. Injection needle penetration should target 2.54 cm or 1.0 in. injection depth. The segregated two-step action separates the required force and speed to permeate the shell from those required for a softer slower needle penetration of the CAM and embryo/amnion. The needle for vaccine delivery can also be designed specifically for that purpose (injection proper) and be of small diameter to minimize trauma to the embryo. A smaller needle and small needle tip design will additionally minimize needle lumen interface and exposure of the vaccine to microbial challenge and sanitation fluids as well as improves sanitation efficacy by limiting total surface area to disinfect. The injection needle should have a minimal effect on the embryo and support structures such as the chorioallantoic blood supply, the inner shell membranes and allantoic and amniotic membranes. The needle should be a small size (gauge) and the correct tip design for piercing the membranes or the embryo. Additionally, the needle should be large enough to deliver the vaccine without causing physical damage to the vaccine. Consideration must be given to the total surface area of the needle that is exposed to the internal contents of the egg and therefore the ease of sanitation. The surface area created by large needle sizes, extended canula openings or side port openings restrict the ability to sanitize the needle. An additional benefit to a small-tip angle is a more precise delivery point. Comparisons between a customized 20 gauge 45° straight cut needle tip and standard 28° B bevel (subcutaneous) tip show both a more consistent delivery site and a greater sanitation efficacy for the customized tip design. The 45° needle tip also minimizes membrane and embryo trauma in ovo as compared to the 28° needle tip, presumably due to cutting action presented by the design for subcutaneous injection.

3.5 Vaccine specifications and delivery in ovo

Vaccines licensed for use in ovo include several cell-associated vaccines comprised of CEFs infected with herpes viruses for Marek's disease (Table 1). These vaccines have been shown to be safe for the developing embryo as well as efficacious against subsequent disease challenge later in life. The vaccines include not only all three serotypes of Marek's disease but also several recombinant vaccines utilizing the serotype 3 herpes virus of turkeys (HVT) as the vector to express immunogenic proteins of other diseases (Newcastle's disease, infectious laryngotracheitis, infectious bursal disease). The Marek's disease vaccines are cell cultures (CEF) with approximately 20% of the cultured cells infected with the herpes virus. The vaccine is stored frozen prior to use in sealed ampules submerged in liquid nitrogen. Once thawed, the vaccine is diluted in special nutritive diluent that not only supports the viability of the cell culture but also provides a pH-sensitive medium (color change) to visually ensure osmotic stability of the vaccine mixture. The mixed vaccine has a moderate cell concentration rarely exceeding ~1.0×10^7 per milliliter of total cells, of which ~2.0×10^6 are infected with vaccine virus. In contrast, human red blood cell concentration is approximately 5.3×10^7 per milliliter. As the Marek's disease-based vaccines are a live cell culture, and the cells are somewhat swollen due to the viral infection, dispensing or pumping of the mixture must be gentle and cause little or no further damage. Successful designs include peristaltic and diaphragm-type pumping systems that simulate kidney dialysis or heart bypass pumps that handle blood. It is also important to consider the required time and pressure created by the pump to complete the injection process if the needle penetrates the embryo proper (versus delivery into the fluids of the amniotic sac). Dispense timing or time required for fluid pumping to initiate and complete dispense out of the needle tip must be coordinated with device programing such that needles are in the egg correctly before dispensing initiates. Similarly, coordination must occur such that needle retraction happens after dispensing is complete. The displacement-type pumping presented by peristaltic devices easily meets pressure requirements for placement of vaccine into embryonic tissue; however, diaphragm pumps and their control mechanisms require unique pressure assurance by design. Normally the diaphragm is driven by a spring-load electronic solenoid valve. The strength of the spring, and therefore the voltage required to open and set the spring tension, is paramount to the ability of the pump to dispense into tissue and against a 'head pressure'. These pressures and therefore the whole pumping mechanism design require thorough intralumenal pressure analysis and testing to ensure no cavitation (formation of bubbles) occurs during fluid movement to remove any risk of damage to the vaccine. Other vaccines licensed for use in ovo include live viral antigens that are not cell-associated (fowl pox, fowl pox recombinants [infectious laryngotracheitis, Newcastle's disease], infectious bursal disease), antigen-antibody complex viruses (IgG plus infectious bursal disease) and protozoal vaccines of *Eimeria* (coccidia) oocysts of combined species (*E. maxima*, *E. acervulina*, *E. tenella*). While not as sensitive to pressure and handling, the live viral and protozoal antigens also require safe and aseptic handling during preparation and administration. It is important to consider the diluent not only as a supportive media, but also the key constituent that must be routinely removed and cleaned effectively from the system to prevent inadvertent contamination.

Important features of the vaccine delivery systems include smooth polished bores and manifolding, high-density plastics (such as Hydex®, Ertalyte® Rynite®, Delrin®, PET), bottom-sealing barb-type fittings (for tubing connections), pharmaceutical grade tubing and robust, accurate, low maintenance operation. Equally important to the design of the pumping system are the material and internal surface characteristics to enhance and support clean-in-place sterilization. Commercial application of vaccine in ovo in a practical sense requires that the system remain closed by design to maintain a sterile internal environment in storage and impervious to environmental contamination during use. Key components of the vaccine delivery system need to remain in place and subjected to 'clean-in-place' sanitation on a daily use basis. Disposable entities such as tubing and needles must also maintain a relatively long duration of use and be easily replaced during planned maintenance during off use period. In considerations for programs of cleaning and sanitation, two key entities exist. One entity is the complete cleaning and disinfection of the vaccine pumping system (before and after use) and the other is the sanitation required between injection sequencing (from egg to egg) as part of the in ovo vaccination process.

3.6 Sanitation for in ovo vaccination

It is imperative that the process of egg to egg or needle/punch sanitation between injections be robust and effectively sterilize the key delivery components of the injection device that interface with each egg/embryo. The sanitation process requires a very short time frame, and the disinfecting agent must have a broad microbial kill spectrum. Successful systems utilize a high concentration of chlorine with buffered pH to maintain a high level of hypochlorous acid. In direct 'contaminated egg to live egg' injection studies (bacterial challenge modeling), needle/punch disinfection with 2500–5000 ppm chlorine (with pH of 7.0–8.0) utilizing a ~6 s injection interval can successfully address a bacterial challenge of $1 \times 10^{5-7}$ CFU per milliliter of egg content. The cycle time requirement (~6 s) between injection is the primary restriction of the process and limits the possible candidates for needle/punch disinfection. Kill time for many disinfectants with or without additional organic challenge normally takes minutes, much less seconds. Another important aspect is addressing and minimizing organic buildup. Without addressing the entire lower portion of the injection device, calcium 'dust' or organic debris from 'exploding' rotten eggs created or liberated during egg punch can build up and deposit/contaminate the exterior of the punch and inner areas of the location device. Targeted dispense of the needle sanitation fluid to only the injection needle is not adequate. The sanitation process should permit cyclic rinsing and cleansing/disinfection of these adjacent areas. Small holes in the upper regions of the punch within the needle sanitation fluid pathway or cyclic rinsing of the complete lower portions of the devices can accomplish this additional sanitation requirement.

The cleaning and sterilization of the vaccine delivery system of in ovo vaccination devices present a unique challenge (Denoya et al., 2010). When clean and sterile, the vaccine system cannot contain any residual chemicals or biofilm formation which may negatively affect the vaccine. The cleaning process must focus detergent activity upon the removal of primarily carbohydrates (sugars) that make up the diluent but must also address certain levels of proteins (albumen, chick embryo fibroblasts) and inadvertent contaminants (bacteria, fungi, virus) that may have been introduced by mistake. The disinfection steps cannot utilize any residual chemicals, and storage of the equipment must consider probable microbial challenges in the harsh environment of the hatchery. Catastrophic results have been observed when pathogenic bacteria or viruses have been inadvertently introduced into vaccine or vaccine pathway utilized for in ovo application. It is important to consider that these accidents, while rare, do occur. The vaccine clean-in-place system must be designed and be able to overcome the mistake during every cleaning cycle. A vaccine contamination is not apparent until 2-3 days after application, so the clean-in-place system must completely sterilize the vaccine delivery system before and after every use. Any level of biological or chemical 'carryover' from day-to-day production process is not acceptable for successful in ovo veterinary medical devices used for vaccination of embryos. Chemicals typically utilized by clean-in-place in ovo vaccination devices are chlorine-based detergents and sanitizers that leave no residual chemical. Volumes and exposure times must be adequate to remove not only the residual vaccine and diluent but also all chemicals used in the cleaning cycles. Different pumping systems require different cleaning programs, as the pumps themselves may be included in the vaccine pathway or the pump may be mechanically separated (peristaltic). Regardless of pumping type for vaccine dispense through the needles, clean-in-place systems require control by a separate dedicated pump system to adequately address pressures, pathways and possible low flow areas of dispense systems. During each step of the cleaning cycle (each chemical), the solutions should replace the dead volume of the total vaccine pathway at least ten times. It is important to note that sterile saline is utilized to remove storage chemicals (isopropyl alcohol) just prior to the addition of vaccine. Dead volume of delivery system should be replaced at least ten times as target volume for sterile saline to retain sterile and chemical free pathway. Overnight and short-term storage should include system fill and fluid exposures to the entire internal surface via isopropyl alcohol. Isopropyl alcohol is preferred as it leaves no residual upon evaporation, is easily removed with sterile saline, scavenges free chlorine, has minimal effect on most plastics and stainless steel and presents an effective bacteriostatic hurdle within the clean interior of the delivery systems.

3.7 Hatchery hygiene and biosecurity

Additional sanitation requirements to support the vaccination process include heightened awareness and cleanliness of the hatchery, and more specifically the hatch baskets and hatcher compartments. The hole in the egg from vaccination procedure is not sealed and represents a possible entry point for microbial contamination after the process. This is most apparent with fungi, specifically *Aspergillus* species. Other areas of focus for fungal control in the hatchery include the large ventilation equipment (HVAC, heating units, evaporative cooling units) that supply air to the entire facility as well as the systems utilized to control pressure of the exhaust systems (plenums). Priority can be placed upon the ventilation for the hatcher and incubators (hallways), the egg storage area and the dirty exhaust from processing and hatcher compartments. The control of hatch dander or fluff in exhaust systems is a priority such that minimal recirculation to fresh air inlets is accomplished. Plenums and misting exhaust ducts are usually required to capture and collect the airborne contamination within the building for routine waste handling and cleaning. It is important to understand the risks of poor-quality hatching eggs being introduced into incubation and the production environment. These eggs represent a significant long-term negative influence and microbiological burden due to high probability and incidence of horizontal transmission of bacteria and fungus throughout the various environments of the hatchery. There exists a threshold of microbial challenge which can influence the quality of all production parameters, including in ovo vaccination.

Biosecurity of the hatchery work areas as well as product (egg/chick/equipment) flow through the hatcher is integrated into the hatchery ventilation systems and programs. Clean areas such as clean basket storage, egg storage and incubation hallways should maintain positive air pressure environments and designated dirty areas should consistently have negative pressures that supplement 'clean to dirty' biosecurity of the facility. Limited compromises of counterflow (dirty to clean) may happen as well as minimizing shared work areas function to reduce environmental risks of the hatchery operation. Additional attention to detail includes the water quality that supplies various functions in the hatchery. Humidification units are of primary importance as they can serve as dispersion source for high levels of bacteria through contaminated water supply. The use of steam humidification for humidity in air supplied to incubator and hatcher hallways greatly improves the quality of the humidity, it also serves to eliminate bacteria from the water source. For equipment that utilizes spray for cooling incubation and hatching compartments, it is recommended that the water be treated with reverse osmosis (RO) to greatly reduce the mineral content and to chlorinate the RO water prior to use to prevent subsequent contamination downstream of the treatment (water storage accumulation tank). It is recommended to evaluate main water supply and treat accordingly. In the hatchery, many end-use applications (water hoses, pressure washers, utility sink, laboratory faucet, etc.) are contaminated with environmental and fecal organisms. Routine chlorine 'shock' or drain/dry programs need to be enforced along with automatic pH and disinfectant application to incoming water supplies.

Qualitative and quantitative microbial sampling should be completed prior to the initiation of in ovo vaccination at any given facility (hatchery). Additionally, routine microbial monitoring should be completed (typically on a semiannual basis) to assess ongoing risks associated with quality of the air, water and hatchery environment. Table 2 is a summary of environmental assessments completed in commercial hatcheries during the late 1990s prior to the incorporation of in ovo vaccination. Approximately 20% of the hatcheries

Table 2 Percentage of 250 commercial hatcheries with *Aspergillus* contamination[a]

Level of *Aspergillus*[b]	Hatchers	Hatcher ventilation	Hatcher hall	General ventilation	Incubators (setters)	Miscellaneous[c]
	(%)					
None	63	56	46	68	45	53
Low	23	18	27	15	30	28
Moderate	6	10	8	7	10	10
High	8	16	19	10	15	9

[a] Data from Williams et al. (1994) and King (1995).
[b] Low levels = 1-3 colonies per plate; moderate levels = 4-10 colonies per plate; high levels = >10 colonies per plate.
[c] Egg room, vaccine prep, roof, etc.

evaluated required remediated cleaning and disinfection prior to incorporation of in ovo vaccination technology due to high level of fungal challenge found in the hatcher hall and ventilation systems.

3.8 Importance of compartmental delivery for in ovo vaccination

Detailed evaluations were conducted to better understand in ovo vaccine delivery site and subsequent efficacy to Marek's disease vaccination (Wakenell et al., 2002). The studies clearly showed preferred compartments or sites of vaccine delivery within the egg and subsequent efficacy against disease challenge. Marek's disease vaccine requires placement into the amnion or the embryo body during the in ovo vaccination process. Cell-associated vaccines placed into the waste compartment (allantois) or directly onto the air cell membrane showed very reduced or no efficacy against subsequent disease challenge. Research conducted by Jochemsen and Jeurissen (2002) investigated and determined that localization and uptake of substances injected in ovo and particularly to the amnion were characterized by the delivery site as well as the substance to determine its final localization. The amniotic delivery of particulate colloidal carbon and microspheres resulted in the substances entering the embryo by the mouth (imbibed) and ingested in the intestinal and respiratory tract. These substances reached the lungs via the trachea and the bronchi and were absorbed by the air capillaries. Substances were also absorbed by the bursa of Fabricius (retrograde absorption). Viral substances (infectious bursal disease vaccine) were also found in many different organs after delivery into the amnion.

In contrast to the amnion, the allantois supports embryonic development primarily as a waste reservoir and the very important function of water balance and control of water loss. The chorioallantoic blood supply supporting the allantois serves a respiratory function (CO_2, O_2) across the shell and air cell membrane, but also works in conjunction with water balance and waste compartmentalization. The resulting allantoic fluid is acidic and thus will negatively affect cell cultures such as Marek's disease vaccine. The allantoic fluid environment serves an important antimicrobial barrier as it is found along the exterior of the inner egg along the eggshell surface. Similarly, the air cell membrane is a significant physical barrier, and the cell-associate (MD) vaccine cannot cross directly into the vascular bed or access the amnion or embryo. Molecules with a size below ~10 000 Å and presented in liquid solutions can pass directly into the chorioallantoic blood supply when placed upon the air cell membrane (examples include many antibiotics). It is important to note that in the past, antibiotics were routinely incorporated into in ovo vaccination. While not necessarily required, the use of antibiotics with in ovo vaccination represented a very controlled and efficacious delivery route. In the last 4–5 years, we have seen the elimination of all antibiotics important to human medicine used in animal production, including the use in ovo. Work by Phelps (1995) showed compartmental absorption levels of gentamicin following delivery to the air cell, amnion, embryo and allantois (Fig. 2). These data highlight the differences and importance of delivery to the individual compartments found in the egg in late-stage embryonic development.

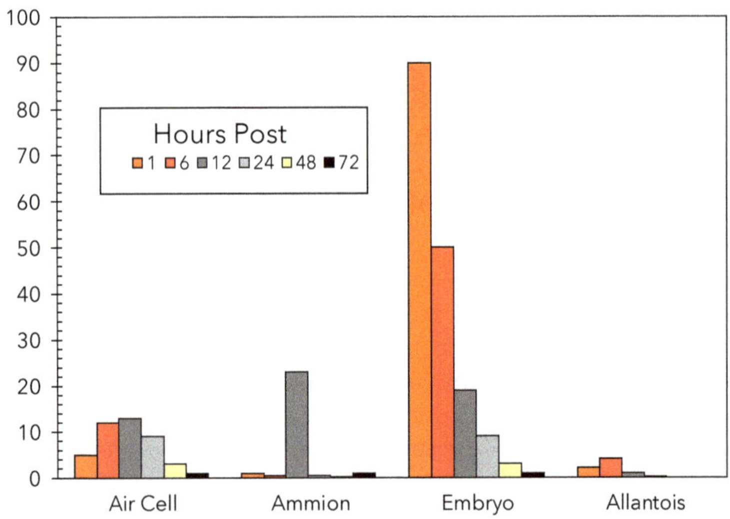

Figure 2 Plasma gentamicin levels (µg) following in ovo administration of 1 mg gentamicin into specific compartments of day 18 embryonated broiler eggs. Source: Phelps, P., 'The In Ovo Administration of Antibiotics into Broiler Eggs at Transfer', PhD dissertation, North Carolina State University, 1995.

Compartmental delivery of Marek's disease vaccine in ovo has been thoroughly evaluated; however, the application of other vaccine viruses to various compartments of the egg is not as well understood. Evaluations with infectious bursal disease vaccines have shown that while many strains are safe and licensed for in ovo use, certain intermediate plus and 'hot' strains of the vaccine virus are not safe when given in ovo unless combined into an antibody-antigen complex vaccine (Bursaplex®, Bursammune®[5], Magniplex®). Without the antibody-antigen complex, the more aggressive bursal disease vaccine viruses, when given in the egg, invade the bursal tissue and replicate to the degree that the tissue is functionally destroyed, severely affecting the humoral arm of immunity for the life of the animal. The bursal tissue of the embryo is still immature and maternal antibody levels are lower than those found at hatch. In addition, bursal-derived immunocompetent cell migration to secondary tissue (bone marrow, spleen) is not fully complete until post-hatch. We know immune development of chickens is not complete at hatch, but many of the necessary tools are in place for a protective immune response to certain vaccines on day 18 of incubation. During late-stage embryo development at injection/transfer (day 18/19), the bone marrow, spleen, bursa of Fabricius, liver and thymus are populated with functioning immune cells, although less than the full complement of T and B cells. B cells are still undergoing extensive growth and development in the bursa. T-cell migration from the thymus occurs in three waves, with the first two waves occurring prior to hatch and the third wave occurring during the first week after hatch. Destruction of the bursa prior to these migrations can be seen with certain IBD vaccines and viruses.

Similarly, certain fowl pox vaccines are safe and efficacious (licensed), but others can cause issues when given in ovo (Williams et al., 2010). The problems are exacerbated with the use of chick embryo origin pox viruses as well as in ovo vaccine delivery timing that is early on day 18 of incubation, presumably applying most of the vaccine into the amnion rather than the embryo body. Efficacy and safety studies show that dosage control (viral units) as well as virus type (origin, culture media) and injection timing should be considered in design and use of fowl pox and pigeon pox vaccine viruses for in ovo use (Avakian et al., 1999). Application should be limited to tissue culture origin pox viruses at log 2-3 concentration given late on day 18 or early day 19 of incubation. Avian reoviruses (tenosynovitis) have been shown to be efficacious against subsequent disease challenge; however, the vaccines are immunosuppressive when given in ovo to chickens with low or no maternal immunity (Guo et.al., 2004). The safety of the reovirus vaccines can be enhanced using an antibody

[5] Bursamune® is registered trademark of Boehringer Ingelheim, Ingelheim am Rhein, Germany.

complex; however, no commercial vaccines are available in this design or for use in ovo.

3.9 Immunity as a function of vaccine specifications

Without vaccination in ovo, the immune system of the chicken develops only after hatch (Lehtonen et al., 1989; Mast and Goddeeris, 1999; Schierman and Ordskog, 1961; Seto, 1990) and only transient T cell responsiveness occurs during the first few weeks of life (Lowenthal et al., 1994). Expression of MHC-1, activation of T cells and humoral immune responses are low or nonexistent at hatch and develop within the first weeks of life. The poor immune response is not necessarily due to the lack of T cells but rather due to the cells inability to complete the steps necessary for cell activation (cytokine secretion, receptor expression) (Lowenthal et al., 1994). More specifically, receptors that mediate interferon types 1 and 3 transcripts are delayed and develop gradually (Karpala et al., 2012). It is also important to consider the role of maternal antibodies during the critical first weeks of life, as they enable growth and maturity of immunocompetent organs and spread of immunocompetent cells to secondary sites within the chicken without overt compromise of the tissues due to viral or infective challenge. One of immunology's central tenets is the concept of neonatal tolerance. Immunization of neonates may result in clonal deletion with subsequent tolerance. It has been shown that animals in the prenatal state are able to mount a protective, lasting immune response to antigens or pathogens administered during neonatal life (Bot et al., 1997; Ridge et al., 1996; Sarzotti et al., 1996; Sharma, 1987a).

Work done at the College of Veterinary Medicine in Raleigh, NC (Gimeno et al., 2015a,b, 2018) showed that the vaccination of chicken embryos at day 18 of incubation with herpesvirus of turkeys (HVT) rendered chickens more immunocompetent at hatch as measured by an increase in activation of cell phenotypes (MHC-I+, CD3 plus MHC-II+) to levels comparable to non-vaccinates at 10–14 days of age. Additionally, the lymphocytes of the chickens vaccinated in ovo responded better to Con A in vitro as well as improved response against antigens unrelated to the HVT antigen (phyto-hemagglutinin subunit L (PHA-L) and Keyhole limpet hemocyanin (KLH)). It is important to note that administration of other Marek's disease vaccines in ovo (SB-1, CVI-988) did not have the same effect on cell phenotypes of the spleen (Gimeno and Cortes, 2011; Gimeno et al., 2015b). The mechanisms through which the herpesvirus of turkeys enhances and matures the immune competence of the chicken embryo are not completely understood; however, more recent work suggests that increased transcripts of interferon gamma (IFN-ɤ) and toll-like receptor 3 (TLR-3) in the lung and spleen may be the mechanism that accelerates immune maturation.

Viral replication after in ovo vaccination is evident with all live viruses whether delivered to the embryo or the amnion and viral replication contributes not only to the subsequent immune response but also to the safety of vaccine when delivered to the embryo. Differences exist with regard to viral replication within vaccine types and these differences are related to the pathogenesis of the individual strains. However, even with limited viral replication after vaccination in ovo, CVI-988 elicits a very good immune response, especially against highly virulent Marek's disease isolates (Zhang and Sharma, 2001). The type of virus, the concentration and the delivery site as a function of embryo compartmentalization and maturity contribute to replication rate but more importantly to both safety and efficacy of the vaccine and is unique for any given virus. Many vaccine viruses are not safe given in ovo, for example the respiratory viruses of Newcastle's disease, infectious bronchitis and infectious laryngotracheitis. Their replication rates and invasive pathogenesis cause embryo morbidity and mortality even with repeated passages and isolations designed to reduce virulence. As noted earlier, the safety of the specific virus utilized for in ovo vaccination is paramount and cannot be understated. HVT is unique in that it accelerates immune maturation, but also replicates at a high rate when administered to the embryo (Abdul-Carrem et al., 2008; Sharma, 1985, 1987b, 1989), it is safe at excessively high doses (25X) and it is safe and efficacious across a large delivery window of development (day 14 to day 19 of incubation). The use of HVT as the viral backbone for recombinant or vector-based vaccines presents a unique vaccine construct, as the HVT virus is a large virus suitable for gene insertion, the virus replicates at a high rate, the insertion does not significantly affect replication and expression, the protein expression resulting from the insertion are intact and exact, and most importantly from a regulatory and ethical stance, the virus is not readily shed from the bird in feather dander to the environment. While the vaccine virus can be found in the feather pulp of the developing feather of the young chicken after vaccination, the vaccine virus for HVT in both recombinant and non-recombinant form does not shed horizontally via follicular dander release. This characteristic of the virus in vivo improves the safe application and acceptance of these genetically modified viruses for vaccine use.

When we consider the overall enhancement of immune competence and earlier protection against Marek's disease afforded by in ovo vaccination of chickens, we can see why the vaccination process in commercial industry applications results in improvements to parameters such as feed conversion, mortality and production efficiencies (Gildersleeve et al., 1993; Sarma et al., 1995; Ricks et al., 1998; Williams and Villalobos-Chaves, 2000; Bruzual et al., 2005). Feed efficiencies during the growing period can be directly related to better immune status at any given time, but we should also consider the energy requirements utilized during maturation. 'In the egg' maturation of immune competence does not involve feed consumption; however, maturation and immunocompetence developed during the first weeks of life for birds vaccinated at hatch will have a feed cost variable.

3.10 Asepsis of vaccine mixing and handling

While strictly defined with standard procedures and basic in concept, mixing and handling of vaccine for in ovo use is challenged primarily by repeated execution and shear volume requirements. It is also a challenge due to the frozen glass ampule design for Marek's disease vaccines and the limits and challenges that the procedure to thaw and open the ampules present to aseptic technique. The glass ampules require rapid thaw process, typically in a temperature-controlled water bath. After thawing, the ampules are dried and the top is snapped or broken off, opening the contents to atmosphere. The concentrated vaccine is then removed from the ampule with a needle and syringe and transferred to an intravenous-type bag of diluent. It is important to note the utilization of the IV bag. Not only is sterility easily maintained through mixing procedures via injection ports, the collapse of the bag as the vaccine is utilized during injection requires no venting or air introduction. Glass bottles require ventilation (due to negative back pressure) and should not be utilized for in ovo vaccination due to contamination risks. The diluent also contains pH-sensitive dye (phenylalanine red) that remains red at the desired neutral pH = 7.0. If diluent contamination occurs during storage prior to use, the waste products and enzymes produced by the contaminating bacteria will change the pH of the fluid and the color of the diluent will change (yellow = acidic pH <6.5, maroon = basic pH > 7.5).

Many procedures outlined from vaccine manufacturers recommend rinsing of the vials, thus further increasing the risk of contamination of the open glass ampules utilized for Marek's disease vaccines. It is advised to review these recommendations in light of contamination risk versus reward of recovering less than 2% of vaccine during nominal ampule rinse. Several critical control points exist during the mixing process and involve limiting inadvertent environmental contamination. The vaccine preparation area or laboratory should be a strictly controlled environment with high-quality sanitation and hygiene. In many high-quality incubation facilities, the mixing procedures are completed in a laboratory area supplemented with a work area beneath a laminar flow hood to improve biosecurity. Daily, weekly and monthly programs of sanitation should be enforced for the laboratory environment as well as limiting entry to only designated and trained personnel.

Inadvertent bacterial contamination of vaccines for in ovo use has been traced primarily to the thaw bath water and subsequent poor handling technique in drying and opening the ampules. Capillary action of thaw bath water on the exterior of the ampule at opening can result in transmission of the thaw bath water into the interior of the ampule and bacterial contamination of the vaccine at that point. It is a routine recommendation and practice to use a low level of chlorine (100–150 ppm Cl) in the thaw bath water. It is also advised to eliminate direct handling of the ampules while they are in the bath (utilize use of spaghetti strainer or perforated ladle). Additionally, a new clean paper towel (not cloth) should be utilized with every mixing session as well as fresh needle and syringe. It is important to consider other types of contaminations such as respiratory vaccines (viral or bacterial). Segregated work areas for preparation of vaccines for in ovo and spray vaccines applied at day of hatch are required to ensure no sharing of thaw bath containers or needle/syringes. Inadvertent contamination of vaccine for in ovo use can result in catastrophic morbidity and mortality for the developing embryos such that prevention is a requirement.

There exist several types of licensed vaccines for in ovo use that are not frozen in sealed glass ampules. These vaccines are presented in concentrated dehydrated form or as a refrigerated liquid contained in a sealed glass bottle with a rubber septum. These types of vaccines are much more user-friendly container and ensure asepsis due to the sterile entry made possible by the rubber septum.

While the vaccine mixing and handling process requires a certain level of skill and attention to detail, there also exists the challenge of sheer volume of vaccine and harsh environment of application. A hatchery incubating 1.2 million eggs per week will require approximately 20 800-ml bags of vaccine prepared each day (4 days/week). Each bag will require handling four to ten frozen Marek's vaccine ampules through the thawing and diluting process to provide the final solution of vaccine for use. On an annual basis, the procedure demands attention to detail and safe handling up to 50 000 ampules as well as safely attaching over 4000 individual bags of vaccine to injection devices, with no mistakes. Routine training and re-training programs are an important part of success as well as routine evaluations and monitoring of the adherence to protocol. Monitoring and training should include microbial sampling of mixed vaccine during normal run days representative of applications.

3.11 Egg handling during injection/transfer

As aforementioned, the window for timing of vaccination in ovo is chronologically set to occur during the late stages of development between day 17.5 and 19 days of incubation. Considerations include the optimal time for removal of the eggs from the incubator as part of the individual incubator design and operation as well as relative embryo development or stage. Multi-stage incubators can be divided into two basic types: one whereby the eggs of each age or batch are contained in a trolley, and the second type whereby the eggs are placed by individual incubator flat into fixed turning racks secured inside the incubator. Single-stage incubation utilizes trolleys, so for most practical purposes they can be treated as a 'trolley' system. The trolley system allows for whole racks of eggs (5000-8000) to be removed at one time with the ability to move these eggs directly to an injection/transfer device for in ovo vaccination. While requiring a certain amount of labor to move the trolley, the trolley systems are more readily adapted to further automation for 'de-stacking' these trolleys and 're-stacking' the full hatch baskets during the process (more information later). The fixed rack systems require individual handling of each flat of eggs to remove them from incubation. This is obviously more labor intensive, and these types of incubation may process their eggs by double handing (placing incubated eggs into a 'farm' trolley, moving the trolley to injection/transfer device) or by removing eggs from the fixed rack positions and conveying eggs directly from the incubator to the injection/transfer device. It is important to consider that the removal process must be managed and controlled with attention to detail as to egg cracking and managing embryonic temperature. The embryos are exothermic and individually produce approximately 190 mW (0.65 BTU) of heat at this stage of incubation. Egg shell temperatures should target 100–101°F (not exceed 102.0°F) upon removal from the incubator and caution should be taken to stage the trolleys in horizontal airflow while they await the injection/transfer process. Overheating is the primary concern during the injection/transfer process.

Optimal timing for removal of eggs is different for the two multi-stage incubator designs. It is important to consider that the metabolic mass of eggs that are being removed represent a significant heat supply for that incubator. Removing the oldest eggs are an important step in the continual incubation process for all the remaining eggs and stages still in the incubator. The trolley systems recommend removal and transfer of eggs to the hatching compartments on day 18–18.5 of development. Extending the incubation cycle results in excessive heat management issues as the embryonic mass (same day of development) is localized into the trolley and position in the incubator. Earlier removal will conversely affect the incubation with cooler incubation challenges (Brake and Williams, 2002). Fixed rack incubation results in a broader spread of embryonic heat within the incubator as the six different developmental stages found within the machine are spread uniformly across the entire internal environment. Typically, these incubators perform optimally with transfer occurring after day 18.5 of incubation but before day 19 and 2–4 h. Their embryonic development is slightly behind trolley-based incubators in strict chronological versus developmental stage comparison.

Embryonic development and population dynamics should be determined for both incubation systems to set final protocol and adjust 'start-to-finish' scheduling for the entire process day at the commercial hatchery. Utilizing embryonic stage of development for optimizing in ovo vaccination should be cognizant of population curves associated with eggs from young, prime and older breeder hens. Minimal development target delineated via egg breakout should be the embryo in the position to hatch with the head under the right wing and the stalk of the yolk physically entering the abdomen (no visible intestines at navel). The maximum development target would be 5-15% internally pipped with less than 1% external pip when injection process is complete. During the final scheduling of injection/transfer timing, considerations include total incubation time (eggs from older hens/younger hens > prime hens), egg set schedule (begin incubation) and general microbial load (eggs from older hens/younger hens > prime hens). For these purposes, eggs are usually processed on any given day in the order of prime, then young and finally eggs from older hens last.

Supportive quality assurance of the injection/transfer process for in ovo vaccination includes coordination of hatcher and hatch basket sanitation such that all baskets and hatchers are clean and dry, preferably pre-warmed to receive the eggs immediately after vaccination. The filters of the ventilation system (HVAC) and handling units also require sanitation and maintenance prior to introduction of vaccinated eggs into the hatching environment. As previously mentioned, fungal challenges to the hatching environment are a primary concern and focus of control in the 'everyday' processes of the hatchery utilizing in ovo vaccination. The vaccination process may take place in the hatcher hallway itself, in a designated injection/transfer room (preferred) or in an adjacent hallway. The environment for the device and process should be clean and dry at start with adequate air exchange to ventilate for embryonic needs (8-12 ft^3/min or 13.6-20.3 m^3/h per 1000 eggs) while maintaining a neutral or slightly positive room pressure. This is especially important to consider if the vaccination process is conducted in the hatcher hallway. Optimal environmental control (temperature, pressure) should be established when the first eggs are vaccinated and placed into the initial hatcher. The hatcher hallways for both single- and multi-stage incubation are typically held at 78-82°F with slightly positive air pressure (1.25-3.74 Pa or 0.005-0.015 in. of water draft) (Barnwell, 1998; Jung and Magrans, 1997; Martin, 2019). These conditions should be established and maintained during the vaccination procedures as the complete process may take several hours and the environment controlling air supplied to the hatchers must be controlled. Upon completion of the vaccination process, the hallway environment is normally cleaned of any organic debris (wash down) and care should be taken to not introduce excess water into the hatcher or adjacent incubator environments. Disinfectant fogging is also commonly utilized after hallway cleanup to minimize the fungal and bacterial challenge to the hatching environment as supplied by the air of the hallway.

3.12 Automation and in ovo vaccination

The global acceptance and utilization of vaccination in the egg has opened the poultry production environment at the breeder farm and hatchery to further automation and labor efficiencies. Vaccination by individual handling and injection of day-old chicks after hatch created a significant delay and holding period between actual hatch time and transport to the farm. With the incorporation of in ovo vaccination, chick handling after hatch could realize significant improvements through automation. Downstream processing efficiencies include rapid chick separation from unhatched eggs, hatch basket/trolley washing, chick counting, spray vaccination and chick sorting and boxing for transport to farms. The efficiencies in after-hatch processing has significantly reduced the time between removal of the birds from the hatcher environment to transport and placement on the grow-out farm with access to food and water. It is important to realize that the early and uniform access to feed and water for any given flock significantly improves feed efficiencies and time (days) to achieve desired body weights and size (Peebles et al., 2017). It is typical of most production programs at present to require only 2-4 h between removal of chicks from the hatcher to placement on the farm in the United States. The industry has also looked at improved efficiencies at the breeder farm and incubation processes and how they interact. New incubator flat designs are larger and are one- or two-piece design per tray or level in the incubation rack. This makes egg handling more efficient and more easily adapted to robotic handling of the full incubation trays during injection/transfer. Multipurpose trolleys are designed to hold eggs from the initial loading at the breeder farm completely through incubation until removal of the flats for injection/transfer. There has been an increase in the construction and conversion to single-stage incubation due to the inherent efficiencies and technological advances presented by these technologies.

In more direct application to the in ovo vaccination process, automated unloading of incubation flats from incubator trolleys and loading them robotically onto in ovo injection devices has developed primarily over the last 8–10 years. Optically controlled robotic functions have advanced considerably over the last decade and have provided automation equipment specifically designed to handle manipulations required for individual incubator flat and trolley designs as they are utilized with in ovo vaccination. Additionally, robotic devices can repeatedly remove vaccinated eggs from the incubation trays following vaccination and place them accordingly in stacks of hatching baskets ready for placement in the hatcher compartments. The automation of these processes is unique to each type of incubation flat and presents unique challenges in their application; however, most are continually quality controlled and produce more predictable and consistent results with the elimination of human error. Automated production can result in up to 60 000–70 000 eggs/h vaccinated and transferred to hatchery baskets with 2–3 operators handling eggs through automation devices and one operator managing vaccine and in ovo device quality control. The labor savings alone between production utilizing hand vaccination after hatch and in ovo vaccination can exceed 500 man-hours/week. Additional benefits to bird health and growth due to lower stress during handling after hatch as well as shorter times to access feed and water combine to improve feed to body weight gain efficiencies.

4 Additional interactions and considerations for in ovo vaccination

4.1 Egg candling for viability

In ovo vaccination devices have advanced in their application processes to include candling technologies to determine embryo viability prior to injection. Candling or embryo viability determination has enabled targeted delivery of vaccine only to viable eggs as well as the ability to remove the non-viable eggs prior to injection and transfer. Removal of non-viable eggs reduces the microbial challenge presented by the nutrient potential for bacterial and fungal entities associated with these eggs. Additional benefits from candling include information as to individual flock status at injection/transfer with regard to viability (predictive analytics) and the removal of these eggs also presents an improvement in airflow and environmental conditions of the individual hatch baskets for the chicks during the hatching period. Candling can be accomplished with visible light or infrared light for simple determination of developmental stages or can be accomplished with more elaborate viability detection such as heartbeat or pulse oximeter determinations. The in ovo vaccination devices utilized throughout the industry routinely utilize viability detection for vaccine delivery (>90%) and additional non-viable egg removal is included in approximately 80% of production globally.

4.2 Chick handling and welfare

As aforementioned, in ovo vaccination has eliminated the need for the individual handling of day-old chicks for injectable vaccinations after hatch, and this in itself reduces stress on the hatched chick, thus improving welfare status. Chick handling after hatch involves removal of the stacks of hatching baskets from the hatcher compartments and manual transport to central area of hatchery for manual or automated chick shell separator (baskets stacked upon dolly with wheels or use of pallet jack). Welfare considerations are primarily chick comfort as to appropriate temperature and fresh air supply (10–25 cfm per 1000) while in the basket stacks and during separation processes. The separation of chicks from unhatched egg and shell residue requires handling the individual hatch baskets. The automated process gradually inverts the basket onto roller or shaker bars that allow the chick to fall through the gaps between the bars while the unhatched eggs and eggshells are carried into an offal vacuum system. Humane handling includes maceration of all unhatched eggs. The live chicks are conveyed into a clean room (chick room) via belted conveyors for subsequent counting and sorting into handling boxes for spray vaccination and accumulation for transport to grow-out farms. To complete these processes in a repeatable rapid process, it is important in the design and execution to have a uniform presentation of the hatch baskets and control of the environmental conditions where the process is completed. The separation process includes automated devices similar in concept to other agricultural food-handling mechanization that sorts vegetables from organic debris during packaging. Air handling and temperature control are important to maintain the animals' body temperature and control organic debris (e.g. chick fluff, unhatched eggs) liberated during the process. Obvious challenges are evident as the safety and welfare of the day-old chicks require attention to their physical well-being and comfort and the fact that we are handling live somewhat fragile animals. Special attention to the mechanics of speeds, conveying belts and surfaces as well as drop heights required for isolating individual animals for counting and quality control. These equipment operate in relatively harsh and organic laden environments and must be robust to withstand rigorous cleaning and disinfection routines performed daily after processing. The devices must also be somewhat impervious to organic buildup as biofilms and microorganisms cannot build up within the animal contact and waste-handling entities of the equipment.

4.3 Labor efficiencies

As discussed earlier in Section 3.12, labor savings in direct comparisons between vaccination of day-old chicks and in ovo vaccination significantly favor the latter. Many individual complexes reduced labor requirements in the hatchery by more than 500 man-hours/week with the incorporation of in ovo vaccination to replace vaccination by hand after hatch. These efficiencies are not however the only efficiencies that can be attributed to the in ovo vaccination processes. With chicks receiving their vaccinations before hatch, chick handling after hatch became a great opportunity to improve efficiencies. Automated shell separators, hatch basket or tray wash equipment and robotic de-stack and stacking devices replaced manual labor required to perform the after-hatch processes. Efficiencies include labor reduction, but also eliminated many of the health and injury claims related to repetitive motion disorders associated with hand vaccination and handling the hatch baskets during chick and shell separation. Similarly, the handling efficiencies of hatching eggs and incubator functions became more attuned to the egg-handling possibilities prior to and after incubation. Efficiencies at the breeder farm and hatching egg production facilities include larger single-tray incubation flats more readily adaptable to automation (versus smaller multi-flat designs). Many incubator companies now offer dual-purpose trolleys for hatching eggs that can be loaded on the farm, transported to the hatchery and then utilized in the incubation process. The cycle will repeat once the trolleys are unloaded during injection/transfer, cleaned and sanitized and then returned to the farm. Prior designs required two different sets of trolleys, one for use at the farm and transport to the hatchery and the second set only utilized in incubation. Eggs required transfer from farm rack to incubation rack, thus doubling the amount of egg handling. The double handling not only required more labor but also increases the level of cracked or damaged eggs. Arguably, both improved egg handling and after-hatch automation could have occurred without in ovo vaccination; however, both efficiencies would not be as important and require longer-term return on investment associated with commercial poultry production.

4.4 Quality assurance

In ovo vaccination is not a simple veterinary medical application, and thus requires shared microbial responsibilities in assurance to standards and operating procedures. The quality assurance aspect begins prior to the incorporation of the technology with a basic understanding of the challenges of hatchery ventilation and work areas. As reviewed in Section 3.6, an evaluative risk analysis is recommended prior to incorporation of in ovo vaccination. The analysis is focused upon quality control of the ventilation system and the overall sanitation programs of the hatchery. More detailed quality control programs become part of the hatchery's standard operating procedures once in ovo vaccination is embraced. These include greater attention to detail with cleaning and disinfection programs and routine microbiological evaluations to support the execution of these programs. The mechanical aspects of all in ovo vaccination devices require routine maintenance and functional operations should be ensured by appropriate technically trained personnel. Quality control should include routine and periotic evaluations as to compliance to protocols for setup and shutdown procedures as well as operator functions to be completed during vaccine application. These include appropriate training and educational exchanges to ensure confident operation of the equipment. Technical evaluations to ensure that biological and mechanical interface is correct include evaluations of embryonic stage of development at injection/transfer time, vaccine site of injection delivery in ovo and determinations to ensure quality of vaccine preparation and application through the device (microbial sampling, live/dead cell counting and vaccine concentration).

5 Conclusion and future trends

The discovery of the virus that causes Marek's disease in 1967 opened the door for the development of a preventative strategy, with the first vaccines available in the early 1970s. Continued research and development with Marek's disease led to the discovery and expansion of the science of in ovo vaccination in the early 1980s, and this continues today. Engagement and collaboration between the poultry industry, academia and biotechnical scientists at Embrex brought in ovo vaccination to commercial reality during the 1990s. Veterinary medical in ovo injection devices now bring the vaccination technology to more than 35 countries with a growing list of more than 35 different approved and licensed vaccines. Globally, more than 24 billion broiler chicken eggs are vaccinated in ovo annually. The technology has enabled adjacent evolutions of candling systems, automation for chick and egg handling as well as a required influence on improved sanitation and environmental controls at the hatchery and breeder farm level.

The application of egg injection technology is not simply a matter of delivering biologically active antigens to the developing embryo. The technique must be sterile and provide precise execution through the process. The details and specific aspects of avian incubation must be coordinated with the timing of egg injection and transfer of eggs from the incubator to the hatching environment to maximize the potential benefits. Each hatchery has specific microbiological criteria that must be controlled and optimized to incorporate in ovo vaccination and maximize the potential benefits from egg injection, while working with a real-world situation of people, programs and costs. With respect to the injectable antigens and vaccines, they must be safe for the embryo, as well as efficacious in establishing a superior immunological response to defend against disease challenge and cannot interfere with each other's efficacy.

The future holds clear challenges to develop more safe and efficacious vaccines for in ovo use; however, the technology has clear advantages as outlined herein. We will see further development of recombinant vaccines for in ovo use utilizing the HVT virus construct, and these families of vaccines will continue to provide tailored programs for disease prevention of the broiler chicken. The adjacent technologies of candling, egg handling and hatchery automation will also advance in their predictive and applied use during the vaccination process as well as support the efficiencies required in poultry production. We will see improvements in sanitation and quality control to address bacterial and fungal disease challenges of the environment and within the production of hatching eggs. Vaccines targeting bacterial and protozoal diseases will also be developed and become available as reliance on antibiotic treatment and use will continue to wain in the animal health industry. Prevention of disease requires controlled exposure such as vaccination provides. In ovo vaccination will continue to be the most accurate and uniform application route for cell-associated vaccines in support of broiler production for the global poultry industry.

6 Where to look for further information

As briefly reviewed in the final paragraph of Section 5, there will be future developments in vaccines at commercial level through animal health industry as well as research in many universities globally. Additionally, there are current governmental agencies (e.g. USDA) conducting research involving in ovo vaccine applications both internally and through university settings.

7 References

Abdul-Carrem, M. F., Hunter, D. B., Lambroune, M. D., Read, L. R., Parviszi, P. and Sharif, S. (2008), Expression of cytokine genes following pre- and post-hatch immunization of chickens with herpesvirus of Turkeys, *Vaccine*, 26(19), 2369-2377.

Avakian, A., Singbeil, B., Poston, R., Grosse, D., Klein, D., Whitfill, C. and Tripathy, D. (1999), Safety and efficacy of fowl and pigeon pox vaccines administered in ovo to SPF and broiler embryos, Proceedings of the 48th Western Poultry Disease Conference, Vancouver, CA, 56-60.

Barnwell, R. (1998), Recognizing common hatch ventilation problems, *World Poult.*, 14, 11.

Bot, A., Antohi, S. and Bona, C. (1997), Immune response of neonates elicited by somatic transgene vaccination with naken DNA, *Front. Biosci.*, April, 173-188.

Brake, J. T. and Williams, C. J. (2002), Timing egg transfer to improve hatchability, Proceeding Broiler/Hatchery Conference. Delmarva Poultry Industry.

Bruzual, J. J., Padilla, J. and Campos, R. (2005), Field trial conducted in Peru to evaluate the route of vaccine application against Marek's and infectious bursal disease (Burcell-HVT) in broiler chickens. In Ovo versus Subcutaneous. Proceedings of the For 54th Western Poultry Disease Conference, Vancouver, BC Canada.

Denoya, C., Dawson, E., Sessoms, D., Baumstein, M. and Shabushnig, J. (2010), A rapid microbiological assay to monitor the effectiveness of a vaccine injector sanitization following a microbial challenge procedure. *Am. Pharm. Rev.*, 13, 4.

Gildersleeve, R. P., Hoyle, C. M., Miles, A. M., Murray, D. L., Ricks, C. A., Secrest, M. N., Williams, C. J. and Womack, C. L. (1993), Developmental performance of an egg injection machine for administration of Marek's disease vaccine, *J. Appl. Poult. Res.*, 2(4), 337-346.

Gimeno, I. M. and Cortes, A. L. (2011), Chronological study of cytokine transcription in the spleen and lung of chickens after vaccination with serotype 1 Marek's disease vaccines, *Vaccine*, 29(8), 1583–1594.

Gimeno, I. M., Cortes, A. L., Faiz, N. M., Barbosa, T. and Villalobos, T. (2015a), Evaluation of factors influencing efficacy of vaccine strain CVI988 against Marek's disease in meta-type chickens, *Avian Dis.*, 59(3), 400–409.

Gimeno, I. M., Cortes, A. L., Barbosa, T., Villalobos, T. and Pandiri, A. R. (2015b), In ovo vaccination with HVT hasten Maturation of chicken embryos immune responses in specific Pathogen Free Chickens (SPAFAS), *Avian Dis.*, 59, 375–383.

Gimeno, I. M., Glaize, A. and Cortes, A. L. (2018), Effect of Marek's disease vaccines on interferon and toll like receptors when administered in ovo, *Vet. Immunol. Immunopathol.*, 201, 62–66.

Guo, Z. Y., Giambrone, J. J., Liu, Z., Dormitorio, T. V. and Wu, H. (2004), Effect of in ovo administered reovirus vaccines on immune responses of specific-pathogen-free chickens, *Avian Dis.*, 48(1), 224–228.

Jochemsen, P. and Jeurissen, S. H. M. (2002), The localization and uptake of in ovo injected soluble and particulate substances in the chicken, *Poult. Sci.*, 81(12), 1811–1817.

Jung, S. and Magrans, R. (1997), Hatchability depends on adequate ventilation, *Zootecnica Int.*, December 19–21.

Karpala, A. J., Bagnaud-Baule, A., Goossens, K. E., Lownthal, J. W. and Bean, A. G. (2012), Ontogeny of the interferon system in chickens, *J. Reprod. Immunol.*, 94(2), 169–174.

King, S. F. (1995), The basics about *Aspergillus*, *Inovoject Customer Serv. Newsl.*, 3(5), 4–5.

Lehtonen, L., Vainio, O. and Toivanen, P. (1989), Ontogeny of alloreactivity in the chicken as measured by mixed lymphocyte reaction, *Dev. Comp. Immunol.*, 13(2), 187–195.

Lowenthal, J. W., Connick, T. E., McWaters, P. G. and York, J. J. (1994), Development of T cell immune responsiveness in the chicken, *Immunol. Cell Biol.*, 72(2), 115–122.

Manders, T. T. M., Matthijs, M. G. R., Veraa, S., van Eck, J. H. H. and Landman, W. J. M. (2021), Success rates of inoculation of the various compartments of embryonated chicken eggs at different incubation days, *Avian Pathol.*, 50(1), 61–77.

Martin, S. (2019), Hatchery Audits. Webinar Series from Scott Martin, LLC., scottmartinllc @att.net.

Mast, J. and Goddeeris, B. M. (1999), Development of immunocompetence of broiler chickens, *Vet. Immunol. Immunopathol.*, 70(3–4), 245–256.

Miles, A. M., Williams, C. J., Womack, C. L., Murray, D. L. and Gildersleeve, R. P. (1992), Commercial broiler studies of Marek's disease vaccination in ovo. Proceedings of the XIX WPC Amsterdam, the Netherlands (vol. 1), 320–322.

Parkhurst, C. R. and Mountney, G. J. (1995), *Poultry Meat and Egg Production*, Boston, MA: Springer Publishing.

Peebles, E. D. (2018), In ovo applications in poultry: a review, *Poult. Sci.*, 0, 1–17.

Peebles, E. D., Barbosa, T. M., Cummings, T. S., Dickson, J., Womack, S. K. and Gerald, P. D. (2017), Comparative effects of in ovo versus subcutaneous administration of the Marek's disease vaccine and pre-placement holding time on the post-hatch performance of Ross 708 broilers1,2,3, *Poult. Sci.*, 96(5), 1071–1077.

Phelps, P. (1995), The in ovo administration of antibiotics into broiler eggs at transfer, Ph.D. Dissertation, North Carolina State University.

Ricks, C. A., Avakian, A., Bryan, T., Gildersleeve, R. P., Haddad, E., Illich, R., King, S., Murray, L. M., Phelps, P., Poston, R., Whitfill, C. and Williams, C. J. (1998), In ovo vaccination technology, *Adv. Vet. Med.* 41, 495–515.

Ridge, J. P., Fuchs, E. J. and Matzinger, P. (1996), Neonatal tolerance revisited: turning on newborn T cells with dendritic cells, *Science*, 271(5256), 1723–1726.

Sarma, G., Greer, W., Gildersleeve, R. P., Murray, D. L. and Miles, A. M. (1995), Field safety and efficacy of in ovo administration of HVT +SB-1 bivalent Marek's disease vaccine in commercial broilers, *Avian Dis.*, 39(2), 211–217.

Sarzotti, M., Robbins, D. S. and Hoffman, P. M. (1996), Induction of protective CTL responses in newborn mice by a nurine retrovirus, *Science*, 271(5256), 1726–1728.

Schierman, L. W. and Ordskog, A. W. (1961), Relationship of blood type to histocompatibility in chickens, *Science* 134(3484), 1008–1009.

Seto, F. (1990), Immune responsiveness of neonatal chicks and immunocyte precursors from unprimed juvenile chicken donors, *Poult. Sci.*, 69(7), 1103–1109.

Sharma, J. M. (1985), Early pathogenesis of turkey herpesvirus in chickens inoculated as embryos, Proceedings of the International Symposium on Marek's Disease (Calnek and Spencer Editors), AAAP, Kennett Square, PA, 462–468.

Sharma, J. M. (1987a), Delayed replication of Marek's disease virus following in ovo inoculation during late stages of embryonal development, *Avian Dis.*, 31(3), 570–576.

Sharma, J. M. (1987b), Embryo vaccination of chickens with turkey herpesvirus: characteristics of the target cell of early viral replication in embryonic lung, *Avian Pathol.*, 16(4), 567–579.

Sharma, J. M. (1989), In situ production of interferon in tissues of chickens exposed as embryos to turkey herpesvirus and Marek's disease virus, *Amer. Jour. Vet. Res.*, 50, 842–852.

Sharma, J. M. and Burmester, B. R. (1982), Resistance to Marek's disease at hatching in chickens vaccinated as embryos with turkey herpesvirus, *Avian Dis.*, 26(1), 134–149.

Sharma, J. M. and Graham, C. K. (1982), Influence of maternal antibody on efficacy of embryo vaccination with cell-associated and cell-free Marek's disease vaccine, *Avian Dis.*, 26(4), 860–870.

Sharma, J. M. and Burmester, B. R. (1984), Disease control in avian species by embryonal vaccination, US Patent No. 4,458,630, July 10, 1984.

Sharma, J. M., Lee, L. F. and Wakenell, P. S. (1984), Comparative viral, immunologic, and pathologic responses of chickens inoculated with herpesvirus of turkeys as embryos or at hatch, *Am. J. Vet. Res.*, 45(8), 1619-1623.

Sharma, J. M. and Witter, R. L. (1983), Embryo vaccination against Marek's disease with serotypes 1, 2 and 3 vaccines administered singly or in combination, *Avian Dis.*, 27(2), 453–463.

Wakenell, P. S., Bryan, T., Schaeffer, J. L., Avakian, A., Williams, C. J. and Whitfill, C. E. (2002), Effect of in ovo vaccine delivery route on herpesvirus of turkeys/SB-1 efficacy and viremia, *Avian Dis.*, 46(2), 274–280.

Williams, C. J., Griffin, C. L., Klein, D. R., Sorrell, W. R., Secrest, M. N., Miles, A. M., Gildersleeve, R. P. and Brake, J. T. (1994), A microbiological survey of broiler hatcheries in North America – the prevalence of *aspergillus* and other deuteromycetes, *Poult. Sci.*, 73(Suppl. 1), 166.

Williams, C. J. and Villalobos-Chaves, T. (2000), Field testing of the application of Marek's disease vaccination (HVT) to broiler chickens in costa Rica, In: ovo versus sub-cutaneous. Proceedings of the For XXI World's Poultry Congress, Sixth International Symposium on Marek's disease. Montreal CA, Aug 20-24, 2000.

Williams, S. M., Smith, J. A., Garcia, M. C., Brinson, D., Kiupel, M. and Hofacre, C. (2010), *Severe Histiolymphocytic and Heterophilic Bronchopneumonia as a Reaction to In Ovo Fowlpox Vaccination in Broiler Chicks*, American College of Veterinary Pathologists publication.

Witter, R. L. and Lee, L. F. (1984), Polyvalent Marek's disease vaccines: safety, efficacy and protective synergism in chickens with maternal antibodies, *Avian Pathol.*, 13(1), 75–92.

Zhang, Y. and Sharma, J. M. (2001), Early post hatch protection against Marek's disease in chickens vaccinated in ovo with a CVI988 serotype 1 vaccine, *Avian Dis.*, 45(3), 639–645.

Published by Burleigh Dodds Science Publishing Limited, 2023.

Part 2

Incubation

Chapter 9

Incubator design for poultry eggs: principles and techniques

Ron Meijerhof, Poultry Performance Plus, The Netherlands

1 Introduction

Incubation is a delicate process that requires good control of the environment around the egg. Ensuring the requirements of the embryo are met prevents a loss of hatchability (Lourens and Middelkoop, 2000) and inferior chick quality, which can be even more costly as it can influence performance later in life (Hulet et al., 2000). How are the conditions to meet these requirements in a commercial incubator, containing many thousands of eggs, fulfilled? A starting poing is the requirements of the embryo during incubation.

Factors enabling the embryo to develop into a good-quality chicken include:

- Temperature: a constant temperature inside the egg (the so-called embryo temperature). The temperature of the shell is usually measured at the equator of the egg and should be approximately 37.8°C (100°F) from the start until internal pipping. Once the embryo starts hatching, the internal temperature will rise due to an increase in heat production.

http://dx.doi.org/10.19103/AS.2022.0118.13

- Position of the egg: the embryos orientate themselves via gravity and need to end up with their head at the side of the air cell, in order to allow them to use the air in the air cell to begin lung ventilation. Therefore, the air cell needs to be elevated compared with the centre of the egg, usually achieved by placing the eggs with the blunt end up.
- Turning: eggs need to be turned, especially during the first half of incubation, over a 45° angle to both sides. This is important for the development of extra-embryonic tissues.
- Moisture loss: the egg needs to lose water during incubation to create a big enough air cell for the embryo to inflate its air sacs after internal pipping. Moisture loss should be, on average, 12–14% until external pipping. Moisture loss (weight loss) is usually measured between egg setting and 18 days of incubation (transfer from setter to hatcher) in chickens and should be 11%–13%, on average, with a minimum preferably over 10%. The same percent of water loss is also required in non-chicken species by their appropriate age at transfer. Both moisture, oxygen (O_2) and carbon dioxide (CO_2) must move across the eggshell through the pores in the shell. The resistance of the shell against this moisture loss can be expressed as the conductance of the shell.
- O_2 supply: the embryo needs to take up O_2 to oxidise carbohydrates but especially fat from the yolk.
- CO_2 removal: the metabolism of the embryo produces CO_2, which must be removed from the egg content.

The requirements of the embryo during incubation are rather simple; however, creating these conditions in a commercial incubator is not always easy (Owen, 1991). A complicating issue is that influencing one factor often impacts other factors. Let's take a closer look at the different requirements to identify optimal conditions for the embryos.

2 Temperature, heat production and heat transfer

The most important factor during incubation is the temperature that the embryo experiences, i.e. the temperature inside the egg. This temperature is important for a number of reasons (discussed in more detail in Chapter 10). One of the main reasons is that the embryo reacts as a poikilotherm, meaning that its metabolism is correlated with temperature. If the temperature of the embryo rises, its metabolism and the requirements for nutrients and O_2 also increase.

A key question to address is if the embryo changes in such a way that incubator design has to be adapted. Genetic progress in broilers is rapid, which may also influence the embryo. If we compare growth rates and metabolism of embryos described by Romanoff in 1960 with more recent work, we see that the embryos reach a higher body weight earlier in the process. Modern embryos at 14 and 18 days are approximately 40% bigger than the lines described in 1960 by Romanoff, although the egg size has increased by approximately 10%. When we compare the heat production of embryos measured during that period (Romijn and Lokhorst, 1960) with the heat production reported more recently (Nangsuay et al., 2015), there is a small difference of 5-8%, which could be caused by differences in measuring methods or in initial conditions such as egg weight and incubation temperature. As the maximum heat output will be limited by the amount of O_2 that can pass through the shell, this is not likely to change in the near future. Research by Nangsuay et al. (2015) showed that modern broilers have approximately 15% more heat production than layers with the same size eggs, suggesting that there is a genetic difference and perhaps a difference over the years due to genetic development. However, the same research also showed that modern broiler lines of a young breeder age have 15% less heat production as an embryo compared with eggs from the same lines at an older age with identical egg size. This indicates that it might not be genetic but that the environment (egg composition, conductance) may be responsible for the difference. However, although the development of genetic potential in broilers is very quick, the metabolic changes during incubation seem to progress much slower, so we probably do not have to give too much consideration to future genetic development.

Embryos are extremely sensitive to temperature, even a shift of a degree Celsius in eggshell temperature can result in metabolic alterations as shown by Lourens et al. (2007) and Nangsuay et al. (2021); however, birds in nature do not appear to precisely control temperature. This is probably because the embryo in the egg can partly direct its blood flow (Tzschentke and Rumpf, 2011), and hence can use the gradient in temperature between the breast of the mother hen and the bottom of the nest to influence its internal temperature.

Modern commercial incubators do not have the option of providing a temperature gradient to the embryo to allow it to regulate its own temperature, as the air surrounding the egg creates an equal temperature for the top and bottom of the egg. This means that a very accurate temperature in the egg must be achieved, which is a challenge. As the temperature inside the egg is the result of a balance between heat production by the embryo and heat loss from the egg to air, the influence of the conditions around the egg on the heat transfer from the egg to air (and vice versa if warming up the eggs at the start of incubation) needs to be considered.

There are several factors that influence the heat transfer from the egg to air and vice versa including: the difference in temperature between the egg and air, the air velocity over the egg and the loss of heat via water evaporation.

- If there is a difference in temperature between the eggshell and air, heat will be transferred from the medium with the higher temperature towards the medium with the lower temperature. The higher the difference in temperature is, the higher the heat transfer will be.
- Air velocity will accelerate this heat transfer. A higher air velocity will increase the heat transfer and will therefore help to accelerate the levelling of temperature between the egg and air. A difference in air flow makes a huge difference in heat transfer. Theoretical calculations show that if an egg produces enough heat to make the egg content 2°C warmer than the air at an air velocity of 0.5 m/s, an increase to 2 m/s in air velocity will result in an egg being only 1°C warmer than the air. The theoretical background is given in Meijerhof and van Beek (1993) and Lourens et al. (2011). Measurements in commercial machines confirm this, as illustrated in Fig. 1 (Lourens, 2001).
- Water evaporation costs energy and this will cool the egg. It doesn't matter if this water is produced by natural moisture loss of the eggs or by the sprayer to influence relative humidity (RH) in the machine, the evaporation of 1 g of water costs 2.26 kJ of energy. To get an idea of how much energy evaporation costs: the evaporation of 1 g of water costs approximately 5 times more energy than to bring the temperature of that 1 g of water up from 0°Cto 100°C.

RH is often expected to also influence heat transfer, but this is a misconception. RH influences the heat capacity of air, meaning that 1 m^3 of air with a higher RH requires more energy to go up 1°C in temperature than 1 m^3 of air with a lower

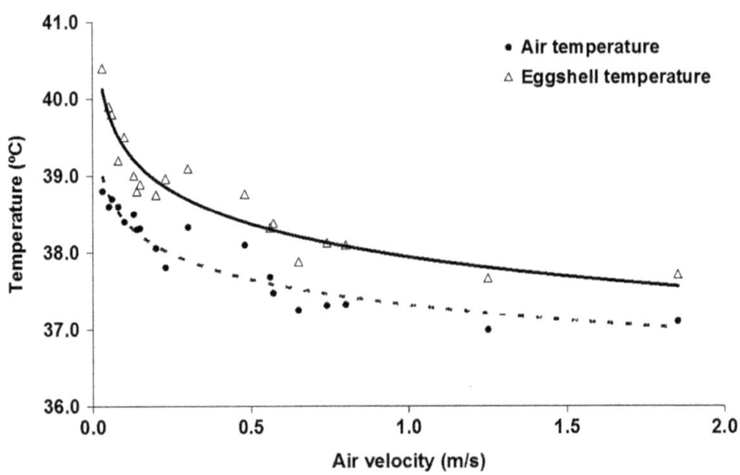

Figure 1 Effect of air velocity at a given machine temperature of 37.2°C on air (5 cm from the egg) and eggshell temperature in a commercial incubator at 18 days of incubation (Lourens, 2001).

RH. This is because the water in the air also needs to be heated up, which costs extra energy. Consequently, the same amount of air used for ventilating a machine can take up more energy when it is more humid, if it is warmed up from the temperature at the inlet to the machine temperature. However, this doesn't influence the heat transfer from the egg to air, it just means that less air is needed for the same amount of heat removal by ventilation. The influence of this should not be overestimated: to increase the temperature of air from 36 to 37°C requires 4.5 kJ/kg of air at an RH of 60%, whereas at 30% RH only 3.3 kJ/kg of air is needed.

3 Positioning and turning of eggs

In commercial incubators, the eggs are usually placed in incubator trays with the point down and air cell up. This allows the egg to be turned over an angle in both direction. As the embryo orientates itself on gravity, placing the eggs with the point down will force the embryo to position itself at the end of incubation with the head towards the air cell, allowing it to find the air cell when internal pipping starts. Failing to place the eggs in this position (upside down placement) will reduce the hatchability by approximately 10–20%, as several embryos will be unable to start lung ventilation and obtain sufficient energy to open the eggshell and hatch.

However, the mother hen doesn't place the eggs point down in the nest but orientates them horizontally. So how does the embryo find the air cell in that position, as there is a 50/50 chance that the head will be at the air cell end? The shape of an egg is not fully elliptic but is thicker towards the air cell than towards the point, which means that, when placed in a horizontal position, the egg will turn with the air cell slightly up. This effect is, for instance, used when eggs are automatically packed: when the egg is moved over a set of rollers it will move in the direction of the blunt end, enabling the machine to determine the point from the blunt end. As the air cell enlarges during incubation, this will make the blunt end relatively light and the side with the air cell will come up even more, allowing the embryo to determine where the air cell will be.

Although commercial incubators always place the eggs point down, there are some potential benefits to placing them horizontally, as the mother hen does. First of all, upside down placement due to poorly recognised egg shapes or inexperienced staff will be less problematic. Research by Kaltofen (1955) has shown that eggs placed in a horizontal position hatch well, as long as their horizontal position is not fixed but the eggs are allowed to get into a position with the air cell being slightly higher than the point. Another potential advantage is that turning the eggs can be achieved by rolling the eggs over a short distance, which will require less space. This has the advantage that the orientation of the eggs towards the flow of air can be better controlled and kept more constant, which might result in a more uniform air flow and therefore a more uniform egg temperature. However, practical aspects such as handling the eggs at placement and at transfer from setter to transfer tray will be more complicated and might outweigh the advantages.

The embryo requires turning, preferably over a 45° angle, but a minimum 38° angle, to both sides. This is to facilitate the formation of subembryonic membranes. These membranes are completed around days 11 and 12, which means turning can be stopped after approximately 12 days of incubation.

In a traditional system, i.e. turning the trays over at an angle, the turning can influence the air flow over the eggs. When the trays and eggs are in a specific position towards the air flow, often the eggs on one side of the tray will differentiate in temperature from the eggs on the other side. When the eggs are turned, this temperature distribution will become equalised over time (Romijn and Visschedijk, 1952). This is why we often see that more frequent turning is beneficial for temperature control. Nowadays, turning once every hour is widely accepted as sufficient, although some experiments indicate that a more frequent turning can be beneficial (Elibol and Brake, 2003). In his experiments, Kaltofen (1955) used small machines that did not turn the trays in the machines, but turned the whole machine instead. In this way, the effect of turning itself without the effect that turning has on air distribution and velocity could be investigated. He concluded that a turning frequency of once an hour is adequate, which could indicate that turning is for not only membrane formation but also temperature distribution.

Placing the trays horizontally after 10–11 days can also influence the temperature in the eggs. A stack (trolley) of trays in a turned position will have less space between the trays than a horizontal track, but more space between the stacks (trolleys). Although it depends on the type of machine (place of the ventilator, position of the trolleys, resistance against air flow by other components such as radiators) in general, the eggs in the centre of the trays will experience a higher air velocity when the trays are in a horizontal position than in a turned position, influencing the heat transfer of the eggs.

4 Ventilation: managing moisture, CO_2 and O_2

Ventilation is required to remove moisture, heat and CO_2 and provide O_2. These three factors require different ventilation rates based on the properties and demands of the embryo. We can calculate the properties of ventilation for each of these factors.

Let us start with RH. Eggs need to lose water during incubation, in order to create a big enough air cell for the embryo to start lung ventilation after internal pipping. This amount is approximately 12–14% on average over the incubation period, which means on average about 0.6% per day. The driving force for this moisture loss is the water vapour pressure deficit, the difference in water vapour pressure over the shell, in combination with the eggshell conductance, which is the resistance of the eggshell and membranes against gas exchange. The water vapour pressure is a function of temperature and RH. This means that inside the egg it is a function of temperature, as we can estimate the RH inside an egg as 100%, whereas outside the egg it is determined by temperature and RH. As the temperature during incubation inside and outside of the egg is typically constant and fixed, in practice it means that the moisture loss is a function of RH in the machine and the conductance of the eggs.

As the eggs need to lose on average 12–14% of moisture until internal pipping, a machine of 100 000 eggs, each weighing 65 g, should lose on average between 1.6 L/h and 1.9 L/h ((100 000 eggs × 65 g × 12% or 14%)/20 days/24 h). If we calculate using 1.8 L/h on average, we can calculate the cooling effect of the evaporation of that water on the eggs. The evaporation of water costs 2.26 mJ/L, so 1.8 L of water costs 4.1 mJ.[1] The specific heat of eggs is 3.4 kJ/kg per°C, so the cooling of 6500 kg of eggs will cost 22.1 mJ/°C. The evaporation of 1.8 L costs 4.1 mJ, so the eggs will be cooled with 22.1/4.1 = 0.18°C. This means that without heat production, the eggs will be slightly under the air temperature. As this cooling is uniformly spread over all the eggs, it is a factor that can be easily dealt with.

However, if we want to keep the RH at a certain level and we need to spray in order to maintain that humidity, it is a different story. This water also needs to be evaporated and the energy for that will be mainly supplied by the eggs, which means they will cool down. However, the distribution of the water spray will often not be sufficiently equalised to cool all the eggs in the machine at the same rate. The water will be distributed with the air flow created by the ventilator. This means that some of the eggs that are close to the sprayer will not only experience a high air velocity that will cool them, but also receive a high amount of water that needs to be evaporated. These eggs will be cooled more than eggs further away from the ventilator, contributing to an uneven spread of eggshell temperature in the machine.

A higher level of ventilation will remove more moisture out of the machine, resulting in an increased amount of spraying if the humidity is set at a fixed level. This is one of the reasons why, in most machines, an increase in ventilation rate contributes to an unevenness in eggshell temperature.

Not spraying in the machine means that the humidity will go down when ventilation is increased towards the end of incubation. As the humidity regulates the moisture loss of the eggs, the control point for optimum humidity is the amount of weight loss of the eggs during incubation. As long as the weight loss

[1] Unit of energy: the calculations in this chapter use the term joule (J) for energy. 1 watt equals 1 joule per second (J/s). 1000 joules is 1 kilo joule (1 kJ), 1,000,000 joules is 1 mega joule (1 mJ).

is sufficient, and therefore the air cell is big enough, the RH has no influence on the embryo. In older machines a sprayer is needed for additional cooling of the air, which increases the RH. In modern machines, the cooling is usually sufficient to remove the heat, which means spraying is not necessary, resulting in a more equal eggshell temperature throughout the machine.

This concept is used in so-called non-linear moisture loss. By keeping the ventilation very low in the first week of incubation, the RH will go up and the moisture loss will be limited. The sprayers are not used and will therefore not cool the eggs at the start. At the end of incubation, when more ventilation is needed, the sprayers remain off and the RH will drop, resulting in a higher moisture loss. In this way, the total moisture loss at the end of incubation is sufficient without using the sprayer, which results in a more stable temperature control.

If we know the temperature and RH of the incoming air and the air in the machine, we can calculate, using the Mollier diagram, how much air is needed to remove the moisture produced by the eggs. If the incoming air is 24°C and 60% RH, 1 m^3 of air contains 13 g of water. If the air in the machine is 37.5°C and 55% RH, it contains 24.7 g of water per m^3. This means that each m^3 of air passing through the machine takes up 24.7–13 = 11.7 g of water. If the eggs are losing 1.8 L (1800 g) of moisture per hour to achieve sufficient moisture loss, we can calculate that the air volume needed to remove that moisture will be 1800/11.7 = 154 m^3/h. Of course, if we change the conditions of the incoming air or the conditions in the machine, this value will change, but as the mentioned conditions are not uncommon in hatcheries, it gives a starting point.

4.1 Ventilation: CO_2 and O_2

Ventilation is also needed for the removal of CO_2 and the supply of O_2. In normal situations, the control of CO_2 will determine the ventilation level more than the O_2 supply. As CO_2 production and O_2 consumption are related in a ratio of 1:1 with the heat production of the embryo, we can calculate the embryonic heat production if the CO_2 production is known and vice versa. At the start of incubation, the embryo is so small that the heat production (and with it the CO_2 production) is negligible, but at the end of incubation (day 18) the heat production is approximately 0.15 W/egg, which results in a CO_2 production of 18.5 L/h/1000 eggs. Half-way through the incubation process (day 9), the embryo is much smaller and produces less heat, about 0.02 W/egg, resulting in a production of CO_2 of 2.7 L/h/1000 eggs.

Knowing the CO_2 level of the incoming air and the maximum level required in the machines, the ventilation rate needed to control the CO_2 level can be calculated. If the incoming air contains 500 ppm CO_2 and the maximum CO_2 level in the machine is set at 4000 ppm, we can add 3500 ppm to every m^3 of air. One litre per m^3 (1000 L) is 1000 ppm, 3500 ppm is 3.5 L of CO_2 per m^3. As the eggs are producing 2.7 L CO_2 per 1000 eggs per hour at 9 days, the required ventilation is 2.7/3.5 = 0.77 m^3/h/1000 eggs. At 18 days, the CO_2 production of the eggs is 18.5 L/h/1000 eggs, so to maintain a level of 4000 ppm, with 500 ppm in the incoming air, a ventilation of 18.5/3.5 = 5.3 m^3/h/1000 eggs is needed. This means that in our imaginary machine of 100 000 eggs, the ventilation for CO_2 at 9 days of incubation will require approximately 77 m^3 of ventilation, while at 18 days it will be 530 m^3 of ventilation per hour to keep the required levels of CO_2 in the machine.

5 Ventilation: managing heat

Ventilation also removes heat. Air that enters the machines is lower in temperature and, therefore, in energy content than air that is removed, so each m^3 removes heat. How much heat is removed by ventilation? If we again calculate using air of 24°C and 60% RH coming into the machine, according to the Mollier diagram, the air contains 45 kJ/m^3, where air of 37.5°C and 55% RH holds 85 kJ/kg, a difference of 40 kJ/m^3. At 18 days, the embryos produce 0.15 W of energy, so our machine with 100 000 embryos is producing 15 kW of energy. A watt equals a J/s, so 100 000 embryos producing 0.15 W produce 100 000 × 0.15 J/s × 3600 s = 54 mJ/h. As each m^3 of air takes out 40 kJ, we need to ventilate 54 mJ/40 kJ = 1350 m^3/h at 18 days of incubation if we can only rely on ventilation to remove heat. The actual amount depends on the quantities of the incoming and outgoing air, as the energy content depends on the temperature and RH. If the incoming air is 27°C and 60% RH, each m^3 of air passing through the machine can hold less energy and ventilation will be almost 1700 m^3/h.

For multi-stage incubation (or for a total setter room with various stages of incubation), the ventilation requirements for heat removal are much lower, as the average embryo is much smaller and produces less heat. If we calculate with an embryo of, on average, 9 days that produces, on average, 0.02 W of energy or 2 kW per 100 000 eggs, the heat production is 100 000 × 0.02 J/s × 3600 s = 7.2 mJ/h. If each m^3 of air still takes out 40 kJ of energy, the amount of air that we need is 7.2 mJ/40 kJ = 180 m^3/h.

What becomes the first limiting factor for ventilation: CO_2, moisture or heat removal? Especially in single-stage incubation, the requirements of ventilation for the different parameters change over time. In Fig. 2, the development of the required ventilation for the different parameters is given for a machine of 100 000 eggs.

This figure shows that until day 8, the ventilation is mainly for RH control, but after that the ventilation is required for heat removal. The ventilation for CO_2 (and O_2) is always lower than for energy; however, this is calculated for a situation where the RH is kept constant, and the produced energy is removed only by ventilation, but this is not the case in a practical situation.

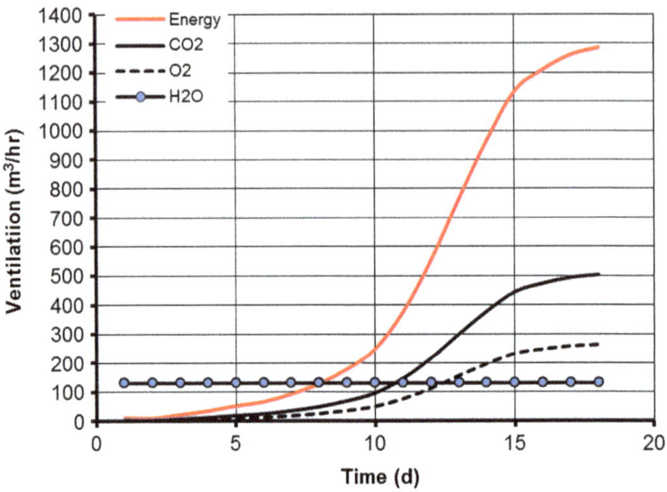

Figure 2 Theoretical ventilation levels during incubation of 100 000 eggs in a machine, based on heat exchange, and the control of CO_2, O_2 and RH (after Meijerhof and Lourens, 2018).

6 Cooling techniques

Ventilation enables the removal of heat, in addition to evaporation and cooling coils. If cooling coils are installed in a machine to remove heat, ventilation can be reduced to the level that is needed for the second limiting factor, i.e. CO_2 control. Heat is usually removed by cooling water that travels through pipes or radiators in the machine. If we know how much heat needs to be removed, we know how much water should run through the pipes with a certain temperature difference (delta *T*). Water has a specific heat capacity of 4.184 kJ/l/°C, so for every °C of increase in temperature, it takes up 4.184 kJ/L.

Let us follow the steps that help us to remove the heat based on our imaginary machine of 100 000 eggs, 100% fertility, at 18 days of incubation, with a heat production of 0.15 W/egg. The incoming air into the machine is 24°C and 60% RH (45 kJ/m^3), the air in the machine is 37.5°C and 55% RH (85 kJ/m^3).

The eggs produce 0.15 W × 100 000 × 3600 = 54 mJ/h. As stated, if we only remove the produced heat by ventilation we would need to ventilate 1350 m^3/h, because every m^3 of air takes out 85–45 = 40 kJ, and 54 mJ/40 kJ = 1350 m^3/h. But the eggs are losing moisture as well, and that moisture needs to be evaporated, which costs energy. We have estimated the moisture loss of all the eggs to be, on average, 1.8 L/h. The evaporation of a litre of water takes 2.26 mJ, so the natural evaporation of the eggs takes 1.8 × 2.26 = 4.068 mJ to remove. This means that the ventilation does not have to take out 54 mJ, but only 50 mJ. This reduces the amount of ventilation needed from 1350 to 50 mJ/40 kJ = 1250 m^3/h. When the RH increases or decreases, the moisture loss will change and with it the cooling effect of evaporation, which gives a starting point.

However, what happens if we limit the ventilation and only ventilate to keep the CO_2 level at 4000 ppm. If we reduce the ventilation to the level required for CO_2 control, we have to ventilate 530 m^3/h for our machine containing 100 000 eggs. With that air, heat will also leave the machine. If the incoming air is 24°C and 60% RH, 40 kJ would be removed with each m^3 of air that went through the machine. That means that when ventilating at 18 days for the requirement of CO_2, we also remove 530 m^3 × 40 kJ is 21.2 mJ/h. Our 100 000 embryos were producing 0.15 W, so 100 000 × 0.15 × 3600 = 54 mJ/h. Our evaporation from the eggs took out 4 mJ/h, and our minimum ventilation for CO_2 takes out 21.2 mJ/h, so the cooling coils should achieve 50–21.2 = 28.8 mJ/h. If the difference in temperature between incoming and outgoing water in the cooling coils is 10°C, it would require 28.8 mJ/(10 × 4.184 kJ) = 688 L of water/h.

However, the sprayers also remove heat as they evaporate water. If we ventilate 530 m^3/h for CO_2, and the incoming air is 24°C and 60% RH, we calculated that every m^3 of air will remove 11.7 g of water, or 530 × 11.7 = 6.2 L of water per hour. The eggs are producing 1.8 L/h, so we are 6.2–1.8 L short to keep the RH up. This means that the sprayer will have to produce 4.4 L of water, which has to be evaporated together with the water from the eggs, both contributing to cooling of the eggs. It takes 2.26 mJ to evaporate a litre of water, so 4.4 L/h will cool with 9.9 mJ/h. It means that the total cooling effect of the evaporation in the machine will be roughly 14 mJ; 4 mJ/h will come from evaporation of the eggs and 10 mJ from evaporation of the water from the sprayer.

When we add this to the total cooling needed, we come to the following calculation for eggs at 18 days of incubation:

- 100 000 fertile eggs of 18 days produce 54 mJ/h.
- Ventilation of 530 m^3/h takes out 21.2 mJ/h.
- Evaporation of water from the eggs takes out 4 mJ/h.
- Evaporation of water from the humidifier takes out 10 mJ/h.
- To be taken out by the cooling coils is: 54–21.2–4–10 = 18.8 mJ/h.
- To achieve that using water with a delta *T* of 10°C, we need 18.8 mJ/(10 × 4.184 kJ) = 450 L/h.

So, when we run the machines under these conditions, the amount of cooling that we need is limited. However, it requires quite a lot of spraying, and water from the sprayers is usually not very well distributed over the eggs in the machine. If we do not want to use the sprayer (and with it accept that the RH goes down), 10 mJ/h that is now removed by the evaporation of the water from the humidifier needs to be taken out by the cooling coils. The calculation then becomes:

- (54 mJ/h-21.2 mJ/h-4 mJ/h)/(10 × 4.184 kJ) = 700 L/h.

Therefore, with the ventilation on a level to keep the CO_2 to a maximum of 4000 ppm, we need 450 L of cooling water per hour in the machine if we keep the RH at 55% and 700 L/h if we switch off the sprayers and let the RH drop to the balance created by the ventilation for CO_2.

7 Incubator machine design issues

These equations show that the design of a machine is highly dependent on what we want to achieve. Do we need to ventilate for RH, for CO_2 or for heat? Do we want to avoid spraying? Do we want to keep the RH at a constant level with limited ventilation? Do we want to ventilate at a level that keeping the RH at the required level is sufficient to remove all the heat, as the sprayers will have to work hard to achieve a substantial amount of cooling? If we want to switch off or limit the sprayers but the machine does not have enough cooling capacity, the result will be overheating.

Although RH has no direct influence on temperature, the manner in which it is related to evaporation, ventilation and cooling makes it an important factor that can influence the results. This is why, especially in the past, RH has been seen as very important, as the machines didn't have enough cooling capacity to take over the function of the sprayer. This is true in a setter and even more in a hatcher, as the hatching chicks produce twice the amount of heat as an egg of 18 days, which makes cooling even more important.

So, what should a setter look like? There is no single answer to this question as designers will be able to find different solutions for the same problem, but with the parameters described we can estimate some of the boundaries for machine design.

First of all, it is pretty safe to assume that new machines will be based on single-stage concepts. The multi-stage concept, with an equal temperature from start to end, does not provide enough possibilities to control the eggshell temperature with the required accuracy. Although it is theoretically possible to create an air velocity high enough to get all eggs to air temperature, practically, it would be very difficult.

Temperature control in the machine should be based on the internal temperature of the egg, or more practically, on the eggshell temperature.

This can be done by placing infrared sensors or contact thermometers between the eggs. The uniformity of the temperature between different eggs and different positions within the machine is problematic. The biological variation in embryos, egg size, line, breed, etc. will contribute to variation in egg temperature; hence, adequate measurements will be needed. To obtain a more predictable indication of the heat loss at different positions, an egg-shaped thermometer with an internal heat source could be used. However, to use this to control the incubation process in a practical situation, the heat production of specific eggs in the machine should be known precisely at every moment of incubation to calculate the required air temperature resulting in the desired egg temperature.

The number one challenge for machine design will be to obtain an even egg or eggshell temperature for all (fertile) eggs in the machine, which is not easy to achieve, as it requires uniform air temperature, air velocity and evaporation. Complicating factors will be that differences in egg size, line, breed, etc. will contribute to variations in egg temperature. Establishing an equal air temperature in the machine will be a function of mixing the air thoroughly, which most machines are unable to do. A point to note is that the intake of fresh air is (usually) colder than the air in the machine, which can create potential cold spots. Creating an equal air velocity over all the eggs is much more challenging and requires a controlled pressure difference at different spots in the machine. A ventilator just moving the air will probably not result in an even air velocity, as the resistance of the trays and the openings between the trolleys will contribute to an unequal and uncontrolled pressure difference. Turning makes the distances between trays and trolleys variable and the egg size will also have an influence, which will complicate the airflow. To create an equal air velocity in a traditional turning system, the air will have to move through a tunnel-type system, creating an equal pressure on all the spots. However, this will have consequences for the ventilator's capacity to create enough pressure difference for an equal distribution.

The last factor is the spraying in the machine. As water droplets are not easily distributed equally through the machines, it will be difficult to get an even evaporation and therefore an even eggshell temperature if the sprayer is used. Therefore, a better option is to limit or not use the sprayer, which requires more cooling capacity and an adequate ventilation system. Systems that do not spray but let the moisture evaporate from a surface will still create local cooling, although the cooling will be more uniform. Systems that apply humidity in an evaporated form are beneficial. However, spraying water in the air inlet, corridor or air-handling unit will decrease the unevenness in the machine but will still create a non-uniform temperature, as the droplets need to take up energy before they evaporate, which will usually be done by the eggs in the machine. If humidity is added to the air-handling unit, additional heating will be needed to evaporate it and prevent cold spots in the machines.

Limited ventilation and spraying require adequate cooling. For hygienic reasons, flat surfaces are used for cooling but, as they create more laminar airflow than turbulent airflow, the exchange of heat (or cold) between a flat surface and the air can be challenging. A good design that considers the flow of air and the exchange of heat from cooling coils to air is beneficial.

Using a gradient of temperature over the eggs (as a mother hen does) would probably require a less fine-tuned temperature system but will require some technical adjustments and complicated technical solutions. Some smaller-scale incubators use this concept with good results, but it is quite challenging to upgrade such an idea to a commercial concept that can be used in the machine capacities required in modern hatcheries.

8 Hatcher design issues

After transfer (18 days), the embryos are in the so-called plateau phase: the embryo has reached its maximum O_2 uptake and, therefore, heat production. This means that the air temperature does not need to be reduced anymore, it can stay at the level required to keep the eggshell temperature at the optimal 100.0–100.5°F (38°C). Once internal pipping starts, the embryo will have more access to O_2 and will start moving within the egg to prepare for hatching, and as a consequence, the temperature of the eggs will rise to approximately 102.5°F (39°C). This is a natural process; however, if the eggshell temperature does not rise, the embryo might be too slow and lethargic to open the shell. Once the shell is opened, more O_2 becomes available and the embryo will (hopefully) hatch. Once the chick is hatched and dried, the heat production is approximately 0.3–0.35 W, so about twice as high for a day-old chick when compared with an egg at 18 days. If the birds are getting feed (early feeding), the heat production increases rapidly to 0.6 W or even more.

A point of concern during hatching is the air velocity over the chicks. Although this air velocity will help cool the chicks, they can feel uncomfortable when experiencing high air velocity, especially during the process of drying. As far as the author is aware, there is no literature available on the influence of air velocity on behaviour during the hatching phase, but observations have suggested that with air velocities above 0.5 m/s at chick level, the birds start to move away from the airflow and appear to become more stressed.

In a commercial hatchery, hatching traditionally takes place in a specialised hatcher. Nowadays, we see more 'alternative' hatching systems being used, especially in Western Europe. These systems are often motivated by the intention to feed the chicks immediately after hatch (early feeding). This can be done using specialised hatchers with feed and water available during hatching, or by bringing the fertile eggs, after 18 days of incubation, into the broiler house and letting them hatch in the presence of feed and water.

An important aspect during the hatching process is bacterial contamination. The high temperature, high humidity and presence of organic material create an optimal environment for bacterial growth, which can cause problems with omphalitis (navel/yolk sac infection) in the first days after hatching. Although the presence and growth of bacteria in the hatcher environment are difficult to avoid, the reduction of bacterial growth and the reduction of the transmission of bacteria from one chick to another is a factor to be considered in the design and operation of a hatcher.

9 Managing the hatching process

The period between the first and last chick hatching is called the hatch window and usually takes 24 h to approximately 36 h. When the hatch window is longer, it is usually because the last chicks are taking longer to hatch, and with it, there is a risk that chick quality, navel quality and hatchability will decline. In addition, the chance that chicks become dehydrated due to overheating is increased.

When the hatching process starts, the birds become more active and there is an increase in production of heat and CO_2. Knowing the heat and CO_2 production of the eggs/chicks, we can calculate the ventilation dimensions for the hatchers. Typically, hatchers are much smaller than setters, but for a good comparison, we can use the properties of our imaginary setter and calculate an imaginary hatcher with 100 000 fertile egg incubators. At 19 days, the heat of the embryo is almost identical to the level at 18 days, 0.16 W (vs. 0.15 W at 18 days). This means the CO_2 production goes up to 1947 L/h in our machine, and to keep the CO_2 level at 4000 ppm we need to ventilate 564 m^3/h, if the incoming air holds 500 ppm CO_2.

However, this changes rapidly once the chicks start internally pipping and hatching. The heat production increases to 0.25 W at 20 days and 0.35 W at 21 days. The CO_2 production increases accordingly, and to keep the CO_2 level in the machine at 4000 ppm we need to ventilate 881 m^3/h at day 20 and 1234 m^3/h at day 21.

The production of CO_2 increases during the hatching process, along with moisture loss. Every chick that hatches produces approximately 3 g of water, which needs to be evaporated. This evaporation will help the chick to cool down from the exhausting hatching process but will also increase the RH, depending on the ventilation rate. After a certain period of time, the RH becomes constant and then starts to drop again. This drop indicates that the last chicks have hatched and are drying, and the ventilation is removing more and more moisture without new moisture being added by newly hatched chicks.

During the period of internal and external pipping, the embryo is very sensitive to a drop in its internal temperature, as it will slow down the hatching process resulting in a dragging hatch, a long hatch window, increased problems with chick and navel quality, and eventually a reduced hatch. This means that care must be taken to regulate the level of cold air and spraying. In particular, if a large quantity of relatively cold incoming air reaches the eggs during the hatching process, the last chicks might be too slow to hatch in time, creating more problems with navel closure. To avoid this, the ventilation is often kept at a low level during the hatching phase. As a consequence of this limited ventilation, the RH and CO_2 levels in the hatchers will go up, which is why high RH and high CO_2 levels are associated with a relatively fast and good hatch.

The increase in CO_2 level and RH during hatching is a result of the moisture and CO_2 production of the embryo and chick, and is a natural indicator of the hatching process. It is not a stimulus for hatching, but a consequence of limited ventilation. This can be observed when alternative hatching systems, such as on-farm hatching, are used. The volume of a broiler house is much more than a hatcher and the natural ventilation through cracks and doors is much higher. As a result, both CO_2 and RH levels are much lower. However, if the temperature is controlled well, the hatching results are comparable and sometimes even better than in traditional hatchers, indicating that CO_2 and RH levels in traditional hatchers are the result of the limited ventilation that we apply, and not the trigger for the hatching process.

When the CO_2 level is set at a maximum, as is done in the setters, the ventilation will increase substantially during the hatching process as the CO_2 production of the embryos and chicks increases. With that increase in ventilation, a substantial amount of heat is also removed. Introducing high levels of ventilation to remove the CO_2 often results in cold spots, as the machine takes in substantial amounts of relatively cold air. To avoid these cold spots, the ventilation is often kept at a minimum level during hatching but this has consequences for cooling of the eggs. We have calculated for the setters that in a machine with 100 000 eggs at 18 days of incubation, the eggs produce 54 mJ/h. With the ventilation needed to keep the CO_2 at 4000 ppm, 21.2 mJ is removed through ventilation and 4 mJ/h through evaporation of the eggs. During the hatching process, the heat production increases due to embryo activity, from 0.16 W at day 19 to 0.25 W at day 20 and 0.35 W at day 21. When we calculate this per hour, it means a heat production of 57.6 mJ/h at day 19, 90 mJ/h at day 20 and 126 mJ/h at day 21. This heat also has to be removed, but when the ventilation is not increased (until all the chicks are hatched) it has to be removed either through evaporation or the cooling coils.

One would expect that the capacity of the cooling coils has to be increased dramatically as the heat production increases rapidly, but in this process, we are helped by Mother Nature. Each chick that hatches is evaporating 3 g of water, which is a significant cooling force. In our machine of 100 000 eggs, it would mean the evaporation of 300 L of water if all the chicks are hatching. The evaporation of 300 L of water costs 300 × 2.26 mJ = 678 mJ, spread out over a period of approximately 36 h. On average this will be 18.8 mJ/h, but of course the hatch will not be evenly spread over the hatch window, so the evaporation of water by the chicks that are hatching will also not be evenly spread. Once the last chick has hatched, the evaporation of the extra water and with it the extra cooling stops, but as we can then increase the ventilation without the risk of the last hatching eggs getting too cold, we can increase the heat loss with the extra ventilation. On top of that, the chicks are losing water through their normal breathing, which is approximately 1-2 g per 24 h. This means for 100 000 chicks approximately 4–8 L/h, i.e. a cooling rate of 9–18 mJ/h.

During the hatching period, the situation changes rapidly. If we calculate for an imaginary hatcher with 100 000 fertile eggs (to make the situation comparable with our imaginary setter), we get the following situations on consecutive days (assuming no spraying is used):

Day 18: the last day in the setter, just before transfer:

- Total heat production (0.15 W/egg): 54 mJ/h
- Ventilation of 530 m^3/h (4000 ppm CO_2) takes out 21.2 mJ/h
- Heat loss by evaporation of eggs: 4 mJ/h
- Required cooling capacity: 54–21.2–4 = 28.8 mJ/h

Day 19: first day in the hatcher, some eggs internally pipped:

- Total heat production (0.16 W/egg): 56 mJ/h
- Ventilation of 564 m^3/h (4000 ppm CO_2) takes out 22.5 mJ/h
- Heat loss by evaporation of eggs: 4 mJ/h
- Required cooling capacity: 56–22.5–4 = 29.5 mJ/h

Day 20: hatching in full progress:

- Total heat production (0.25 W/egg): 90 mJ/h
- Ventilation of 564 m^3/h (fixed, CO_2 increases) takes out 22.5 mJ/h
- Heat loss by evaporating eggs (half of the eggs still closed) 2 mJ/h
- Heat loss by evaporation of hatched and drying chicks: 18.8 mJ/h
- Required cooling capacity: 90–22.5–2–18.8 = 46.7 mJ/h

Day 21: all chicks have hatched and are dry, ready to be pulled, no additional spraying:

- Total heat production (0.35 W/chick): 126 mJ/h
- Ventilation of 1234 m^3/h (4000 ppm CO_2) takes out 49.4 mJ/h
- Heat loss by evaporation of breathing chicks: 9 mJ/h
- Required cooling capacity: 126–49.4–9 = 67.6 mJ/h

What happens if we humidify the air?

These calculations show that the situation during the hatching phase changes rapidly, and that especially during the last day(s), a significant amount of cooling is needed. We often see that the RH during and after hatching is kept at a relatively high level, which requires humidification of the air. This will support the cooling process if that water is evaporated in the machines. As the air volume is high, especially after the hatch, it requires a significant amount of humidification and evaporation.

Let us assume that we use the same conditions for incoming and outgoing air as in the setters. If the incoming air is 24°C and 60% RH, 1 m^3 of air contains 13 g of water. If the air in the machine is 37.5°C and 55% RH, it contains 24.7 g of water per m^3, so each m^3 takes up 24.7-13 = 11.7 grams of water. With 1234 m^3/h of ventilation, the air removes 14.4 L of water per hour. Part of that water will originate from the chicks and part from the sprayer. We estimated the chicks lose 4–8 L/h, so an additional 6–10 L need to be brought in each hour by the sprayer. The evaporation of 10 L of water costs 22.6 mJ/h, so the cooling coils do not have to take out 67.6 mJ but only 67.7-22.6 = 45.1 mJ/h.

Day 21: all chicks have hatched and are dry, ready to be pulled, additional spraying (55% RH):

- Total heat production (0.35 W/chick): 126 mJ/h
- Ventilation of 1234 m^3/h (4000 ppm CO_2) takes out 49.4 mJ/h
- Heat loss by evaporation of breathing chicks: 9 mJ/h
- Heat loss by evaporation of the humidifier: 22.6 mJ/h
- Required cooling capacity: 126–49.4–9–22.6 = 45.1 mJ/h

This calculation shows that when the sprayers in hatchers are used to increase the RH artificially, the evaporation of water will support the cooling of the eggs and less cooling capacity will be needed in the hatchers. It is often assumed that hatchers need a high RH to obtain a satisfactory hatch and good chick quality, and perhaps the additional spraying contributes to this because it helps to avoid overheating. However, as in setters, this evaporation is often very local and can, therefore, contribute to unevenness of temperature in the machine. It may be necessary to add extra humidification in situations where the cooling capacity of the machines is limited, but we must realise that there is increased risk of local cooling due to this evaporation. An additional complicating factor is that spraying of (extra) water will stimulate bacterial growth, contributing to bacterial contamination and possibly mortality due to omphalitis in the first week after placement.

10 What should a hatcher look like?

As in a setter, the air velocity in a hatcher should be distributed evenly over all the eggs/chicks. However, different to a setter, the increase in air velocity over the chicks should not be excessive, as they will experience a draft and will try to move over the tray to escape. A maximum air velocity at chick level of 0.5 m/s seems to be effective in a hatcher.

To obtain a uniform air velocity over the eggs and chicks, the machine should be carefully designed with respect to pressure differences created by the ventilators. As a ventilator does more than just mixing the air for uniform air temperature, a uniform and controlled air velocity over the eggs and chicks should be the target to create a uniform and controlled heat loss.

The temperature measurement should not be influenced by the incoming air or the amount of water. Temperature sensors are usually placed in a position with high air velocity. As this is close to the ventilator, it often means that the incoming air and the sprayer will cool the sensor, forcing the machine to increase the heating. The temperature sensor will then show a set value, but in some places the eggs/chicks might get overheated.

As the embryos reach their plateau phase, the heat production will remain relatively constant directly after transfer. This means that the eggshell temperature target should be 100–100.5°F (37.8–38°C). Once internal pipping starts, heat production increases, which is needed for successful hatching. From that moment onwards, the eggshell temperature should be allowed to increase to 102.5°F (39°C).

To avoid slow, dragging hatches and stimulate a short hatch window, direct cold air and/or spraying of water on the eggs should be avoided during the hatching process, i.e. the incoming cold air should be mixed with the air in the machine before it gets to the eggs. If this is not possible, the ventilation should be limited, which will lead to an increase in CO_2 and RH.

Spraying in the hatchers should be avoided as it adds to uneven temperatures and bacterial growth. To avoid spraying, the cooling capacity of the machines should be sufficient to remove the heat without the need for the additional cooling effect of humidifiers.

In addition to these requirements, setters and especially hatchers should be constructed in such a way that they are easy to clean. The cleaning of hatchers is a crucial but time-consuming procedure that needs to be done after every hatch, i.e. on a very regular basis. As there is a lot of organic material produced in a hatcher and they are used in high frequency, the cleaning procedures should be as easy and thorough as possible for optimal results.

11 Where to look for further information

Additional information on embryo requirements and machine design can be found in:

French, N. A. (2002). Managing the Incubation environment in commercial hatcheries to meet the requirements of the embryo. *Avian Poult. Biol. Rev.* 13: 179-185.

French, N. A. (1977). Modelling incubation temperature: the effect of incubator design, embryonic development and egg size. *Poult. Sci.* 76: 124-133.

Meijerhof, R. (2009). Principles and practice of incubator design. *Avian Biol. Res.* 2(1/2): 81-86.

Molenaar, R., Reijrink, I. A. M., Meijerhof, R. and van den Brand, H. (2010). Meeting embryonic requirements of broilers throughout incubation: A review. *Brazilian J. Poult. Sci.* 12(3): 243-250.

Molenaar, R, Hulet, R., Meijerhof, R., Maatjens, C. M., Kemp, B. and van den Brand, H. (2011). High eggshell temperatures during incubation decrease growth performance and increase the incidence of ascites in broiler chickens. *Poult. Sci.* 90(3): 624-632.

Wright, M. L. and Epps, N. A. (1958). Hatchery sanitation. *Can J. Comp. Med. Vet. Sci.* 22(11): 396-399.

12 References

Elibol, O. and Brake, J. (2003). Effect of frequency of turning from three to eleven days of incubation on hatchability of broiler hatching eggs. *Poult. Sci.* 82(3): 357-359.

Hulet, R. M., Gladys, G., Hill, D. and Meijerhof, R. (2000). Embryonic temperature effects on post-hatch performance in broilers. *Avian Poult. Biol. Rev.* 11: 298-299.

Kaltofen, R. S. (1955). Hatching experiments at Beekbergen–Turning the eggs. *Worlds Poult. Sci. J.* 11(3): 204-209.

Lourens, A. (2001). The importance of air velocity in incubation. *World Poult.* 17: 29-30.

Lourens, A., Meijerhof, R., Kemp, B. and van den Brand, H. (2011). Energy partitioning during incubation and consequences for embryo temperature: A theoretical approach. *Poult. Sci.* 90(2): 516-523.

Lourens, S. and Middelkoop, J. H. (2000). Embryo temperature affects hatchability and grow-out performance of broilers. *Avian Poult. Biol. Rev.* 11: 299-301.

Lourens, A., van den Brand, H., Heetkamp, M. J. W., Meijerhof, R. and Kemp, B. (2007). Effects of egg shell temperaure and oxygen concentration on embryo growth and metabolism during incubation. *Poult. Sci.* 86(10): 2194-2199.

Meijerhof, R. and Lourens, S. (2018). Ventilation during incubation is a complex matter. *Poult. World*, February 14.

Meijerhof, R. and van Beek, G. (1993). Mathematical modelling of temperature and moisture loss of hatching eggs. *J. Theor. Biol.* 165(1): 27–41.
Nangsuay, A., Molenaar, R., Meijerhof, R., van den Anker, I., Heetkamp, M. J. W., Kemp, B. and van den Brand, H. (2015). Differences in egg nutrient availability, development, and nutrient metabolism between broiler and layer embryos. *Poult. Sci.* 94(3): 415–423.
Nangsuay, A., Molenaar, R., Meijerhof, R., van den Anker, I., Heetkamp, M. J. W., Kemp, B. and van den Brand, H. (2021). Effects of oxygen concentration during incubation and broiler breeder age on embryonic heat production, chicken development, and 7-day performance. *Animal* 15(9): 1751–7311.
Owen, J. (1991). Principles and problems of incubator design. In: Tullett, S. G. (Ed.), *Avian Incubation*. London: Butterworth-Heinemann, pp. 205–224.
Romanoff, A. L. (1960). *The Avian Embryo: Structural and Functional Development*. New York: MacMillan Company.
Romijn, C. and Lokhorst, W. (1960). Foetal heat production in the fowl. *J. Physiol.* 150(1): 239–249.
Romijn, C. and Visschedijk, A. H. J. (1952). *Onderzoekingen over Kunstmatig Broeden (2e Mededeling), Uitgave* (Vol. V). B.N.
Tzschentke, B. and Rumpf, M. (2011). Embryonic development of endothermy. *Respir. Physiol. Neurobiol.* 178(1): 97–107.

Chapter 10

Understanding the effects of incubator temperature on embryo and post-hatch chick development

R. Michael Hulet, Penn State University, USA; and Michael J. Wineland, North Carolina State University, USA

1 Introduction

Early research on incubation showed that the major factors controlling incubation were temperature, humidity, ventilation, position and turning, egg handling, and sanitation (Romanoff, 1960). While ventilation, egg handling prior to and during incubation, and sanitation have not significantly changed, temperature and humidity have seen significant changes in the incubation process. Temperature is often said to be the most important parameter of incubation. Romanoff (1936) was able to show the effects of different incubation temperatures upon development. Our incubators use a temperature set point (multistage) or program (single-stage) to regulate the temperature that the incubator will run at, although that may not be the temperature that the embryo is experiencing. The temperature of the embryo is also influenced by the other parameters of incubation. Ventilation within the incubator also impacts embryonic temperature by the ability of air movement over the eggs to remove

http://dx.doi.org/10.19103/AS.2022.0118.14

excess metabolic heat from the older embryos and to provide heat to the younger embryos. Humidity can influence temperature of the embryo because of its ability to allow heat transfer between the air and the egg. Additionally, turning can influence embryonic temperature in some incubators by how it allows for proper airflow over the eggs. The negative effects of temperature upon the embryo (hatchability, yolk utilization, and some organ maturation) can also be offset by availability of oxygen to the embryo either by a higher eggshell conductance or increased oxygen level in the incubator (Wineland, M. J., unpublished). Eggshell conductance of an egg is dependent upon the number of pores, diameter of pores, and length of pores (shell thickness) (Rahn et al., 1981).

It is important to understand that the detrimental effect of abnormal temperatures upon the proper development of an embryo is dependent upon two factors. The effect is influenced by the deviation of the abnormal temperature from the optimum and the duration that the abnormal temperature is influencing the embryo. Investigations of temperature effects upon the developing embryo and the hatchlings subsequent performance in the field have been approached by researchers from two directions. Investigators have looked at the detrimental effects of elevated or suboptimal temperatures upon the developing embryo and hatchling performance in commercial operations. Others have looked at how altering the temperature (thermal manipulation) during different periods of incubation could potentially enhance development of different organ systems and thus potentially enhance performance of hatchlings.

In the past 30 years, significant changes in the rate of growth and development of commercial poultry have occurred. Cunningham (2006) reported comparative production data for broiler production between 1980 and 2002. From his data, it was calculated that broilers could achieve a 2 kg body weight in 61 days, while in 2002, broilers were able to achieve the same body weight in 41 days. The genetic increase in body weight gain was accomplished with no change in the 21-day incubation time for the eggs. Therefore, for the life of the bird, embryonic growth in the egg takes on an increasingly important percentage of the total development of the bird. Incubation time now represents 30–40% of the total development time for the embryo/bird, while 20 years ago, incubation time was only 20–25%. Therefore, more emphasis on optimizing the growth and maturity of the developing embryo is needed to be able to maximize post-hatch growth.

2 Single-stage versus multiple-stage incubation

Factors that affect embryo temperature are egg size and means of incubation, such as multiple-stage versus single-stage incubation. Multiple-stage incubation was designed so that the embryo heat of the older embryos could provide heat to the younger embryos, making an economical method of incubation, but not necessarily best environment for the embryos. Typically, multistage incubators have slightly cool eggshell temperatures at the start of incubation, which can result in slow embryo development and poorer chick quality, and, more importantly, high eggshell temperatures at the end of incubation, which may increase late embryo mortality and reduce chick quality. Early investigators (Romanoff and Faber, 1933) showed that lowering the temperature during the latter part of incubation could have positive effects upon embryo growth, less embryonic mortality, and higher quality chicks. Tejeda et al. (2021) presented data that indicated egg tray location within the incubators because of non-uniformity of temperature within an incubator has an impact on embryos and subsequent growout. Selection of the optimum incubation temperature profile is important to achieve greatest hatchability (Swann and Brake, 1990; French, 1997) and chick quality (Wilson, 1991; Decuypere and Michels, 1992). Increased eggshell temperature in the later stages of incubation was found to be related to decreases in hatch time, hatchability, and chick/poult quality (Hulet and Meijerhof, 2000). Additionally, possibilities of greater heat production in the more modern breeds have been implicated. Meijerhof (2000) calculated heat production of high-yielding embryos to be significantly greater than reported in classic broilers. However, it is also possible that the greater egg size in the recent breed crosses could be the contributing factor.

Observations have shown eggshell temperature (EST) variations of 2.2°C within a multistage setter among fertile eggs with some eggs reaching EST of 41.1-41.7°C (Hulet et al., 2007). While problems with machine maintenance, incubator cooling, airflow patterns, or other conditions may cause embryos to overheat (French, 1997; Mauldin and Buhr, 1995), the result is lower hatchability and reduced chick quality (Hulet et al., 2000).

3 Embryonic heat requirements

The difference between late embryonic temperature and incubator air temperature is dependent upon thermal conductivity, which, in turn, is mostly influenced by the air velocity over the eggs (Meijerhof and van Beek, 1993). Significant production of metabolic heat from the embryo starts around day 4. By day 9, embryo temperatures are greater than incubator air temperature. Therefore, removal of heat from the embryo, rather than just distribution of the air within the incubator, becomes a critical factor influencing hatchability (Lourens et al., 2006). The uniformity of the air velocity within the incubator will depend on the ease at which the air is able to pass over the eggshells to dissipate heat and metabolic water from the incubator cabinet.

Physical conditions or design of the incubator can also influence thermal conductivity. For example, failure to properly fasten air-directing curtains in a multistage incubator, incomplete turning, positioning of larger eggs in the incubator, or design of egg trays may result in air passing around the egg mass rather than over the eggshell (Mauldin, 1989).

The embryo has some ability to react to non-optimal incubation temperatures by altering blood flow in the chorioallantoic membrane. During the latter part of incubation, the blood flow through the chorioallantoic membrane will increase when exposed to increased ambient temperature and it will decrease when exposed to decreasing ambient temperatures (Holland et al., 1998; Nichelmann and Tzschentke, 2003).

Lourens et al. (2007) showed that elevated incubation temperature had a greater influence than oxygen concentration during the first week of incubation on growth of the embryo and that oxygen concentration was more important during the third week of incubation. This observation would coincide with the work of Rahn (1981) as the embryo during the latter part of incubation is subjected to the plateau stage of oxygen consumption.

Values of maximal heat production of eggs during incubation published by Romanoff and Faber (1933) have been shown to be much lower than that measured for our modern broiler and layer strains (Hulet and Meijerhof, 2000). The overheating of embryos found in many multistage hatcheries over the last decade has occurred due to different age breeder flocks, various sizes of eggs, and inadequate maintenance of the incubators or hatchery hallways. While standards of hatchability have not changed perceptively, managers with less ability to change incubation environs suffered decreased hatchability and post-hatch performance. For that reason, more emphasis on single-stage incubation that has all eggs incubating at a similar stage of embryonic development has resulted in greater success in meeting the thermal needs of the developing embryo at different embryological stages (Lourens et al., 2006). Hamidu et al. (2018) showed that different genetic lines within a genetics company can demonstrate differences in heat production. During the last week of incubation, the Ross 708 was more sensitive to incubation temperature than the Ross 308 as the Ross 708 showed increased heat production as the embryo temperatures increased and the Ross 308 did not show this increase. Between lines from different genetic companies, it has been shown that there are differences between broiler lines in their efficiency to convert energy to yolk-free body mass (Nangsuay et al., 2015). These investigators questioned if the optimal eggshell temperature should be the same for the two different broiler lines. When comparing Ross 308 to Cobb 500 FF, Nangsuay et al. (2016) proposed that differences in response to a normal and elevated incubation temperature may be due to eggshell conductance which influences availability for critical gas exchange between the embryo and the exterior environment. The ability of the egg to dissipate heat is directly related to egg size. French (1997) stated that as egg mass increases, thermal conductance does not increase proportionally because shell surface area does not increase proportionally to the increase in egg weight. Therefore, larger eggs have greater difficulty losing metabolic heat produced by the embryo.

Janke et al. (2013) demonstrated that different lines of chicken showed different heat production from day 12 to hatch. The layer line had a lower heat production during this period than did the broiler lines (Ross 308 and Ross 508). Additionally, van den Brand et al. (2019) showed that the rooster contributes to the embryos heat production. They demonstrated in two different genetic laying hen crossbred lines where the female line was the same but different rooster lines and the heat production was higher in the AB crossbred when compared to the BB crossbred.

4 Effect of temperatures on hatchling quality and growth

Traditionally, hatchery managers were thought to be responsible only for hatchability and the first 3-day mortality in the brooder house. However, Hulet et al. (2007; Table 1) showed that incubation conditions can significantly affect broiler post-hatch growth to 44 days of age and feed conversion to 21 days of age.

The impact of embryonic temperature on broiler performance has been difficult to measure. Most of the earlier published work on incubation temperature and post-hatch performance reports incubator air temperatures and not the temperature of the embryo (Wilson, 1991). When reading the published work on the effects of different temperatures upon embryonic development, one must consider if the research is reporting incubator set

Table 1 Effects of embryo shell temperature on post-hatch broiler body weights, feed conversion, and percent mortality (Hulet et al., 2007)

	Embryo shell temperature			
Factors[a]	37.5°C	38.6°C	39.7°C	SEM
Body weight (g), 44 days	2213.8[b]	2263.3[c]	2165.7[d]	9.77
Feed conversion, 1-21 days	1.56[b]	1.55[b]	1.60[c]	0.26
Feed conversion, 1-44 days	1.91	1.86	1.87	0.01
Mortality (%), 1-21 days	2.21	2.55	3.99	0.41

[a] Each value is the mean of 8 pens (34 birds/pen).
[b-d] Means within a row with no common superscripts differ significantly ($P < 0.05$).

point or eggshell temperature which is slightly greater than the set point of the incubator during the latter part of incubation. Field observations of decreased hatch times and hatch residue lesions (poor closure of the navel, enlarged yolk sac, etc.) showing the effects of increased temperature (inadequate removal of excess metabolic heat) have led many to believe that maximum heat production from our modern breeds is higher than previously calculated and results in overheating. Whether the difference is egg mass (Lourens et. al., 2006) or whether it is a result of increased metabolic activity, the result is that there is a need to balance our incubation systems to optimize the temperature that best suits the development of the embryo.

Overheated embryos either in the setter or hatcher characteristically show significantly decreased yolk-free chick weight (YFBM), decreased heart size, and heart as a percentage of chick weight when compared to controls (Wineland et. al., 2000a, b). Other hatch residue observations among chicks or poults have been red, bruised hocks, enlarged yolk mass, unhealed navels, increased late dead embryos, malposition of late dead embryos with head over the top of the wing, malposition of late dead embryo with head in the small end, and others (French, 1994; Hulet, et.al., 2007).

Wineland et al. (2000a) reported that week-3 embryonic temperature affected chick heart size as a percentage of body mass. Maatjens et al. (2016b) using different eggshell temperatures beginning at day 15, 17, and 19 found that eggshell temperatures lower than 37.8°C and onward appear to be beneficial with improved yolk-free body mass and organ development.

The finding of lower relative heart size in chicks became a consistent characteristic of overheating and poor chick quality in hatchery dead-embryo and chick evaluations. Other studies have shown that the variation in chick performance was related to decreased yolk-free body mass (YFBM) and increased yolk weight (Lourens et al., 2005). That is, yolk mass was increased, and yolk-free body mass decreased for hatchlings grown under the high versus normal temperature. Willemsen et al. (2011) suggested that the increased yolk size at hatch is because the embryo when exposed to high temperatures shifted their metabolism toward an anerobic state. They showed that these embryos demonstrated higher pCO_2 and lower pO_2 in the air cell as well as increased presence of lactate. A common observation of poor incubation temperature is seen as abnormal absorption and utilization of the residual yolk sac during the hatching process. Dayan et al. (2020) determined the effect of abnormal temperatures of 1.5°C above and below the normal incubation temperature of 37.8°C upon residual yolk sac and upon gene expression. These investigators showed that 37.8°C had the best yolk utilization, while the elevated temperature had the poorest and the cooler temperature was intermediate. They further demonstrated that incubation temperature affected the genes involved in gluconeogenesis and glycogenesis. Other investigators (Molenaar et al., 2013)

Table 2 Body, yolk, and yolk-free body mass for poults from turkey eggs incubated at different eggshell temperatures from 1 to 25 days (Lopez et al., 2018)

Embryo shell temperature treatment[a]	Hatch of total eggs (%)	Body weight (g)	Yolk weight (g)	Yolk-free body mass (g)
99.2-99.4°F	88.23[b]	52.11[b]	3.38[c,d]	48.73[b]
99.34-99.6°F	89.25[b]	52.17[b]	4.48[b]	47.70[b,c]
100°F	84.75[b,c]	49.99[c]	3.87[b,c]	46.10[d]
100.5-101.0°F	81.71[c]	49.22[c]	2.60[d]	46.63[c,d]

[a] Each value is the mean of 8 pens (34 birds/pen).
[b-d] Means within a row with no common superscripts differ significantly ($P < 0.05$).

demonstrated how elevated incubation temperatures increased anaerobic glycolysis because those embryos have a higher metabolic rate.

Determination of the optimal embryo temperature for growth, maturation, nutrient utilization, hatchability, and hatchling quality has received research attention (Lourens et al., 2005; Hulet et al., 2007; and Leksrisompong et al., 2007). While several studies in other avian species have been reported, studies to fine-tune turkey hatchability have not been done. Therefore, a study looking at four different eggshell temperatures (99.2-99.4, 99.4-99.6, 99.8-100.2, and 100.5-101.0°F) from day 1 to day 25 of incubation was conducted. Results show a decrease in hatch of fertile eggs for eggs incubated at eggshell temperatures greater than 100.5°F (Table 2; Lopez et al., 2018). Body weight and yolk-free body mass were also decreased for turkey eggs incubated at these same higher eggshell temperatures. The optimal eggshell temperature throughout incubation for turkey embryos seems to be lower than the optimum EST found for broiler embryos and could be related to the greater turkey egg mass (French, 1997).

Maatjens et al. (2016a) used different eggshell temperatures (35.6°C, 36.7°C, 37.8°C, or 38.9°C) during the last week of incubation on Ross 308 broiler eggs. These investigators showed that poorer chick quality, organ development, and growth performance parameters were seen when using 38.9°C, while the eggshell temperature of 35.6°C demonstrated better or as good of results as a 36.7°C temperature.

5 Incubational thermal manipulation

The use of thermal manipulation has been investigated to improve organ system development and maturation to better handle post-hatch environments. With the genetic selection for growth in broilers (Cunningham, 2006; Havenstein et al., 2003) and turkeys (Christensen et al., 2001) comes a weakness during a seasonal acute environmental temperature increase (Piestun et al., 2011). However, thermoregulation ability (by changing incubation temperature) was altered when young chicks were exposed to short periods of elevated temperature. The birds were able to cope with the higher temperatures and eliminate the potential negative effects (Yahav and Huriwitz, 1996). Incubation allows a more controlled environment to impose a thermal manipulation upon the embryos. Research trials to stress the embryos during the last stages of incubation and see if the hatchlings would thermoregulate when they reached market age were conducted in broilers (Piestun et al., 2013; Tzschentke and Halle, 2009) and turkeys (Krischek et al., 2013). However, no significant differences were found between control and post-hatch heat stress on hatchability, growth, slaughter, or meat quality (Werner et al, 2010).

Janke et al. (2006) demonstrated that short time increases in incubation temperatures during the last 4 days of incubation may be advantageous compared to a control incubation temperature or continuous elevated temperature. The subsequent growth was greatest when the short time increase in temperature was used. Wilsterman et al. (2015) showed that slightly elevated temperature during incubation failed to develop a glucocorticoid (GC) peak, while the normal incubated embryos did show a peak of GC when exposed to a heat challenge. Iqbal et al. (1990) and Al-Rukibat et al. (2017) showed that a thermal manipulation late in embryo development helped the birds to respond to a heat stress later during the growing period. Iqbal et al. (1990) demonstrated 2 groups: one a control group incubated at 37.8°C and a second group exposed to 39°C only between day 11 and day 20 of incubation for 6 h/day when the thyroid gland is maturing. Both groups had been grown under similar conditions with no differences in body weight when an acute heat stress of 38°C for 1 h at 40 days was applied. Both groups experienced reduced body weight, but in the control group the acute heat stress caused a significant elevation of plasma T3 and corticosterone, while the incubation-treated birds exhibited no hormonal changes.

6 Influence of embryonic temperature on organ development

Understanding that embryo and chick/poult development could be influenced by temperature, humidity, conductance, and other environmental factors, many researchers focused their studies to explain some embryonic organ growth and/or maturation at higher and lower incubation temperatures than what was considered optimal (37.5°C). Wijnen et al. (2020a) investigated the use of a lower eggshell temperature during week 3 of incubation. They found that a lower temperature of 36.7°C during the last week stimulated some organ growth in newly hatched chicks, but the performance in growout was not better.

6.1 Bone

Some bone malformations are found in all commercial flocks of chickens and turkeys and averaged around 4%. Observations by Applegate and Lilburn (2002) showed the greatest period of bone growth was days before and just after hatch. As leg problems were also responsible for welfare problems (foot pad dermatitis, hock burns, morbidity, and mortality), incubation conditions were examined (Oviedo-Rondon et al., 2008; Oviedo-Rondo et al., 2009a). Oviedo-Rondo et al. (2009a, b) found that broiler bone development is influenced by incubator temperature, oxygen concentration, and eggshell conductance at the plateau stage in oxygen consumption. Studies showed that proper late-stage incubation and transportation temperatures (37.8°C) could significantly decrease the incidence of leg problems and therefore could reduce welfare problems in high-yielding broilers (Groves and Muir, 2014; Oviedo-Rondo et al., 2009b). Van der Pol et al. (2014) showed that elevated incubation temperatures resulted in decreased tibia femur and metatarsus length.

Guz et al. (2020) investigated the effects of eggshell temperature pattern on tibia development of bone seen at the time of slaughter. Their results indicated that an elevated eggshell temperature (38.9°C) rather the normal 37.8°C temperature during the second week of incubation resulted in better morphological, biophysical, and mechanical tibia characteristics, especially when followed by 37.8°C in the third week of incubation. These investigators did note that although the treatment improved bone development, it did not affect the incidence of leg disorders at slaughter.

6.2 Heart

Christensen et al. (2004) demonstrated in turkey embryos exposed to 36°C, 37°C, 38°C, 39°C temperatures during the last 4 days of incubation, that the elevated temperatures depressed heart weight but not body weight and decreased heart tissue energy metabolism. The depressed heart weights have also been demonstrated by many other investigators (van den Brand, 2019; Leksrisompong, 2007; Wineland et al., 2000 a,b).

6.3 Endocrine

Effects of temperature on two of the endocrine organs that have been investigated are the thyroid and adrenal glands. The hypothalamic-pituitary axis acts on both. The thyroid gland is a very important organ system that becomes functional in the developing embryo sometime between 11 days and 19 days of incubation. This organ is important in the development of other organ systems. The thyroid hormones T3, T4 and T3:T4 ratio was determined in chicken embryos from two breeds (Wineland et al., 2006b). Eggs from one breed exhibited a low-conductance eggshell, while eggs from a second breed exhibited a higher eggshell conductance when all were divided into four groups for each breed and were exposed to 36°C, 37°C, 38°C, or 39°C during the last 4 days of incubation. The hatchlings exposed to the higher temperature demonstrated suppressed T3 and T4 hormone concentrations.

Morita et al. (2016a) in a trial used normal, suboptimal, and elevated incubation temperatures from day 13 to hatching. The low temperature treatment showed lower T3 at hatch and then again at 42 days compared to the control. The high treatment temperature showed lower T3 levels at 21 and 42 days of age compared to the control. It was also shown that the high-temperature treatment preferred a higher ambient temperature during growout. Morita et al. (2016ab) found lower T3 and growth hormone levels in hatchlings incubated at a lower temperature from 13 days onward.

The hypothalamic-pituitary axis also has an impact on the adrenal gland which produces corticosterone (Ruuskanen et al., 2021). Wilsterman et al. (2015) investigated if elevated temperatures during various times of incubation could alter the release of glucocorticoids in response to a heat challenge 3 weeks post-hatch. These investigators demonstrated that following the post-hatch heat challenge, the early elevated incubation temperature decreased the peak level of glucocorticoid compared to the control. Piestun et al. (2009a) demonstrated how thermal manipulation for 12 h/day from day 7 to day 16 of incubation lowered the functional setpoint of the adrenal and thyroid resulting in a lower metabolic rate during embryonic development and at hatch. Moraes et al. (2004) believed that the optimal time for administering thermal conditioning for thermotolerance was during 14–15 days of incubation. They noted that the thermal conditioning did not cause any stress-induced tissue damage. Also, they noted that this thermal conditioning decreased hatchability and delayed the hatching process. Moraes et al. (2003) demonstrated that exposing embryos to a thermal conditioning of 2 h/day from day 13 to day 17 of incubation assisted in the ability of the broiler to cope with heat stress when they were 33–43 days of age. These investigators believed that this thermal manipulation benefited the broilers to decrease thyroid hormone and triglycerides during heat stress.

6.4 Thermoregulation

Thermal manipulation has been shown to alter corticosterone and thyroid hormones. Thermoregulation is controlled by the hypothalamic-pituitary thyroid axis (HPT) in the brain which in turn influences the release of thyroid-stimulating hormone to act on the thyroid gland to release thyroxine (T4), which is converted to the biologically active T3 that is important in controlling thermogenesis. The hypothalamic-pituitary adrenal axis (HPA) also has an impact on corticosterone production (Ruuskanen et al., 2021). Both the HPA and HPT axes can alter the physiology of thermal regulation. Tzschentke and Basta (2002) demonstrated that there is influence on thermoregulation by neuronal and hormonal response when using thermal manipulation.

When birds are outside of their thermoneutral range, they use physiological and/or behavioral mechanisms to help them adapt. Tzschentke and Nichelmann (1999) investigated development of the avian thermoregulatory system. They incubated Muscovy eggs at a control temperature of 37.5°C and at a treatment of 34.5°C during the last week of incubation. The lower incubation temperature resulted in greater heat production in the incubated ducklings and a higher body temperature than control incubated ducklings. They hypothesized that the lower incubation temperature during the last week decreased the thermoregulatory setpoint and thus increasing cold tolerance. Morita et al. (2016a) demonstrated that chicks that were incubated at an elevated temperature from day 13 to hatch preferred a higher ambient temperature during growout. Loyau et al. (2014) using cold incubation temperatures and subsequent cold post-hatch exposure suggested that cyclically cold incubation could alter pathways involved in energy metabolism and antioxidant defense long term.

6.5 Muscle and adipose tissue

Maltby et al. (2004) showed an alteration in muscle development, fiber number, and muscle weight in 16-day turkeys when increasing the incubation temperature for 4 days early in incubation. Similar work in broilers showed increased carcass and meat yields without any harm to meat quality standards (Werner and Wicke (2008); Werner et al. (2010); Krischek et al. (2016); Ipek and Sozcu (2015); Leksrisompong et al. (2009)). Werner et al. (2010) when increasing incubation temperature between day 7 and day 10 by 1°C in Cobb 500 eggs showed no difference in males between the normal and elevated incubation temperatures. Only the normal temperature hens had significantly larger muscle fibers when compared to the other groups. There was also no difference in slaughter and meat quality indices except for the high-temperature males. While elevated temperatures early in incubation was positive for muscle development, the elevated temperatures during the last 4 days while the embryo is experiencing some hypoxia in the plateau stage of oxygen consumption resulted in decrease breast muscle weight and degeneration of muscle fibers (Christensen et al., 2007).

Piestun et al. (2009b) showed that if short daily increases in temperature were provided during the latter part of incubation, there was a stimulatory effect on cell proliferation creating a higher reservoir of myogenic progeny. In a subsequent publication, Piestun at al. (2015) showed that elevated temperature during the middle period of incubation increased the myogenic progeny reservoir in muscle, resulting in increased muscle growth during incubation and post-hatch.

Janisch et al. (2015) examined the effect of 3 treatments, control (37.8°C), lower (36.8°C) and higher (38.8°C) temperatures at 2 periods (7–10 days and 10–13 days) of incubation, which covers the periods of embryonic myogenesis upon day old and 35-day-old broilers. The higher temperature had the greatest weight at 35 days, while the lower temperature had the lowest weight at 35 days. This indicates elevated temperatures during the period of embryonic myogenesis had the greatest effect. Yalcin et al. (2022) demonstrated the effects of an elevated temperature being applied from day 10 to day 14 of development. These researchers showed that at slaughter, the elevated temperature had greater body weight and increased muscle fiber area. Interestingly, their work showed different broiler breeds were affected differently with regard to gene expression in breast muscle. Collin et al. (2007) had conducted a trial using a thermal manipulation during an early and or late period of incubation to see if there was an effect upon thermotolerance and breast muscle characteristics. The results showed that the thermal manipulation had no long-term effects in the chickens, but there was an effect upon breast muscle growth.

Clark et al. (2017) demonstrated that they could influence satellite cells in muscle which develop during the latter part of embryo development. The purpose of these cells is to repair damaged muscle fibers and increase the growth of the muscle fibers in the broiler in the house. Eggs were incubated to day 14 at 37.8°C. During 14–18 days of incubation, the eggs were kept at the control temperature or other groups of eggs were subjected to 39.5°C for 3 or 12 h/day. The chicks were separated into early, mid, or late hatching chicks. The birds were grown out for performance indicators and meat quality. The investigators found the group which experienced the elevated temperature for 12 h had reduced body weight and poorer feed conversion throughout the trial. The early hatched broilers were significantly heavier at processing when compared to the mid- or late-hatched birds. The broilers exposed to the 12-h treatment had reduced number of birds demonstrating moderate or severe myopathies. Additionally, the late-hatched birds had fewer fibrotic and necrotic muscles compared to the early hatched group. Although the following work (Harding et al., 2016) is conducted in cell culture, it is still interesting that using normal, above normal, and below normal temperatures, they investigated effects upon different chicken skeletal muscle. They showed that satellite cells of pectoralis major muscle (anaerobic) are more sensitive than those in femoris muscle (aerobic) in both the hot and cooler conditions. Additionally, in another report by Harding et al. (2015), using increased incubation temperatures in cell culture, these investigators showed that the p. major (anaerobic) satellite cells are more predisposed to adipogenic conversion than the femoris cells which are aerobic.

Thermal manipulation during the late period of incubation in the duck showed that there was an endoplasmic reticulum stress (ERS) in the breast muscle (Li et al., 2017). These investigators believe that the temperature increase inhibited muscle hypertrophy and proliferation at a post-transcriptional level by activation of ERS. Oviedo-Rondon et al. (2016) showed negative effects on early muscle development and on the severity of wooden breast by suboptimal incubation temperatures. The effect of incubation upon muscle myopathies such as woody breast, white striping and spaghetti muscle has been presented in a review by Oveido-Rondon et al. (2020). In this review, the authors suggest that the myopathies can start early in incubation by causing alterations in breast muscle histomorphology of chicken, turkey, and duck.

Both myoblasts and adipocytes originate from stem cells (Bowers and Lane, 2007). Thus, it has been thought that the incubator temperature could alter the development of muscle fibers and adipose tissue. Al-Musawi et al. (2012) suggested that in the broiler, different temperature manipulations could influence mesenchymal stem cells to myogenesis or adipogenesis. Yahav and Brake (2014) speculated that thermal manipulation of the embryo during the first 12 days of incubation affects fat tissue formation as that the time coincides with embryonic adipose tissue formation. Almeida et al. (2016, 2022) used an optimal eggshell temperature and a higher and lower temperature from day 13 onward to see if the temperatures affected fat deposition. They showed that thermal manipulation altered both cervical and abdominal adipocyte size.

6.6 Intestine

Broiler and turkey chicks hatched from eggs incubated at high temperature (39°C) in the last 4 days of incubation had a reduced jejunal maltase and alkaline phosphatase activity compared to chicks from eggs incubated at normal temperature (37.5°C), indicating less mature intestine development (Wineland et al., 2006a, Christensen et al., 2004). Wineland et al. (2006a) also showed the effect of incubation temperature on intestine maturation was greater in a broiler breed that had, on average, eggs with a lower eggshell conductance compared to a breed with a greater eggshell conductance. In a similar trial with turkey eggs during the last 4 days of incubation by Christensen et al. (2004), they also showed that elevated temperatures depressed intestinal maltase and alkaline phosphatase activities.

6.7 Immune system

Oznurlu et al. (2010) conducted a trial where they incubated eggs at 37.8°C until 10 days of incubation. At that time, they divided the eggs into two groups, a control and a treatment 38.8°C for the remainder of incubation. The results showed that the elevated temperature group showed retarded development of both the thymus and bursa of Fabricius. Additionally, the elevated temperature resulted in decreased blood ACP-ase and ANAE-positive lymphocyte levels. These investigators surmised that immunosuppression could be induced by high temperatures during incubation. Other investigators showed similar results in ducks (Liu et al., 2013) when they applied an additional 1°C above normal temperature from embryonic day 11 to day 24. They noted that the increased temperature depressed the bursa of Fabricius index and the spleen index. However, Shanmugasundaram et al. (2018) showed that duck embryos which had been incubated at a 0.5°C higher temperature than the controls from day 1 to day 21 had increases in Heat Shock Protein 70, Major Histocompatibility Complex I, and inflammatory cytokines with an immune response favoring cell-mediated immunity.

Work by Wijnen et al. (2020b) utilized different eggshell temperatures. During week 1, all eggs were incubated at 37.8°C. In week 2, the EST was either 37.8°C (control) or 38.9°C (higher), and in week 3 the EST was either control or 36.7°C (lower). At hatch, histology of bursal follicles, jejunal villi, and crypts as well as heterophil:lymphocyte ratio (H:L) was examined. Chicks received a Newcastle disease vaccination. The higher EST during second week resulted in lower density within bursal follicles, higher Newcastle titers when slaughtered, and a tendency for lower H:L ratio at hatch, suggesting that a higher EST in week 2 during incubation may benefit embryonic immune organ development and post-hatch immunocompetence while the lower eggshell temperature in week 3 showed opposite indications.

7 Incubation temperature effects on incidence of ascites

Molenaar et al. (2011) published results of different EST on the growth, performance, and incidence of ascites in market chickens. High eggshell temperatures (EST; ≥38.9°C) during the second half of incubation are known to decrease the body and organ development of broiler hatchlings when compared with eggs incubated at a normal EST of 37.8°C. Relative heart weights are decreased by a high EST, and this may increase the incidence of metabolic disorders that are associated with cardiovascular development, such as ascites. Molenaar et al. (2011) determined that the effect on the heart (ventricle size) resulted in increased incidence of broilers with ascites. Eggs were incubated at a normal (37.8°C) or high (38.9°C) EST from day 7 of incubation onward. After hatching, the broilers were brooded by EST treatment in pens, at a normal or cold temperature schedule that was applied during the growout period. Hatchability, hatchling quality, body weight, feed conversion ratio, total mortality, mortality associated with ascites, and growth performance were determined. Hatchability was comparable between incubation temperature treatments, percentage of lower quality chicks was greater, yolk-free body mass and heart weights were also less. While body weight was significantly different for growth at 21 and 42 days of age, heart size as a percentage of body weight and percentage incidence of ascites at market age were greater. Total mortality (4.1% greater) and mortality associated with ascites (3.8% greater) were greater compared to the normal EST. High incubation treatment increased the incidence of ascites, which may be related to the reduced right ventricle development at hatch.

Sozcu and Ipekce (2015) utilized both an acute elevated temperature of 39°C for 3 h from day 19-20 or a chronic high temperature of 39°C during 19-20 days to compare incubation temperature. After 6 weeks of growth, the total mortality was 2.5%, 9.2%, and 13.3%, respectively, for the control, acute, and chronic groups. Mortality due to ascites was 2.1%, 8.3%, and 12.9%, respectively.

Additional research has been examining the use of periodic low eggshell temperatures during incubation as a method to reduce the incidence of ascites (Afsarian et al., 2016; 2018). Afsarian et al. (2016) demonstrated that incidence of ascites was reduced when eggs were exposed to 15°C for 1 h on days 11,13,15, and 17, which is the period when thyroid maturation occurs. Afsarian et al. (2018) showed that in addition to the periodic exposure at 15°C for 1 h, an injection of T4 also enhanced the chick quality, productive performance, and reduced incidence of ascites.

8 Temperature effects on integument system

One factor that has positive animal welfare implications was the effect on feathering. More dense and faster feathering has the potential to protect the skin from damage and prevent bacterial exposure to the skin of the market chicken or breeder chicken. Scott et al. (2015) and Ipek and Sozcu (2016) conducted studies using different methods, but both showed improved effects on using late incubation high EST can result in greater feather density and, therefore, better welfare status of the birds.

Da Costa et al. (2015) examined the effects of incubation temperature and eggshell conductance on duck footpad development. Since incubation temperature drives the rate of embryonic development and thus potentially influence the oxygen needs by the embryo, the two parameters can exert a synergistic effect upon the embryo. The results of their work showed that oxygen availability and incubation temperature could affect the development of the footpad skin which could lead to foot pad dermatitis. Morita et al. (2016b) using three incubation temperatures from day 13 onward showed differences in vascularity in the skin and skin thickness depending upon which incubation temperature was used.

9 Conclusion

In summary, control of embryo temperature is critical for our current strains of commercial poultry. There are differences in metabolic heat production between strains, broilers and layers, old and young breeder flocks, and larger and small eggs. Incubation by strain could be used in single-stage incubation to control EST during different stages of embryonic development. Incubation can affect not only hatchability but also post-hatch broiler growth performance. Incubation temperature has been shown to effect development of bone, heart, thyroid, muscle, intestine, immune system, and integument. Therefore, incubation research has moved from merely setter/hatcher temperatures to optimizing conditions for embryo/hatchling growth and development with influences on animal welfare, disease resistance, thermotolerance, organ development, incidence of ascites, disease resistance, and others. Meeting the needs of the embryo and hatchling at different stages of development is critical for optimal results. It is conceivable that in the next 5–10 years, the hatcheries with single-stage incubation will be utilizing thermal manipulation to optimize the performance of the hatchlings in the field.

10 Where to look for further information

There are many good journals available to do a literature search. Future work is going to have to be directed specifically at the different breed crosses and not generalize. The different breeds and the male crosses on the female will respond differently to temperature. When conducting future experiments, we should use eggshell temperatures as a standard tool as different incubator types have different airflow over the eggs as well as air speed. It is also critical to report age of breeder, egg size and storage length of eggs.

11 References

Afsarian,O., Shahir, M. H., Akhlaghi, A., Lotfolahian, H., Hoseini, A. and Lourens, A. 2016. Periodical low eggshell temperatures during incubation and post hatch dietary arginine supplementation: Effects on performance and cold tolerance acquisition in broilers. *Poultry Science* 95(10):2427–2434.

Afsarian, O., Shahir, M. H., Lourens, A., Akhlaghi, A., Lotfolahian, H., Hoseini, A. and Mousavi, N. 2018. Eggshell temperature manipulations during incubation and in ovo injection of thyroxine are associated with decreased incidence of cold-induced ascites in broiler chickens. *Poultry Science* 97(1):328–336.

Almeida, A. R., Morita, V. S., Matos Junior, Joao B., Sgavioli, S., Vicentini, T. I. and Boleli, I. C. 2022. Long-lasting effects of incubation temperature during fetal development on subcutaneous adipose tissue of broilers. *Frontiers in Physiology* 13:article 913496.

Almeida, V. R., Morita, V. S., Sgavioli, S., Vicentini, T. I., Castiblanco, D. M. C. and Boleli, I. C. 2016. Incubation temperature manipulation during fetal development reduces adiposity of broiler hatchlings. *Poultry Science* 95(2):316–324.

Al-Musawi, S. L., Stickland, N. C. and Bayol, S. A. M. 2012. In ovo temperature manipulation differentially influences limb musculoskeletal development in two lines of chick embryos selected for divergent growth rates. *The Journal of Experimental Biology* 215(9):1594–1604.

Al-Rukibat, R. K., Al-Zghoul, M. B., Hananeh, W. M., Al-Natour, M. Q. and Abu-Basha, E. A. 2017. Thermal manipulation during late embryogenesis: effect on body weight and temperature, thyroid hormones, and differential white blood cell counts in broiler chickens. *Poultry Science* 96(1):234–240.

Applegate, T. J. and Lilburn, M. S. 2002. Growth of the femur and tibia of a commercial broiler line. *Poultry Science* 81(9):1289–1294.

Bowers, R. R. and Lane, M. D. 2007. A role for bone morphogenetic protein-4 in adipocyte development. *Cell Cycle* 6(4):385–389.

Christensen, V. L., McMurtry, J. P., Donaldson, W. E. and Nestor, K. E. 2001. Incubation temperature affects plasma insulin-like growth factors in embryos from selected lines of turkeys. *Poultry Science* 80(7):949–954. doi: 10.1093/ps/80.7.949.

Christensen, V. L., Wineland, M., Grimes, J., Oviedo, E., Mozdziak, P. S., Ort, D. and Mann, K. 2007. Effect of incubator temperature and oxygen concentration at the plateau stage in oxygen consumption on turkey embryo muscle growth and development. *International Journal of Poultry Science* 6(6):406–412.

Christensen, V. L., Wineland, M. J., Yildirum, I., Ort, D. T. and Mann, K. M. 2004. Incubator temperature and oxygen concentration at the plateau stage affect cardiac health of turkey embryos. *Journal of Animal and Veterinary Advances* 3:52–65.

Clark, D. L., Walter, K. G. and Velleman, S. G. 2017. Incubation temperature and time of hatch impact broiler muscle growth and morphology. *Poultry Science* 96(11):4085–4095.

Collin, A., Berri, C., Tesseraud, S., Requena Rodin, F. E., Skiba-Cassy, S., Crochet, S., Duclos, M. J., Rideau, N., Tona, K., Buyse, J., Bruggeman, V., Decuypere, E., Picard, M. and Yahav, S. 2007. Effects of thermal manipulation during early and late embryogenesis on thermotolerance and breast muscle characteristics in broiler chickens. *Poultry Science* 86(5):795–800.

Cunningham, D. L. 2006. Cash flow estimates for contract broiler production in Georgia: A 20-year analysis. Available at: http://pubs.caes.uga.edu/caespubs/pubcd/B1228.htm.

Da Costa, M. J., Oviedo-Rondón, E. O., Wineland, M. and Jeffrey, D. 2015. Effects of eggshell conductance and incubation temperatures on duck footpad development. *Journal of Applied Poultry Research* 24(4):536–546.

Dayan, J., Reicher, N., Melkman-Zehavi, T. and Uni, Z. 2020. Incubation temperature affects yolk utilization through changes in expression of yolk sac tissue functional genes. *Poultry Science* 99(11):6128–6138.

Decuypere, E. and Michels, H. 1992. Incubation temperature as a management tool: A review. *World's Poultry Science Journal* 48(1):28–38.

French, N. A. 1994. Effects of temperature incubation on the gross pathology of turkey embryos. *British Poultry Science* 35(3):363–371.

French, N. A. 1997. Modeling incubator temperature: The effects of incubator design, embryonic development, and egg size. *Poultry Science* 76(1):124–133.

Groves, P. J. and Muir, W. I. 2014. A meta-analysis of experiments linking incubation conditions with subsequent leg weakness in broiler chickens. *PLoS ONE* 9(7):e102682. doi: 10.1371/journal.pone.0102682.

Guz, B. C., Molenaar, R., de Jong, I. C., Kemp, B., van Krimpen, M. and van den Brand, H. 2020. Effects of eggshell temperature pattern during incubation on tibia characteristics of broiler chickens at slaughter age. *Poultry Science* 99(6):3020–3029.

Hamidu, J. A., Torres, C. A., Johnson-Dahl, M. L. and Korver, D. R. 2018. Physiological response of broiler embryos to different incubator temperature profiles and maternal flock age during incubation. *Poultry Science* 97(8):2934–2946.

Harding, R., Halevy, O., Orna, Yahav, Shlomo and Velleman, S. 2016. The effect of temperature on proliferation and differentiation of chicken skeletal muscle satellite cells from different muscle types. *Physiological Reports* 4:e12770.

Harding, R. L., Clark, D. L., Halevy, O., Coy, C. S., Yahav, S. and Velleman, S. G. 2015. The effect of temperature on apoptosis and adipogenesis on skeletal muscle satellite cells derived from different muscle types. *Physiological Reports* 3(9):e12539.

Havenstein, G. B., Ferket, P. R. and Qureshi, M. A. 2003. Growth, livability, and feed conversion of 1957 versus 2001 broilers when fed representative 1957 and 2001 broiler diets. *Poultry Science* 82(10):1500–1508.

Holland, S., Hochel, J., Burmeister, A., Janke, O. and Nichelmann, M. 1998. A method for measuring deep body temperature in avian embryos. *Journal of Thermal Biology* 23(2):123–129.

Hulet, R., Gladys, G., Hill, D., Meijerhof, R. and El-Shiekh, T. 2007. Influence of egg shell embryonic incubation and boiler breed flock age on post hatch growth performance and carcass characteristics. *Poultry Science* 86:408–412.

Hulet, R. M. and Meijerhof, R. 2000. *Energy Production in Our Commercial Broiler during Incubation. WPSA World Group (Reproduction).* Hall, Oxford, UK: St. Edmunds, 151–155.

Ipek, A. and Sozcu, A. 2015. The effects of high setter and hatcher temperatures during incubation on slaughter weight and carcass yield in broilers. *Turkish Journal of Veterinary and Animal Sciences* 39:450–454.

Ipek, D. and Sozcu, A. 2016. The effects of eggshell temperature fluctuations during incubation after performance to 70 days of age. *Poultry Science* 95:1296–1303.

Iqbal, A., Decuypere, E., Abd El Azim, A. and Kuhn, E. R. 1990. Pre- and post-hatch high temperature exposure affects the thyroid hormones and corticosterone response to acute heat stress in growing chicken (*Gallus domesticus*). *Journal of Thermal Biology* 15(2):149–153.

Janisch, S., Sharifi, A. R., Wicke, M. and Krischek, C. 2015. Changing the incubation temperature during embryonic myogenesis influences the weight performance and meat quality of male and female broilers. *Poultry Science* 94(10):2581–2588.

Janke, O., Tzschentke, B. and Boerjan, M. 2013. Comparative investigations of heat production and body temperature in embryos of modern chicken breeds. *Avian and Poultry Biology Reviews* 15(3):191–196.

Janke, O., Tzschentke, B. and Halle, I. 2006. Does variation in incubation temperature increase broiler chicken performance. In: *Proceedings of the European Poultry Conference, Verona (on CD).*

Krischek, C., Gerken, M. and Wicke, M. 2013. Effect of a higher incubation temperature between day 9 and 12 on growth and meat quality characteristics of turkeys. *British Poultry Science* 54(1):5–11.

Krischek, C. J. S., Jaiseh, W., Naralalllabh, P. and Brunner, K. 2016. Altered incubation temperatures between embryonic day 7 and 13 influences the weights and the mitochondrial respiratory and enzyme activities in breast and leg muscle of broiler embryos. *Molecular Reproduction and Development* 88:71–78.

Leksrisompong, N., Romero-Sanchez, H., Plumstead, P. W., Brannan, K. E. and Brake, J. 2007. Broiler incubation 1. Effect of elevated temperature during late incubation on body weight and organs of chicks. *Poultry Science* 86(12):2685–2691.

Leksrisompong, N., Romero-Sanchez, H., Plumstead, P. W., Brannan, K. E., Yahav, S. and Brake, J. 2009. Broiler incubation. 2. Interaction of incubation and brooding temperatures on broiler chick feed consumption and growth. *Poultry Science* 88(6):1321–1329.

Li, X., Qiu, J., Liu, H., Wang, Y., Hu, J., Gan, X. and Wang, J. 2017. Long term thermal manipulation in the late incubation period can inhibit breast muscle development by activating endoplasmic reticulum stress in duck (*Anasplatyrhynchos domestica*). *Journal of Thermal Biology* 70(B):37–45.

Liu, J., Yan, X., Li, Q., Wang, G., Liu, H., Wang, J., Li, L., Du, X., Han, C. and He, H. 2013. Thermal manipulation during the middle incubation stage has a repressive effect on the immune organ development of Peking ducklings. *Journal of Thermal Biology* 38(8):520–523.

Lopez, J. C., Kitto, L. and Hulet, R. M. 2018. Effect of eggshell temperature on survival rate, development at hatch and 7day growth. *Journal of Applied Poultry Research* 27(2):249–252.

Lourens, A., Brand, H. V. D., Heetcamp, M. J. W., Meijerhof, R. and Kemp, B. 2007. Effects of eggshell temperature and oxygen concentration on embryo growth and metabolism during incubation. *Poultry Science* 86(10):2194–2199.

Lourens, A., Brand, H. V. D., Meijerhof, R. and Kemp, B. 2005. Effect of eggshell temperature during incubation on embryo development, hatchability, and posthatch development. *Poultry Science* 84(6):914–920.

Lourens, A., Molenaar, R., van den Brand, H., Heetkamp, M. J. W., Meijerhof, R. and Kemp, B. 2006. Effect of egg size on heat production and the transition of energy from egg to hatchling. *Poultry Science* 85(4):770–776.

Loyau, T., Collin, A., Yenisey, C., Crochet, S., Siegel, P. B., Aksit, M. and Yalcin, S. 2014. Exposure of embryos to cyclically cold incubation temperatures durably affects energy metabolism and antioxidant pathways in broiler chickens. *Poultry Science* 93(8):2078–2086.

Maatjens, C. M., van Roovert-Reijrink, I. A. M., van den Anker, I., Engel, B., van der Pol, C. W., Kemp, B. and van den Brand, H. 2016a. Temperature during the last week of incubation. II. Effects on first week broiler development and performance. *Poultry Science* 95(9):2136–2144.

Maatjens, C. M., van Roovert-Reijrink, I. A. M., Engel, B., van der Pol, C. W., Kemp, B. and van den Brand, H. 2016b. Temperature during the last week of incubation. I. Effects on hatching pattern and broiler chicken embryonic development. *Poultry Science* 95(4):956–965.

Maltby, V., Somaiya, A., French, N. A. and Strickland, N. G. 2004. In ovo temperature manipulation influences post-hatch muscle growth in the turkey. *British Poultry Science* 45(5):1–8.

Mauldin, J. M. 1989. An analysis of reproductive efficiency in Georgia hatcheries. General Report Number 113. Extension Poult. Sci. Dept. University of Georgia, Athens, GA.

Mauldin, J. M. and Buhr, R. J. 1995. What is really in your incubator? *International Hatchery Practice* 9:13–23.

Meijerhof, R. 2000. *The Influence of Incubation on Chick Quality and Broiler Performance. WPSA World Group (Reproduction)*. Hall, Oxford, UK: St. Edmunds, 167–176.

Meijerhof, R. and van Beek, G. 1993. Mathematical modeling of temperature and moisture loss of hatching eggs. *Journal of Theoretical Biology* 165(1):27–41.

Molenaar, R., Hulet, R., Meijerhof, R., Maatjens, C. M., Kemp, B. and van den Brand, H. 2011. High eggshell temperatures during incubation decrease growth performance and increase the incidence of ascites in broiler chickens. *Poultry Science* 90(3): 624–632.

Molenaar, R., van den Borne, J. J. G. C., Hazejager, E., Kristensen, N. B., Heetcamp, M. J. W., Meijerhof, R., Kemp, B. and van den Brand, H. 2013. High environmental temperature increases glucose requirement in the developing chicken embryo. *PLoS ONE* 8(4):e59637.

Moraes, V. M. B., Malheiros, R. D., Bruggeman, V., Collin, A., Tona, K., Van As, P., Onagbesan, O. M., Buyse, J., Decuypere, E. and Macari, M. 2003. Effect of thermal conditioning during embryonic development on aspects of physiological responses of broilers to heat stress. *Journal of Thermal Biology* 28(2):133–140.

Moraes, V. M. B., Malheiros, R. D., Bruggeman, V., Collin, A., Tona, K., Van As, P., Onagbesan, O. M., Buyse, J., Decuypere, E. and Macari, M. 2004. The effect of timing thermal conditioning during incubation on embryo physiological parameters and its relationship to thermotolerance in adult broiler chicken. *Journal of Thermal Biology* 29(1):55–61.

Morita, V. S., Almedida, V. R., Matos Junior, J. B., Vicentini, T. I., van den Brand, H. and Boleli, I. C. 2016a. Incubation temperature alters thermal preference and response to heat stress of broiler chickens along the rearing phase. *Poultry Science* 95(8): 1795–1804.

Morita, V. S., de Almeida, V. R., Matos, J. B., Vicentini, T. I., van den Brand, H. and Boleli, I. C. 2016b. Incubation temperature during fetal development influences morphophysiological characteristics and preferred ambient temperature of chicken hatchlings. *PLoS ONE* 11(5):e0154928.

Nangsuay, A., Meijerhof, R., van den Anker, I., Heetkamp, M. J. W., De Souza Morita, V. S., Kemp, B. and van den Brand, H. 2016. Effect of breeder age, broiler strain, and eggshell temperature on development and physiological status of embryos and hatchlings. *Poultry Science* 95(7):1666–1679.

Nangsuay, A., Meijerhof, R., van den Anker, I., Heetkamp, M. J. W., Kemp, B. and van den Brand, H. 2015. Development and nutrient metabolism of embryos from two modern broiler strains. *Poultry Science* 94(10):2546–2554.

Nichelmann, M. and Tzschentke, B. 2003. Efficiency of thermoregulatory control elements in precocial avian embryos. *Avian and Poultry Biology Reviews* 14(1):1–19.

Oviedo-Rondon, E. O., Cordova, H. A., Saldana, B., Sarsour, A., Bernal-Arango, L., Mesquita, M., Fansanaro, R., Wineland, M., Borst, L. and Barnes, J. 2016. Effects of genetic line, incubation profiles and gender on woody breast and white striping at 57 d of age. In: *World's Poultry Congress CD Proceedings*, Beijing, China, September 5–9.

Oviedo-Rondon, E. O., Small, J., Wineland, M. J., Christensen, V. L., Mosdziak, P. S., Koci, M. D., Funderbunk, S. L. V., Ort, D. and Mann, K. M. 2008. Broiler embryo bone development is influenced by incubator temperature, oxygen concentrations and egg shell conductance at the plateau stage in oxygen consumption. *British Poultry Science* 50(5):634–640.

Oviedo-Rondon, E. O., Velleman, S. G. and Wineland, M. J. 2020. The role of incubation conditions in the onset of avian myopathies. *Frontiers in Physiology* 11:545045 doi: 10.3389/fphys.2020.545045:3389/fphys.2020.

Oviedo-Rondon, E. O., Wineland, M. J., Funkerburk, S., Small, J., Cutchin, H. and Mann, M. 2009a. Incubation conditions affect leg health in large, high yielding broilers. *Journal Applied Poultry Science* 18:640–646.

Oviedo-Rondon, E. O., Wineland, M. J., Small, J., Cutchin, H., McElroy, A., Barri, A., and Martin, S. 2009b. Effect of incubation temperature and chick transportation conditions on bone development and leg health. *Journal Applied Poultry Science* 18:671–678.

Oznurlu, Y., Celik, I., Telatar, T. and Sur, E. 2010. Histochemical and histological evaluations of the effects of high incubation temperature on embryonic development of thymus and bursa of Fabricius in broiler chicks. *British Poultry Science* 51(1):43–51.

Piestun, Y., Druyan, S., Brake, J. and Yahav, S. 2013. Thermal manipulations during broiler incubation alter performance of broilers to 70 days of age. *Poultry Science* 92(5):1155–1163.

Piestun, Y., Halevy, O., Shinder, D., Ruzal, M., Druyan, S. and Yahav, S. 2011. Thermal manipulations of broiler embryos – The effect on thermoregulation and development during embryogenesis. *Poultry Science* 88:2677–2688.

Piestun, Y., Halevy, O. and Yahav, S. 2009a. Thermal manipulation of broiler embryos - The effect on thermoregulation and development during embryogenesis. *Poultry Science* 88(12):2677–2688.

Piestun, Y., Harel, M., Barak, M., Yahav, S. and Halevy, O. 2009b. Thermal manipulations in late-term chick embryos have immediate and longer term effects on myoblast proliferation and skeletal muscle hypertrophy. *Journal of Applied Physiology* 106(1):233–240.

Piestun, Y., Yahav, S. and Halevy, O. 2015. Thermal manipulation during embryogenesis affects myoblast proliferation and skeletal muscle growth in meat type chickens. *Poultry Science* 94(10):2528–2536.

Rahn, H. 1981. Gas exchange of avian eggs with special reference to turkey eggs. *Poultry Science* 60(9):1971–1980.

Rahn, H., Christensen, V. L. and Edens, F. W. 1981. Changes in shell conductance, pores, and physical dimensions of egg and shell during the first breeding cycle of turkey hens. *Poultry Science* 60(11):2536–2541.

Romanoff, A. L. 1936. Effects of different temperatures in the incubator on the prenatal and postnatal development of the chick. *Poultry Science* 15(4):311–315.

Romanoff, A. L. 1960. *The Avian Embryo. Structural and Functional Development*. New York, NY: Macmillan.

Romanoff, A. L. and Faber, H. A. 1933. Effect of temperature on the growth, fat, and calcium metabolism and mortality of the chick embryo during the latter part of incubation. *Journal of Cellular and Comparative Physiology* 2(4):457–466.

Ruuskanen, S., Hsu, B. Y. and Nord, A. 2021. Endocrinology of thermoregulation in birds in a changing climate. *Molecular and Cellular Endocrinology* 519:111088.

Scott, J., Da Costa, M. J. and Oviedo-Rondon, E. O. 2015. Incubation temperature profiles affect broiler feathering. *Journal of Applied Poultry Research* 24(1):49–57.

Shanmugasundaram, R., Wick, M. and Lilburn, M. S. 2018. Effect of embryonic thermal manipulation on heat shock protein 70 expression and immune system development in Pekin duck embryos. *Poultry Science* 97(12):4200–4210.

Sozcu, A. and Ipekce, A. 2015. Acute and chronic eggshell temperature manipulations during hatching term influence hatchability, broiler performance, and ascites incidence. *Poultry Science* 94(2):319–327.

Swann, G. S. and Brake, J. 1990. Effect of incubation dry-bulb and wet-bulb temperatures on time of hatch and chick weight at hatch. *Poultry Science* 69(6):887–897.

Tejeda, O. J., Meloche, K. J. and Starkey, J. D. 2021. Effect of incubator tray location on broiler chicken growth performance, carcass parts yields, and the meat quality defects wooden breast and white striping. *Poultry Science* 100(2):654–662.

Tzschentke, B. and Basta, D. 2002. Early development of neuronal hypothalamic thermosensitivity in birds: Influence of epigenetic temperature adaptation. *Comparative Biochemistry and Physiology. Part A, Molecular and Integrative Physiology* 131(4):825–832.

Tzschentke, B. and Halle, I. 2009. Influence of temperature stimulation during the last 4 days of incubation on secondary sex ratio and later performance in male and female broiler chicks. *British Poultry Science* 50(5):634–640.

Tzschentke, B. and Nichelmann, M. 1999. Development of avian thermoregulatory system during the early postnatal period: Development of the thermoregulatory set-point. *Ornis Fennica* 76:189–198.

van den Brand, H., van de Kraats, S. J. F., Sozcu, A., Joerissen, R., Heetkamp, M. J. W., van den Anker, I., Ooms, M. and Kemp, B. 2019. Both the rooster line and incubation temperature affect embryonic metabolism and hatchling quality in laying hen crossbreds. *Poultry Science* 98(6):2632–2640.

van der Pol, C. W., van Roovert-Reijrink, I. A. M., Maatjens, C. M., van den Anker, I., Kemp, B. and van den Brand, H. 2014. Effect of eggshell temperature throughout incubation on broiler leg bone development. *Poultry Science* 93(11):2878–2883.

Werner, C. and Wicke, M. 2008. Increasing the temperature at a certain period during the incubation of broiler eggs improves the carcass and breast yields without a negative impact on the meat quality. *Arch Tierz Drummerstorf* 51:11–16.

Werner, C., Wicke, C., Fliebert, F. and Wicke, M. 2010. Increasing the incubation temperature between embryonic day 7 and 10 has no influence on the growth and slaughter characteristics as well as meat quality of broilers. *Animal* 4(5):810–816.

Wijnen, H. J., Molenaar, R., van Roovert-Reijrink, I. A. M., van der Pol, C. W., Kemp, B. and van den Brand, H. 2020a. Effects of incubation temperature pattern on broiler performance. *Poultry Science* 99(8):3897–3907.

Wijnen, H. J., van den Brand, H., Lammers, A., van Roovert-Reijrink, I. A. M., van der Pol, C. W., Kemp, B. and Molenaar, R. 2020b. Effect of eggshell temperature pattern during incubation on primary immune organ development and broiler immune response later in life. *Poultry Science* 99(12):6619–6629.

Willemsen, H., Li, Y., Willems, E., Franssens, L., Wang, Y., Decuypere, E. and Everaert, N. 2011. Intermittent thermal manipulations of broiler embryos during late incubation and their immediate effect on the embryonic development and hatching process. *Poultry Science* 90(6):1302–1312.

Wilson, H. R. 1991. Physiological requirements of the developing embryo: Temperature and turning. In: Tullet, S. G. (Ed.), *Avian Incubation*. Northampton: Butterworth-Heinemann, 145–156.

Wilsterman, K., Mast, A. D., LuuMark, T. H. and Haussman, M. F. 2015. The timing of embryonic exposure to elevated temperature alters stress endocrinology in domestic chickens (*Gallus domesticus*). *General and Comparative Endocrinology* 212:10–16.

Wineland, M. J., Mann, K. M., Fairchild, B. D. and Christensen, V. L. 2000a. Effect of high and low incubator temperatures at different stages of incubation upon the broiler embryo. *Poultry Science* 79(Suppl. 1):S180.
Wineland, M. J., Mann, K. M., Fairchild, B. D. and Christensen, V. L. 2000b. Effect of different setter and hatcher temperatures upon the broiler embryo. *Poultry Science* 79(Suppl. 1):S180.
Wineland, M. W., Christensen, V. L., Yildrum, I., Fairchild, B. D., Mann, K. M. and Ort, D. T. 2006a. Incubator temperature and oxygen concentration at the plateau stage in oxygen consumption affects intestinal maturation of broiler chicks. *International Journal of Poultry Science* 5(3):229–240.
Wineland, M. J., Christensen, V. L., Yildrum, I., Fairchild, B. D., Mann, K. M. and Ort, D. T. 2006b. Incubator environment interacts with genetic line of broiler at the plateau stage to affect embryo thyroxine and triiodothyronine concentrations. *International Journal of Poultry Science* 5:714–722.
Yahav, S. and Brake, J. 2014. Chick embryogenesis: A unique platform to study the effects of environmental factors on embryo development. *Journal of Stem Cells* 9(1):17–37.
Yahav, S. and Huriwitz, S. 1996. Induction of thermotolerance in male chickens by temperature conditioning at an early age. *Poultry Science* 75(3):402–406.
Yalcin, S., Aksit, M., Ozkan, S., Hassanzadeh, M., Bilgen, G., Helva, I. B., Izzetoglu, G. T., Buyse, J. and Yılmaz, M. C. 2022. Effect of temperature manipulation during incubation on body weight, plasma parameters, muscle histology, and expression of myogenic genes in breast muscle of embryos and broiler chickens from two commercial strains. *British Poultry Science* 63(1):21–30.

Chapter 11

Understanding the effects of light on embryo and post-hatch chick development

Israel Rozenboim, Liron Dishon, Natalie Avital-Cohen and Joanna Bartman, The Hebrew University of Jerusalem, Israel; and Ronen Shviki, BeyondEdge Inc, USA

1 Introduction

The role of light in biological activities associated with avian growth and reproduction is very well known. Light quality can be defined by three criteria: (1) pattern of light–dark exposure, photoperiod; (2) light intensity/brightness (W/m^2 and lux or foot-candle, respectively); and (3) spectral composition of the light source (Mobarkey et al., 2010). In meat-type birds, photoperiod is involved mainly in feeding activity by providing more feeding time to birds. Several studies indicated the role of brightness (measured by lux or foot-candle) on meat-type birds. High brightness (64.8 lux) reduces the growth rate in broilers (Barott and Pringle, 1951); Cherry and Barwick (1962) stated that light brightness beyond 10.8 lux probably depresses growth. No data is available on the effect of light intensity per se (W/m^2) on the growth of meat-type birds. Light spectra also affect growth in meat-type birds. Broilers raised under blue light (BL) or green light (GL) gained significantly more weight than birds reared under red light (RL) or white light (WL), whereas feed conversion and mortality were not affected (Wabeck and Skoglund, 1974; Lauber et al., 1961; Ookawa, 1970; Foss et al., 1972). In addition, we found that the effect of monochromatic light on the growth and development of broilers is age-related; i.e. GL stimulates growth and muscle development at an early age while BL furtherly stimulates growth in alder birds (Rozenboim et al., 1999; Halevy et al., 1998).

http://dx.doi.org/10.19103/AS.2022.0118.15

The only light source for chickens in environmentally controlled houses is an artificial one. Thus, targeted illumination manifested by source, spectra, intensity, brightness and regimen of light supplementation became major factors, known as light quality, in modern poultry management (Andrews and Zimmerman, 1990).

2 The role of monochromatic light on growth of meat-type birds

2.1 The effect of post-hatch monochromatic photostimulation on growth and development of broilers

Light spectra affect growth in broilers. Broilers raised under blue or green fluorescent lamps gained significantly more weight than birds reared under RL or WL, whereas feed conversion and mortality were not affected (Wabeck and Skoglund, 1974). There are, however, conflicting reports regarding the effect of monochromatic light on birds' growth. Whereas GL was found to stimulate growth (Wabeck and Skoglund, 1974; Lauber et al., 1961; Ookawa, 1970; Foss et al., 1972), mature female Japanese quail had lower body weight (BW) when reared under GL and BL as compared to those reared under RL or WL (Woodard et al., 1969). Turkeys grew better under BL until 16 weeks of age, but by the 24th week those reared under RL and WL were heavier (Leighton and Mason, 1976). Pink light depressed chicken hatching weights with reversal by WL at 4 weeks of age (Tamimie, 1967).

GL advanced early chick-embryo growth, as did BL (Lauber, 1975) and ultraviolet (UV) light (Coleman and McDaniel, 1976). The possible causes of contradictory findings were probably due to variability in light sources (Goodman, 1983), methods of measuring light doses (Blum, 1964), species, breed, gender and age of experimental birds (Ookawa, 1970; Foss et al., 1972; Woodard et al., 1969).

In order to eliminate some of these controversies, a light-emitting diode (LED) was used for broiler photostimulation. LED is a pure monochromatic light source. Early studies conducted in our laboratory (Rozenboim et al., 1999) on broiler post-hatch chicks have shown an acceleration in growth rate in birds photostimulated by GL and BL during incubation (Fig. 1), this acceleration in growth rate was also associated with an increase in breast muscle weight. These early findings demonstrated that pure monochromatic photostimulation affects the growth rate of broiler chicks. GL enhanced growth rate significantly more than WL and RL 3 days after hatch. Growth was also enhanced by BL, but the onset of this effect was at an older age. These results agreed with Wabeck and Skoglund (1974), who demonstrated it in broilers, and Phogat et al. (1985) who found it in quails.

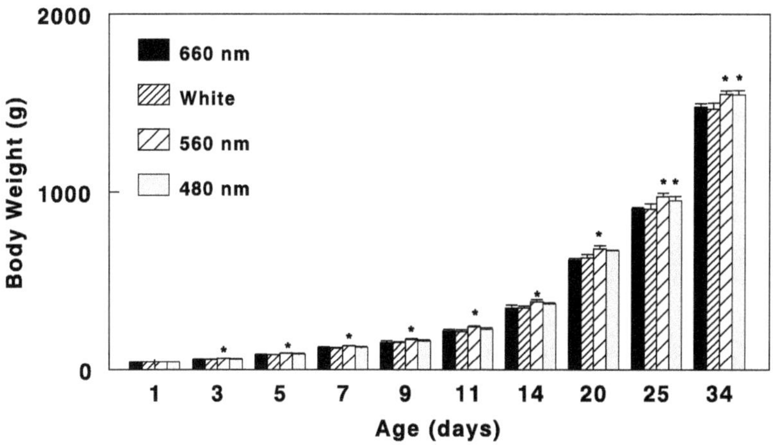

Figure 1 Body weight of male broiler reared under different light spectra. Each group of treated birds was exposed to 480, 560, 660 nm, or white light from 1 day old until termination of experiment at 35 days of age. Data marked with * are significantly different from control ($P < 0.05$) (Rozenboim et al., 1999).

Chicks reared under GL and BL, by 5 days of age, had 2-2.4 times as much breast muscle satellite cells compared to white and red photostimulated chicks. Since satellite cells are members of myoblast family, any rise in their level may indicate stimulation of muscle growth, and a possible mechanism explaining the elevation in BW associated with green and blue monochromatic photostimulation (Halevy et al., 1998).

We later tested the effect of green–blue photostimulation combinations on the growth and development of broiler males. A significant increase in relative cumulative growth rate was observed at 4 days of age in birds reared under GL compared with WL-reared birds. Switching of GL birds at 10 or 20 days of age to BL caused a further acceleration in this parameter until 46 days of age. Shifting of birds from BL to GL at 10 days of age caused a reduction in cumulative growth rate at 15 days of age; however, the growth rate was elevated by the end of the experiment at 46 days of age. Furthermore, shifting of BL-reared birds to GL at 20 days of age caused an increase in their cumulative growth rate until the end of the experiment (Rozenboim et al., 2004).

2.2 The effect of in ovo monochromatic photostimulation on growth and development of broilers

In commercial hatcheries, broiler eggs are set in the dark; however, several studies have shown that photostimulation accelerates embryonic development. Early studies showed that stimulation with WL increases BW in avian embryos, including broilers, White Leghorn chickens, turkeys and quail (Shutze et al.,

1962; Siegel et al., 1969; Cooper, 1972; Walter and Voitle, 1972; Coleman and McNabb, 1975). Moreover, embryos that had been exposed to light hatched earlier than those kept in the dark. WL stimulation of White Leghorn and broiler eggs accelerated hatching by approximately 1 day (Seigel et al., 1969; Walter and Voitle, 1972; Coleman and McDaniel, 1976). A similar acceleration has been observed in turkey embryos, with no effect on hatchability or BW on hatching day (Fairchild and Christensen, 2000), suggesting embryonic photostimulation as a tool for shortening the hatching period. A recent study has shown that broiler embryos stimulated with fluorescent GL hatch earlier and are heavier at hatch than those kept in the dark (Shafey, 2002). However, the positive effects on embryonic development and hatching observed in these studies could have been due to an effect of overheating, in addition to the illumination effect (Romanoff, 1960), because the eggs were incubated under continuous illumination. In addition, none of these studies analyzed the effect of pure monochromatic light on embryonic development.

In the first study, we investigated the effect of monochromatic GL on egg core temperatures (Fig. 2). Although LED lamps are considered 'cold' lamps, a preliminary study conducted in our lab demonstrated early hatching (6 h) compared to dark incubated birds. This observation is a result of electromagnetic radiation degradation to heat at the target area (egg core). Further, it led us to test egg core temperature related to in ovo photostimulation with several wavelengths. Photostimulation of broiler eggs with monochromatic GL caused a significant, time-dependent elevation in yolk temperature (Fig. 2a). The highest elevation was detected after 25 min of GL photostimulation and was 0.24% of the initial temperature, corresponding to 0.25°F. Figure 2b demonstrates the high correlation between the photostimulation period and the elevation in yolk temperature, suggesting that 1 min of photostimulation increases yolk temperature by 0.01% of its initial value (approximately 0.01°F). In eggs illuminated for periods longer than 15 min, the temperature did not decline to zero-time temperatures, even 60 min after the light had been turned off, suggesting the need for intermittent lighting. Setting the eggs under an intermittent light regimen (15 min on, 15 min off) eliminated the rise in egg yolk temperature, and, as a result, prevented early hatch associated with in ovo photostimulation (Fig. 2c) (Rozenboim et al., 2004).

GL photostimulation enhances embryonic BW, calculated as the percentage of egg weight, and percentage of pectoralis muscle weight (Fig. 3) (Rozenboim et al., 2004). The positive effect of light stimulation on muscle growth preceded its effect on BW in embryos and was evident on almost all days during incubation, suggesting a specific effect of GL on muscle growth (Halevy et al., 2006).

At hatch (day 0), BW of the chicks did not significantly differ between the two in ovo-treated groups. However, during the first-week post-hatch, BW of chicks that were in ovo photostimulated with GL was significantly higher than

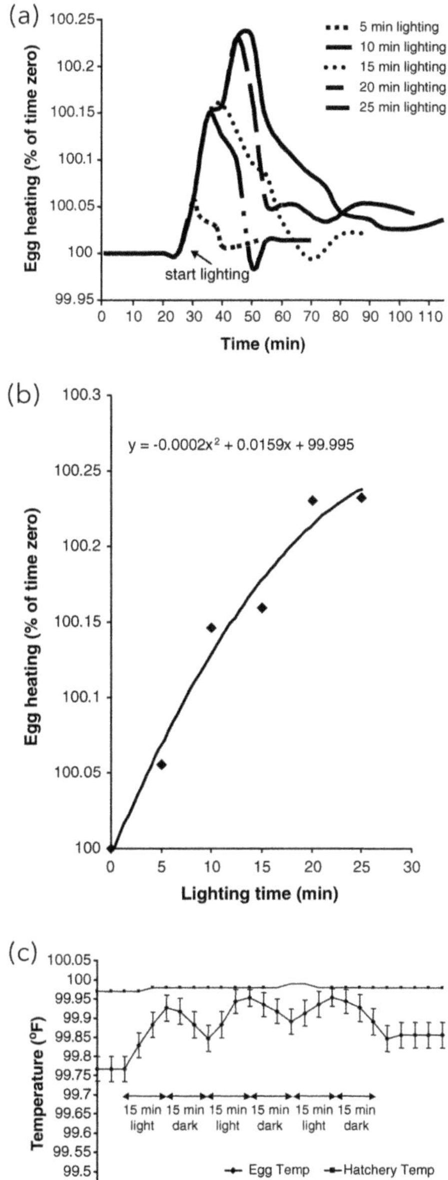

Figure 2 In ovo GL photostimulation affects the core temperature of incubated eggs. (a) Time-course effect of lighting period on egg temperature and (b) correlation between lighting period and temperature elevation. Data are presented as a percentage of temperature at time zero. (c) Egg core and hatchery temperature changes by intermittent light (15 min on and 15 min off). Results are means ± SEM (Rozenboim et al., 2004).

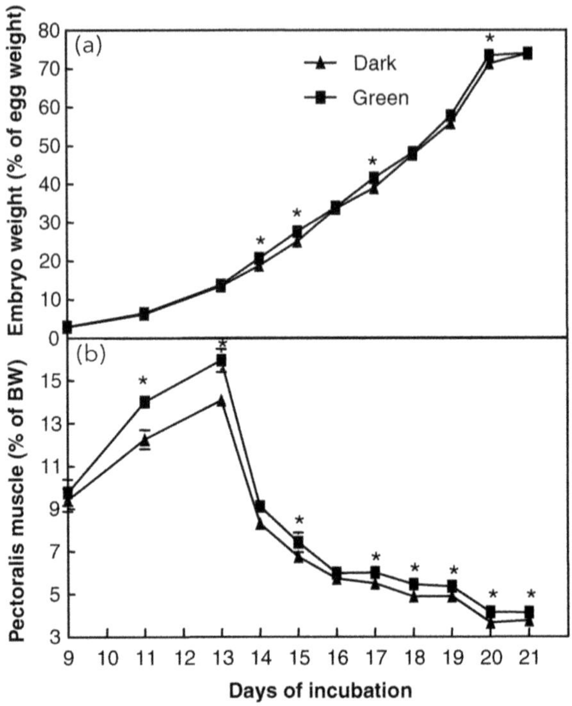

Figure 3 Percentage BW of egg weight (a) and pectoralis muscle weight as a percentage of BW (b) of chick embryos incubated under green light or kept in the dark on various days during incubation. Results are means ± SEM and asterisks mark significant differences between treatments ($P < 0.05$) (Rozenboim et al., 2004).

that of those hatched in the dark (Fig. 4a). The pectoralis muscle weight as the percentage of BW of the chicks that hatched under GL was significantly higher than that of the chicks that hatched in the dark, both on hatching day and at 6 days of age (Fig. 4b) (Rozenboim et al., 2004).

Skeletal muscle fibers are formed during embryogenesis and continue to enlarge postnatally until the mature size has been reached. This postnatal myofiber growth entails an increase in myofiber protein accretion and in the number of myofiber nuclei (reviewed by Allen et al., 1979; Edgerton and Roy, 1991). The primary source of these additional myofiber nuclei is the satellite cells, myogenic stem cells situated on the surface of the myofiber between the myofiber plasmalemma and its covering basement membrane (Mauro, 1961; Hawke and Garry, 2001; Zammit and Beauchamp, 2001).

Satellite cells are first detected in muscles during the late stages of fetal development, when the myofiber basement membrane is formed (Yablonka-Reuveni, 1995). At birth, skeletal muscle nuclei consist of a high percentage of proliferating satellite cells. However, at the end of the growth phase, the

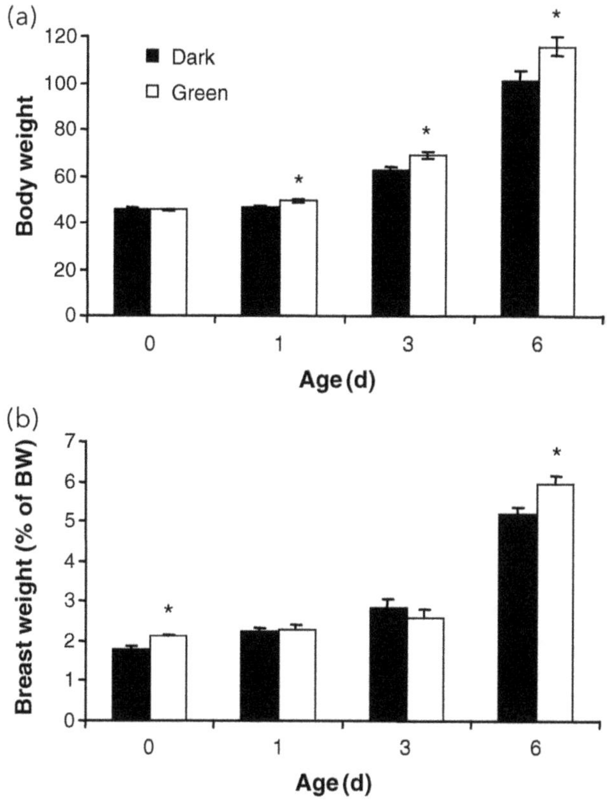

Figure 4 Body weight (a) and pectoralis muscle weight as a percentage of BW (b) of chicks incubated under green light or kept in the dark on various days post-hatch. Results are means ± SEM and asterisks mark significant differences between treatments ($P < 0.05$) (Rozenboim et al., 2004).

number of satellite cells decreases to less than 5% of total muscle fiber nuclei, and at this stage, most of the satellite cells become quiescent (Hawke and Garry, 2001). The quiescent satellite cells can rapidly enter the cell cycle in response to various muscular stresses ranging from work overload to major muscle trauma.

Proliferation and differentiation of myoblasts is regulated by the muscle-specific basic helix-loop-helix (bHLH) family of transcription factors (reviewed in Weintraub, 1993; Ludolph and Konieczny, 1995; Naya and Olson, 1999). Upon satellite cell activation, the muscle-specific transcription factors are expressed in a sequential pattern with Myf5 and MyoD being expressed in the proliferating progeny followed by myogenin expression as the cells enter differentiation (Smith et al., 1994; Yablonka-Reuveni and Rivera, 1994; Cornelison and Wold, 1997; Cooper et al., 1999; Yablonka-Reuveni et al., 1999; Kästner et al., 2000; Yablonka-Reuveni and Paterson, 2001).

What are the cellular and molecular basis underlying skeletal muscle developments in embryos and post-hatch chicks in response to in ovo GL illumination?

In a collaborative study, we found evidence for the mechanism behind the elevation in body weight and muscle growth associated with in ovo GL photostimulation.

1 The number of cells per gram of muscle was higher in the green group than in the dark group on *days 1* and *3* of age ($P < 0.05$). In both groups, the number of cells peaked on *day 3* and then declined; however, the decline was slightly greater in the green group (Fig. 5) (Halevy et al., 2006).
2 A stimulatory effect of in ovo GL illumination on muscle cell proliferation was further observed by immunohistochemical staining for proliferating cell nuclear antigen (PCNA), a marker for cell proliferation, in muscle sections (Fig. 6) (Halevy et al., 2006).
3 Expression levels of Pax7 and Myogenin in the early post-hatch period are elevated by in ovo green lighting (Fig. 7). Pax7 levels were higher in the green group than in the dark group on all days examined. The largest differences were observed on days 3 and 6 of age, reaching up to ninefold. Myogenin levels were significantly higher, up to fourfold, in the green group than in the dark one at hatch and on day 1. Myogenin levels did not differ between the two groups on day 3, but on day 6 post-hatch, these levels were significantly lower in the green versus dark groups. The higher levels of Pax7 on *day 3* and on *day 6* (the day on which a decline in muscle cell number was observed in both groups) in the green versus dark group support the presence of a higher reservoir of myogenic progeny resulting from in ovo GL illumination. The importance of Pax7 as an indicator of the proliferation capacity of satellite cells has been recently shown in turkey poults that were held from food early post-hatch; Pax7 expression was dramatically reduced in parallel to a decline in cell proliferation (Halevy et al., 2006). Our earlier findings that the percentage of pectoralis muscle of total body weight in in ovo illuminated chicks is significantly higher than that in chicks incubated in the dark at all post-hatch days analyzed (through *day 42*; 32), together with higher Pax7 levels (i.e. higher reservoir of quiescent/proliferative satellite cells), imply an enhanced hypertrophy potential of muscle because of in ovo GL illumination.
4 We have shown a higher expression of growth hormone (GH) receptor mRNA in satellite cells derived from GL illuminated chicks (Fig. 8) (Halevy et al., 1998). In addition, plasma GH levels, and IGF-I levels in muscle tissue,

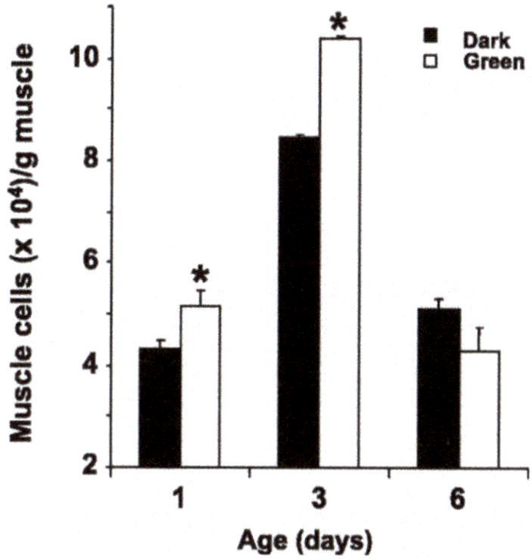

Figure 5 Number of muscle cells per gram of pectoralis muscle from dark and green group chicks on various days post-hatch. *$P < 0.05$ (Halevy et al., 2006).

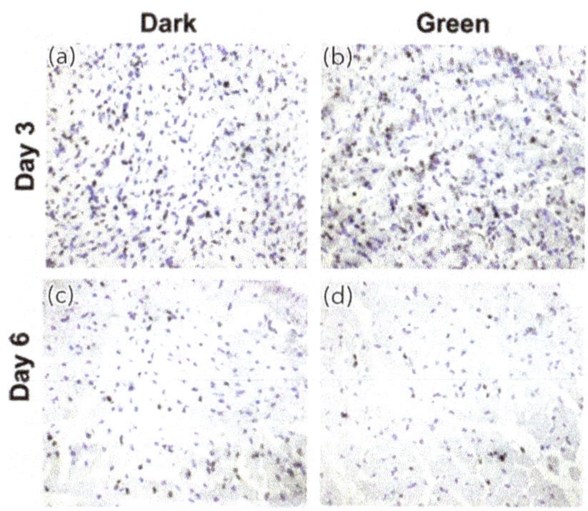

Figure 6 Representative proliferating cell nuclear antigen (PCNA) staining of breast muscle cross sections prepared from dark (a and c) and green (b and d) groups at 3 and 6 days of age (Halevy et al., 2006).

were higher in the green group relative to the dark one in early post-hatch (Halevy, O., Piestun Y., Hoizman, O., and Rozenboim, I., unpublished data). GH induces myogenic cell proliferation (Halevy et al., 1996; Sadowski et al., 2001), whereas IGF-I induces proliferation, differentiation and fusion of muscle cells (Engert et al., 1996; Florini et al., 1996): another possible explanation for the in ovo photostimulation effect on adult myoblasts, at least while they populate the muscle during late embryogenesis.

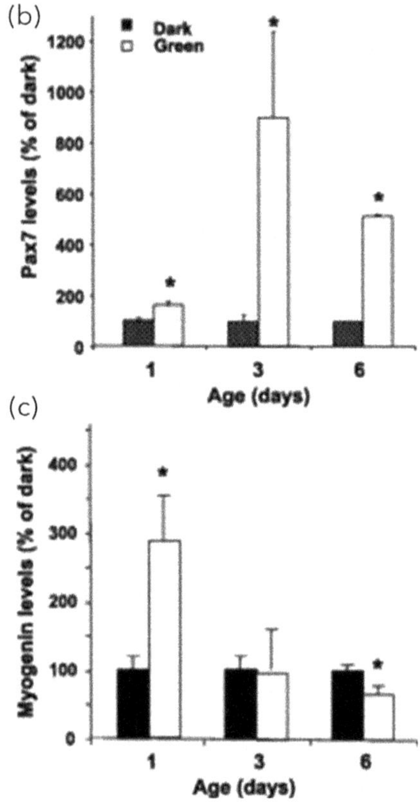

Figure 7 Pax7 and myogenin levels in the pectoralis muscle are upregulated in response to in ovo green light stimulation. Pectoralis muscle samples were removed in parallel from individual chicks from both groups at various days of age (Halevy et al., 2006).

3 The effect of in ovo green light photostimulation on somatotrophic axis activity in embryos and post-hatch

In several studies, we explored the effect of in ovo GL photostimulation on the somatotropic axis; first we compared eggs photostimulated with GL from E0-E21 to dark control eggs (Fig. 9). For plasma GH levels, blood sampling was drawn every other day from ED10 to ED18 from the chorioallantoic vein, and in ED20 (hatching day) from the jugular vein (Fig. 10).

In ovo GL photostimulation caused a significant increase in plasma GH levels (Fig. 10) from ED14 to ED18 and on ED14 and ED20 (hatching d). In addition, plasma GH levels were elevated during the second half of incubation for both treatment groups between ED16 and ED20 ($P < 0.05$) (Dishon et al., 2017).

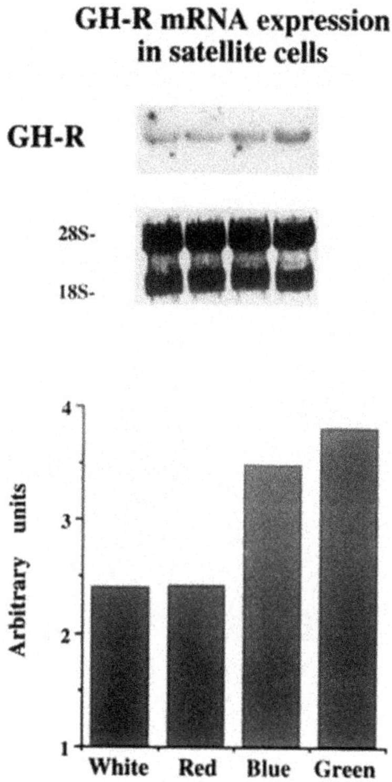

Figure 8 GH-R mRNA expression in satellite cells of broiler males photostimulated after hatch with white, red, blue and green monochromatic lights (Halevy et al., 1998).

In addition, the hypothalamus, pituitary, liver and breast muscle were taken for gen mRNA expression of the somatotropic axis (Figs. 11–14).

In ovo monochromatic GL photostimulation caused a significant increase in hypothalamic GHRH mRNA on ED16 and ED20 (Fig. 11), as well as a significant increase in liver GHR mRNA between ED16 and ED18 (Fig. 12), and liver IGF-1 mRNA between ED14 and ED18 (Fig. 13), while increasing breast muscle IGF-1 mRNA only on ED14 (Fig. 14) (Dishon et al., 2017).

Following this study, we investigated the effect of in ovo GL photostimulation during several incubation periods on in ovo and post-hatch somatotropic axis activity and post-hatch growth (Dishon et al., 2021a,b).

The effect of in ovo-selected GL photostimulated periods on embryonic plasma GH level is presented in Fig. 15. In general, plasma GH levels increased from ED12 until hatch. All GL treatments caused an elevation in plasma GH levels from ED14. On the day of hatch, the GD18–20 group, as well as GD15–20 and GD16–20 groups, and the positive control (GD0–20) showed similar results.

Figure 9 Green light (560nm, at intensity of 0.1 watts/m2) in ovo photostimulation is conducted by LED lamps above the eggs position in the setter. Light intensity was provided intermittently to avoid over heating of egg corr.

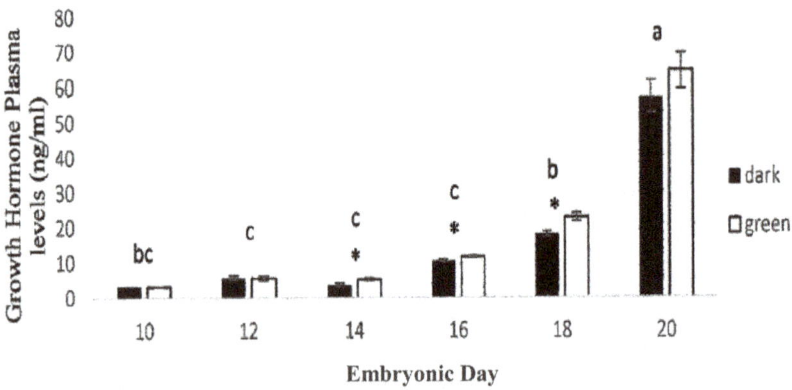

Figure 10 Plasma GH levels, between ED10 and ED20, of broilers incubated in the dark or under monochromatic GL. Data are presented as mean ± SEM. Means with asterisks show a significant difference between the two groups on the same day ($P < 0.05$). Means of days with no common letter differ significantly ($p < 0.05$) (Dishon et al., 2017).

Hypothalamic GHRH mRNA expression increased from ED14 until hatch in all treatment groups (Fig. 16). On ED12 and from ED16 until hatch, all treatment groups showed an increase in GHRH mRNA gene expression compared to the negative control group.

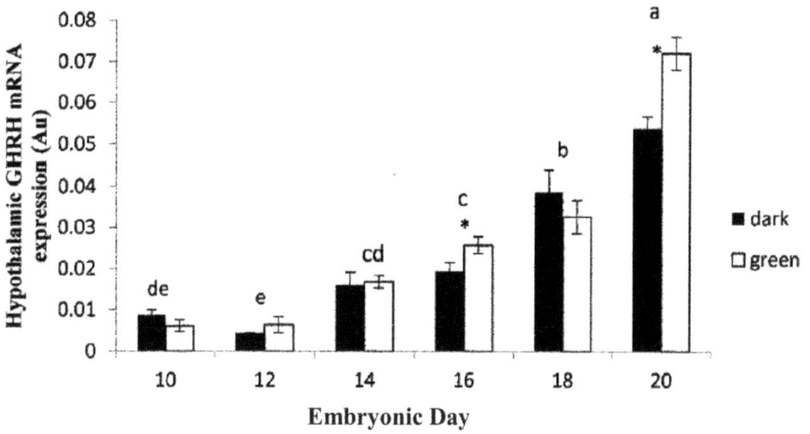

Figure 11 Hypothalamic GHRH mRNA expression, between ED10 and ED20, of broilers incubated in the dark or under monochromatic GL. Data are presented as mean ± SEM. Means with asterisks show a significant difference between the two groups on the same day ($P < 0.05$). Means of days with no common letter differ significantly ($P < 0.05$) (Dishon et al., 2017).

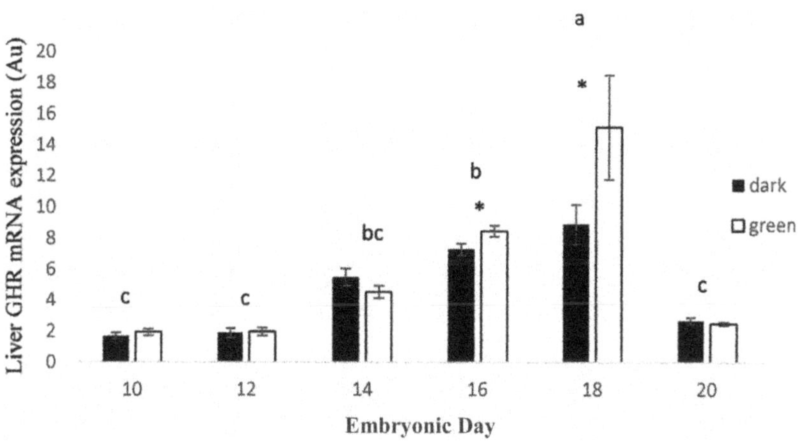

Figure 12 Liver GHR mRNA expression of liver GHR, between ED10 and ED20, of broilers incubated in the dark or under monochromatic GL. Data are presented as mean ± SEM. Means with asterisks show a significant difference between the two groups on the same day ($P < 0.05$). Means of days with no common letter differ significantly ($P < 0.05$) (Dishon et al., 2017).

On the day of hatch, the GD18–20 group, as well as the GD15–20 and GD16–20 groups, showed a similar elevation of GHRH mRNA expression as the positive control group (GD0–20). Liver GHR mRNA expression increased from ED12 to a peak on ED18 and then decreased to ED20 (Fig. 16). The different treatment groups all showed an increase in GHR mRNA expression

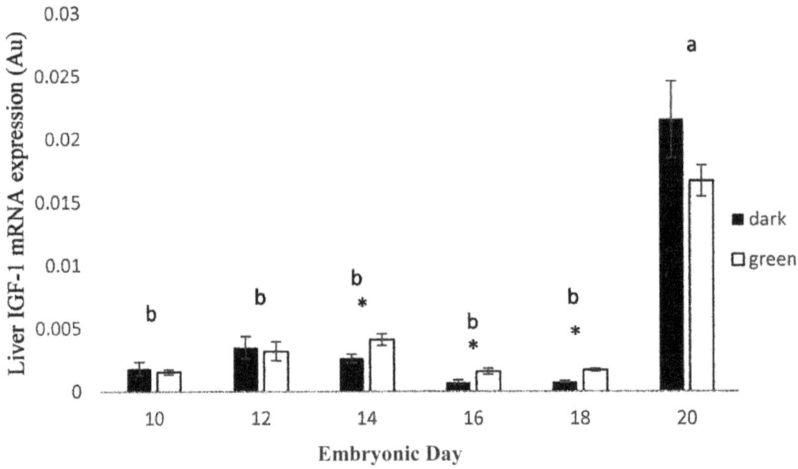

Figure 13 Liver IGF-1 mRNA expression, between ED10 and ED20, of broilers incubated in the dark or under monochromatic GL. Data are presented as mean ± SEM. Means with asterisks show a significant difference between the two groups on the same day ($P < 0.05$). Means of day with no common letter differ significantly ($P < 0.05$) (Dishon et al., 2017).

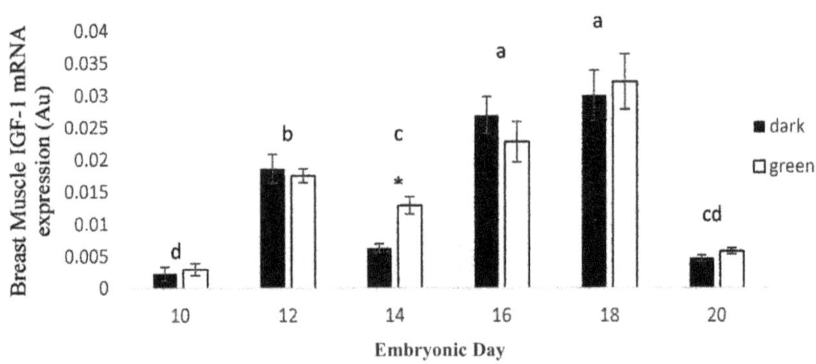

Figure 14 Breast muscle mRNA expression, between ED10 and ED20, of broilers incubated in the dark or under monochromatic GL. Means with asterisks show a significant difference between the two groups on the same day ($P < 0.05$). Means of day with no common letter differ significantly ($P < 0.05$) (Dishon et al., 2017).

compared with the negative control group. On the day of hatch, the GD18-20 treatment group, as well as GD16-20 group, showed a similar elevation of GHR mRNA expression as the positive control group (GD0-20). Liver IGF-1 mRNA expression was low during the incubation period but was elevated on ED20 (day of hatch). On the day of hatch, all treatment groups showed significantly higher IGF-1 mRNA expression compared with the negative control, with no significant difference between the light treatments (Fig. 16).

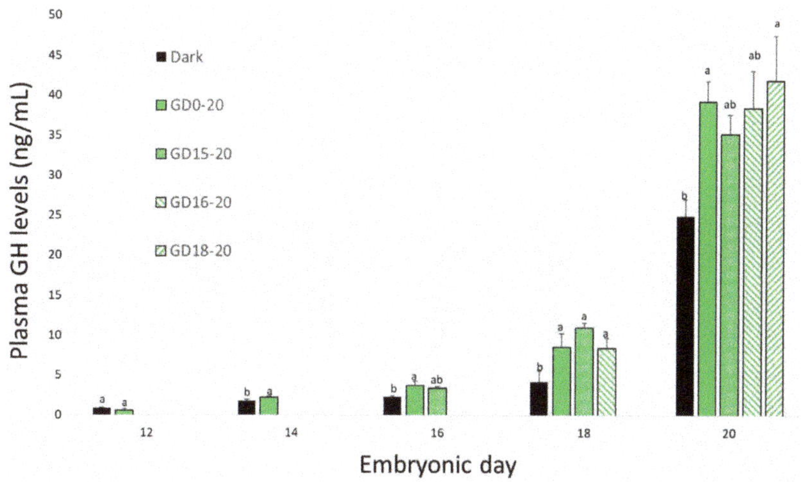

Figure 15 Plasma growth hormone levels, between ED12 and ED20, of broiler embryos incubated in the dark as a negative control (dark) or under monochromatic GL from different days of incubation to the day of hatch (GD0-20, GD15-20, GD16-20 and GD18-20). Means of the groups on a specific day with different lowercase letters differ significantly ($P < 0.05$) (Dishon et al., 2021a).

In a follow-up experiment, we further investigated the post-hatch effect of in ovo GL photostimulation on somatotropic axis manifestation and body and muscle growth (Dishon et al., 2021a,b).

Figure 17 presents post-hatch somatotropic axis activity followed by in ovo GL photostimulation at several embryonic periods.

Plasma GH levels were elevated from day 0 to day 5 and then declined up to day 35. We found that all in ovo GL photostimulated groups had higher plasma GH levels than the dark incubated group. Hypothalamic GHRH mRNA expression on day 0 was elevated in all in ovo GL treatment groups compared to the dark group and by day 35 significant higher hypothalamic GHRH expressions were found in birds in ovo GL photostimulated form on day 0, day 15 and day 18 of incubation compared to the dark incubated group. Similar expressions were found in liver GHR, IGF-1 and muscle IGF-1 post-hatch.

Post-hatch BW (Fig. 18) and breast muscle weights (Fig. 19) clearly demonstrate the beneficial activity of in ovo GL photostimulation of broilers (Dishon et al., 2021b).

Taken together our current findings that in ovo GL photostimulation during the last critical period of the broiler's incubation increased growth and somatotropic axis expression, we suggest that the last incubation days might be considered a photostimulation critical period for broiler somatotropic axis activity. These findings are important, since there is a technical difficulty in photostimulation during incubation. Modern hatcheries are packed and light

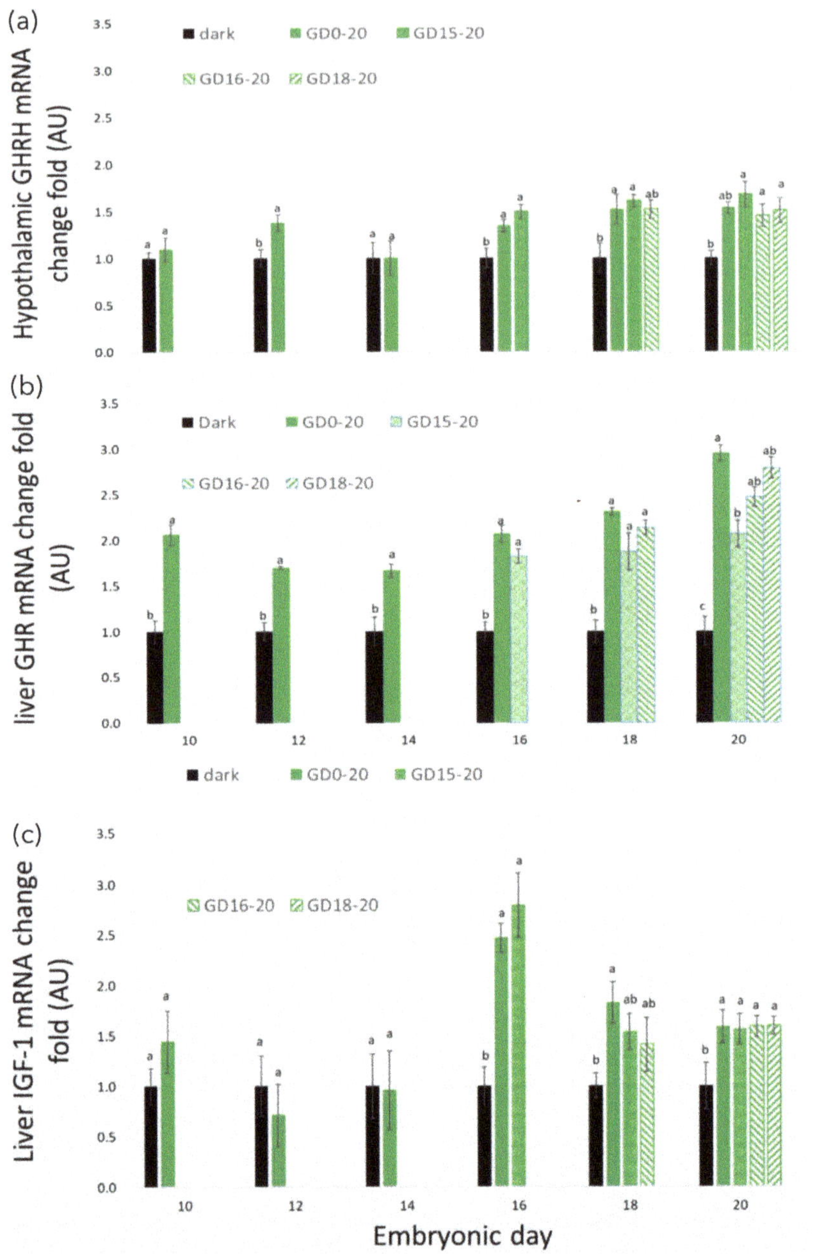

Figure 16 Hypothalamic GHRH (a), liver GHRH (b) and liver IGF-1 (c) mRNA gene expression, between ED10 and ED20, of broiler embryos incubated in the dark as a negative control (dark) or under monochromatic GL from different days of incubation to day of hatch (GD0–20, GD15–20, GD16–20 and GD18–20). Data are presented as mean 6 SEM. Means of the groups on a specific day with different lowercase letters differ significantly (*P* > 0.05) (Dishon et al., 2021a).

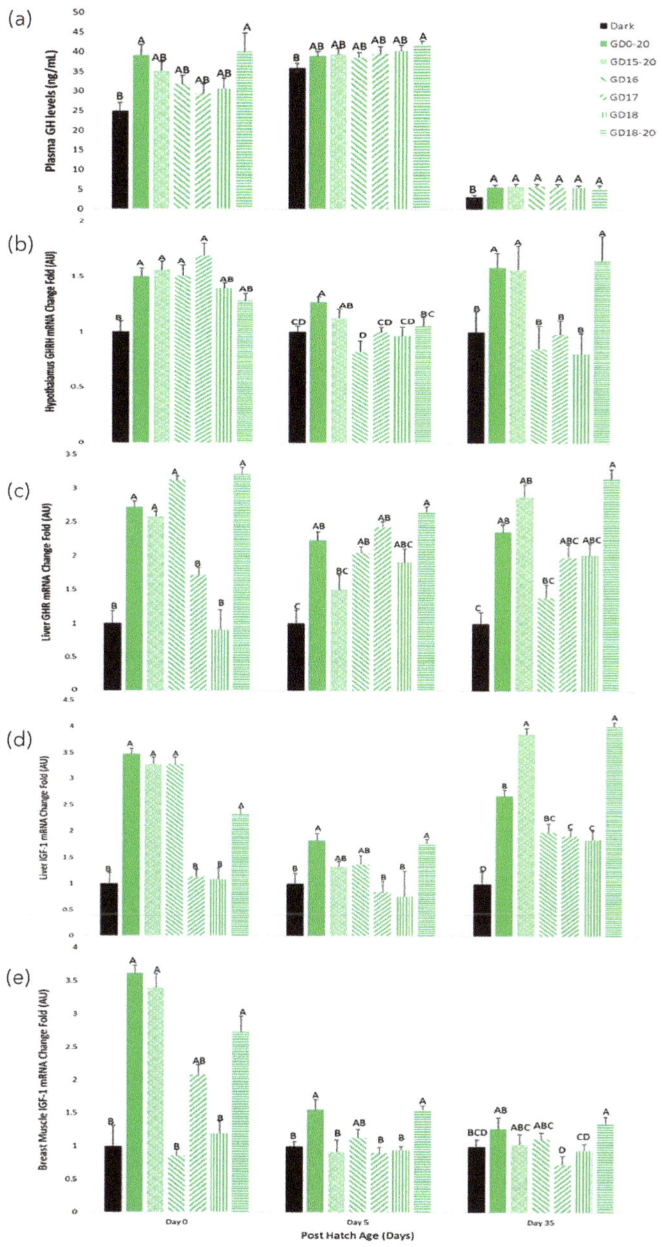

Figure 17 Plasma GH level (a), hypothalamic GHRH (b), liver GHR (c), liver IGF-1 (d) and breast muscle IGF-1 (e) mRNA gene expression, between hatch (day 0) and 35 days of the age of broiler incubated in the dark as a negative control (dark) or under monochromatic GL from different days of incubation to day of hatch (GD0–20, GD15–20, GD16–20 and GD18–20). Data are presented as mean 6 SEM. Means of the groups on a specific day with different lowercase letters differ significantly (*P* > 0.05) (Dishon et al., 2021b).

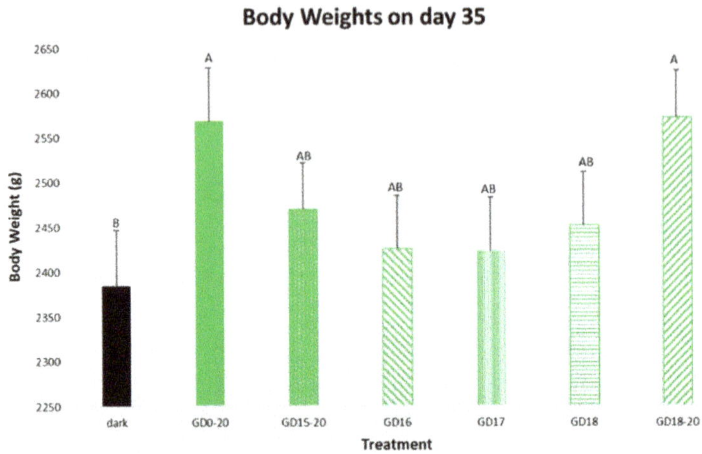

Figure 18 Broiler BW at 35 days of age. Broilers incubated in the dark as a negative control (dark) or under monochromatic green light from different days and periods of incubation to the day of hatch (GD0–20, GD15–20, GD16, GD17, GD18 and GD18–20). Data are presented as means ± SEM. Means of the groups on a specific day with different letters differ significantly ($P < 0.05$) (Dishon et al., 2021b).

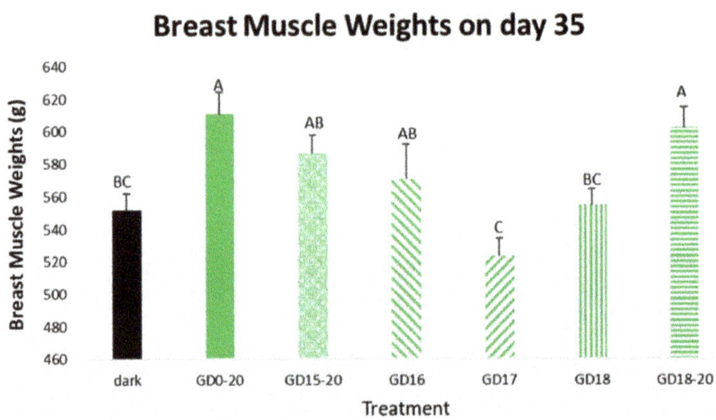

Figure 19 Broiler breast muscle weights at 35 days of age. Broilers incubated in the dark as a negative control (dark) or under monochromatic green light from different days and periods of incubation to the day of hatch (GD0–20, GD15–20, GD16, GD17, GD18 and GD18–20). Data are presented as means ± SEM. Means of the groups on a specific day with different letters differ significantly ($P < 0.05$) (Dishon et al., 2021b).

cannot evenly distribute on the setters. However, at the hatching period (ED18-21), there is a possibility to conduct photostimulation in the hatching baskets that will remain during the entire growth period and will give results similar to GL photostimulation during the entire incubation period. Furthermore, it is possible that the in ovo GL photostimulation effect on plasma GH levels during the last critical stage of incubation (Dishon et al., 2021a,b) may affect the hormonal imprinting process by affecting the responsiveness and quantity of the GH receptors, as found during last stages of incubation and hatching (Dishon et al., 2021a,b) as well as during the post-hatch period in this experiment.

One of the questions arising from these results is whether the changes we discovered may be considered epigenetic alterations/adaptations? Epigenetics refers to the regulation of gene expression with no changes to the DNA sequence (Jiang et al., 2016). Epigenetic adaptation or modification is defined by a prenatal manipulation that generates a physiological memory, which in turn facilitates a better adaptation during the postnatal period (Yahav, 2009; Vinoth et al., 2018). According to Guerrero-Bosgna et al. (2018), a variety of traits can be affected by environmental interaction, through epigenetics. Thus, external changes which affect the pre- and postnatal periods can have enormous effects on the adult phenotype (Guerrero-Bosgna et al., 2018). We found that in ovo photostimulation during the last days of the incubation (GD18–20 group), as well as during the entire embryogenesis period (GD0–20 group), elevated somatotropic axis gene expression during the post-hatch period (even 35 days post-hatch), as well as affected the growth and production of the broilers throughout this period of time. Therefore, we suggest it may be possible that the in ovo GL photostimulation, especially during the entire embryogenesis, or during the last critical stage of incubation, causes an epigenetic adaptation for somatotropic axis expression, which in turn affects the performance of the broilers. However, further studies are required in order to examine and understand whether there are epigenetic modifications during these changes, as well as to examine whether these changes/adaptations can carry on to future generations. Moreover, the in ovo GL photostimulation between ED18 and hatching can become a commercial tool to increase performance in broilers, as it requires GL LEDs only during the hatching process, and increases performance as well as photostimulation from ED0 to hatching.

4 Conclusion

Broilers' meat and meat products supply a very important nutritional source, such as minerals, vitamins, and proteins. The problem is that meat production is not an efficient process. This inefficiency is due to the cost of rearing animals and the process of meat production. Furthermore, increasing the broiler's growth rate and muscle building results in a decrease in the meat's fat content, which in turn may increase its nutritional values (Hays and Preston, 1994).

Many studies examined different external manipulations and their effects on the broiler's growth, production, and quality. One example is the effect of rearing broilers under monochromatic lights. Growing broilers under monochromatic green light increased their body weight as well as cumulative growth rate. In ovo photostimulation of broilers' eggs under monochromatic green light increased the embryos' body weight and breast muscle weights throughout the incubation period, an effect which continued during the first week of the rearing period (Rozenboim et al., 2013). Furthermore, it was found that in ovo monochromatic green light photostimulation increased broilers' body weights until day 42 of the rearing period, as well as increased pectoral muscle weights as a percentage of body weights (Rozenboim et al., 2004).

This phenomenon raised a question whether the endocrine/molecular mechanisms play a role in the effect of those manipulations on broiler's performance. In our studies, we found that growth acceleration was the activation of somatotropic axis. We found that incubation under monochromatic green light from ED0 until hatch elevated different components of the somatotropic axis. These effects are represented by an increase of hypothalamic GHRH mRNA gene expression from ED12 until day 35 of the rearing period, as well as an increase in plasma GH levels from ED14 until day 35 post-hatch, an increase in the liver GHR gene expression from ED10 until day 35 of the rearing period, and an increase in the liver IGF-1 mRNA gene expression from ED16 until day 35 post-hatch (Dishon et al., 2017; Dishon et al., 2018; Dishon et al., 2021; Dishon et al., unpublished). As for the breast muscle IGF-1, the results we found are similar to those of Halevy et al. (2006), which found that incubation under monochromatic green light elevated the muscle IGF-1 gene expression only during the post-hatch period (Halevy et al., 2006). In our experiments, we found that during the embryonic stages, the GL photostimulation elevated the muscle IGF-1 gene expression only on ED14 (Dishon et al., 2017), but it did cause an elevation in the gene expression from hatch until day 35 of the rearing period.

Several papers describe the critical period of different external manipulation and their effects on the chicken embryo. We used in ovo GL photostimulation in different parts of incubation and we found out that photostimulation from ED18 until hatch resulted in elevation in the somatotropic axis activity followed by elevation in post-hatch growth. This finding suggest a tool for the industry to be used,

5 References

Allen, R. E., Merkel, R. A. and Young, R. B. (1979), Cellular aspect of muscle growth: myogenic cell proliferation. *J. Anim. Sci.*, 49(1), 115–127.

Andrews, D. K. and Zimmermann, N. G. (1990), A comparison of energy efficient broiler 255 house lighting sources and photoperiods, *Poult. Sci.*, 69(9), 1471–1479.

Barott, H. G. and Pringle, E. M. (1951), Effect of environment on growth and feed and water consumption of chickens. 4. The effect of light on early growth, *J. Nutr.*, 45(2), 265–274.

Blum, H. (1964), *Photodynamic Action and Diseases Caused by Light*, New York, NY: Hafner.

Cherry, P. and Barwick, M. W. (1962), The effect of light on broiler growth: II. Light patterns, *Br. Poult. Sci.*, 3(1), 41–50.

Coleman, M. A. and McDaniel, G. R. (1976), Light alterated changes in the embroynic age versus incubation age of White Leghorn embryos, *Poult. Sci.*, 55(6), 2483–2485.

Coleman, M. A. and McNabb, R. A. (1975), Photoacceleration of embryonic development in depigmented Japanese quail eggs, *Poult. Sci.*, 54(6), 1849–1855.

Cooper, J. B. (1972), Effect of light during incubation on hatchability of turkey eggs, *Poult. Sci.*, 51(4), 1105–1108.

Cooper, R. N., Tajbakhsh, S., Mouly, V., Cossu, G., Buckingham, M. and Butler-Browne, G. S. (1999), In vivo satellite cell activation via Myf5 and MyoD in regenerating mouse skeletal muscle, *J. Cell Sci.*, 112(17), 2895–2901.

Cornelison, D. D. W. and Wold, B. J. (1997), Single-cell analysis of regulatory gene expression in quiescent and activated mouse skeletal muscle satellite cells. *Dev. Biol.*, 191(2), 270–283.

Dishon, L., Avital-Cohen, N., Malamud, D., Heiblum, R., Druyan, S., Porter, T. E., Gumulka, M. and Rozenboim, I. (2017), In-ovo monochromatic green light photostimulation enhances embryonic somatotropic axis activity, *Poult. Sci.*, 96(6), 1884–1890.

Dishon, L., Avital-Cohen, N., Zaguri, S., Bartman, J., Heiblum, R., Druyan, S., Porter, T. E., Gumulka, M. and Rozenboim, I. (2018), In-ovo green light photostimulation during different embryonic stages affect somatotropic axis, *Poult. Sci.*, 97, 1998–2004.

Dishon, L., Avital-Cohen, N., Zaguri, S., Bartman, J., Heiblum, R., Druyan, S., Porter, T. E., Gumulka, M. and Rozenboim, I. (2021a), In ovo green light photostimulation during the late incubation stage affects somatotropic axis activity, *Poult. Sci.*, 100(2), 467–473.

Dishon, L., Avital-Cohen, N., Zaguri, S., Bartman, J., Heiblum, R., Druyan, S., Porter, T. E., Gumulka, M. and Rozenboim, I. (2021b), The effect of selected *in ovo* green light photostimulation periods on post-hatch broiler growth and somatotropic axis activity. *Poult. Sci.*, 100(8), 101229.

Edgerton, V. R. and Roy, R. R. (1991), Regulation of skeletal muscle fiber size, shape and function, *J. Biomech.*, 24(Suppl. 1), 123–133.

Engert, J. C., Berglund, E. B. and Rosenthal, N. (1996), Proliferation precedes differentiation in IGF-I-stimulated myogenesis, *J. Cell Biol.*, 135(2), 431–440.

Fairchild, B. D. and Christensen, V. L. (2000), Photostimulation of turkey eggs accelerates hatching times without affecting hatchability, liver or heart growth, or glycogen content, *Poult. Sci.*, 79(11), 1627–1631.

Florini, J. R., Ewton, D. Z. and Coolican, S. A. (1996), Growth hormone and the insulin-like growth factor system in myogenesis, *Endocr. Rev.*, 17(5), 481–517.

Foss, D. C., Carew, L. B. Jr. and Arnold, E. L. (1972), Physiological development of cockerels as influenced by selected wavelengths of environmental light, *Poult. Sci.*, 51(6), 1922–1927.

Goodman, G. (1983), Physiological responses of the cockerel to light quality: growth and sexual development. Doctoral thesis, The Hebrew University of Jerusalem, Jerusalem Israel.

Guerrero-Bosagna, C., Morisson, M., Liaubet, L., Rodenburg, T. B., de Haas, E. N., Košťál, Ľ. and Pitel, F. (2018), Transgenerational epigenetic inheritance in birds. *Environ. Epigenet.*, 4(2), dvy008.

Halevy, O., Biran, I. and Rozenboim, I. (1998), Various light source treatments affect body and skeletal muscle growth by affecting skeletal muscle satellite cell proliferation in broilers. *Comp. Biochem. Physiol., Part A Mol. Integr. Physiol.*, 120(2), 317-323.

Halevy, O., Nagler, A., Levi-Schaffer, F., Genina, O. and Pines, M. (1996), Inhibition of collagen type I synthesis by skin fibroblasts of graft versus host disease and scleroderma patients: effect of halofuginone, *Biochem. Pharmacol.*, 52(7), 1057-1063.

Halevy, O., Piestun, Y., Rozenboim, I. and Yablonka-Reuveni, Z. (2006), In ovo exposure to monochromatic green light promotes skeletal muscle cell proliferation and affects myofiber growth in posthatch chicks, *Am. J. Physiol. Regul. Integr. Comp. Physiol.*, 290(4), R1062-R1070.

Hawke, T. J. and Garry, D. J. (2001), Myogenic satellite cells: physiology to molecular biology, *J. Appl. Physiol.*, 91(2), 534-551.

Hays, V. W. and Preston, R. L. (1994), In: *Low-Fat Meats Design Strategies and Human Implications Food Science and Technology*, 13-33. Academic Press.

Jiang, Y., Chen, L., Zhang, H., Zhang, Q., Chen, W., Zhu, J. and Song, D. (2016), Two-dimensional Co_3O_4 thin sheets assembled by 3D interconnected nanoflake array framework structures with enhanced supercapacitor performance derived from coordination complexes, *Chem. Eng. J.*, 292, 1-12.

Kästner, S., Elias, M. C., Rivera, A. J. and Yablonka-Reuveni, Z. (2000), Gene expression patterns of the fibroblast growth factors and their receptors during myogenesis of rat satellite cells. *J. Histochem. Cytochem.*, 48(8), 1079-1096.

Lauber, J. K. (1975), Photoacceleration of avian embryogenesis, *Comp. Biochem. Physiol. A Comp. Physiol.*, 51(4), 903-907.

Lauber, J. K., Shutze, J. V. and McGinnis, J. (1961), Effects of exposure to continuous light on the eye of the growing chick. *Proc. Soc. Exp. Biol. Med.*, 106, 871-872.

Leighton, A. T. and Mason, J. P. (1976), Environmental factors affecting growth performance of turkeys, *Am. Soc. Agric. Eng.*, 76, 4508.

Ludolph, D. C. and Konieczny, S. F. (1995), Transcription factor families: muscling in on the myogenic program, *F.A.S.E.B. J.*, 9(15), 1595-1604.

Mauro, A. (1961), Satellite cell of skeletal muscle fibers, *J. Biophys. Biochem. Cytol.*, 9(2), 493-495.

Mobarkey, N., Avital, N., Heiblum, R. and Rozenboim, I. (2010), The role of retinal and extra-retinal photostimulation in reproductive activity in broiler breeder hens, *Domest. Anim. Endocrinol.*, 38(4), 235-243.

Naya, F. J. and Olson, E. (1999), MEF2: a transcriptional target for signaling pathways controlling skeletal muscle growth and differentiation, *Curr. Opin. Cell Biol.*, 11(6), 683-688.

Ookawa, T. (1970), Effects of bilateral optic enucleation on body growth and gonad in young male chicks, *Poult. Sci.*, 49(1), 333-334.

Phogat, S. B., Aggarwal, C. K. and Chopra, S. K. (1985), *Effect of Red and Green Lights growth of quail. Indian J. Poult. Sci.* 20, 126-128.

Romanoff, A. L. (1960), The avian embryo. Structural and functional development. The avian embryo, *Struct. Funct. Develop.*

Rozenboim, I., Biran, I., Chaiseha, Y., Yahav, S., Rosenstrauch, A., Sklan, D. and Halevy, O. (2004), The effect of a green and blue monoch romatic light combination on broiler growth and development, *Poult. Sci.*, 83(5), 842-845.

Rozenboim, I., El Halawani, M. E., Kashash, Y., Piestun, Y. and Halevy, O. (2013), The effect of monochromatic photostimulation on growth and development of broiler birds, *Gen. Comp. Endocrinol.*, 190, 214-219.

Rozenboim, I., Robinzon, B. and Rosenstrauch, A. (1999), Effect of light source and regimen on growing broilers, *Br. Poult. Sci.*, 40(4), 452–457.

Sadowski, C. L., Wheeler, T. T., Wang, L. H. and Sadowski, H. B. (2001), GH regulation of IGF-I and suppressor of cytokine signaling gene expression in C2C12 skeletal muscle cells, *Endocrinology*, 142(9), 3890–3900.

Shafey, T. M. (2002), Eggshell 322 conductance, embryonic growth, hatchability and embryonic mortality of broiler breeder eggs dipped into ascorbic acid solution, *Br. Poult. Sci.*, 43(1), 135–140.

Shutze, J. V., Lauber, J. K., Kato, M. and Wilson, W. O. (1962), Influence of incandescent and coloured light on chicken embryos during incubation, *Nature*, 196, 594–595.

Siegel, P. B., Isakson, S. T., Coleman, F. N. and Huffman, B. J. (1969), Photoacceleration of development in chick embryos, *Comp. Biochem. Physiol.*, 28, 753–758.

Smith, C. Kn., Janney, M. J. and Allen, R. E. (1994), Temporal expression of myogenic regulatory genes during activation, proliferation, and differentiation of rat skeletal muscle satellite cells, *J. Cell. Physiol.*, 159(2), 379–385.

Tamimie, H. S. (1967), Light exposure of incubating eggs and its influence of the growth of chicks. I. Brooding chicks under different light regimens, *Comp. Biochem. Physiol.*, 21(1), 59–63.

Vinoth, A.,Thirunalasundari, T., Shanmugam, M., Uthrakumar, A., Suji, S. and Rajkumar, U. (2018), Evaluation of DNA methylation and mRNA expression of heat shock proteins in thermal manipulated chicken, *Cell Stress Haperones*, 23, 235–252.

Wabeck, C. J. and Skoglund, W. C. (1974), Influence of radiant energy from fluorescent light sources on growth, mortality, and feed conversion of broilers, *Poult. Sci.*, 53(6), 2055–2059.

Walter, J. H. and Voitle, R. A. (1972), Effects of photoperiod during incubation on embryonic and post-embryonic development of broilers, *Poult. Sci.*, 51(4), 1122–1126.

Weintraub, H. (1993), The MyoD family and myogenesis: redundancy, networks, and thresholds, *Cell*, 75(7), 1241–1244.

Woodard, A. E., Moore, J. A. and Wilson, W. O. (1969), Effect of wave length of light on growth and reproduction in Japanese quail (Coturnix japonica), *Poult. Sci.*, 48(1), 118–123.

Yablonka-Reuveni, Z. (1995), Development and postnatal regulation of adult myoblasts, *Microsc. Res. Tech.*, 30(5), 366–380.

Yablonka-Reuveni, Z. and Paterson, B. M. (2001), MyoD and myogenin expression patterns in cultures of fetal and adult chicken myoblasts, *J. Histochem. Cytochem.*, 49(4), 455–462.

Yablonka-Reuveni, Z. and Rivera, A. J. (1994), Temporal expression of regulatory and structural muscle proteins during myogenesis of satellite cells on isolated adult rat fibers, *Dev. Biol.*, 164(2), 588–603.

Yablonka-Reuveni, Z., Seger, R. and Rivera, A. J. (1999), Fibroblast growth factor promotes recruitment of skeletal muscle satellite cells in young and old rats, *J. Histochem. Cytochem.*, 47(1), 23–42.

Yahav, S. (2009), Alleviating heat stress in domestic fowl: different strategies, *World's Poult. Sci. J.*, 65(4), 719–732.

Zammit, P. S. and Beauchamp, J. R. (2001), The skeletal muscle satellite cell: stem cell or son of stem cell? *Diff. Res. Biol. Diver.*, 68(4–5), 193–204.

Chapter 12

Understanding the effects of humidity/air composition on embryo and post-hatch chick development

E. David Peebles, Mississippi State University, USA

1 Introduction

Incubation conditions influence embryo growth and viability, which subsequently affect egg hatchability. Along with temperature, environmental factors in the incubator that are of primary importance for optimum hatchability and chick quality include humidity and vital gas [carbon dioxide (CO_2) and oxygen (O_2)] levels. Because the movements of gases across the eggshell occur by diffusion, avian embryos are dependent on the composition of the surrounding air, with their development and hatching success being dependent on the ambient partial pressures of O_2, CO_2, and water vapor. The influences of the incubational environment can extend beyond embryonic development into the post-hatch period with ensuing effects on chick survival, quality, and performance. Various studies have been conducted to establish humidity and vital gas levels during incubation that will optimize the embryonic and post-hatch livability and development of poultry. The prospects for the pragmatic commercial use of these incubational regimens, as well as the physiological bases for their observed effects, are explored in this chapter.

http://dx.doi.org/10.19103/AS.2022.0118.16

2 Physical function of the eggshell as a respiratory organ for the developing avian embryo

2.1 Eggshell porosity

Different integrated physical or structural properties of the avian eggshell determine its porosity, and the transport of water and the vital gases (O_2 and CO_2) across the eggshell occurs by their diffusion through the pores of the eggshell when in their gaseous phases (Paganelli, et al., 1975). These integrated eggshell properties subsequently predispose the proficiencies by which water vapor and the vital gases diffuse and are exchanged between the exterior and interior of the embryonated egg. The pores of the eggshell extend through the shell proper, which is composed of calcite crystals, and terminate on two underlying membrane layers. The membranes overlay and make immediate contact with the vascularized chorioallantois during the first 5-6 days of incubation (Fig. 1). In conjunction with the continual loss of water, the exchange of O_2 and CO_2 occurs across the chorioallantois during the prenatal period (Rahn et al., 1979). The conductance of the eggshell to gases, as governed by its porosity, and differences in the partial pressure of

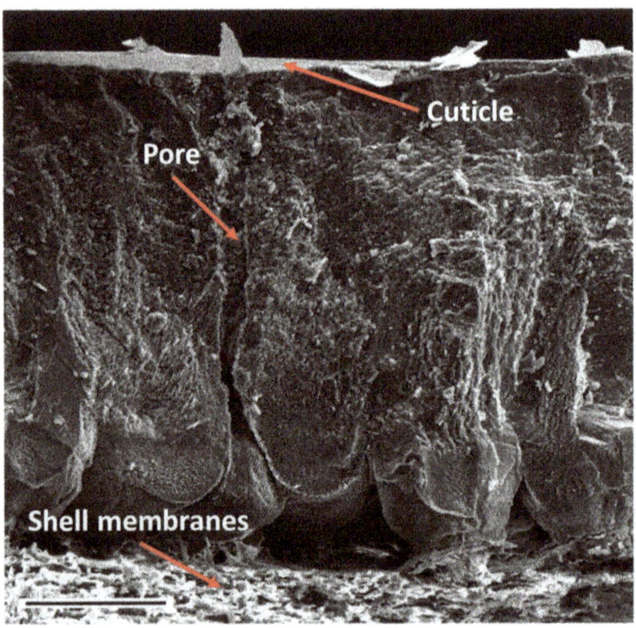

Figure 1 Scanning electron microscope micrograph depicting a longitudinal section of an avian eggshell. The cuticle, pore, and shell membranes are indicated (scale bar = 100 μm, image magnification 200×). The chorioallantois will form underneath the shell membranes.

gases between the outside and inside of the egg contribute to the control of gas exchange across the eggshell. It has been summarized that the flux of gas across the eggshell is determined by the product of its effective conductance and the partial pressure gradient of the gases between the inner surface of the eggshell and the ambient air surrounding the eggshell (Visschedijk, 1980). The passage of water vapor and the vital gases through the pores of the eggshell proper and underlying membranes allows for their exchange in the circulation and tissues of the developing avian embryo (Paganelli, 1991). Consequently, embryonic and post-hatch chick development are significantly influenced by the transfer of heat and the exchange of O_2, CO_2, and water between the internal and external environments of the egg (Boleli et al., 2016). Meir and Ar (1990) have noted that an increase in the rate of egg mass-specific O_2 consumption by ostrich embryos 1 day prior to external pipping is associated with their active movement inside the egg in accommodation for a change in their means of O_2 uptake during the shift from chorioallantoic to pulmonary respiration. Furthermore, regional gas tensions in the air spaces within the egg are also a consequence of the ratio of diffusive conductance to the perfusion of blood in the chorioallantois and can change with location on the eggshell (Paganelli et al., 1988; Paganelli, 1991).

2.2 Water vapor conductance and incubational egg weight loss

Water vapor, O_2, and CO_2 share common pathways through the pores that traverse the eggshell (Paganelli et al., 1978), and the shell proper itself has the primary influence on the partial-pressure differences that exist for these gases between the internal and external environments of the egg (Rahn et al., 1979). Moreover, variations in water vapor conductance reflect proportional variations in the conductance values of O_2 and CO_2 (Paganelli et al., 1975). The determination of water vapor conductance is the easiest to measure because it can be calculated from the rate of water loss from an egg, which equates to its weight loss, and egg weight is reciprocally and equally affected by the uptake of O_2 and the loss of CO_2. Therefore, water vapor conductance is commonly used to also represent the functional conductance of an eggshell to O_2 and CO_2. In brief, the physiological function of the eggshell as an embryonic respiratory component is more specifically best described in terms of its water vapor conductance since water vapor conductance incorporates the interactions of these gases with the eggshell's physical properties (Peebles and McDaniel, 2004). The water vapor partial pressure gradient that exists between the inside and outside of the egg and which is determined by the difference in the levels of humidity between the egg's internal and external environments is affected by the temperature of the egg and the local barometric pressure.

The determination of water vapor conductance is a function of the division of the rate of water loss from the egg by the water vapor gradient across the shell at a particular temperature and barometric pressure (Ar et al., 1974; Tullett, 1984; Peebles and McDaniel, 2004; Pulikanti et al., 2011b). However, Rokitka and Rahn (1987) have reported that there may be regional differences in the rates water vapor conductance over the entire eggshell surface. For example, eggshell water vapor conductance over the air cell may be higher by a factor of 1.6 in comparison to the other regions of the eggshell (Rokitka and Rahn, 1987). It should be noted that embryonic metabolism and hatchability are impaired if eggs that are laid at a low altitude are subsequently incubated at a high altitude (Visschedijk, 1980). A reduction in the atmospheric barometric pressure, associated decrease in O_2 tension, and an increase in eggshell conductance or functional porosity, leading to an excessive loss of CO_2 and water are the major causes of this impairment (Visschedijk, 1991).

The loss of egg weight in association with the loss of water during incubation is known to influence embryogenesis and the post-hatch nutrient utilization, metabolism, growth, and development characteristics of broiler chicks (Peebles et al., 2005). A negative correlation has been observed between the average daily percentage egg weight loss of broiler hatching eggs from 10.5 days to 18.5 days of incubation and the body weight of chicks relative to set egg weight through 12 h post-hatch (Pulikanti et al., 2012b). Pulikanti et al. (2012b) have more specifically suggested that a higher eggshell water vapor conductance adjusted for egg weight results in increased metabolism of the broiler embryo presumably due to increased O_2 uptake, which then increases the successive growth and rate of yolk sac absorption of the chicks through 3 days post-hatch. Influences of the water vapor conductances of eggshells on various physiological attributes of broilers can extend into the middle and late post-hatch grow-out periods (Pulikanti et al., 2013). Pulikanti et al. (2013) further observed that adjusted eggshell water vapor conductance values were positively correlated with breast muscle weight on day 48 post-hatch. Because the incubational weight loss of broiler hatching eggs through the first 10.5 days of incubation has been shown to be negatively correlated with hatchling body weight and to influence day 49 processing yield (Peebles et al., 2014), it should be closely monitored, particularly during the first half of incubation.

2.3 Influences of incubational temperature and airflow

Changes in egg temperature can be affected by changes in the rate of water loss from an egg during incubation. Because the physical change of water from a liquid to vapor requires heat, an increase in the loss of water would result in an increase in the loss of heat from an egg due to evaporative cooling (French, 2009b). An increase in evaporation rate promoted by an increase in the rate of water vapor conductance can help reduce the effects of increased embryonic metabolic heat production on egg temperature (Meir and Ar, 1990). Nevertheless, other studies have specifically focused on incubation temperature and airflow rate as factors that contribute to the control of water vapor and vital gas exchange across the eggshell (Robertson, 1961b; French, 1997, 2009a). Wilson (1991) indicated that small incremental deviations from 37.0°C to 38.0°C dry bulb incubational temperature can significantly impact embryo development and subsequent hatchability. Nevertheless, it has been emphasized that incubator air temperature and the internal temperature of the egg can be distinctly different, as the temperature that the developing embryo experiences is not only dependent on incubator temperature but also influenced by the transmission of heat between the external and internal environments of the egg and the production of metabolic heat by the metabolizing embryo (French, 1997, 2009a). French (2009a) has precisely indicated that the temperature that the developing embryo inside the egg experiences should be considered as the real or true incubation temperature. The temperature of the external environment of an egg influences heat transfer to or from the egg, and the rise in temperature of the air passing over the egg is inversely proportional to the rate of airflow or airflow volume (French, 2009a). Nevertheless, airflow rate itself can be effectively used to regulate the external temperature of an egg without directly affecting its moisture content, because the rate of airflow over an egg has a negligible effect on its rate of water loss (Kaltofen, 1969; Spotila et al., 1981). When maintaining an optimum air temperature around eggs, a lower air speed is required when the spacing between eggs is increased (French, 1997).

2.4 Measurement of internal egg temperature

Pulikanti et al. (2011a) reported that temperature transponders could be inserted into the air cells of broiler hatching eggs between 12 days and 14 days of incubation without negatively influencing eggshell porosity or causing physiological stress to the growing embryo. Peebles et al. (2012) have suggested that the use of transponders in the air cells of embryonated eggs circumvents the confounding effects that the thermal properties of the eggshell, as well as the flow of air across the shell, may impose on the temperature that the embryo actually experiences. Moreover, Pulikanti et al. (2012a) observed that transponders in the air cells of embryonated eggs detected minute internal temperature fluctuations and recorded mean temperatures that were consistently higher than those of non-embryonated eggs as well as the external microenvironments surrounding the eggs between 13 days and 18 days of incubation. Using transponders, Peebles et al. (2012) have likewise concluded that in comparison to eggshell surface temperatures, air cell temperatures were higher and closer to those of actual broiler embryo temperatures previously reported in other studies. At a 75% survival rate, it has also been concluded that temperature transponders could be successfully implanted in the air cells of broiler hatching eggs at 10.5 days of incubation to determine internal egg temperature for the subsequent accurate calculation of eggshell water vapor conductance unadjusted or adjusted for set weight (Pulikanti et al., 2011b, 2012a,b), as well as their water vapor conductance constants (Pulikanti et al., 2012a,b).

The measurement of internal egg temperature is the most accurate and, therefore, the most ideal means to assess embryo temperature; however, it is not practical in commercial operations (French, 2009a). Sotherland et al. (1987) have suggested that the measurement of eggshell temperature is more pragmatic and has been shown to be closely related to internal egg temperature. Using eggshell temperature to measure embryonic temperature (Lourens et al., 2005; Joseph et al., 2006), it has been reported that deviating from a constant eggshell temperature of 37.8°C can result in 10% differences in the yolk-free body mass and hatchability of broiler hatchlings (Lourens et al., 2005). Therefore, because eggshell temperature is affected by embryonic metabolic heat production, and because embryonic heat production can be affected by the amount and efficiency of energy utilization by the embryo, changes in eggshell temperature can subsequently impact embryogenesis, hatchability, and chick quality (Meijerhof, 2002; Lourens et al., 2011).

The topics discussed in this section have described various aspects that are the bases for understanding how humidity and the O_2 and CO_2 concentrations of the air within incubators affect embryo development, hatchability, and post-hatch chick development. The techniques by which the concentrations and exchange or flux of water, O_2, and CO_2 between the external and interior environments of the embryonated egg are measured and the environmental factors that affect their values have also been considered. The following sections describe in more detail the individual effects of humidity, O_2, and CO_2 on embryo and post-hatch chick development.

3 Humidity

3.1 Egg water content and its production and loss during embryonic development

Water is the dominant constituent in eggs, comprising approximately 75% of the internal contents of an egg. The concentrations of water in the albumen, yolk, and eggshell are 88.8%, 47.5%, and 1.0%, respectively (Romanoff and Romanoff, 1949; Vieira and Moran, 1999). A partial loss of this water must occur throughout incubation for embryonic development to be properly achieved

Published by Burleigh Dodds Science Publishing Limited, 2023.

(Barbosa et al., 2013). As summarized by Davis et al. (1988), the total internal water content of an egg during embryonic development is affected by the loss of water through the pores of the eggshell (Paganelli, 1980) and the production of metabolic water by the oxidation of yolk lipids (Ar and Rahn, 1980). Considering the smaller portion of water that is lost to the external environment by its diffusion across the eggshell, the water content of the embryo is increased by its absorption from the extra-embryonic tissues including the amnion, allantois, and yolk sac (Davis et al., 1988; Ar, 1991b). Furthermore, the accretion of metabolic water as a byproduct of the metabolism of lipids by the embryo, which increases with development (Ar, 1991a), is responsible for approximately 8–13% of the water content of the embryo (Ar, 1991b). Not only does the loss of water during incubation have an important association with the loss of heat that is produced by the metabolically active embryo, but it is necessary for the formation of the egg air cell or chamber (Rahn and Ar, 1980; Visschedijk et al., 1980), which will possess a volume of gas that is positively related to the volume of water that is lost from the egg (Durojaye et al., 2018). A substantial air cell with an adequate volume and a well-proportioned supply of vital gases is critical for the subsequent hatching success of the embryo (Rahn et al., 1979; Ar and Rahn, 1980; Ar, 1991b; Barbosa et al., 2013; Uçar et al., 2021). Most importantly, it is essential that the air cell volume is large enough to support the embryo's initial breathing activities and to allow for the adequate filling of its respiratory system, including its lungs and air sacs. The volume of the air cell space needed to accommodate the pipping and subsequent hatching processes is approximately 14% of the total egg volume (Rahn et al., 1976).

3.2 Relative humidity

As described by Cormick (2021), absolute humidity is the amount of water contained in air (g/m^3) regardless of temperature, whereas relative humidity (RH) is the volume of moisture that air can hold at a given temperature and is expressed as a percentage. At a given RH, air at a cold temperature holds less moisture, whereas at a higher temperature it holds more moisture (Cormick, 2021). Relative rather than absolute humidity has been commonly used as the mode of measurement in hatcheries. Because the absolute humidity of the air surrounding the egg modulates the rate of water loss through the pores of the eggshell (Ar and Sidis, 2002), the RH of the incubational environment is a primary means by which the internal water content of the egg can be influenced and altered (Davis et al., 1988). The rate of water loss from an egg increases as the RH of the air surrounding it decreases. Therefore, to augment the proper development of the embryo, the RH of the air that surrounds a fertile hatching egg should be monitored and controlled (Decuypere et al., 2001).

3.3 Egg water loss requirement

Interspecific and intraspecific egg water loss during incubation can vary substantially (Walsberg, 1980). However, it has been well established that prior to pipping, eggs laid by poultry species must lose between 12% and 14% of their fresh weight at set by the loss of water through the pores of their eggshells to hatch successfully (Ar, 1991a). Lundy (1969) has further stated that peak hatching success can be achieved in domestic fowl when total egg water loss equals 10–12% of initial egg weight. In the review article by Barbosa et al. (2013), it was summarized that the RH of an incubator containing broiler hatching eggs should be set between 56% and 60% to allow for a 12% loss of set egg weight through 18 days of incubation and that maintaining a weight loss range within 11–13% during the first 18 days of incubation led to a better rate of hatch than did weight losses that were lower. When considering the effects of altitude changes on the water vapor conductance of eggs, Visschedijk (1991) has noted that in order to compensate for an increase in the water vapor conductance rate of eggs at higher altitudes, the water vapor gradient across the eggshell must be reduced to allow water loss to remain similar to that at lower altitudes (sea level).

3.4 Effects of relative humidity on hatchability and hatchling quality

A 5% change in humidity results in a 1% difference in moisture loss by 18 days of incubation (Cormick, 2021). Although the incubation of broiler breeder eggs of similar weight at either a 33% or 50% RH did not affect the absolute yolk sac weights of hatchlings, Tullett and Burton (1982) observed that, in conjunction with a lower incubational egg weight loss, their body weights were higher when they were incubated at the higher RH. If the RH during incubation is too low, embryonic moisture loss will be excessive, which will affect hatchability and will cause hatchlings to be dehydrated and to have lower body weight. Conversely, if the RH during incubation is too high, an early hatch of mushy chicks may result (Barbosa et al., 2013). The effects of these improper RH settings can further extend into the post-hatch growing period resulting in sub-optimal performance and decreased processing yield (Bruzual et al., 2000b; Pulikanti et al., 2012b; Peebles et al., 2014). Adjustment of an RH in the hatcher that is higher than that in the setter has been used to facilitate the hatching process of chicks. Upon examining the effects of setter and hatcher humidities on the hatchability of Bobwhite quail eggs, Wilson and Dugan (1992) concluded that previous recommendations for setter and hatcher RH levels of 55.4% and 75.0%, respectively, were satisfactory for optimum hatchability. Nevertheless, after conducting studies on the effects of incubational RH during the last 5 days of incubation, Bruzual et al. (2000b) summarized that despite the lack of an effect on hatchling body weight, optimal performance was achieved overall by incubating eggs from young broiler breeder hens at an RH of 53% rather than at 43% or 63%.

3.5 Effects of relative humidity on embryo metabolism and embryonic and post-hatch chick development

Although varying RH outside a 42–70% range was not found by Barott (1937) to significantly affect the time of hatch of chicks, when set at 42% or 70%, it decreased embryonic energy metabolism and thereby compromised their growth. Barott (1937) concluded that the metabolic activity of the embryos occurred at a higher level at an RH of 60%. Lundy (1969) later reported that the optimal RH range is wide and can be between 40% and 70%, but Robertson (1961a) has concluded that RH should be approximately 50% to maximize hatchability. Robertson (1961b) further noted that within a 40–70% range in RH, hatchability and hatchling body weight, which reflect embryonic growth, were not significantly affected. However, it was found that the two endpoints of that RH range inversely altered the rate of egg weight loss, thereby potentially disrupting embryo metabolism at various stages of development, with the subsequent consequence of increasing embryonic mortality. In that report, it was also indicated that egg size may be a confounding factor in determining an optimum RH, with larger eggs requiring a lower RH. Because both incubation RH and temperature not only affect the loss of heat from the embryonated egg but also affect its rate of water loss, eggs that are incubated at a high or low RH may need to be incubated at a different incubation temperature to maintain the same embryo temperature when incubated in a more normal (55–60%) RH range (Molenaar et al., 2010; van der Pol et al., 2013; Boleli et al., 2016). A higher level of hatchability for broiler eggs was obtained by van der Pol et al. (2013) when they were incubated within a 55–60% range of RH and when their eggshell temperature was 37.8°C.

Upon comparing the effects of incubating layer hatching eggs at a 45% or 55% RH, Hamdy et al. (1991) reported that chicks hatched from eggs incubated at 55% RH had higher body weights at the hatch in comparison to those from eggs that had been incubated at the 45% RH. However, this difference ceased to exist by day 2 post-hatch, suggesting that the chicks from eggs incubated at the 45% RH were able to soon adapt physiologically and compensate in growth. Furthermore, although incubational RH treatment did not affect chick feed intake, feed conversion, or rectal temperature through 4 weeks post-hatch, after exposure to 39.0°C for 48 h, fewer chicks in the 45% RH treatment group died in comparison to those in the 55% RH treatment group. Bruzual et al. (2000a) tested the effects of 43%, 53%, and 63% RH incubational settings from the time that broiler hatching eggs were set to the time that the hatch was pulled. With increasing RH, the body weight of the hatchlings was increased, but the time of hatch was not significantly affected. In that study, it was also observed that the percentage hatchability of fertile eggs was highest at 53% RH, whereas the percentage of late dead embryonic mortalities was highest at 63% RH. Therefore, although increasing RH to 63% produced the heaviest hatchling body weight, it had a detrimental effect on embryonic development. Bruzual et al. (2000b) further tested the effects of 43%, 53%, and 63% RH incubational settings during the last 5 days of incubation. Conversely, although the time of hatch was likewise not significantly affected by these same RH settings, no significant differences in hatchling body weight were noted. Restricting these changes in RH to only the last 5 days of incubation, therefore, does not appear to impact broiler hatchling body weight. The effects of RH on hatchling BW may be associated with the rate of yolk uptake from the yolk sac by embryos. Upon comparing the effects of 43%, 53%, and 63% RH incubational environments on yolk utilization in broiler hatching eggs, Burnham et al. (2001) noted that the rate of embryonic yolk uptake was increased by incubating eggs at the 53% RH. The concentration of palmitic acid in the yolk was also reported to be lower at 17 days of incubation at the 53% RH setting.

3.6 Influences of breeder age on the effects of relative humidity on embryonic development

The results of experiments conducted by Vick et al. (1993) have suggested that an incubational RH that produces the best hatch results for broiler breeder eggs is influenced by the age of the breeder hen. A 50% rather than a 58% RH resulted in a higher level of hatchability and a lower early embryonic mortality in eggs as the age of the breeder hen decreased between 28 weeks and 64 weeks. Conversely, the 58% RH resulted in a higher hatchability and a lower late embryonic mortality in eggs laid by hens that were 60 weeks of age. The 50% RH also led to a lower hatchability of eggs from hens that were 66 weeks of age, whereas young hens exhibited a higher level of hatchability. The differential influences of breeder age on the effects of RH on hatchability can be attributed to changes in the diffusivity of the eggshell to water vapor with hen age (Peebles and Brake, 1987; Vick et al., 1993). In a study exploring the influence of incubational RH on the characteristics of broiler embryo progeny from young breeder hens, Peebles et al. (2001) found that changes in RH between 43% and 63% did not affect embryo moisture content and did not have consistent effects on their crude fat and protein contents. However, a reduction in RH to 43% depressed embryogenesis, which may augment the inferior performance of broiler chicks from young breeder hens. Upon testing the effects of 43%, 53%, and 63% RH settings on broiler hatching egg yolk composition, Burnham et al. (2001) observed that the effects of these RH levels on yolk lipid content were influenced by breeder age. More specifically, the percentage of yolk lipid was higher when eggs from 26-week-old breeders were incubated at a 63% rather than a 53% RH. However, yolk lipid content was lower in eggs from 30-week-old breeders when they were incubated at an RH of 43% rather than an RH of 53% or 63%.

To summarize, an approximate 12% loss of the internal water content of an embryonated egg prior to pipping, in conjunction with the formation of an air cell with well-proportioned concentrations of O_2 and CO_2, must occur for the embryo to metabolize nutrients, develop, and hatch successfully. The rate of water loss during incubation is primarily a consequence of the functional porosity of the eggshell and the RH of the air surrounding the egg at a given temperature. In addition, hen species and age, egg size, stage of incubation, and the elevation at which incubation occurs are influential factors that can further modify the rate of water loss. Because RH affects post-hatch chick performance as well as embryogenesis, hatchery managers must carefully monitor incubator RH while considering the various influential factors which can modify the effects of RH.

4 Vital gases

4.1 Oxygen

4.1.1 Eggshell porosity, air cell oxygen tension and the transition from chorioallantois vasculature to pulmonary respiration

Burton and Tullett (1983) predicted the O_2 consumption of embryos based on the diffusion as well as air cell gas composition measurements of O_2. In the report, it was shown that eggshell porosity affects O_2 availability and subsequent embryonic metabolism. Throughout incubation up until external pipping, all embryonated eggs consume a total amount of O_2 equivalent to 90 cm^3 per gram of fresh egg weight (Ar and Rahn, 1978; Tullett, 1984). The events of internal and external pipping are associated with changes in the partial pressure of O_2 in the air cell and various physiological responses of the late-stage embryo. It is known that the O_2 tension in the gas space of the air cell is associated with the metabolic rate of the embryo and that an increase in embryonic metabolism lowers O_2 tension within the air cell (Rahn et al., 1974; Tullett and Deeming, 1982). The tension of O_2 in the air cells of eggs reaches approximately 14% (100 mmHg) immediately prior to internal pipping (Ar and Rahn, 1978; Tullett, 1984). Subsequently, endocrine-induced external pipping and successive hatching occur in response to decreased O_2 concentrations in the air cell (Decuypere et al., 2006; Decuypere and Bruggeman, 2007; Mortola, 2009). When O_2 availability for the embryo becomes limited, its physiological demands increase considerably (Tazawa et al., 1983). The loss of access to O_2 requires a transition to pulmonary ventilation (Vince, 1976: Menna and Mortula, 2002). There is a transition from gas exchange that is facilitated by the chorioallantoic vasculature to pulmonary respiration after internal pipping. The increased availability of O_2 after this transition allows for the predominant use of aerobic fatty acid oxidation to produce energy by the embryo (Moran, 2007).

4.1.2 Interactive effects of eggshell porosity, embryo metabolism and temperature on embryonic oxygen consumption and hatchling quality

As previously discussed, the water vapor conductance of an eggshell is known to be controlled by the temperature of the egg. However, as eggshell water vapor conductance increases with an increase in egg content surface temperature, the O_2 diffusive conductance of the eggshell decreases (Meir and Ar, 1990). Upon examining the interactive effects of incubational O_2 concentration and RH on the hatchability of alligator eggs, Reigh and Williams (2020) concluded that when RH is adequate, a good level of hatching success can be achieved when eggs are incubated under normal ambient O_2 concentrations. Moreover, Deeming and Thompson (1991) have further indicated that eggs can experience higher conductance rates to respiratory gases, such as O_2 and CO_2, when incubated in high humidity environments. Nevertheless, Burton and Tullett (1983) have more specifically noted that high porosity eggshells are able to adequately provide O_2 to late-stage embryos, whereas low porosity eggshells may lead to a reduction in embryonic metabolism. In accordance with an increase in the metabolic demands of the embryo that are associated with its various successive developmental stages, there is an increased demand for gas exchange through the pores of the eggshell (Boleli et al., 2016). French (2009b) has likewise suggested that an increase in eggshell conductance facilitates an increase in embryonic O_2 consumption in association with an increase in their rate of metabolism. Even though the diffusivity of gases through the eggshell is known to be controlled by egg temperature, Lourens et al. (2007) showed that the O_2 concentration of the surrounding air in an incubator during the third week of incubation had a greater effect in determining embryo development than did the eggshell temperature. More specifically, in comparing high (38.9°C) to normal (37.8°C) eggshell temperatures and high (25%) to low (17%) O_2 concentrations, it was found that embryonic heat production was highest when both eggshell temperature and O_2 concentration were highest and that heat production was lowest when eggshell temperature was high but when O_2 concentration was low. Facilitation of the high metabolic demands for increased O_2 availability by the late-stage embryo must, therefore, include a sufficiently high partial pressure of O_2 in an egg's external environment regardless of eggshell temperature and porosity. It was subsequently observed by Lourens et al. (2007) that an increased O_2 concentration increased the yolk-free body weight and length of hatchlings, while residual yolk sac weight was decreased.

4.1.3 Physiological responses of the embryo and hatchling to environmental oxygen concentrations

Diverse physiological responses also occur to increase the gas exchange demands of the late embryo and hatchling that include increases in circulating hematocrit, hemoglobin, and red blood cell count values (Morita et al., 2009; Tazawa et al., 2011, 2012). Özge et al. (2006) tested the hypothesis that O_2 supplementation could improve the survival rates of broiler breeder embryos during the late stage of incubation. Although the time of hatch and circulating red blood cell, packed cell volume, and hemoglobin values were not affected by subjecting the eggs to 23% O_2 concentrations between 18 days and 21 days of incubation, an improvement in embryonic survival rate, in association with a decrease in late embryonic mortality, were observed. The hatchlings were also heavier and had higher blood glucose concentrations in response to the O_2 supplementation. The use of supplemental O_2 as high as 23% may, therefore, be a useful means by which to meet the higher O_2 demands of late-stage embryos. Christensen et al. (1997) tested the hypothesis that providing an O_2-enriched environment to embryonated turkey eggs between 25 days and 28 days of incubation would increase the survival rate of the embryos from genetic lines selected for growth or egg production when compared to their respective random-bred controls. They observed significant interactions between O_2 treatment and genetic line for embryonic survival, heart growth, and hepatic glycogen concentration. It was concluded that the two types of genetic selection resulted in a diminishment in the response of the embryos to O_2 enrichment, which may be attributed to decreases in their eggshell conductances and their altered metabolisms. In a later related study by Christensen et al. (1999), in which the metabolism of carbohydrates and lipids was examined, it was confirmed that embryos selected for growth or egg production are unable to adequately respond to elevated environmental O_2 partial pressures by normally modifying the metabolism of their energy sources, thereby leading to an increase in embryonic mortality during the plateau stage of O_2 consumption that exists during the latter stage of incubation.

An adequate influx of O_2 is vital to the metabolic function of embryos, as the partial pressure of CO_2 increases in the blood and extraembryonic fluids during the period of incubation (Boutilier et al., 1977). In enclosed environments, higher rates of embryonic respiration during the later stages of incubation can increase deviations from normal atmospheric O_2 tensions (Deeming and Thompson, 1991). However, higher porosity eggshells that allow for increased O_2 influx through the eggshell to the embryo counteract the natural accumulation of CO_2 in the blood and extraembryonic fluids as incubation progresses. This effect is manifest in the report by Burton and Tullett (1983) showing that a decreased availability of O_2 to a developing embryo, due to the low porosity of an eggshell, is a major factor that influences embryo weight late in incubation. The atmosphere consists of approximately 21% O_2 at sea level (Stock and Metcalfe, 1984; Weekley and Bland, 2021). Walsberg (1980) has noted that the fractional O_2 concentration in the typical nests of birds is like that in the general atmosphere but can decline during incubation from 20.9% to about 20.0–20.3%. Changes in the O_2 levels in an incubator can affect hatching results (Okur, 2019). It has been further observed that hatchability and subsequent performance can be adversely affected by O_2 levels that are either too low (<17%) or too high (25%) (Stock and Metcalfe, 1984; Lourens et al., 2007; Molenaar et al., 2010). During the early stages of incubation, chick embryos are very sensitive to O_2 deprivation; however, embryos display an improved tolerance to acute hypoxia as their age increases. The embryo also becomes more tolerant to hyperoxia as its age increases between the middle and late periods of incubation (Taylor et al., 1956; Taylor and Kreutziger, 1965, 1966; Onagbesan et al., 2007). Nevertheless, upon testing the effects of progressively increasing atmospheric O_2 concentrations between 15% and 50% on chick hatchability, Barott (1937) observed that a 21% level of O_2 produced the best hatch results.

4.1.4 Influences of altitude on the effects of atmospheric oxygen concentrations on the embryo and hatchling

Although the O_2 concentration in air at any altitude is 21%, there is a near-linear decrease in the atmospheric partial pressure of O_2 as altitude increases (Weekley and Bland, 2021). This is a result of a reduction in barometric pressure as altitude increases (Visschedijk, 1991). A hypoxic environment, due to a reduction in the partial pressure of O_2 that occurs at high altitudes (approximately 600 m above sea level), can cause increased early embryonic mortality (Şahan et al., 2011), reduced organ growth (Bagley and Christensen, 1989), and reductions in hatchability (Francis et al., 1967; Şahan et al., 2011) and hatchling weight (Şahan et al., 2011; Boleli et al., 2016). Boleli et al. (2016) more specifically summarized that the adverse effects of hypoxia on embryonic development during the first half of incubation are related to increased chorioallantoic development and vascularization, whereas its effects during the second half of incubation are related to the compensatory responses of the organs. After assessing the effects of incubational O_2 concentrations at high altitudes on the hatchability of chicken and turkey eggs in the studies conducted by Ells and Morris (1947) and Meshew (1949), Davis (1955) summarized that the hatchability of chicken eggs incubated at a high altitude (2195 m) could be increased with the use of supplemental O_2 concentrations as high as 25%, which equated to a 19% O_2 concentration at sea level. However, it was also suggested that more rapid genetic improvements in hatchability may be made possible by natural selection in an un-supplemented environment at high altitudes and that a higher level of hatchability in an environment without supplemental O_2 may be a partial result of a reduction in the variation of O_2 concentrations that is experienced between trials in which supplemental O_2 is used.

Broiler embryos in eggs incubated at high altitudes, in which hypoxic conditions existed, have been observed by Şahan et al. (2011) to show increased plasma triiodothyronine (T_3) and thyroxine (T_4) concentrations and higher T_3:T_4 ratios. The effect of high altitude carried over into the newly hatched chicks, which also exhibited higher circulating T_3:T_4 ratios. Şahan et al. (2011) noted that newly hatched chicks that had been incubated at a high altitude exhibited higher hemoglobin and plasma hematocrit values. Chan and Burggren (2005) examined the effects of continuous hypoxia (15% O_2) exposure on the development of various organs, including the chorioallantoic membrane, at several successive critical intervals of development in chicken embryos. The differential effects that hypoxia had on the various organs were dependent on the developmental stage of the embryo. More specifically, although chorioallantoic membrane mass was not affected by hypoxia during the early and middle developmental stages, a remarkable compensatory increase in its mass was observed by day 18 of development to mitigate the physiological effects of the hypoxic condition. Furthermore, eye mass and beak length were reduced in middle development, while the masses of the liver, brain, heart, kidneys, stomach, intestines, and skeletal long bones were not affected at any developmental stage.

In summary, eggshell porosity and the availability of O_2 in the external environment, in conjunction with changes in air cell O_2 partial pressure, affect the metabolism, O_2 consumption, induction of pipping, and transition from chorioallantois vasculature to pulmonary respiration in embryos. Moreover, the impacts of O_2 availability on embryo development and its physiological responses depend on the stage of embryogenesis and the altitude at which incubation occurs.

4.2 Carbon dioxide

4.2.1 Effects of changes in carbon dioxide production by the developing embryo on incubator carbon dioxide concentrations and the influences of temperature

Increased levels of CO_2 early in incubation are known to cause a more rapid acidification and liquefaction of the albumen, and development of the sub-embryonic fluid (Benton and Brake, 1996; Bruggeman et al., 2007). Tona et al. (2013) have further noted that the albumen pH of layer hatching eggs decreases more rapidly than that in broiler hatching eggs. The normal production of CO_2 by the embryo is 0.7 times the uptake of O_2 (Visschedijk, 1991). In association with the output of CO_2, a byproduct of embryonic metabolism under standard incubation conditions, the CO_2 concentration in a single-stage incubator can gradually increase from 0.05% at the beginning of incubation to approximately 0.3–0.5% at the end of incubation (Özlü et al., 2019). Because the CO_2 concentration in the air surrounding eggs in a nest is linearly related to the production of CO_2 by the embryo (Romanoff and Romanoff, 1967; Walsberg, 1980), the CO_2 concentration in nests peaks near the end of incubation, but it is when the embryo is less sensitive to elevated levels (Lundy, 1969). Deeming and Thompson (1991) have further stated that increased rates of metabolic CO_2 production, in conjunction with the higher respiratory rates of late-stage embryos, can cause a more intense divergence from normal atmospheric CO_2 partial pressures in the air within an incubator. However, Tona et al. (2013) have also hypothesized that reductions in elevated incubator temperature and CO_2 concentrations resulting from the increased production of heat and CO_2 by late-stage embryos can be achieved when the eggs are incubated at standard ventilation rates. Lourens et al. (2006) found that when standard incubator temperature was decreased by 0.30°C for 1 h, embryonic CO_2 production was initially increased by 0.5%, but then decreased thereafter. Conversely, when machine temperature was increased, CO_2 production initially decreased by 0.4%, but then increased thereafter. The embryos, therefore, displayed an initial inverse response followed by a more long-term direct response to the short-term change in incubational temperature. It was surmised that the changes in CO_2 production were not due solely to the level of embryonic heat production, but that alterations in the flow of blood in the chorioallantois that followed changes in incubation temperature affected heat transfer and the rate of CO_2 diffusion. Lourens et al. (2006) further concluded that the ability of the embryo to liberate CO_2 may limit its development under higher temperatures.

4.2.2 Effects of changes in incubator carbon dioxide concentrations on embryogenesis and hatchability

Carbon dioxide levels ranging between 0.1% and 0.5% are customarily used in the incubation of poultry eggs (Onagbesan et al., 2007). Hatching results are affected by changes in CO_2 concentration in an incubator (Okur, 2019). Despite the possible confounding influences of various environmental factors during experimentation, it has been suggested that incubational CO_2 levels above 1.0% can adversely affect hatching success in domestic fowl (Lundy, 1969). In a series of studies (Taylor et al., 1956; Taylor and Kreutziger, 1965, 1966), it was shown that the effects of hypercapnia on the hatchability of chicks were related to the period of incubation in which they occurred in a setter unit. For instance, hatchability was reduced when CO_2 levels were equal to or above 1.0% between days 1 and 4 (Taylor et al., 1956), 3.0% between days 5 and 8 (Taylor and Kreutziger, 1965), and greater than 6.0% between days 9 and 12 (Taylor and Kreutziger, 1966). The increased tolerance of embryos to hypercapnia as they age during incubation can be partially attributed to the different buffering systems that they possess (Onagbesan et al., 2007). Nevertheless, embryos in low conductance eggshells not only can experience low water loss and O_2 uptake levels, but an accumulation of CO_2 in the egg can lead to a decrease in their blood pH levels (French and Tullett, 1991). The processes of internal and external pipping are influenced by CO_2 partial pressure changes in the air cell. A CO_2 tension of approximately 6% (40 mmHg) is reached in the air cell just before the initiation of internal pipping (Ar and Rahn, 1978; Tullett, 1984). Increased CO_2 concentrations in the air cell are associated with hormonally stimulated external pipping and the complete emergence of hatchlings from the eggshell (Decuypere et al., 2006; Decuypere and Bruggeman, 2007; Mortola, 2009). Nevertheless, it has been suggested in other studies that not only may high CO_2 concentrations not be detrimental to embryonic development and subsequent hatchability, but that higher levels of CO_2 can be beneficial to embryonic development and to further improve hatchability and to stimulate an early hatch response (Bruggeman et al., 2007; Tona et al., 2007). The results that were observed were alleged to be related to the conservation of energy in association with a greater uptake of O_2 in response to physiological adaptations including increases in circulating erythrocyte numbers and a general improvement in cardiovascular development (Tazawa et al., 2002; Decuypere et al., 2006; Verhoelst et al., 2011).

4.2.3 Potential benefits of elevated atmospheric carbon dioxide concentrations during incubation on embryogenesis, hatchability and post-hatch performance

It has been shown that upon exposing broiler hatching eggs to hypercapnic conditions, through gradual increases in CO_2 levels from 0.7% to 1.5% during the first 10 days of incubation, that embryogenesis was accelerated, and hatchability was improved (De Smit et al., 2008). Accelerated embryo growth was likewise observed by Carlea et al. (2012) when they exposed Cobb 500 broiler hatching eggs to 0.85% CO_2 concentrations during the first half of incubation. Willemsen et al. (2008) concluded that a gradual increase in incubator air CO_2 concentration up to 1.0% during the first 10 days of incubation decreased embryonic mortality and subsequently increased fertile egg hatchability by lowering embryo malposition incidence. However, although higher CO_2 levels increased broiler body weight through 2 weeks post-hatch, it had no consistent effects on body weight at slaughter age. Conversely, Fernandes et al. (2014) showed in a subsequent study that when broiler hatching eggs were exposed to graded levels of CO_2 between 0.4% and 1.0% during the first 10 days of incubation, that although the circulating heterophil:lymphocyte ratios of the birds were increased at 42 days of post-hatch age, it did not affect their post-hatch livability, performance, heart and liver weights, or heart characteristics. These same results were observed even when they were subjected to fluctuating temperatures between 35 days and 42 days of age. Gildersleeve and Boeschen (1983) have indicated that CO_2 levels between 0.3% and 1.5% early in incubation improve hatch, and De Smit et al. (2006) have also indicated that levels in that same range early in incubation can also stimulate embryonic growth. In comparison to low (0.2%) CO_2 concentrations, when high (1.0%) CO_2 concentrations were applied during only the hatching phase (beginning on day 19) of incubation, higher relative heart and lung weights were observed at 12 h after hatch, thereby indicating that when applied during the hatching phase, high CO_2 levels in an incubator may only exert a temporary physiological effect during the early post-hatch period (Maatjens et al., 2014a). In a subsequent study by Maatjens et al. (2014b), in which the same experimental design was employed, it was reported that the high CO_2 concentration resulted in a lower blood pH and hepatic glycogen concentration in embryos that had been internally pipped.

4.2.4 Physiological responses of embryos and hatchlings to elevated atmospheric carbon dioxide concentrations and the influences of genotype

Increased plasma T_3 and T_4 concentrations and higher T_3:T_4 ratios have been observed in broiler embryos in eggs incubated under hypercapnic conditions at high altitudes. Higher circulating T_3:T_4 ratios in response to the high altitude also subsequently occurred in hatchlings (Şahan et al., 2011). Tona et al. (2013) conducted experiments to determine the effects of incubator CO_2 concentration on the physiological variables of chicken eggs from broiler and layer genotypes exhibiting different growth trajectories. It was reported that increasing incubator CO_2 concentration during the first 10 days of incubation differentially affected the growth trajectories and the physiological variables of late-stage embryos. It was also noted that the higher CO_2 level during incubation led to higher plasma T_3 and corticosterone concentrations in both genetic lines at the time of internal pipping. However, the higher CO_2 concentration resulted in a shorter length of incubation in only the slower-growing genetic line. Buys et al. (1998) further explored the effects of different incubational CO_2 levels (0.2% and 0.4%) on the hatch results and physiological characteristics of embryo and post-hatch chicks from ascites-sensitive and ascites-resistant broiler lines. At the 0. 2% but not the 0.4% level, the ascites-resistant line hatched earlier than the ascites-sensitive line. Embryos from both lines had higher plasma T_3 concentrations when incubated under a 0.4% rather than a 0.2% CO_2 concentration. In the post-hatch period, lower ascites mortalities occurred in chicks that hatched from eggs that were incubated at the 0.4% level in comparison to the 0.2% level, and heart right ventricular to total ventricular ratios were higher in the 0.2% CO_2 treatment group in comparison to the 0.4% CO_2 treatment group.

To summarize, incubational temperature and the increased production of CO_2 by developing embryos can affect atmospheric CO_2 concentrations in incubators. However, although increased incubational CO_2 concentrations can affect hatching success, late-term embryos become more tolerant to hypercapnia. Nevertheless, elevated CO_2 concentrations during the first half of incubation may be used to stimulate embryo growth and improve hatch. These effects are mediated by various physiological responses and are influenced by altitude and bird genotype.

5 Conclusion

To achieve a successful hatch, an approximate 12% loss of the internal water content of an embryonated egg must occur prior to pipping in conjunction with a well-formed air cell at the large end of the egg containing a correct proportion of O_2 and CO_2. The rate of water loss is primarily a product of an eggshell's functional porosity and the RH, at a given temperature, of the air immediately surrounding an egg. While carefully monitoring the incubational environment with these factors in mind, it is also necessary to consider other confounding factors previously described in this chapter that can further modify the rate of water loss from the embryonated egg.

Knowing the need for the adequate elimination of CO_2 and the supply of O_2 for the developing embryo, it can be summarized that maintenance of the vital gas concentrations across the eggshell by proper adjustments of their concentrations in the internal incubator environment during the setter and hatcher phases of incubation is necessary to meet the embryo's physiological requirements. The physiological needs and adjustments of the embryo in response to O_2 availability include its metabolism, pipping activity, and transition from vascular to pulmonary respiration. Like that of RH, other modifying factors that can affect the impact of O_2 availability on embryogenesis include the stage of development of the embryo and the altitude at which incubation occurs. In contrast to the required uptake of O_2, the elimination of CO_2 as a metabolic byproduct of the developing embryo is also required. A subsequent accumulation of CO_2 in an incubator can occur with an increased growth of the embryo. However, after taking into consideration the modifying effects of altitude and bird genotype, the practice of elevating CO_2 concentrations during the first half of incubation may be used to aid in stimulating embryonic growth and improving hatchability.

To understand the subsequent effects of air humidity and vital gas composition in an incubator on embryo and post-hatch chick development, it is necessary to have a comprehensive knowledge of the interactive effects of the physiological function of the eggshell and the modifying influences of the various physical factors in the incubational environment. With this knowledge, hatchery managers will be better equipped to optimize productivity in their hatcheries.

6 Where to look for further information

It is recommended that researchers consult current company guidelines for the incubation of their specific hatching eggs. This would include following their recommended procedures for monitoring incubational egg weight loss for the proper adjustment of incubational conditions necessary to achieve an optimal weight loss prior to pipping. Further information can also be obtained from updated extension articles published by a university Poultry Science department. Future research should focus on determining fine incubational regimen adjustments, including those for temperature and humidity, that could be made at various periods during incubation to better accommodate the changing growth and metabolic rates of embryos of modern strain hatching eggs.

7 References

Ar, A. (1991a). Egg water movements during incubation. In: Tullett, S. G. (Ed.), *Avian Incubation*, Butterworth-Heinemann, London, UK, pp. 157–173.

Ar, A. (1991b). Roles of water in avian eggs. In: Deeming, D. C. and Ferguson, M. W. J. (Eds.), *Egg Incubation: Its Effects on Embryonic Development in Birds and Reptiles*, Cambridge University Press, New York, pp. 229-243.

Ar, A., Paganelli, C. V., Reeves, R. B., Greene, D. G. and Rahn, H. (1974). The avian egg: water vapor conductance, shell thickness, and functional pore area, *Condor* 76(2), 153–158.

Ar, A. and Rahn, H. (1978). Interdependence of gas conductance, incubation length and weight of the avian egg. In: Piiper, J. (Ed.), *Respiratory Function in Birds, Adult and Embryonic*, Springer, Berlin, pp. 227–236.

Ar, A. and Rahn, H. (1980). Water in the avian egg: overall budget of incubation, *Am. Zool.* 20(2), 373-384.

Ar, A. and Sidis, Y. (2002). Nest microclimate during incubation. In: Deeming, D. C. (Ed.), *Avian Incubation: Behaviour, Environment, and Evolution*, Oxford University Press Inc., New York, pp. 143-160.

Bagley, L. G. and Christensen, V. L. (1989). Comparisons of turkey embryos incubated in tenuous or dense gas environments–II. Organ growth, *Comp. Biochem. Physiol. A Comp. Physiol.* 93(2), 451–454.

Barbosa, V. M., Rocha, J. S. R., Pompeu, M. A., Martins, N. R. S., Baião, N. C., Lara, L. J. C., Batista, J. V. M. S. P. and Leite, R. C. (2013). The effects of relative humidity and turning in incubators machines on the incubation yield and chick performance, *Worlds Poult. Sci. J.* 69(1), 89–98.

Barott, H. G. (1937). Effect of temperature, humidity, and other factors on hatch of hens' eggs and on energy metabolism of chick embryos, USDA Technical Bulletin No. 553, pp. 1–45.

Benton, C. E. and Brake, J. (1996). The effect of broiler breeder flock age and length of storage on egg albumen during early incubation, *Poult. Sci.* 75(9), 1069–1075.

Boleli, I. C., Morita, V. S., Matos, Jr., J. B., Thimotheo, M. and Almeida, V. (2016). Poultry egg incubation: integrating and optimizing production efficiency, *Rev. Bras. Cienc. Avic.* 18, 1–16, 10.1590/1806-9061-2016-0292.

Boutilier, R. G., Gibson, M. A., Toews, D. P. and Anderson, W. (1977). Gas exchange and acid-base regulation in the blood and extraembryonic fluids of the developing chicken embryo, *Respir. Physiol.* 31(1), 81–89.

Bruggeman, V., Witters, A., De Smit, L., Debonne, M., Everaert, N., Kamers, B., Onagbesan, O. M., Degraeve, P. and Decuypere, E. (2007). Acid-base balance in chicken embryo (*Gallus domesticus*) incubated under high CO_2 concentrations during the first ten days of incubation, *Respir. Physiol. Neurobiol.* 159(2), 147–154.

Bruzual, J. J., Peak, S. D., Brake, J. and Peebles, E. D. (2000a). Effects of relative humidity during incubation on hatchability and body weight of broiler chicks from young breeder flocks, *Poult. Sci.* 79(6), 827–830.

Bruzual, J. J., Peak, S. D., Brake, J. and Peebles, E. D. (2000b). Effects of relative humidity during the last five days of incubation and brooding temperature on performance of broiler chicks from young broiler breeders, *Poult. Sci.* 79(10), 1385–1391.

Burnham, M. R., Peebles, E. D., Gardner, C. W., Brake, J., Bruzual, J. J. and Gerard, P. D. (2001). Effects of incubator humidity and hen age on yolk composition in broiler hatching eggs from young breeders, *Poult. Sci.* 80(10), 1444–1450.

Burton, F. G. and Tullett, S. G. (1983). A comparison of the effects of eggshell porosity on the respiration and growth of domestic fowl, duck and turkey embryos, *Comp, Biochem. Physiol.* 75A, 167–174.

Buys, N., Dewil, E., Gonzales, E. and Decuypere, E. (1998). Different CO_2 levels during incubation interact with hatching time and ascites susceptibility in two broiler lines selected for different growth rate, *Avian Pathol.* 27(6), 605–612.

Carlea, L., Miclea, V., Zahan, M. and Miclea, I. (2012). The effect of supplemental carbon dioxide in chicken incubation with eggs from heavy breeder parents. *J. Anim. Sci. Biotechnol.* 45, pp. 136–141.

Chan, T. and Burggren, W. (2005). Hypoxic incubation creates differential morphological effects during specific developmental critical windows in the embryo of the chicken (*Gallus gallus*), *Respir. Physiol. Neurobiol.* 145(2-3), 251–263.

Christensen, V. L., Donaldson, W. E. and Nestor, K. E. (1997). Effects of an oxygen-enriched environment on the survival of Turkey embryos between twenty-five and twenty-eight days of age, *Poult. Sci.* 76(11), 1556–1562.

Christensen, V. L., Donaldson, W. E. and Nestor, K. E. (1999). Effect of supplemental oxygen on blood plasma organic acids within embryos from selected lines of turkeys, *Poult. Sci.* 78(11), 1601–1605.

Cormick, J. (2021). Factors affecting incubation in high humidity regions, *Int. Hatch Prac.* 35, 12–13.

Davis, G. T. (1955). Influence of oxygen concentration on hatchability and on selecting for hatchability, *Poult. Sci.* 34(1), 107–113.

Davis, T. A., Shen, S. S. and Ackerman, R. A. (1988). Embryonic osmoregulation: consequences of high and low water loss during incubation of the chicken egg, *J. Exp. Zool.* 245(2), 144–156.

De Smit, L., Bruggeman, V., Debonne, M., Tona, J. K., Kamers, B., Everaert, N., Witters, A., Onagbesan, O., Arckens, L., De Baerdemaeker, J. and Decuypere, E. (2008). The effect of non-ventilation during early incubation on the embryonic development of chicks of two commercial strains differing in ascites susceptibility, *Poult. Sci.* 87(3), 551–560.

De Smit, L., Bruggeman, V., Tona, J. K., Debonne, M., Onagbesan, O., Arckens, L., De Baerdemaeker, J. and Decuypere, E. (2006). Embryonic development plasticity of the chick: increased CO_2 during early stages of incubation changes developmental trajectories during prenatal and postnatal growth, *Comp. Biochem. Physiol. A* 145(2), 166–175.

Decuypere, E. and Bruggeman, V. (2007). The endocrine interface of environmental and egg factors affecting chick quality, *Poult. Sci.* 86(5), 1037–1042.

Decuypere, E., Onagbesan, O., DeSmit, L., Tona, K., Everaert, N., Witters, A., Debonne, M., Verhoelst, E., Buyse, J., Hassanzadeh, M. and De Baerdemaeker, J. (2006). Hypoxia and hypercapnia during incubation of chicken eggs: effects on development and subsequent performance, *Worlds Poult. Sci. J.* 62, 486–487.

Decuypere, E., Tona, K., Bruggeman, V. and Bamelis, F. (2001). The day-old chick: a crucial hinge between breeders and broilers, *Worlds Poult. Sci. J.* 57(2), 127–138.

Deeming, D. C. and Thompson, M. B. (1991). Gas exchange across reptilian eggshells. In: Deeming, D. C. and Ferguson, M. W. J. (Eds.), *Egg Incubation: Its Effects on Embryonic Development in Birds and Reptiles*, Cambridge University Press, New York, pp. 277–284.

Durojaye, O. A., To, S. D. F., Elliott, K. E. C., Gerard, P. D. and Peebles, E. D. (2018). Broiler hatching egg air cell: air cell size profile and its relationship to shell temperature and weigh loss in Ross 708 broiler hatching eggs between 3 and 12 days of incubation, *RRJAAS* 7, 104–112.

Ells, J. B. and Morris, L. (1947). Factors involved in hatching chicken and Turkey eggs at high elevations, *Poult. Sci.* 26(6), 635–638.

Fernandes, J. I. M., Bortoluzzi, C., Esser, A. F. G., Contini, J., Stokler, P. and Faust, D. (2014). Performance of broilers submitted to high CO_2 levels during incubation combined with temperature fluctuations at late post-hatch, *Rev. Bras. Cienc. Avic.* 16(3), 285–290.

Francis, D. W., Bernier, P. E. and Hutto, D. C. (1967). The effect of altitude on the hatchability of chicken eggs, *Poult. Sci.* 46(6), 1384–1389.

French, N. A. (1997). Modeling incubation temperature: the effects of incubator design, embryonic development, and egg size, *Poult. Sci.* 76(1), 124–133.

French, N. A. (2009a). The critical importance of incubation temperature, *Avian Biol. Res.* 2(1–2), 55–59.

French, N. A. (2009b). Incubation and hatching. In: Hocking, P. M. (Ed.), *Biology of Breeding Poultry, Poultry Science Symposium Series* (vol. 29), CAB International, Wallingford, UK, pp. 206–223.

French, N. A. and Tullett, S. G. (1991). Variation in the eggs of various poultry species. In: Tullett, S. G. (Ed.), *Avian Incubation*, Butterworth-Heinemann, London, UK, pp. 59–77.

Gildersleeve, R. P. and Boeschen, D. P. (1983). The effects of incubator carbon dioxide level on turkey hatchability, *Poult. Sci.* 62(5), 779–784.

Hamdy, A. M. M., Henken, A. M., Van Der Hel, W., Galal, A. G. and Abd-Elmoty, A. K. (1991). Effects of incubation humidity and hatching time on heat tolerance of neonatal chicks: growth performance after heat exposure, *Poult. Sci.* 70(7), 1507–1515.

Joseph, N. S., Lourens, A. and Moran, E. T., Jr. (2006). The effects of suboptimal eggshell temperature during incubation on broiler chick quality, live performance, and further processing yield, *Poult. Sci.* 85(5), 932–938.

Kaltofen, K. S. (1969). The effect of air movements on hatchability and weight loss of chicken eggs during artificial incubation. In: Carter, T. C. and Freeman, B. M. (Eds.), *The Fertility and Hatchability of the Hen's Egg*, Oliver and Boyd, Edinburgh, Scotland, UK, pp. 177–190.

Lourens, A., Meijerhof, R., Kemp, B. and van den Brand, H. (2011). Energy partitioning during incubation and consequences for embryo temperature: a theoretical approach, *Poult. Sci.* 90(2), 516–523.

Lourens, A., van den Brand, H., Heetkamp, M. J. W., Meijerhof, R. and Kemp, B. (2006). Metabolic responses of chick embryos to short-term temperature fluctuations, *Poult. Sci.* 85(6), 1081–1086.

Lourens, A., van den Brand, H., Heetkamp, M. J. W., Meijerhof, R. and Kemp, B. (2007). Effects of eggshell temperature and oxygen concentration on embryo growth and metabolism during incubation, *Poult. Sci.* 86(10), 2194–2199.

Lourens, A., van den Brand, H., Meijerhof, R. and Kemp, B. (2005). Effect of eggshell temperature during incubation on embryo development, hatchability, and posthatch development, *Poult. Sci.* 84(6), 914–920.

Lundy, H. (1969). A review of the effects of temperature, humidity, turning and gaseous environment in the incubator on hatchability of hen's eggs. In: Carter, T. C. and Freeman, B. M. (Eds.), *The Fertility and Hatchability of the Hen's Egg*, Oliver and Boyd, Edinburgh, Scotland, UK, pp. 143–176.

Maatjens, C. M., Reijrink, I. A. M., Molenaar, R., van der Pol, C. W., Kemp, B. and van den Brand, H. (2014a). Temperature and CO_2 during the hatching phase. I. Effects on chick quality and organ development, *Poult. Sci.* 93(3), 645–654.

Maatjens, C. M., Reijrink, I. A. M., van den Anker, I., Molenaar, R., van der Pol, C. W., Kemp, B. and van den Brand, H. (2014b). Temperature and CO_2 during the hatching phase. I. Effects on chicken embryo physiology, *Poult. Sci.* 93(3), 655–663.

Meijerhof, R. (2002). Design and operation of commercial incubators. In: *Practical Aspects of Commercial Incubation*, Ratite Conference Books, Lincolnshire, UK, pp. 41–46.

Meir, M. and Ar, A. (1990). Gas pressures in the air cell of the ostrich egg prior to pipping as related to oxygen consumption, eggshell gas conductance, and egg temperature, *Condor* 92(3), 556–563.

Menna, T. M. and Mortula, J. P. (2002). Metabolic control of pulmonary ventilation in the developing chick embryo, *Respir. Physiol. Neurobiol.* 130(1), 43–55.

Meshew, M. H. (1949). The use of oxygen in the hatching of chicken and turkey eggs at high altitudes, *Poult. Sci.* 28(1), 87–97.

Molenaar, R., Meijerhof, R., van den Anker, I., Heetkamp, M. J., van den Borne, J. J., Kemp, B. and van den Brand, H. (2010). Effect of eggshell temperature and oxygen concentration on survival rate and nutrient utilization in chicken embryos, *Poult. Sci.* 89(9), 2010–2021.

Moran, Jr., E. T. (2007). Nutrition in the developing embryo and hatchling, *Poult. Sci.* 86(5), 1043–1049.

Morita, V. S., Boleli, I. C. and Cargnelutti Filho, H. (2009). Hematological values and body, heart and liver weights in male and female broiler embryos taken from eggs of young and old breeders, *Braz. J. Poult. Sci.* 11, 19–27.

Mortola, J. P. (2009). Gas exchange in avian embryos and hatchlings, *Comp. Biochem. Physiol. A Mol. Integr. Physiol.* 153(4), 359–377.

Okur, N. (2019). Effects of incubator carbon dioxide and oxygen levels, and egg weight on broilers' hatchability of fertile eggs, *Braz. J. Poult. Sci.* 21(3). DOI: 10.1590/1806-9061-2019-1038.

Onagbesan, O., Bruggeman, V., De Smit, L., Debonne, M., Witters, A., Tona, K., Everaert, N. and Decuypere, E. (2007). Gas exchange during storage and incubation of avian eggs: effects on embryogenesis, hatchability, chick quality and post-hatch growth, *Worlds Poult. Sci. J.* 63(4), 557–573.

Özge, A., Şahan, Ü., İpek, A., Aydin, C. and Bayraktar, H. (2006). Effects of oxygen supplementation on embryonic survival, hematological parameters and plasma glucose level of broiler chicks, *Arch. Geflügelk* 70, 64–68.

Özlü, S., Uçar, A., Banwell, R. and Elibol, O. (2019). The effect of increased concentration of carbon dioxide during the first 3 days of incubation on albumen characteristics, embryonic mortality and hatchability of broiler hatching eggs, *Poult. Sci.* 98, 771–776.

Paganelli, C. V. (1980). The physics of gas exchange across the avian eggshell, *Am. Zool.* 20(2), 329–338.

Paganelli, C. V. (1991). The avian eggshell as a mediating barrier: respiratory gas fluxes and pressures during development. In: Deeming, D. C. and Ferguson, M. W. J. (Eds.), *Egg Incubation: Its Effects on Embryonic Development in Birds and Reptiles*, Cambridge University Press, New York, pp. 261–275.

Paganelli, C. V., Ackerman, R. A. and Rahn, H. (1978). The avian egg: in vitro conductances to oxygen, carbon dioxide, and water vapor in late development. In: Piiper, J. (Ed.), *Respiratory Function in Birds, Adult and Embryonic*, Springer, Berlin, pp. 212–218.

Paganelli, C. V., Ar, A., Rahn, H. and Wangensteen, O. D. (1975). Diffusion in the gas phase: the effects of ambient temperature and gas composition, *Respir. Physiol.* 25, 247–258.

Paganelli, C. V., Sotherland, P. R., Olszowka, A. J. and Rahn, H. (1988). Regional differences in diffusive conductance/perfusion ratio in the shell of the hen's egg, *Respir. Physiol.* 71(1), 45–55.

Peebles, E. D. and Brake, J. (1987). Eggshell quality and hatchability in broiler breeder eggs, *Poult. Sci.* 66(4), 596–604.

Peebles, E. D., Burnham, M. R., Gardner, C. W., Brake, J., Bruzual, J. J. and Gerard, P. D. (2001). Effects of incubational humidity and hen age on embryo composition in broiler hatching eggs from young breeders, *Poult. Sci.* 80(9), 1299–1304.

Peebles, E. D., Keirs, R. W., Bennett, L. W., Cummings, T. S., Whitmarsh, S. K. and Gerard, P. D. (2005). Relationships among prehatch and posthatch physiological parameters in early nutrient restricted broilers hatched from eggs laid by young breeder hens, *Poult. Sci.* 84(3), 454–461.

Peebles, E. D. and McDaniel, C. D. (2004). *A Practical Manual for Understanding the Shell Structure of Broiler Hatching Eggs and Measurements of Their Quality, Mississippi Agricultural and Forestry Experiment Station*, Bulletin 1139, April, 2004, pp. 1–16.

Peebles, E. D., Pulikanti, R., Zhai, W. and Gerard, P. D. (2014). Relationships of incubational hatching egg characteristics to posthatch body weight and processing yield in Ross × Ross 708 broilers, *J. Appl. Poult. Res.* 23(1), 34–40.

Peebles, E. D., Zhai, W. and Gerard, P. D. (2012). Comparative evaluation of air cell and eggshell temperature measurement methodologies used in broiler hatching eggs during late incubation, *Poult. Sci.* 91(7), 1536–1541.

Pulikanti, R., Peebles, E. D., Bennett, L. W., Zhai, W. and D. Gerard, P. (2013). Physiological relationships of the middle and late post-hatch performance of broilers to their embryo and eggshell characteristics, *J. Poult. Sci.* 50(4), 375–380.

Pulikanti, R., Peebles, E. D. and Gerard, P. D. (2011a). Physiological responses of broiler embryos to in ovo implantation of temperature transponders, *Poult. Sci.* 90(2), 308–313.

Pulikanti, R., Peebles, E. D. and Gerard, P. D. (2011b). Use of implantable temperature transponders for the determination of air cell temperature, eggshell water vapor conductance, and their functional relationships in embryonated broiler hatching eggs, *Poult. Sci.* 90(6), 1191–1196.

Pulikanti, R., Peebles, E. D., Zhai, W. and Gerard, P. D. (2012a). Determination of embryonic temperature profiles and eggshell water vapor conductance constants in incubating Ross × Ross 708 broiler hatching eggs using temperature transponders, *Poult. Sci.* 91(1), 55–61.

Pulikanti, R., Peebles, E. D., Zhai, W., Bennett, L. W. and Gerard, P. D. (2012b). Physiological relationships of the early posthatch performance of broilers to their embryo and eggshell characteristics, *Poult. Sci.* 91(7), 1552–1557.

Rahn, H. and Ar, A. (1980). Gas exchange of the avian egg: time, structure, and function, *Am. Zool.* 20(2), 477–484.

Rahn, H., Ar, A. and Paganelli, C. V. (1979). How bird eggs breathe, *Sci. Am.* 240(2), 46–55.

Rahn, H., Paganelli, C. V. and Ar, A. (1974). The avian egg: air-cell gas tension, metabolism and incubation time, *Respir. Physiol.* 22(3), 297–309.

Rahn, H., Paganelli, C. V., Nisbet, I. C. T. and Whittow, G. C. (1976). Regulation of incubation water loss in eggs of seven species of terns, *Physiol. Zool.* 49(2), 245–259.

Reigh, R. C. and Williams, M. B. (2020). Effects of relative humidity and elevated oxygen concentrations on egg-hatching success of American alligator, *Alligator mississippiensis*, *J. World Aquacult. Soc.* 51(4), 982–989.

Robertson, I. S. (1961a). Studies on the effect of humidity on the hatchability of hen's eggs. I. The determination of optimum humidity for incubation, *J. Agric. Sci.* 57(2), 185–194.

Robertson, I. S. (1961b). Studies on the effect of humidity on the hatchability of hen's eggs. II. A comparison of hatchability, weight loss and embryonic growth in eggs incubated at 40 and 70% RH, *J. Agric. Sci.* 57(2), 195–198.

Romanoff, A. L. and Romanoff, A. J. (1949). *The Avian Egg*, John Wiley & Sons, New York.

Romanoff, A. L. and Romanoff, A. J. (1967). *Biochemistry of the Avian Embryo*, John Wiley & Sons, New York.

Rokitka, M. A. and Rahn, H. (1987). Regional differences in shell conductance and pore density of avian eggs, *Respir. Physiol.* 68(3), 371–376.

Şahan, Ü, İpek, A., Yilmaz-Dikmen, B., Aydin, C. and Kederlı, E. (2011). Effects of oxygen supplementation in the hatcher at high altitude on the incubation results of broiler eggs laid at low altitude, *Br. Poult. Sci.* 52(3), 388–394.

Sotherland, P. R., Spotila, J. R. and Paganelli, C. V. (1987). Avian eggs: barriers to the exchange of heat and mass, *J. Exp. Zool. Suppl.* 1, 81-86.

Spotila, J. R., Weinheimer, C. J. and Paganelli, C. V. (1981). Shell resistance and evaporative water loss from bird eggs: effects of wind speed and egg size, *Physiol. Zool.* 54(2), 195-202.

Stock, M. K. and Metcalfe, J. (1984). Stimulation of growth of the chick embryo by acute hyperoxia, *Respir. Physiol.* 58(3), 351-358.

Taylor, L. W. and Kreutziger, G. O. (1965). The gaseous environment of the chick embryo in relation to its development and hatchability. 2. Effect of carbon dioxide and oxygen levels during the period of the fifth through the eighth days of incubation, *Poult. Sci.* 44, 98-106.

Taylor, L. W. and Kreutziger, G. O. (1966). The gaseous environment of the chick embryo in relation to its development and hatchability, *Poult. Sci.* 45(5), 867-884.

Taylor, L. W., Sjodin, R. A. and Gunns, C. A. (1956). The gaseous environment of the chick embryo in relation to its development and hatchability, *Poult. Sci.* 35(6), 1206-1215.

Tazawa, H., Akiyama, R. and Moriya, K. (2002). Development of cardiac rhythms in birds, *Comp. Biochem. Physiol. A Mol. Integr. Physiol.* 132(4), 675-689.

Tazawa, H., Andrewartha, S. J. and Burggren, W. W. (2011). Development of hematological respiratory variables in late chicken embryos; the relative importance of incubation time and embryo mass, *Comp. Biochem. Physiol. A Mol. Integr. Physiol.* 159(3), 225-233.

Tazawa, H., Andrewartha, S. J. and Burggren, W. W. (2012). Acute regulation of hematocrit and blood acid-base balance during severe hypoxic challenges in late chicken embryos (*Gallus gallus*), *Respir. Physiol. Neurobiol.* 184(1), 86-96.

Tazawa, H., Visschedijk, A. H. J., Wittman, J. and Piiper, J. (1983). Gas exchange, blood gases and acid-base status in the chick before, during and after hatching, *Respir. Physiol.* 53(2), 173-185.

Tona, K., Everaert, N., Willemsen, H., Gbeassor, M., Decuypere, E. and Buyse, J. (2013). Effects of interaction of incubator CO_2 levels and mixing hatching eggs of different embryo growth trajectory on embryo physiological and hatching parameters, *Br. Poult. Sci.* 54(4), 545-551.

Tona, K., Onagbesan, O., Bruggeman, V., De Smit, L., Figueiredo, D. and Decuypere, E. (2007). Non-ventilation during early incubation and dexamethasone administration during late incubation: 1. Effects of hatching events and physiological hormone levels, *Domest. Anim. Endocrinol.* 33(1), 32-46.

Tullett, S. G. (1984). The porosity of avian eggshells, *Comp. Biochem. Physiol.* 78A, 5-13.

Tullett, S. G. and Burton, F. G. (1982). Factors affecting the weight and water status of the chick at hatch, *Br. Poult. Sci.* 23(4), 361-369.

Tullett, S. G. and Deeming, D. C. (1982). The relationship between eggshell porosity and oxygen consumption of the embryo in the domestic fowl, *Comp. Biochem. Physiol.* 75A, 529-533.

Uçar, A., Özlü, S. and Elibol, L. (2021). Research note: the effects of chick pipping location on broiler live performance, *Poult. Sci.* 100(9), 101381. 10.1016/j.psj.2021.101381.

van der Pol, C. W., van Roovert-Reijrink, I. A. M., Maatjens, C. M., van den Brand, H. and Molenaar, R. (2013). Effect of relative humidity during incubation at a set eggshell temperature and brooding temperature posthatch on embryonic mortality and chick quality, *Poult. Sci.* 92(8), 2145-2155.

Verhoelst, E., De Ketelaere, B., Decypere, E. and De Baerdemaeker, J. (2011). The effect of early prenatal hypercapnia on the vascular network in the chorioallantoic membrane of the chicken embryo, *Biotechnol. Prog.* 27(2), 562-570.

Vick, S. V., Brake, J. and Walsh, T. J. (1993). Relationship of incubator humidity and flock age to hatchability of broiler hatching eggs, *Poult. Sci.* 72(2), 251-258.

Vieira, S. L. and Moran, Jr., E. T. (1999). Effects of egg of origin and chick post-hatch nutrition on broiler live performance and meat yields, *Worlds Poult. Sci. J.* 55(2), 125-142.

Vince, M. A. (1976). Behaviour and aeration of the respiratory system in the domestic fowl embryo, *J. Physiol.* 263(3), 465-474.

Visschedijk, A. H. J. (1980). Effects of barometric pressure and abnormal gas mixtures on gaseous exchange by the avian embryo, *Am. Zool.* 20(2), 469-476.

Visschedijk, A. H. J. (1991). Incubation of eggs at high altitude. In: Tullett, S. G. (Ed.), *Avian Incubation*, Butterworth-Heinemann Ltd., London, UK, pp. 285-291.

Visschedijk, A. H. J., Ar, A., Rahn, H. and Piiper, J. (1980). The independent effects of atmospheric pressure and oxygen partial pressure on gas exchange of the chicken embryo, *Respir. Physiol.* 39(1), 33-44.

Walsberg, G. E. (1980). The gaseous microclimate of the avian nest during incubation, *Am. Zool.* 20(2), 363-372.

Weekley, M. S. and Bland, M. E. (2021). *Oxygen Administration*. StatPearls Publishing, Treasure Island, Fl. Available at: https:www.ncbi.nlm.nih.gov/books/NBK551617/.

Willemsen, H., Tona, K., Bruggeman, V., Onagbesan, O. and Decuypere, E. (2008). Effects of high CO_2 level during early incubation and late incubation *in ovo* dexamethasone injection on perinatal embryonic parameters and post-hatch growth of broilers, *Br. Poult. Sci.* 49(2), 222-231.

Wilson, H. R. (1991). Interrelationships of egg size, chick size, posthatching growth and hatchability, *Worlds Poult. Sci. J.* 47(1), 5-20.

Wilson, H. R. and Dugan, V. P. (1992). The effect of hatcher and setter humidities on the hatchability of bobwhite quail eggs, *J. Appl. Poult. Res.* 1(2), 180-182.

Chapter 13

The role of egg turning in embryo development

Okan Elibol, Ankara University, Turkey

1 Introduction

Hatching chicks is a complex procedure, affected by a great many factors. Since the development of artificial incubation, optimum conditions for hatching have been determined and artificial incubators operate within precise limits. The optimum turning of eggs during artificial incubation has been of interest since wild birds were first observed regularly moving their eggs within the nest.

The main consequences of no egg turning during incubation are reduced hatchability and chick quality, increased embryonic mortality, and greater incidence of malformations and malpositions (Lundy, 1969; Robertson, 1961a, b; Wilson, 1991). Failure to turn eggs during incubation appears to produce worse results for eggs from older flocks than from younger (Elibol and Brake, 2004), due to higher late embryonic mortality and malposition II (head in the small end of the egg) (Fig. 1). This can be attributed to differences in albumen quality, related to flock age (Brake et al., 1997).

Turning has also been found to prevent premature adhesion of embryonic membranes to the inner shell membrane, yolk, or other membranes (Freeman and Vince, 1974; New, 1957), promote optimal development of extra-embryonic (chorioallantois) membranes (New, 1957; Tullett and Deeming, 1987; Tyrrell et al., 1954), and ensure correct orientation of the embryo within the egg prior to hatching (Lundy, 1969; Robertson, 1961b). In addition, failure to turn eggs during incubation delays albumen utilization by the embryo (Deeming, 1989a, b, c;

http://dx.doi.org/10.19103/AS.2022.0118.20

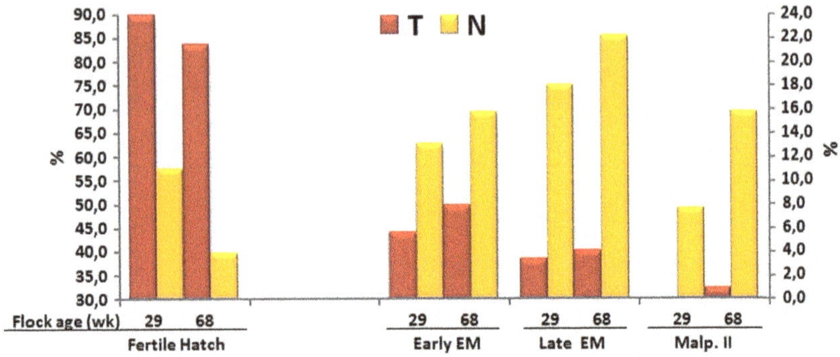

Figure 1 Effect of turning during incubation on percentage fertile hatchability, early embryonic mortality (E0–E6), late embryonic mortality (E18–E21 + pipped), and malposition II of broiler hatching eggs produced by 29- and 68-week-old flocks. Source: Adapted from Elibol and Brake (2004). **T** = Turning during E0–E18; **N** = No turning E0–E18, with trays horizontal to floor.

Randles and Romanoff, 1950), retards area vasculosa growth (Deeming, 1989b), disrupts the proper formation of sub-embryonic fluid (SEF) (Deeming et al., 1987; Deeming, 1989c), decreases the oxygen exchange surface of the chorioallantois (Tazawa, 1980), and reduces embryo growth rate (Deeming et al., 1987; Tullett and Deeming, 1987).

In practice, hatching eggs of most species are generally set on trays with the large end up. They are turned hourly through an angle of about 90° for 18 days (E0–E18) of incubation. Four main aspects of turning have been identified as integral to the ultimate success of incubation:

- Turning during storage.
- Critical periods for turning during incubation.
- Turning frequency during incubation.
- Turning angle during incubation.

These are discussed in the following sections.

2 Turning during storage

It is well documented that increasing the egg storage period reduces hatchability, adversely affects chick quality, and prolongs incubation time. The decline in fertile hatchability associated with prolonged hatching egg storage can be ameliorated by a number of egg management schemes. One of the options used is to turn eggs in the egg store. Turning during egg storage will be covered elsewhere in this book.

In some instances, it may not be possible to manipulate eggs during storage so alternative options are required once incubation is initiated to help recover some of the lost hatching potential. Elibol and Brake (2008) conducted two experiments to examine the effects of length of egg storage and turning frequency during subsequent incubation on fertile hatchability of eggs from older broiler breeders. Hatchability improved with increased turning frequency (96X vs. 24X) during incubation after 14 days of storage but not after 3 days of storage. These data show that the detrimental effects of a long storage period may be ameliorated by an increased turning frequency during subsequent incubation. If there is no opportunity to implement appropriate management during prolonged egg storage, increased turning frequency during subsequent incubation may be an option for older flock eggs.

One possible mechanism for the beneficial effects of incubation turning frequency of eggs from older breeder flocks in eggs stored for extended storage may lie in the juxtaposition of the albumen and shell membranes, because significant changes in albumen pH, albumen thickness, and water loss during extended storage have been documented (Brake et al., 1997). These changes could create an environment that would not be optimally receptive to the growth of the chorioallantoic membrane but may be overcome by an increased frequency of turning.

An associated mechanism may lie in the role of the albumen in buffering or sequestering ammonia, a toxic waste product produced by the blastoderm (Benton and Brake, 2000; Benton et al., 2001). Stern (1991) demonstrated that differences in pH between the yolk and albumen and across the blastoderm were responsible for several membrane-based transport and exchange processes vital to the optimum development of the embryo. During long-term storage, in older flock eggs, the buffering capacity of the albumen must be reduced and the trans-blastoderm difference increased so that an increased turning frequency may allow the blastoderm to interact with a greater quantity of fresh albumen in such a manner as to promote embryonic development (Elibol and Brake, 2008).

3 Critical periods during incubation

Considerable research has been directed at determining the times during incubation when turning eggs is most important to obtain maximum hatchability. In an early study, Card (1926) maintained that eggs turned solely during the first 6 days of incubation hatched almost as well as eggs turned throughout incubation. New (1957) turned eggs twice daily from 4 to 7 days and found that these eggs hatched as well as those turned throughout incubation. Deeming et al. (1987) and Deeming (1989a) agreed with New (1957) that the most critical period for turning was from day 3 to day 7 of incubation. In contrast, Kaltofen

(1961) suggested that frequent turning was the most important during the second week of incubation (7–14 days).

In the study by Elibol and Brake (2004), eggs that were either turned or not turned during the 0–7 day, 8–14 day, and 15–18 day periods of incubation in the first experiment showed that the absence of turning from 0 to 7 days of incubation caused a decrease in fertile hatchability and an increase in all stages of embryonic mortality, as well as incidences of malposition II (head in the small end of the egg). In the second experiment, eggs were either turned or not turned from 0 to 2 days, 3 to 4 days, 5 to 6 days, and 7 to 8 days, in all possible combinations, and the absence of turning from 3 to 8 days or 0 to 2 days (the most critical single 2-day period) alone or in combination with other time periods reduced fertile hatchability and increased embryonic mortality and the percentage of malposition II. They also observed differential effects in the absence of turning during E0 to E2 that primarily increased early embryonic mortality while an absence of turning from E3 to E8 primarily increased late embryonic mortality and malposition II, even though fertile hatchability was similar in both groups (Table 1).

It is common practice to turn eggs for the first 18 days of incubation. It has been reported that ceasing turning after 13 days of incubation (after closure of the chorioallantoic membrane) has no detrimental effect on hatchability (Byerly and Olsen, 1936; North, 1984; Proudfoot et al., 1981). Wilson and Wilmering (1988) found that hatchability was significantly decreased by cessation of egg turning after 10 days but egg turning can be stopped at 16 days of incubation without adversely affecting hatchability. Deeming (2002) found no evidence to suggest that the absence of turning after E15 had any negative effect on chicken embryonic development or hatchability. In the study by Tona et al. (2003), egg turning was stopped after 12, 15, and 18 days of incubation. Hatchability was

Table 1 Effect of an absence of turning during 2-day increments during the first 8 days of incubation on percentage fertile hatchability, early (E0–E6), and late embryonic mortality (E18–E21 + pipped), and malposition II

Days of incubation[1]				Fertile hatch	Embryonic Mortality		
0–2	3–4	5–6	7–8		Early	Late	Malposition II
				---(%)---			
T	T	T	N	91.4[a]	3.60[b]	3.43[c]	0.26[c]
T	T	N	T	90.4[a]	3.95[b]	4.04[c]	0.43[c]
T	N	T	T	86.8[a,b]	5.42[b]	4.30[c]	1.29[b]
N	T	T	T	73.3[c]	13.48[a]	8.22[b]	1.29[b]
T	N	N	N	75.1[c]	4.38[b]	13.04[a]	3.29[a]

Source: Adapted from Elibol and Brake (2004).
[a,b,c] Mean values within columns with no common superscripts differ significantly (*P* > 0.05).
T= Turning during indicated days.
N= No turning during indicated days with trays horizontal to floor.

significantly higher in eggs turned for 18 days than in those turned for 15 days but not significantly different from those turned for 12 days. In addition, Tona et al. (2005) recommended to turn eggs during incubation until at least day 12 or 18 but not stopping turning at day 15 of incubation. In contrast to Tona et al. (2003) and (2005), Elibol and Brake (2004) concluded that the absence of turning after 14 days or 18 days had no effect on hatchability in young (90.1% and 89.7%) and old flock eggs (85.5% and 83.7%). Cessation of turning at E14 may have a beneficial effect on the hatchability of eggs from older flocks in comparison to eggs from younger flocks due to improved machine air flow that may reduce variation in eggshell temperature (Elibol and Brake, 2006b). The importance of turning to control eggshell temperature is more fully discussed elsewhere in this book.

4 Turning frequency during incubation

There has always been interest in determining the optimum frequency for turning eggs during artificial incubation. Olsen (1930, thesis cited by Landauer, 1967) observed that hens moved their eggs during natural incubation about 96 times per day. Olsen and Byerly (1936) reported that turning eggs 48 or 96 times per day during artificial incubation increased hatchability compared with turning eggs 3 times daily. Other authors (Kaltofen, 1955, 1956; Kuiper and Ubbels, 1951) observed that turning eggs 24 times daily resulted in better hatchability than less frequent turning. In a recent study, turning frequencies of 3, 6, and 12 times per day showed significant losses in hatchability compared to a turning frequency of 24 times per day until transfer time (Oliveira et al., 2020).

Elibol and Brake (2003) reported an optimum turning frequency from 3 to 11 days of incubation is 96 times per day for broiler hatching eggs, when compared with turning frequencies of 24 and 48 times per day. However, turning eggs more frequently (e.g. 480 times per day) decreased hatchability (Robertson, 1961a) to less than that of eggs turned 96 times a day. Wilson (1991) concluded that maximum hatchability was achieved with a turning frequency of 96 times per day but that 24 times per day was a practical frequency. Most commercial incubator manufacturers have now designed machines that can be turned 96 times per day. In practice, these are not often used in commercial hatcheries because of concerns about wear on turning mechanisms.

A series of studies was conducted by Elibol and Brake (2006b) to determine if there was a time period during early and middle incubation when turning eggs 96 times per day was more beneficial relative to other time periods. Eggs from flocks at 34 and 37 weeks of age (younger) as well as 59 and 61 weeks of age (older) were subjected to turning frequencies of either 24 (24X) or 96 (96X) times daily from the start of incubation to 8, 10, 12, or 14 days of incubation. There were no differences in overall fertile hatchability,

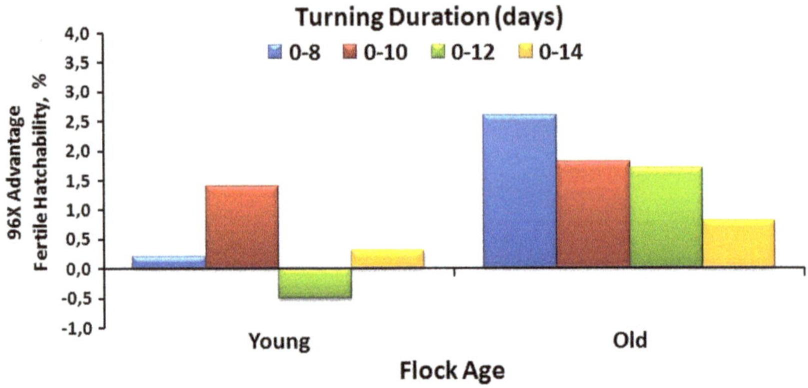

Figure 2 Advantage for 96 vs. 24 turns daily on fertile hatchability of young and old flock eggs. Source: Adapted from Elibol and Brake (2006b).

irrespective of the day of incubation upon which turning was discontinued (the range was 88.9–89.2%), suggesting that turning could be discontinued as early as E8 of incubation without ill effects. On the other hand, turning 96 times daily produced significantly better fertile hatchability than turning 24 times daily only for the older flock eggs, which was due to a lower incidence of late embryonic mortality. Good-quality eggs (younger flocks) benefited less from increased turning frequency during incubation than poor-quality eggs (older flocks) (Fig. 2).

The effects of turning broiler hatching eggs 24 (24X) or 96 (96X) times daily to E3, E9 or E15 of incubation followed by 24X to E18 on broiler embryonic development has been studied. Results suggested that early embryo growth and development were accelerated by 96X turning during early incubation (Lin et al., 2011).

Fasenko et al. (2004) conducted a study in which eggs were set on trays with the small end down (vertically), either turning 24X/d with an angle of 45° (control) or held for 10 min at 45° to the left of vertical, then held in the vertical position for 40 min (egg tray horizontal to the floor), then held for 10 min at 45° to the right of vertical (short turn) until day 18 of incubation. The short turn group had a significantly lower hatchability rate compared to the control group due to a higher percentage of second-grade chicks.

In another study, eggs were incubated from setting to 18 days of incubation with either continuous slow turning through 90° at hourly intervals or normal rapid hourly turning through 90° that took less than 1 min (Ozlu et al., 2008). Turning treatments had no overall effect on fertile hatchability but the incidence of early embryonic mortality was significantly reduced by continuous turning when compared to the normal hourly turning group (Ozlu et al., 2008).

Table 2 Effect of frequency of turning following the absence of turning for the first 2 days of incubation on fertile hatchability, late embryonic mortality, and malposition II

Days of incubation		Fertile hatch	Late Dead	Malposition II
0–2	3–16			
(turns/day)		(%)		
NT	24	80.0	6.0	2.2[a]
NT	96	81.6	5.6	0.8[b]

Source: Elibol et al. (2002).
[a,b] Means that possess different superscripts differ significantly ($P \leq 0.05$).
NT = No Turning.

It is not unusual to accidentally fail to turn hatching eggs in modern incubators, especially immediately after setting. Two experiments were conducted by Elibol et al. (2002) to determine the effect of frequency of turning (96X per day) following a period of no turning during the early period of incubation. After a period of no turning (0–2 days), an increased frequency of turning (96 vs. 24 times daily) significantly reduced the incidence of malposition II. However, no significant effects on fertile hatchability and embryonic mortality were reported (Table 2).

5 Turning angle during incubation

The effect of turning duration and frequency during incubation has been widely investigated. However, very little is known about the effects of the turning angle on hatchability. It is not completely clear what minimum turning angle is required to see the benefits of egg turning.

Funk and Forward (1953) turned eggs at angles of 20°, 30°, 40°, and 45° from vertical and reported increasing fertile hatchability with an increased turning angle, although the difference between 40° and 45° was small. In subsequent research, Funk and Forward (1960) investigated turning angles of 30°, 45°, 60°, and 75° from vertical and found that 45° produced the best results. Wilson (1991) concluded that eggs set small end down should be turned to an angle of 20°–45° from vertical. Cutchin et al. (2009) reported that a decreased turning angle of 15° until day 18 of incubation caused decreased SEF formation, and embryonic size, increased embryonic mortality and an overall hatch of fertile eggs was negatively influenced when compared with a turning angle of 45°.

In the study by Elibol and Brake (2006a), eggs from older flocks were subjected to turning angles (from vertical) of 35°, 40°, and 45°, with a turning frequency of 24 times daily (24X) until the day of transfer. The turning angle had no effect on fertile hatchability. However, the incidence of malpositioned embryos was increased by the 35° angle, compared with both the 40° and 45° angles. In a second experiment, eggs were subjected to turning angles of 35°,

Table 3 The effect of turning angle and daily turning frequency on fertile hatchability, early dead embryos (E0-E7), late dead embryos (E18-E21 + pipped), and malpositioned II embryos with the older flocks in the two experiments.

Turning angle	Turning Frequency (daily)	Experiment 1		Experiment 2	
		Fertile hatch	Malposition II	Fertile hatch	Malposition II
		(%)			
45	24	87.7	0.32[b]	85.8	0.23[b]
35	24	86.2	1.72[a]	84.9	1.40[a]
35	96	-	-	85.3	0.38[b]

Source: Adapted from Elibol et al. (2006a).
[a,b] Means that possess different superscripts differ significantly ($P \leq 0.05$).

with a turning frequency of either 24X or 96X daily, or 45°, with 24X turning. Similar to the first experiment, turning angle and frequency had no effect on fertile hatchability or embryonic mortality, but the incidence of malpositioned embryos was increased by the 35° angle with 24X turning, compared with the 35° angle with 96X turning, and the 45° angle with 24X turning (Table 3). The authors concluded that the incidence of malpositioned embryos was increased by a reduced turning angle, but that this effect was ameliorated by a concomitant increase in turning frequency.

French (1997) reported that increasing the distance between trays by reducing the turning angle could result in an exponential decline in the required air speed in commercial incubators. This might be especially important for larger eggs in commercial machines and to increase incubator capacity and thus reduce incubation cost per chick if the turning frequency was concomitantly increased.

6 Signs of turning failure

Failure to turn eggs during incubation prevents albumen absorption by the embryo, leading to embryos remaining small and covered with albumen. No turning during the early period (0–2 days) primarily increases early embryonic mortality while not turning from three to eight days primarily increases late embryonic mortality. In addition, a lack of turning or the improper turning angle during the first week of incubation increases the number of malpositioned (head in small end of eggs) embryos.

7 Conclusion

Turning has been shown to be fundamental to successful incubation. Turning during incubation involve several parameters including duration, frequency, turning angle and stage of embryonic development requiring turning. Studies on turning during incubation have suggested that the most critical time for turning was during the first week of incubation and an increased rate of turning during early incubation in order to optimize development. In order to maximize the hatch results, turning parameter interact with genotype and breeder age should be considered.

8 References

Benton, C. E. and Brake, J. (2000). Effects of atmospheric ammonia on albumen height and pH of fresh broiler breeder eggs. *Poult. Sci.* 79(11), 1562-1565.

Benton, C. E., Walsh, T. J. and Brake, J. (2001). Effects of presence of a blastoderm on albumen height and pH of broiler hatching eggs. *Poult. Sci.* 80(7), 955-957.

Brake, J., Walsh, T. J., Benton, C. E., Jr., Petitte, J. N., Meijerhof, R. and Peñalva, G. (1997). Egg handling and storage. *Poult. Sci.* 76(1), 144-151.

Byerly, T. C. and Olsen, M. W. (1936). Certain factors affecting the incidence of malpositions among embryos of the domestic fowl. *Poult. Sci.* 15(2), 163-168.

Card, L. (1926). Incubator eggs turned first six days hatch well. *Rep. Illinois Agric. Exp. Stn.* 39, 78.

Cutchin, H. R., Wineland, M. J., Christensen, V. L., Davis, S. and Mann, K. M. (2009). Embryonic development when eggs are turned different angles during incubation. *J. Appl. Poult. Res.* 18(3), 447-451.

Deeming, D. C. (1989a). Characteristics of unturned eggs: Critical period, retarded embryonic growth and poor albumen utilisation. *Br. Poult. Sci.* 30(2), 239-249.

Deeming, D. C. (1989b). Failure to turn eggs during incubation: Development of the area vasculosa and embryonic growth. *J. Morphol.* 201(2), 179-186.

Deeming, D. C. (1989c). Importance of sub-embryonic fluid and albumen in the embryo's response to turning of the egg during incubation. *Br. Poult. Sci.* 30(3), 591-606.

Deeming, D. C. (2002). *Avian Incubation: Behaviour, Environment, and Evolution*. Oxford: Oxford University Press, Inc, p. 440.

Deeming, D. C., Rowlett, K. and Simkiss, K. (1987). Physical influences on embryo development. *J. Exp. Zool. Suppl.* 1, 341-345.

Elibol, O. and Brake, J. (2003). Effect of frequency of turning from three to eleven days of incubation on hatchability of broiler hatching eggs. *Poult. Sci.* 82(3), 357-359.

Elibol, O. and Brake, J. (2004). Identification of critical periods for turning broiler hatching eggs during incubation. *Br. Poult. Sci.* 45(5), 631-637.

Elibol, O. and Brake, J. (2006a). Effect of egg turning angle and frequency during incubation on hatchability and incidence of unhatched broiler embryos with head in the small end of the egg. *Poult. Sci.* 85(8), 1433-1437.

Elibol, O. and Brake, J. (2006b). Effect of flock age, cessation of egg turning, and turning frequency through the second week of incubation on hatchability of broiler hatching eggs. *Poult. Sci.* 85(8), 1498-1501.

Elibol, O. and Brake, J. (2008). Effect of egg position during three and fourteen days of storage and turning frequency during subsequent incubation on hatchability of broiler hatching eggs. *Poult. Sci.* 87(6), 1237-1241.

Elibol, O., Turkoglu, M., Peak, S. D. and Brake, J. (2002). Malpositions induced by the absence of turning during early incubation can be reduced by turning eggs four times hourly thereafter. *Poult. Sci.* 81(Suppl. 1), 15.

Fasenko, G. M., Montgomery, J. E., O'Dea, E. E. and Martin, K. (2004). Establishing if the duration of egg turning effects the hatchability of broiler breeder eggs. IPSF meeting abstracts. *Poult. Sci.* 83, 1777.

Freeman, B. M. and Vince, M. A. (1974). *Development of the Avian Embryo*. London, UK: Chapman and Hall.

French, N. A. (1997). Modeling incubation temperature: The effects of incubator design, embryonic development, and egg size. *Poult. Sci.* 76(1), 124–133.

Funk, E. M. and Forward, J. E. (1953). *The Effect of Angle of Turning Eggs During Incubation on Hatchability*. University of Missouri, College of Agriculture, Agricultural Experiment Station.

Funk, E. M. and Forward, J. (1960). The relation of angle of turning and position of the egg to hatchability of chicken eggs. *Poult. Sci.* 39(3), 784–785.

Kaltofen, R. S. (1955). Hatching experiments at Beekbergen-turning the eggs. *Worlds Poult. Sci. J.* 11(3), 204–209.

Kaltofen, R. (1956). Het bruederij onderzoek te Beekbergen. *Landbouwvoorlichting* 13, 544–550.

Kaltofen, R. (1961). The effect of fertility rate and frequency of turning on embryonic mortality during artificial incubation. *Anim. Breed Abstr*, 106–107.

Kuiper, J. and Ubbels, P. (1951). A biological study of natural incubation and its application to artificial incubation. In: *9th World's Poultry Congress*, Paris, France, pp. 105–112.

Landauer, W. (1967). *The Hatchability of Chicken Eggs as Influenced by Environment and Heredity*. Agricultural Experiment Station Monograph 1.

Lin, Y. M., Brake, J., Yahav, S. and Elibol, O. (2011). Effects of turning frequency during higher incubation temperature on broiler embryonic development. *Poult. Sci.* 90, 10–18.

Lundy, H. (1969). A review of the effects of temperature, humidity, turning, and gaseous environment in the incubator on the hatchability of hen's eggs. In: Carter, T. C. and Freeman, B. M. (Eds.), *The Fertility and Hatchability of the Hen's Egg*. Edinburgh, UK: Oliver and Boyd, pp. 143–176.

New, D. A. T. (1957). A critical period for the turning of hens' eggs. *J. Embryol. Exp. Morphol.* 5(3), 293–299.

North, M. (1984). Factors affecting hatchability. In: *Commercial Chicken Production Manual*. Westport, CT: AVI Publishing Company, Inc.

Oliveira, G. D. S., Dos Santos, V. M., Rodrigues, J. C. and Nascimento, S. T. (2020). Effects of different egg turning frequencies on incubation efficiency parameters. *Poult. Sci.* 99(9), 4417–4420.

Olsen, M. W. (1930). *Influence of Turning and Other Factors on the Hatching Power of Hens Eggs*. Iowa: Iowa State College.

Olsen, M. W. and Byerly, T. C. (1936). Multiple turning and orienting eggs during incubation as they affect hatchability. *Poult. Sci.* 15(1), 88-95.

Ozlu, S., Sariyuz, K. U., Elibol, O. and Brake, J. (2008). Effect of continuous turning during incubation on embryonic mortality and hatchability of broilers. *Avian Biol. Res.* 1, 131–131.

Proudfoot, F. G., Hulan, H. W. and McRae, K. B. (1981). The effect of transferring hen eggs from turning to stationary trays after 13 to 20 days of incubation on subsequent hatchability and general performance. *Poult. Sci.* 60(2), 302–306.

Randles, C. A. and Romanoff, A. L. (1950). Some physical aspects of the amnion and allantois of the developing chick embryo. *J. Exp. Zool.* 114(1), 87–101.

Robertson, I. S. (1961a). The influence of turning on the hatchability of hens' eggs I. The effect of rate of turning on hatchability. *J. Agric. Sci.* 57(1), 49–56.

Robertson, I. S. (1961b). The influence of turning on the hatchability of hens' eggs II. The effect of turning frequency on the pattern of mortality, the incidence of malpositions, malformations and dead embryos with no somatic abnormality. *J. Agric. Sci.* 57(1), 57–69.

Stern, C. (1991). The sub-embryonic fluid of the domestic fowl and its relationship to the early development of the embryo. In: Tullett, S. G. (Ed.), *Avian Incubation*. London, UK: Butterworth-Heinemann, pp. 81–30.

Tazawa, H. (1980). Adverse effect of failure to turn the avian egg on the embryo oxygen exchange. *Respir. Physiol.* 41(2), 137–142.

Tona, K., Onagbesan, O., Bruggeman, V., Mertens, K. and Decuypere, E. (2005). Effects of turning duration during incubation on embryo growth, utilization of albumen, and stress regulation. *Poult. Sci.* 84(2), 315–320.

Tona, K., Onagbesan, O., De Ketelaere, B., Decuypere, E. and Bruggeman, V. (2003). Effects of turning duration during incubation on corticosterone and thyroid hormone levels, gas pressures in air cell, chick quality, and juvenile growth. *Poult. Sci.* 82(12), 1974–1979.

Tullett, S. G. and Deeming, D. C. (1987). Failure to turn eggs during incubation: Effects on embryo weight, development of the chorioallantois and absorption of albumen. *Br. Poult. Sci.* 28(2), 239–249.

Tyrrell, D. A. J., Tamm, I., Forssman, O. C. and Horsfall, F. L., Jr. (1954). A new count of allantoic cells of the 10-day chick embryo. *Proc. Soc. Exp. Biol. Med.* 86(3), 594–598.

Wilson, H. (1991). Physiological requirements of the developing embryo: Temperature and turning. In: Tullett, S. G. (Ed.), *Avian Incubation*. London, UK: Butterworth-Heinemann, pp. 145–156.

Wilson, H. R. and Wilmering, R. F. (1988). Hatchability as affected by egg turning in high density plastic egg flats during the last half of incubation. *Poult. Sci.* 67(5), 685–688.

Chapter 14

Poultry embryo development and skeletal growth

Edgar Orlando Oviedo-Rondón[1], North Carolina State University, USA

1 Introduction

Leg and locomotion problems are among the most prevalent and expensive diseases in growing meat-poultry species worldwide (Gocsik et al., 2017; Guevara-Torres et al., 2021). Bone developmental defects can cause pain, lameness, and immobility that can affect access to feed and water, causing welfare and health issues and increasing culling and total mortality (Granquist et al., 2019; Çapar Akyüz and Onbaşılar, 2020; Rath and Durairaj, 2022). Locomotion issues may reduce carcass quality and increase food safety issues at the processing plant (Granquist et al., 2019). Many leg defects do not depend only on the development of individual bones; the alignment and symmetry between collateral bones, muscles, tendons, and other organs like the liver and kidneys or the whole immunological system could be implicated (Guo et al., 2019; Li et al., 2022). Asymmetry of leg bones and tibiotarsal deformities can affect the development of the vertebral column (Droual et al., 1991) and contribute to vertebral damage and infections (Borst et al., 2017; Braga et al., 2018; Chen et al., 2018). Ossification issues can affect cartilage and bone alignment and impact tendon and muscle development (Bradshaw et al., 2002; Oviedo-Rondón et al., 2010; Pines and Reshef, 2015; Oviedo-Rondón et al., 2020; Rath and Durairaj, 2022). Lack of proper ossification, vascularization, and development can affect bone porosity (Williams et al., 2004) and make bones

[1] Edgar Orlando Oviedo-Rondón, Prestage Department of Poultry Science, North Carolina State University, USA (edgar_oviedo@ncsu.edu)

http://dx.doi.org/10.19103/AS.2022.0118.08

more susceptible to infections (Wideman, 2016). Many of these developmental issues start very early in life, and several research reports indicate they could have an origin during embryo development. This chapter will describe the impact of incubation factors on embryo skeletal development that has long-term effects in the life post-hatch.

Embryos of all avian species can exhibit phenotypic (De Smit et al., 2006; Branum et al., 2016) and metabolic (Loh et al., 2004; McNabb, 2006) developmental plasticity to respond to modifications in environmental conditions. The phenotypic plasticity of an embryo comprises all physiological or metabolic interactions of the genotype with its environment to continue development under suboptimal or adverse conditions (Burggren et al., 2016). Several incubation factors such as temperature (Van der Pol et al., 2014), oxygen (O_2) and carbon dioxide (CO_2) concentrations (Oviedo-Rondón et al., 2008a, b), and light (Van der Pol et al., 2017; Güz et al., 2021; Shah and Özkan, 2022; Yalçin et al., 2022) can affect bone developmental plasticity in broilers, turkeys (Oviedo-Rondón et al., 2008a, b) and ducks (Da Costa et al., 2016).

Most of the research has been focused on the effects of temperature (Hammond et al., 2007; Oviedo-Rondón et al., 2008a, b; Shim and Pesti, 2011; Van der Pol et al., 2014; Oznurlu et al., 2016; Aygün and Narinç, 2016; Güz et al., 2020; Agyekum et al., 2022; Sözcü et al., 2022; Shah and Özkan, 2022). However, under commercial incubator conditions effects of temperature can be concomitant with reduced ventilation, high humidity and CO_2, and low O_2 concentrations in different positions or microenvironments within the machine (Van Brecht et al., 2005; El-Hanoun et al., 2012; Barbosa et al., 2013; Boleli et al., 2016; Okur et al., 2017).

The average eggshell temperature (EST) during incubation is mainly affected by the air velocity and temperature, flow direction, and the metabolic heat of the embryo (French, 1997; Tazawa and Whittow, 2000; Özcan et al., 2010; Lourens et al., 2005). Consequently, all incubation factors may play a role in skeletal development. It has been demonstrated that any variation in incubation parameters, especially the long-term high temperatures and low O_2 tension during the last incubation phases in the hatchers, may negatively affect almost all tissues and organ development, including bones (Leksrisompong et al., 2007; Oviedo-Rondón et al., 2008a b; Molenaar et al., 2011a, b; Boleli et al., 2016; Wijnen et al., 2020a). The harmful effects on yolk-absorption, growth and maturation of the gastrointestinal tract (Wineland et al., 2006a; Leksrisompong et al., 2007) and immune system (Yeboah et al., 2019; Wijnen et al., 2020b) can indirectly affect skeletal development post-hatch and susceptibility to failures in ossification or infection.

It is also known that embryo reactions to environmental factors can be influenced by the maternal and genetic factors related to the transfer of nutrients, growth-factors, hormones, eggshell properties, and other physical egg traits (Oviedo-Rondón et al., 2008a, b; Eusebio-Balcazar et al., 2014; Yair et al., 2017; Muir and Groves, 2018, 2019). The optimum incubation parameters for all embryos in a machine are not the same due to interactive effects of breeder genetics and hen age, egg composition and yolk size, eggshell and albumen properties, and egg storage time and temperature of storage, among other factors (Hamidu et al., 2007, 2011, 2018; Molenaar et al., 2011a, b; Nangsuay et al., 2017; Okur et al., 2017).

The impact of modifications in embryo bone development has been observed during the life post-hatch in live performance, bone size and weight, mineralization, tibial dyschondroplasia (TD), leg blemishes, lameness, and locomotion issues. The objective of this chapter is to present scientific evidence indicating the importance of incubation conditions on skeletal development, leg health, and locomotion. The first section will describe embryo bone development in avian species, its regulation, and the interaction effects of the environment. This information will help the reader to understand the results reported in the literature. The second portion of the chapter will describe some literature documenting the effects of incubation factors on bone development.

2 Embryo bone development

2.1 Early development

Bone development starts very early in embryogenesis (Pines and Reshef, 2015). The formation of long bones like the tibia and femur seems to have similar initial development as flat bones like calvaria in the skull (Caplan, 2007; Pines and Reshef, 2015; Rath and Durairaj, 2022). However, lateral plate mesoderm-derived cells (originating from the mesoderm germ layer) form long bones, and neural crest (originating from the ectoderm germ layer) and cephalic mesoderm-derived cells form skull bones (Wang et al., 2020). After mesenchymal condensation, long bones and skull bones develop using distinct ossification patterns. Endochondral ossification predominates in long bones, and a cartilaginous intermediate is first generated to form a growth plate and, finally, bone. In contrast, most skull or flat bones are formed from intramembranous ossification.

The cellular regulation of endochondral bone formation at the molecular level requires cartilage and bone-encoding genes. Bone-encoding genes include collagen type I (col1a1), bmp-2, and bmp-6, and the cartilage-encoding genes include sox9, collagen type II (col2a1), collagen type X (col10a1), and Indian Hedgehog (Ihh), among others. The bone-encoding gene col1a1 is involved in the formation of bone extracellular matrix, whereas runt-related transcription factor 2 (runx2) is a master transcriptional regulator for osteoblast differentiation. The cartilage-encoding gene, col2a1, is exclusive to endochondral ossification and a major extracellular component of the cartilaginous template (Pines and Reshef, 2015). The knowledge of these genes is essential to understand the effects of incubation factors on bone and its effects in life post-hatch.

On embryo day (ED) 3, it is possible to observe the formation of limb buds. At ED 4, mesenchymal cells are committed to osteogenic lineage. Osteogenic progenitor cells are observed before the formation of the chondrogenic core that is formed hours later. These osteogenic progenitor cells are a group of four to six cells in a stacked conformation around a prechondrogenic core of cells. Osteogenic cell development is governed by a complex balance between the intrinsic capacities of these cells in the context of extrinsic information and signaling (Caplan, 2007). This relation is vital to understand how these cells can be affected by external factors during incubation. The cartilaginous core is replaced initially by vascular and marrow tissues and later bone. The relationships between the osteogenic collar and the chondrogenic core provide an environment that stimulates the cartilage's further differentiation into hypertrophic cartilage and eventually renders this core replaceable by vascular and marrow tissue (Pines and Reshef, 2015; Rath and Durairaj, 2022).

The first bony layer is vascularized, and vascularization is obligatory to start mineralization or during the whole bone development. Therefore, the vascular system is essential in transporting ions in solution and distributing solid minerals toward the embryonic chicken bone. After initial vascularization, the second layer of bone is deposited around the capillaries to start a bony trabecula. During this phase, a relationship between osteoblast cells and capillaries is established per life (Pines and Reshef, 2015).

Osteoblasts secrete collagen and bone matrix proteins, forming a complex interwoven network known as osteoid. Production of osteoid starts around ED 6. The secretion of osteoid is oriented away from the adjacent capillary. This osteoid is eventually mineralized. Osteoblasts are connected as a continuous monolayer of cells held together in close-packed conformation with their faces entirely in contact with their neighbors to produce large-diameter collagen fibrils specific to osteoid. Osteoblasts and chondrocytes produce first collagen type I around ED 5.5 or ED6, at day ED 7.5 to ED 8 collagens I and II, and later collagen type X, which is necessary for ossification. By ED 7.5 it is possible to observe the mineralization of osteoid.

Embryonic long bones are forming and continuously resorbing bone in distinct locations. Osteoblasts deposit new bone at the periosteal or external side, and osteoclasts resorb bone at the inner surface or endosteal side. Osteoclasts dissolve bone minerals by acidifying the underlying bone area and decomposing the collagenous bone matrix (Teitelbaum, 2000). Bone resorption products are removed from the resorption site via transcytosis and released at the osteoclast's basolateral side. Thus, active osteoclasts must be able to intracellularly store and transport large amounts of resorbed bone minerals (Kerschnitzki et al., 2016).

2.2 Growth of long bones

The longitudinal growth of long bones is regulated in the growth plate, an epiphyseal cartilaginous area generated by epiphyseal chondrocytes (Pines and Reshef, 2015; Kerschnitzki et al., 2016). These chondrocytes differentiate and proliferate to create resting, proliferative, and hypertrophic zones. The onset of growth plate chondrogenesis occurs at ED 4 when dividing undifferentiated mesenchymal cells in the limb bud condensate. This cartilaginous tissue is vascularized during this process and initiates growth plate formation. During this process, the Wnt and Hedgehog signaling pathways and transforming growth factor (TGF-) play fundamental roles in regulating the growth plate. These cellular signaling mechanisms of genes and gene products, mostly proteins, convert one kind of signal into another in a cascade of reactions. This transduction process regulates and directs the transcriptional and posttranscriptional processes of embryonic cell development. Phases of skeletogenesis, such as terminal chondrocyte maturation and joint formation, mainly depend on Wnt signaling (Enomoto-Iwamoto et al., 2002).

Chondrocytes in the growth plate have a high capacity to proliferate and produce daughter cells, but new cells must be supplied to the growth plate to maintain their function. Stem/progenitor cells may reside in several regions, including the resting zone, the border between the epiphyseal bone and the growth plate, and Ranvier's groove located in the perichondrium adjacent to the growth plate. These cells can have limited stemness, or ability to perpetuate or give rise to differentiated cells, after the rapid growth phase, which might be a reason for the insufficient reparative ability of the growth plate and susceptibility to apoptosis or failures during embryo development (van der Eerden et al., 2003; Pines and Reshef, 2015; Kerschnitzki et al., 2016; Rath and Durairaj, 2022).

Chondrocytes in each zone of the growth plate differ in morphology, secretion of extracellular matrix components, enzyme activity, and expression of hormone receptors. Chondrocytes of the proliferative zone synthesize collagen type II, and hypertrophic chondrocytes produce collagen type X. The production of TGF- in chondrocytes precedes a marked increase in type II collagen mRNA expression in transitional chondrocytes, suggesting a role for TGF- in the induction of extracellular matrix synthesis. Hypertrophic chondrocytes also start the production of alkaline phosphatase and calcification. Additionally, hypertrophic chondrocytes produce noncollagenous phosphorylated matrix proteins such as osteopontin (OPN) and bone sialoprotein and express parathyroid hormone receptors and the parathyroid hormone-related protein (PTHrP). The PTHrP and the Ihh are potent inhibitors of hypertrophic chondrocyte differentiation that act in a negative feedback loop to regulate endochondral bone formation. The thyroid hormone regulates these processes. Failures in the maturation process to hypertrophic chondrocytes lead to TD, characterized by lower expression of all proteins, enzymes, and hormone receptors previously listed (Kerschnitzki et al., 2016).

The hypertrophic zone chondrocytes are the terminally differentiated chondrocytes. Hypertrophic chondrocytes are responsible for the endochondral ossification that regulates the longitudinal growth of bone. In this ossification process, hypertrophic chondrocytes produce a kind of platform mineralized by their extracellular matrix, which supports bone formation by osteoblastic cells before they undergo apoptosis. Hypertrophic chondrocytes also promote vascular invasion at the chondro-osseous junction. This is the junction between calcified and non-calcified cartilage matrices, a critical process for recruiting osteoblast and osteoclast progenitors. A limited number of chondrocytes within the growth plate are themselves generated from stem-cell-like progenitor cells called chondro-progenitors and directly transform in considerable numbers into the osteogenic lineage in developing bone by the Hedgehog and Wnt mitogen and morphogenic signaling (van der Eerden et al., 2003; Pines and Reshef, 2015; Kerschnitzki et al., 2016).

From ED 14 to ED 19, chicken long bones roughly double their length, thickness, and the total amount of bone mineral (Yair et al., 2012). This requires massive transport of calcium and phosphorous derived from the yolk through the yolk sac membrane and the eggshell through the chorioallantoic membrane's circulatory system (Yair and Uni, 2011). Stressful incubation conditions can affect this rapid and delicate process causing failures in bone development. Long bone development reaches the fastest growth rate days before and after hatch (van der Eerden et al., 2003). This period during internal and external pipping in the hatching process has significant metabolic changes for the embryo, including transitioning from chorioallantoic to lung respiration with transitory hypoxia (Rahn, 1981; Tazawa and Whittow, 2000; Mortola and Labbè, 2005) that impacts skeletal development.

Like most poultry species, embryos from avian precocial species always face hypoxia during the hatching process. This hypoxia occurs during the phase of maximum O_2 consumption before pipping, called the plateau stage of O_2 consumption (Rahn, 1981; Mortola and Labbè, 2005). However, management and machine factors can prolong or make this phase more severe, causing detrimental effects on the development of almost all organs and tissues (Tona et al., 2001; Mortola and Labbè, 2005; De Smit, 2006). During this late incubation phase, the eggshell conductance, frequently nominated as

G, becomes very important for embryo respiration (Rahn, 1981; Tazawa and Whittow, 2000; Mortola and Labbè, 2005; Oviedo-Rondón et al., 2008a, b, 2009a, b). The ***G*** is the eggshell's capacity to passively transfer gases from the environment to the embryo, which can also be measured indirectly by the loss of metabolic water and gases from inside the egg to the external environment (Rahn, 1981; Tazawa and Whittow, 2000). Eggs may vary on ***G*** depending on species, genetics, breeder age, and egg size (Rahn, 1981; Visschedijk et al., 1985; Gualhanone et al., 2012; Nangsuay et al., 2017). More severe hypoxia can occur in larger embryos from bigger eggs laid by older hens with more considerable O_2 requirements. Severe hypoxia can be observed in embryos from some broiler genetic lines that lay eggs with reduced ***G*** or in ducks and geese with thicker eggshells, cuticles, and lower pore density (Oviedo-Rondón et al., 2008b; El-Hanoun et al., 2012; Da Costa et al., 2016).

2.3 Hormonal regulation

Studying the effects of incubation on hormonal regulation of bone development is another way to understand the impacts of some environmental factors during embryo development. The thyroid hormone controls endochondral and intramembranous ossifications (Shao et al., 2006). In longitudinal bone growth, thyroid hormones regulate the critical transition between growth plate chondrocytes into hypertrophic cells in the growth plate. These effects are direct in skeletal cells and indirect, primarily via the growth hormone (GH)/insulin-like growth factor-1 (IGF-I) axis, to control the linear skeletal growth and maturation in the growth plates of long bones. The GH and IGF-I are known to stimulate chondrocyte proliferation in epiphyseal plates (Gouveia et al., 2018). Stressful conditions can alter the thyroid hormone function during the last phase of incubation (Wineland et al., 2006b)

Melatonin also has been related to stimulating bone development. Melatonin seems to regulate mechanisms to maintain symmetry in bones and adequate endochondral ossification. The effect can also be direct and indirect by stimulation or inhibition of hypothalamic GH and IGF-I release. Melatonin, GH, and IGF-I release peaks depend on the dark periods. Melatonin also stimulates osteoblast formation (Yalçin et al., 2022).

2.4 Developmental fluctuating asymmetry

Bilateral symmetry is one of the standard features of vertebrates during skeletal development (Pines and Reshef, 2015). It is crucial for the gait and locomotion of bipedal animals (Radzak et al., 2017), including poultry species (Oviedo-Rondón, 2008). Symmetry and alignment of avian leg bones are pivotal for normal gait, post-hatch proper bone development, ossification, bone strength, and avoiding leg health issues (Oviedo-Rondón et al., 2017, 2018; Guo et al., 2019). Nevertheless, embryo developmental plasticity can lead to significant asymmetry in long bones by the end of the incubation process (Oviedo-Rondón et al., 2008a, b).

Minor deviations from perfect symmetry are called fluctuating asymmetry and arise whenever there is a lack of perfect adaptation to the environment (Møller, 2006). It is important to consider that avian embryos commonly exhibit more asymmetry than other vertebrates. This asymmetry can be detected since the moment of gastrulation by observing the cranial end or tip of the primitive streak or Hensen's node. Organogenesis of avian species is highly asymmetric in the heart, the gut, the mesentery, the oviduct (only the left ovary and oviduct persist), and even the brain and eyes (Monsoro-Burq and Levin, 2018). However, increased limb asymmetry is unnatural and may cause several bone and locomotion issues (Sanotra et al., 2001; Radzak et al., 2017; Oviedo-Rondón et al., 2018). Møller et al. (1999) reported that the asymmetry of long bones in chickens had been correlated with TD, worsening gait scores, and tonic immobility.

There are statistical issues with using the values of fluctuating asymmetry as an indicator of genetic selection or stress. Consequently, relative asymmetry (RA) is calculated as the absolute difference in the length, weight, or paired morphological traits divided by the mean of the same feature. Yalçin and Siegel (2003) reported that temperature during the first week of incubation could increase the RA of long bones. Pre-hatch stress either by maternal deposition of corticosterone in the egg (Eriksen et al., 2003) or suboptimal incubation conditions (Oviedo-Rondón et al., 2008a, b, 2009b; Da Costa et al., 2016; Sözcü et al., 2022) can impair embryonic skeletal development by increasing fluctuating or RA in chickens, turkeys, and ducks.

3 Incubation factors that affect skeletal development

The impact of incubation conditions on embryo skeletal development has been reported several times in the past three decades but probably gained more attention in the past two. In 2014, Groves and Muir published a metanalysis of experiments linking incubation temperature with subsequent leg weakness in broiler chickens using the large body of research studies already available. These authors emphasized the search for an ideal incubation temperature profile to improve bone development. Groves and Muir (2014) developed mathematical models using scientific papers to predict variations in bone ash at hatch and later leg weakness. However, commercial hatcheries have implemented optimum incubation temperature profiles during the past decade to support each embryonic stage. Commercial incubation profiles consider the differences in egg traits and embryos among genetic lines and breeder age. In practice, all incubation factors constantly interact, and machine temperature may not be the only parameter to optimize seeking better embryo development (Lourens et al., 2007; French, 2009).

This section will review the effects of temperature, O_2, CO_2, ventilation, and light regimes within incubators on bone development. Additionally, research studies should evaluate the effects of incubation treatments at hatch and in growing birds close to market age. To understand better the impact of incubation factors is important to include parameters such as gene and protein expression in specific bone and cartilage tissues, bone morphology, ossification, bone ash, density, bone strength, RA of bone weight, length and density, incidence and severity of leg defects, locomotion kinematic parameters or gait scores, and incidence of bone infection. The responses of developmental plasticity at hatch may not always indicate the best quality for bones in life post-hatch. Evaluation of only one bone may not answer the effects on locomotion or leg defects such as valgus or twisted legs (Guo et al., 2019; Li et al., 2022), and all leg bones should be studied.

3.1 Temperature

Temperature can affect bone development during the whole incubation process, but effects on skeletal development vary depending on the time, length, and severity of temperature variation. The most frequent deviations from optimum temperatures occur during incubation's first and last days. However, mechanical failures due to electronic sensor calibration, ventilation equipment, heating or cooling equipment, and turning issues can cause small increments during other periods in microenvironments within a setter machine (French, 2009; Okur et al., 2017). Aside from affecting embryonic development, incubation temperature also affects broiler growth, and on occasions, effects on feed utilization have been reported (Oviedo-Rondón et al., 2009a; Hamidu et al., 2007, 2011, 2018).

Early incubation temperatures changing from 36.5 to 39°C for a short period (2–3 days) may not always affect slaughter weight and feed conversion ratio; however, it may affect muscle and bone development (Hammond et al., 2007; Oviedo-Rondón et al., 2008a, b, 2009a, b). Muir and Groves (2018) concluded that slow start incubation from 37.2°C at ED1 reaching 37.8°C EST at ED 13 resulted in higher hatchability with more late-hatched chicks and higher bone ash. But it is necessary to be cautious when interpreting bone results at this early age post-hatch. More bone ash is not always positive for the life post-hatch. For example, Yalçin et al. (2007) observed that low (36.9°C for 6 h/day) or high (39°C 6 h/day) incubation temperatures from ED 0 to 8 increased TD (14.4% and 12.8%) in broilers at 49 days of age compared to control (5.0%).

Nevertheless, ash and tibia calcium were increased only when incubating eggs at a lower temperature. At hatch, chicks from these treatments had growth plates with more collagen II, OPN, gene expression, wider proliferative zone, and higher number of chondrocytes per column than in chickens from the control treatment with optimum incubation. The higher temperature early in life increased heat-shock protein 90 (Hsp90). These changes in chondrocyte differentiation, bone ash, length, protein gene expression, or TD incidence were not observed when incubation temperatures were modified from ED 10 to 18.

Modifications of incubation temperatures during other periods have also been evaluated. These situations may occur under commercial conditions due to failures in electronic machine control or ventilation issues (French, 2009). Shim and Pesti (2011) evaluated in three experiments the effects of increasing (38.5°C) or decreasing (36.5°C) the incubation temperature at 1°C for 3 days between ED 4 to 7 in combination with diets deficient in calcium or phosphorus to induce rickets. In three experiments, it was possible to observe differences in hatching time with the logical delay in the hatch when temperatures were reduced and faster hatch using higher temperatures. Variable results in hatchability and body weight were observed among these three experiments with the same temperature treatments. These authors could not detect interactions between incubation temperature (ED 4 to 7), hatch time, and bone abnormalities such as TD and rickets. Presumably, incubation temperature in that period (ED 4 to 7) may not affect long bone development during this period. Incubation temperature modifications at any age affect the hatch window and the time chicks spend in the hatcher before removing them to be processed and delivered to the farm. Werner et al. (2010) studied the effects of modifying incubation temperature (37.5 *vs.* 38.5°C) from ED 7 to 10 at hatch and after day 36 post-hatch in Cobb 500 chickens. These authors also concluded that increasing incubation temperature in this period (ED 7 to 10) did not impact the hatch, growth, slaughter, and meat quality characteristics.

In contrast, Oznurlu et al. (2016) demonstrated that exposing embryos to high temperatures (38.8°C), one degree higher than the control treatment after ED 10, negatively affected skeletal development. This heat stress during incubation caused a reduction in the relative tibia weights and length, a narrow proliferative zone in the growth plate, thicker hypertrophic areas, and lower alkaline phosphatase during embryo development and at hatch. Additionally, Morita et al. (2020) observed that incubation temperature from ED 13 until hatching affected femoral and tibiotarsal morphological characteristics, mineral composition, and mechanical properties at slaughter age (42 days). The incubation treatments evaluated were constant cold (36°C), control (37.5°C), or hot (39°C) machine temperatures, which cause EST of 37.4 ± 0.1°C, 37.8 ± 0.15°C, and 38.8 ± 0.3°C, respectively. Shah and Özkan (2022), using 38.8°C for 6 h per day from ED 11 to 16, observed early hatching time, reduced chick quality, and decreased relative bursa and tibia weights, while the heart and sternum weights increased. These reports collectively indicated that incubation temperature after ED 10 might affect skeletal development.

Several reports have indicated that low or high EST (36.7 or 38.4–39°C EST, respectively) during the last week of embryogenesis lowered hatchability, broiler growth rate, body weight, and feed intake at slaughter age, reduced immunity, and increased metabolic issues such ascites, and mortality rate (Molenaar et al., 2011a, b; Maatjens et al., 2014, 2016; Sözcü and İpek, 2015; Okur et al., 2017; Yeboah et al., 2019; Wijnen et al., 2020b). It has also been shown that high incubation temperature (38–39°C EST) reduces tibia weight and increases RA of leg weights in broiler chicks and turkeys, affecting growth plate maturation, which may have implications on TD incidence (Yalçin et al., 2007; Oviedo-Rondón et al., 2008a, b, 2009a, b), or in bone morphology and strength at processing age (Aygün and Narinç, 2016; Güz et al., 2020; Agyekum et al., 2022).

During the plateau stage of O_2 consumption (Tazawa and Whittow, 2000), suboptimal temperatures and low O_2 concentrations interact to affect negatively femur, tibia, and shank development, growth, maturation, and RA at hatch time in turkeys (Oviedo-Rondón et al., 2008a) and broilers (Oviedo-Rondón et al., 2008b). Elevated temperatures reduced the expression of collagen type X and TGF- and the chondrocyte density in the resting and hypertrophic zones of the growth plate of broilers and turkeys. The adverse effects of elevated incubation temperatures or low O_2 concentration were more evident in broilers from genetic lines with low eggshell conductance (Oviedo-Rondón et al., 2008a, b). Suboptimal temperatures during early and late incubation can negatively affect the development of muscle (Oviedo-Rondón et al., 2020) and gastrocnemius tendon (Oviedo-Rondón et al., 2010). Hypoxia in the last incubation phase can worsen the harmful effects of elevated EST on these other components of the musculo skeletal system critical for locomotion.

Under commercial conditions, single-stage incubators could provide optimum environmental conditions to embryos in each developmental stage, compared to multi-stage machines with constant temperatures and relative humidity. In single-stage incubator machines it is possible to control CO_2, ventilation, turning, and humidity according to embryo development. Still, the most significant change is the temperature profiles that start higher and finish lower than in a multi-stage machine. These modifications can affect embryo skeletal development.

Investigating differences between single and multi-stage incubation, Oviedo-Rondón et al. (2009a) conducted six experiments under industrial poultry production conditions, and measured live performance, leg defects, and locomotion of large Ross-708 high-meat-yield broilers at 56 days of age.

Machines from both incubation systems were operated to achieve optimum hatchability, and no significant differences in these parameters were observed. In each experiment, 19 600 (summer flocks) to 20 600 (winter flocks) chickens from each incubator treatment were placed per house and two replicate houses were used per incubation treatment within the six farms evaluated. At 56 days of age, 200 broilers per house were individually weighed and evaluated for leg defects and gait scores for a total of 4800 observations in 24 houses. At 56 days of age, female broilers hatched in single-stage machines were 41 grams heavier than those hatched in multi-stage machines. A higher percentage of broilers with crooked toes (0.8% *vs.* 0.1%), valgus (36.8% *vs.* 26.6%), or mild lameness (gait scores 1 and 2) were observed in broilers incubated in multi-stage incubators. In summary, incubation conditions under commercial conditions can affect growth and some leg health and locomotion parameters.

The early hatching time, wide hatch window, and prolonged time spent in the hatcher may affect skeletal development due to stressful conditions with high temperature, humidity, and extended periods to access feed and water. Shim and Pesti (2011) attributed the lack of incubation treatment effects on bone development and bone disorders post-hatch in their experiments to the variability in hatching time caused by their treatment. They modified the setter temperature 1°C above or under their standard (37.5°C) only between ED 4 to 7, which is different from previous studies (Oviedo-Rondón et al., 2008a, b, 2009b) used for their discussion. Oviedo-Rondón et al. (2008a, b) used only incubation temperature treatments between ED18 to 21 and never earlier, as mistakenly Shim and Pesti (2011) reiterated in their discussion. Oviedo-Rondón et al. (2009a) used incubation profiles in single-stage incubation compared with multi-stage machines or simulating incubation profiles (Oviedo-Rondón et al., 2009b). These temperature profiles (Oviedo-Rondón et al., 2009a) are more subtle modifications during the early periods, and temperatures in the hatcher phase are not comparable with the treatment evaluated by Shim and Pesti (2011). In Oviedo-Rondón et al. (2008a, b, 2009b) studies, bone development, were evaluated as soon as the chick down was dry and not waiting until all chicks in the treatment had hatched. In this way, it is possible to differentiate between hatching time, length of stay in the hatcher, and actual effects of incubation on skeletal development at the hatch.

However, hatching time and hatch window may be essential for bone development. Groves and Muir (2017) reported that earlier hatching times predisposed Cobb broiler chickens to TD. Hopcroft et al. (2020) concluded that removing chicks from the hatcher within 30 h of their hatch is likely to benefit their femoral mineralization. They considered that more than 30 h post-hatch in the hatcher could cause leg problems. Most well-managed single-stage machines or multi-stage systems can obtain hatch windows shorter than 24 h. Hopcroft et al. (2020) also addressed the importance of giving access to feed and water to reduce leg problems. In commercial hatcheries, early feeding depends not only on the hatch window but also on the logistics of chick processing and transportation. Nevertheless, early access post-hatch to feed and water could always be necessary for hatchling health (Wijnen et al., 2022).

3.2 Oxygen, carbon dioxide and ventilation

During early incubation, single-stage machines are closed. Due to egg metabolism and respiration, CO_2 concentrations increase gradually up to 0.5% (5000 ppm) or even 0.75% around 5 days of incubation, and relative humidity also increases to 65% or 70%. Non-ventilated incubators containing eggs of 45 and 60-week-old breeders can reach up to 1.5% and 1% CO_2 at ED10, respectively (De Smit et al., 2006). Consequently, the O_2 tends to reduce to 19.5% (De Smit et al., 2006; Everaert et al., 2007). These levels of CO_2 early in life have positive effects on early chick growth, thyroid hormones, and health (Liu et al., 2022), even with eggs stored for up to 18 days (Bilalissi et al., 2022). Multi-stage incubators usually operate with a maximum of 0.3-0.4% CO_2 and 20.5-20.7% O_2 (at sea level) because they have continuous air exchange.

The content of O_2 in the air is close to 21%. However, altitude and atmospheric pressure affect the number of O_2 molecules per air volume. The partial pressure or effective O_2 can quantify these changes. The atmospheric pressure at sea level is 760 mm Hg or 101 kPa, and it falls as altitude increases. At 1000 meters above sea level (m.a.s.l.), atmospheric pressure reduces to 733 mm Hg or 97.7 kPa, and at 1500 m.a.s.l. it is 720 mm Hg or 96 kPa. Lower atmospheric pressure as altitude increases causes reduced partial pressure of O_2. Partial pressure is the pressure exerted by one of the gases in a mixture if it occupies the same volume. However, the partial pressure of O_2 is affected not only by altitude but also by temperature, water vapor, or relative humidity within the incubators. At sea level, the O_2 partial pressure of O_2 is 159 mm Hg, and the effective O_2 is 20.9%, but at 1000 or 1500 m.a.s.l., the partial pressure of O_2 reduces, and the effective O_2 is 18.6% or 17.3%, respectively, when expressing O_2 as an equivalent to sea level O_2%. The reduced effective O_2 can cause hypoxia. Consequently, hypoxia, as it is evaluated under research conditions, can occur in some commercial poultry hatcheries.

Hypoxia or reduced O_2 and hypercapnia or excessive CO_2 in the bloodstream cause different effects in embryonic tissues (Decuypere et al., 2006; Liu et al., 2022). Their effects depend on the degree of severity and the time of development at which the embryo is exposed to them. Oxygen uptake is related to the surface area vasculosa in early embryonic angiogenesis and the growth rate of the chorioallantoic membrane (Tazawa and Whittow, 2000). The development of these extraembryonic membranes has been related to hypoxia and/or hypercapnia during the first 5 days of incubation (Liu et al., 2022) and egg-turning frequency. Optimum development of these tissues, early hypoxia, and/or hypercapnia may reduce the hatch window and improve hatchability. The momentaneous hypoxia and hypercapnia at the end of the incubation period close to pipping are a stimulus for hatching events and potentially can improve cardiovascular and pulmonary functions (Liu et al., 2022; Bilalissi et al., 2022). However, a chronic situation could be detrimental to many tissues and cause several developmental and metabolic diseases (Oviedo-Rondón et al., 2020).

Chan and Burggren (2005) concluded that long bones were not affected by hypoxia (15% effective O_2) at any period between days 1 and 18 of incubation. Nevertheless, Smith et al. (2013) demonstrated that embryos exposed to hypoxic conditions during early embryonic development (ED 3 to 6) had cellular oxidative stress and apoptosis of neural crest progenitor cells in the skull bones. In this experiment, embryos were exposed to O_2 concentrations from 7% to 19% and were compared to embryos that maintained normoxic, 21% O_2 conditions. Consequently, severe hypoxia can affect bones early in life. But, it is important to remember that commercial setter machines generally contain at least 16% effective O_2 even when hatcheries are located at 2130 m.a.s.l.

On the other hand, hypoxia is typically harmful to bone development in the last phase of incubation. Hypoxia is always a stimulus for angiogenesis and vascularization in the growth plate, but this response is only observed if the birds are adapted genetically to hypoxic conditions. Commercial chickens may not be adapted to hypoxic conditions, and their growth plates can suffer more during the plateau stage of O_2 consumption from ED 17.5 to ED20. These genetic differences were demonstrated in research employing commercial and Tibetan chickens at high altitudes (3000 m.a.s.l.). Huang et al. (2017a, 2018) reported the importance of genes encoding the hypoxia-inducible factors (HIFs), vascular endothelial growth factor A (VEGFA), and VEGF receptors in the onset of TD. These authors compared gene expression of HIFs and VEGFA of commercial broiler chickens with chickens from a native poultry breed from the Tibetan plateau (2600–4500 m.a.s.l.) adapted to hypoxic conditions (14.4–14.7% effective O_2) that do not exhibit leg problems. This group used hypoxic conditions and thiram (tetramethyl thiuram disulfide), a fungicide used in agriculture. Thiram induces TD by reducing vascularization in the bone growth plate. These authors concluded that commercial chickens had lower expression of HIFs, VEGFA, and VEGF receptors in the tibia growth plates than Tibetan chickens under hypoxia and toxicity caused by thiram. The same group (Huang et al., 2017b) demonstrated that O_2 supplementation in Tibetan chickens to provide an environment with 21% O_2 upregulated the same genes and enhanced the development of the tibia growth plate in these chickens.

Under industrial conditions, failures in ventilation in hatchers, and lack of negative pressure in plenum rooms, can reduce air exchange during the last incubation phase (French, 1997; Boleli et al., 2016). Large incubators may have fan speed variations, changes in turning angles of egg trays that reduce airflow over the eggs (French, 1997; Barbosa et al., 2013), and airflow direction (Van Brecht et al., 2005). The most critical period for bone development can be when all embryos in a machine have the fastest growth rate, higher metabolic heat production, and consequently more O_2 demand and CO_2 production during the plateau of O_2 consumption (Rahn, 1981; Wineland et al., 2006a, b; Christense et al., 2007). In these conditions, there is a potential hypoxia between ED 17.5 and 19. Embryos from eggs stored for more than 5 days can have longer hypoxia in this phase due to a delay in development and the extended period between internal and external pipping (Tona et al., 2001; Goliomytis et al., 2015). This could explain outbreaks of leg issues when chickens come from eggs stored for more than 10 days (Oviedo-Rondón and Wineland, 2011).

Looking at bone development during the last 3 days of incubation in hatchers with four O_2 levels between 17% and 23%, Oviedo-Rondón et al. (2008b) observed that hypoxia (17%) had lighter and shorter femurs, tibias, and shanks. Chicks at 21% O_2 had the longest tibias, increasing O_2 to 23% reduced the tibia length but increased the shank length. Chicks of a genetic line with low ***G*** exposed to 19% O_2 had the shortest femurs, and chicks from a line with high ***G*** had the highest RA. In a third experiment, chicks coming from hatchers with 23% O_2 had heavier femur and tibias than 19% and 21% O_2 and longer tibias than 21%. Chicks, at 21% O_2, had the heaviest shanks. In two experiments, hypoxia increased the RA of tibia weights. In experiments with turkey eggs (Oviedo-Rondón et al., 2008a), the poult femur and tibia weights relative to yolk-free body weight were depressed, and RA of shank length increased by incubation at 19% O_2. In a second experiment, the tibia and shanks were lighter in turkey poults incubated at 17% O_2 independently of incubation temperature, and shanks were lighter and thinner when the machine temperature was 39°C from ED 18 to 21. The O_2 concentrations below 21% should be avoided during the last incubation phase for proper bone development.

Hypoxia during incubation can affect more duck, goose, and turkey embryos due to the thicker cuticle and lower relative *G* (El-Hanoun et al., 2012; Da Costa et al., 2016). The peak mortality in geese before hatching is frequently attributed to the proper transition to pulmonary respiration (Salamon, 2020). During an investigation of the fatigued walking condition (FWC) of Pekin ducks, Da Costa et al. (2016) observed the adverse effects of reduced *G*, or hypoxia, in bone development, leg problems, and locomotion. The FWC is observed while loading and unloading Pekin ducks to the processing plant. At market age, ducks are generally herded to the trucks that transport them to the processing plant. The flock runs quickly from the house to each truck floor through inclined ramps. However, during the route in the inclined ramps, some ducks show incapacity to walk to and from the truck, sitting while other ducks run over them, increasing the incidence of scratches and, consequently, condemnations of carcasses. Incubation factors were evaluated as possible causing factors. Ducklings from eggs with reduced *G* had more residual yolk, shorter tarsus bones, and reduced tarsus, tibia, and femur weights relative to free-yolk mass at the hatch. Elevated temperatures increased the RA of tibia and femur weights. During the grow-out, ducks subjected to elevated temperatures during incubation had higher mortality (4.80% *vs.* 3.15%) and culling rates due to leg problems (2.70%) than those incubated at standard temperatures (1.35%). But, ducks from eggs with reduced *G* had the highest culling rate (3.2%) due to leg problems. Bone relative weights at 35 days were increased by reduced *G*. However, ducks with the FWC had higher tarsus RA, and their bones were weaker and had less tibia ash than normal walking ducks. Incubation impacted bone and also heart development in the duck FWC. It is not surprising that hypoxia during incubation can have long-term effects on the health of birds. Skeffington et al. (2020) showed that severe hypoxia (13.5–14% O_2) during incubation (0 to 19 days) caused hypertension, cardiovascular dysfunction, and cardiac wall remodeling in hens 6 months post-hatch. The hypoxia and elevated temperatures during incubation also damaged muscle development in these ducks (Oviedo-Rondón et al., 2020).

3.3 Light

There is some evidence that lighting during the incubation can affect embryo and post-hatch bone growth and leg health in broilers. The mechanisms have been linked mainly to the dependency on releasing GH after dark periods. The IGF-I and melatonin should also be involved when a circadian rhythm has been developed after a photoperiod. However, in studies of poultry incubation related to bone development, the role of these last two hormones has not been established. Some results are inconsistent due to the significant variability in the protocols evaluated. In three studies, Van der Pol et al. (2017, 2019a, b) used diverse photoperiods using 500 lux white LED during the whole incubation. The first study (Van der Pol et al., 2017) reported that continuous lighting (24L) had more epiphyseal plate abnormalities and higher femur mineral density than complete darkness. But, complete darkness had more osteomyelitis. No effects were observed on gait scores or other bone parameters. In a subsequent study, Van der Pol et al. (2019a) observed that complete darkness (24D) during incubation which is the most common current condition, resulted in higher femur length, higher femur, and tibia weight, and width at hatch than 16L:8D or 24L. In the third experiment, Van der Pol et al. (2019b) observed that broiler chickens from eggs exposed to continuous lighting (24L) during incubation had lower femur weight and tibiotarsus length, cortical and medullary area of the tibiotarsus, and second moment of the area around the minor axis indicating weaker bones, and a higher incidence of TD at 35 days of age. The GH plasma concentration at hatching was higher for 24D than 12L:12D or 24L. In both experiments, plasma melatonin and IGF-I did not differ among treatments, or the values were difficult to interpret. In summary, complete darkness during incubation was still better than lighting, and probably only 12L:12D provided some stimulus for bone development that was unclear in the data presented. However, in a recent study, Güz et al. (2021) did not find any significant effect of green LED light on tibia bone parameters when they used a 16L:8D photo schedule.

Yalçin et al. (2022) evaluated the effects of continuous light (24L), 12 h of light, followed by 12 h of darkness (12L:12D), or continuous darkness (24D) throughout incubation of broiler chicken eggs on the development and strength of leg bones, and the role of GH, IFG-I, and melatonin in bone development. The 24L resulted in lower embryonic ossification at ED 13 and 14, lower femur length, and lower tibiotarsus weight, length, cortical area, bone strength, and mean cortical thickness at hatching on D0 compared to 12L:12D, primarily. The continuous light (24L) also had adverse long-term effects, with lower femur weight and tibiotarsus length, cortical and medullary area of the tibiotarsus, weaker bones, and a higher incidence of TD. The GH level at D0 was higher for 24D than for 12L:12D, with 24L intermediate, but plasma melatonin and IGF-I did not differ between treatments. In conclusion, in this experiment, 24L during incubation also had a detrimental effect on embryonic leg bone development and post-hatch leg bone strength compared to 24D and 12L:12D. In contrast, the light-dark rhythm of 12L:12D may have a stimulating effect on leg health.

The development of the whole skeleton size may improve with light stimulation. In a second experiment, Shah and Özkan (2022) added a lighting schedule 16L:8D together with the previously discussed incubation temperature (EST, 38.8°C 6 h/day ED 11 to 16). The lighting schedule and temperature stimulation improved chick length at hatch and chick quality, but no other effects in long bones were observed.

3.4 Egg storage and SPIDES effects on bone

Egg storage is a common practice in commercial poultry production. Eggs stored for more than 5 days lose viability, negatively affecting embryo development, increasing incubation times, and affecting chick quality and live performance post-hatch (Uddin and Hamidu, 2014). In parent, grandparent, and layers, storage can exceed 2 weeks. Chicks from stored eggs can have lower triiodothyronine, thyroxine, glutathione peroxidase activity, total antioxidant capacity, and IgY concentrations (Uddin and Hamidu, 2014; Goliomytis et al., 2015; Ebeid et al., 2017).

The detrimental effects of egg storage can be minimized by reducing pre-storage incubation, storage at 15°C, egg turning, and increasing CO_2 levels, but probably the most effective method tested so far is applying short incubation periods during egg storage, known as SPIDES (see also Chapter 6 of this book). Ebeid et al. (2017) demonstrated that pre-incubation for 6 or 8 h enhanced hatching results, improved thyroid function and antioxidative properties, reduced lipid peroxidation, and elevated humoral immunity in chicks hatched from eggs stored for 14 days. All these aspects could be positive for skeletal development.

Dymond et al. (2013) evaluated the effects of SPIDES in eggs stored for 21 days. SPIDES, in this case, consisted of four 4-h preincubations at 4-to-5-day intervals during storage. The SPIDES treatment minimizes early and late embryo mortality restoring hatchability from 71% in the non-treated stored eggs to 84%. The chick quality at hatch was equivalent to un-stored eggs, and the growth rate at 4 weeks was better than chickens from un-treated stored eggs. In body composition, evaluated with DEXA, it was possible to observe that SPIDES chicks had increased bone mass content relative to the other treatments. At 4 weeks, chickens from non-treated stored eggs had reduced bone, fat, and lean masses. In contrast, chickens from eggs subjected to SPIDES had better bone, fat, lean mass, and increased bone density than chickens from stored and non-stored treatments. In conclusion, SPIDES seems to improve the skeletal development of embryos from stored eggs, and these effects can be observed weeks post-hatch.

3.5 Effects of incubation on vertebral development and chondronecrosis with osteomyelitis

Vertebral column developmental issues and cordal spine injuries could be associated with embryo development (Bradshaw et al., 2002; Braga et al., 2018; Çapar Akyüz and Onbaşılar, 2020). Scoliosis has been associated with tibiotarsal deformities (Droual et al., 1991), which are misaligned bones with increased RA. The RA of long bones at hatch may impact the early development post-hatch of the vertebral column because irregular movements and lameness (Møller et al., 1999; Sanotra et al., 2001) as an adaptation of gait to RA of long bones can force the free-thoracic vertebra, T4-T5 (Oviedo-Rondón, 2008). Frictions in the T4 and T5 vertebra cartilage can cause osteochondrosis dissecans (OCD) lesions (Borst et al., 2017; Chen et al., 2018). As discussed before, several authors have reported correlations between worse gait scores, immobility, and RA.

Borst et al. (2017) described the pathogenesis of enterococcal spondylitis (ES) in natural and experimental infections with *Enterococcus cecorum*. This group observed OCD lesions, as variably large cartilaginous clefts in the free-thoracic vertebra, since the first week of life. These clefts were often partially filled with thrombocytes, fibrin, and erythrocytes, and in some birds, cocci consistent with *E. cecorum* colonized and lined these clefts. By week 3, the inflammation targeted the cartilage clefts infected with bacteria. The inflammation caused bone remodeling and compression of the overlying spinal cord, forming the characteristic ES lesion that frequently causes broiler paralysis. A positive correlation ($r = 0.65$) was observed between OCD severity scores and *E. cecorum* infection. All ES-positive chickens had the highest OCD scores (3 or 4).

Reevaluating the data presented by Chen et al. (2018), higher long bone RA and OCD lesions were observed ($P < 0.001$) already at hatch when suboptimal incubation was used. The suboptimal incubation had EST below 37.6°C during the first 3 days and above 38.4°C during the last 3 days of incubation by not adjusting the machine temperatures. These constant incubator temperatures are the most common situation in multi-stage incubation systems. Incubation conditions during embryo development may affect bone, cartilage, and immunity development (Yeboah et al., 2019; Wijnen et al., 2020b). All these factors can make chickens from suboptimal incubation more susceptible to ES lesions.

Wijnen et al. (2022) reported that high EST (38.9°C) during the second week of incubation decreased the systemic resilience of chickens to *E. coli* infection induced at day 8 of age. Consequently, cellular immunocompetence against bacteria can be impaired by suboptimal incubation and may increase the susceptibility to vertebral and bone infections. The effects of incubation temperature on chondronecrosis, osteonecrosis, and femoral head necrosis should be evaluated considering the previously discussed reports. Other incubation factors may also affect infectious bone diseases. Van der Pol et al. (2017) observed that the continuous darkness during incubation, as commercially practiced nowadays, had more chondronecrosis and osteomyelitis than 16L:8D.

Olkowski et al. (2019) discussed the recent emergence of cervical scoliosis and torticollis that has been observed not only in chickens but also in newly hatched chicks and non-hatched embryos. Chick incidence at hatch ranges from 0.1% to more than 1%, and in fully developed embryos that failed to hatch, about 4% to 5%. This can be another skeletal developmental problem to investigate from the incubation management standpoint.

3.6 Stressful conditions during transportation to the farm

The production of high-quality chicks includes the chick management post-hatch and transportation to the farm. Perinatal stressful conditions can harm bone development during this period of fast elongation. Oviedo-Rondón et al. (2009b) evaluated the effects of low temperature during the early incubation, high temperature during late incubation, and stressful transportation conditions with no ventilation from the hatchery to the farm for 4 h. Transportation stress caused by lack of ventilation increased chick box temperatures to almost 38°C. Bone development, gait scores, and leg health disorders were assessed at 41 days of age. The late high temperatures reduced the relative weight of femurs and shanks at hatch and, at 41 days, increased the incidence of crooked toes and, together with transportation stress, increased broiler lameness (gait score 2). The transportation stress caused a greater incidence of twisted legs and immobility. The increased RA caused by stressors in the perinatal period caused skeletal development issues. Avoiding stress for chicks in the hatchers and immediately after hatch may minimize leg health issues.

4 Conclusion

Skeletal development starts early during embryogenesis. All incubation factors may affect skeletal development with important impacts on its development post-hatch. Temperature, O_2, and CO_2 concentrations during incubation impact embryo endochondral ossification, bone elongation, RA, and incidence of TD, valgus, twisted legs, and other bone and leg issues in broilers. Incubation temperatures also affect other organs and systems that can also be related to post-hatch skeletal development, like muscle, tendons, the liver, kidneys, immunity, and the gastrointestinal tract and their functions. Modifications of temperatures or gas concentrations have the most significant impacts when applied during the first 3 days of incubation or near the plateau stage of O_2 consumption 3 days before hatch. White cool light during incubation affects bone development. However, continuous lighting has had harmful effects during incubation and after hatching. Lighting 12L:12D during incubation seems to have a mild positive effect on bone development. Still, it is not proven to cause significant bone development modifications or reduce leg health issues. Monochromatic lights have not had positive effects on bone development. Prolonged egg storage (> 7 days) can diminish embryo livability, chick quality, broiler performance, and skeletal health. SPIDES treatment can minimize the detrimental effects of egg storage and improve skeletal development. Suboptimal incubation could cause increased RA of long bones in chicks. This effect can contribute to the genesis of OCD lesions that are highly correlated with susceptibility to infections with *E. cecorum* and ES lesions. Finally, stressful conditions in the perinatal period affect skeletal development. Transportation of chicks to the farm is a hatchery responsibility. Heat stress and lack of ventilation during transportation can increase twisted leg incidence and gait score issues in broilers. Incubation is critical for skeletal development, welfare, and health in poultry.

5 Acknowledgements

The author would like to express his gratitude to all colleagues who have collaborated with research projects on leg health, bone development, locomotion, incubation, and embryo development in the past 18 years. I want to give special recognition to Dr. Michael Wineland and in memory of Dr. John T. Brake, my long-term collaborators. I also want to acknowledge the sponsorship and collaboration of Cobb-Vantress Inc., Aviagen Turkeys, the US Poultry and Egg Association, Chick Master Company, Mountaire Farms, Perdue Farms, and several other poultry companies that have participated in multiple research projects.

6 Where to look for further information

Readers interested in learning more about avian embryo skeletal development should read the book chapters Incubation Physiology written by Tazawa and Whittow (2000); Poultry Bone Development and Bone Disorders, written by Pines and Reshef (2015), and Avian Bone Physiology and Poultry Bone Disorders written by Rath and Durairaj (2022) in the book Sturkie's Avian Physiology from Academic Press.

7 References

Agyekum, G., Okai, M. A., Tona, J. K., Donkoh, A. and Hamidu, J. A. (2022). Impact of incubation temperature profile on chick quality, bone, and immune system during the late period of incubation of Cobb 500 broiler strain, *Poult. Sci.* 101(9), 101999. doi:10.1016/j.psj.2022.101999.

Aygün, A. and Narinç, D. (2016). The effects of thermal manipulations during embryogenesis of broiler chicks on the growth of embryo and skeletal traits, *AIP Conf. Proc.* 1726(1), 020015. doi:10.1063/1.4945841.

Barbosa, V. M., Rocha, J. S. R., Pompeu, M. A., Martins, N. R. S., Baiao, N. C., Lara, L. J. C., Batista, J. V. M. S. P. and Leite, R. C. (2013). The effects of relative humidity and turning in incubators machines on the incubation yield and chick performance, *Worlds Poult. Sci. J.* 69(1), 89–98. doi:10.1017/S0043933913000081.

Bilalissi, A., Meteyake, H. T., Kouame, Y. A. E., Oke, O. E., Lin, H., Onagbesan, O., Decuypere, E. and Tona, K. (2022). Effects of pre-incubation storage duration and nonventilation incubation procedure on embryonic physiology and post-hatch chick performance, *Poult. Sci.* 101(5), 101810. doi:10.1016/j.psj.2022.101810.

Boleli, I. C., Morita, V. S., Matos, J. B., Jr., Thimotheo, M. I. and Almeida, V. R. (2016). Poultry egg incubation: Integrating and optimizing production efficiency, *Braz. J. Poult. Sci.* doi:10.1590/1806-9061-2016-0292.

Borst, L. B., Suyemoto, M. M., Sarsour, A. H., Harris, M. C., Martin, M. P., Strickland, J. D., Oviedo, E. O. and Barnes, H. J. (2017). Pathogenesis of enterococcal spondylitis caused by *Enterococcus cecorum* in broiler chickens, *Vet. Pathol.* 54(1), 61–73. doi:10.1177/0300985816658098.

Bradshaw, R. H., Kirkden, R. D. and Broom, D. M. (2002). A review of the aetiology and pathology of leg weakness in broilers in relation to welfare, *Avian Poult. Bio. Rev.* 13(2), 45–103.

Braga, J. F. V. (2018). Vertebral osteomyelitis in broilers: A review, *Br. J. Poult. Sci.* 20, 605–616. doi:10.1590/1806-9061-2017-0690.

Branum, S. R., Tazawa, H. and Burggren, W. W. (2016). Phenotypic developmental plasticity induced by preincubation egg storage in chicken embryos (*Gallus gallus domesticus*), *Physiol. Rep.* 4(4), e12712. doi:10.14814/phy2.12712.

Burggren, W. W., Santin, J. F. and Rojas Antich, M. (2016). Cardio-respiratory development in bird embryos: New insights from a venerable animal model, *R. Bras. Zootech.* 45, 11. doi:10.1590/s1806-92902016001100010.

Çapar Akyüz, H. and Onbaşılar, E. E. (2020). Non-infectious skeletal disorders in broilers, *Worlds Poult. Sci. J.* 76(3), 611–623. doi:10.1080/00439339.2020.1759388.

Caplan, A. I. (2007). Bone development, *Ciba Found. Symp. 136-Cell and Molecular Biology of Vertebrate Hard Tissues: Cell and Molecular Biology of Vertebrate Hard Tissues: Ciba Found. Symposium 136*. Chichester, UK. John Wiley & Sons, Ltd., 3–21.

Chan, T. and Burggren, W. (2005). Hypoxic incubation creates differential morphological effects during specific developmental critical windows in the embryo of the chicken (*Gallus gallus*), *Respir. Physiol. Neurobiol.* 145(2–3), 251–263. doi:10.1016/j.resp.2004.09.005.

Chen, L. R., Suyemoto, M. M., Sarsour, A. H., Cordova, H. A., Oviedo-Rondón, E. O., Barnes, H. J. and Borst, L. B. (2018). Prevalence and severity of osteochondrosis of the free thoracic vertebra in three modern broiler strains and the Athens Canadian Random Bred control broiler, *Avian Pathol.* 47(2), 152–160. doi:10.1080/03079457.2017.1388499.

Christense, V. L., Winelan, M. J., Grimes, J. L., Oviedo, E. O., Mozdziak, P. S., Ort, D. T. and Mann, K. M. (2007). Effect of incubator temperature and oxygen concentration at the plateau stage in oxygen consumption on turkey embryo muscle growth and development, *Int. J. Poult. Sci.* 6(6), 406–412. doi:10.3923/ijps.2007.406.412.

Da Costa, M. J., Oviedo-Rondón, E. O., Wineland, M. and Jeffrey, D. (2016). Pathogeny of fatigued walking condition in Pekin ducks, *Avian Dis.* 60(4), 731–738. doi:10.1637/11292-100315-RegR.

De Smit, L., Bruggeman, V., Tona, J. K., Debonne, M., Onagbesan, O., Arckens, L., De Baerdemaeker, J. and Decuypere, E. (2006). Embryonic developmental plasticity of the chick: Increased CO_2 during early stages of incubation changes the developmental trajectories during prenatal and postnatal growth, *Comp. Biochem. Physiol. A Mol. Integr. Physiol.* 145(2), 166–175. doi:10.1016/j.cbpa.2006.06.046.

Decuypere, E., Onagbesan, O., De Smit, L., Tona, K., Everaert, N., Witters, A., Debonne, M., Verhoelst, E., Buyse, J., Hassanzadeh, M. and De Baerdemaeker, J. (2006). Hypoxia and hypercapnia during incubation of chicken eggs: Effects on development and subsequent performance, *12th Eur. Poultr. Conference*, Verona, Italy, 10–14.

Droual, R., Bickford, A. A. and Farver, T. B. (1991). Scoliosis and tibiotarsal deformities in broiler chickens, *Avian Dis.* 35(1), 23-30. doi:10.2307/1591290.

Dymond, J., Vinyard, B., Nicholson, A. D., French, N. A. and Bakst, M. R. (2013). Short periods of incubation during egg storage increase hatchability and chick quality in long-stored broiler eggs, *Poult. Sci.* 92(11), 2977–2987. doi:10.3382/ps.2012-02816.

Ebeid, T. A., Twfeek, F. A., Assar, M. H., Bealish, A. M., Abd El-Karim, R. E. and Ragab, M. (2017). Influence of pre-storage incubation on hatchability traits, thyroid hormones, antioxidative status and immunity of newly hatched chicks at two chicken breeder flock ages, *Animal* 11(11), 1966–1974. doi:10.1017/S1751731117000738.

El-Hanoun, A. M., Rizk, R. E., Shahein, E. H. A., Hassan, N. S. and Brake, J. (2012). Effect of incubation humidity and flock age on hatchability traits and posthatch growth in Pekin ducks, *Poult. Sci.* 91(9), 2390–2397. doi:10.3382/ps.2011-02075.

Enomoto-Iwamoto, M., Kitagaki, J., Koyama, E., Tamamura, Y., Wu, C., Kanatani, N., Koike, T., Okada, H., Komori, T., Yoneda, T., Church, V., Francis-West, P. H., Kurisu, K., Nohno, T., Pacifici, M. and Iwamoto, M. (2002). The Wnt antagonist Frzb-1 regulates chondrocyte maturation and long bone development during limb skeletogenesis, *Dev. Biol.* 251(1), 142–156. doi:10.1006/dbio.2002.0802.

Eriksen, M. S., Haug, A., Torjesen, P. A. and Bakken, M. (2003). Prenatal exposure to corticosterone impairs embryonic development and increases fluctuating asymmetry in chickens (*Gallus gallus domesticus*), *Br. Poult. Sci.* 44(5), 690–697. doi:10.1080/00071660310001643660.

Eusebio-Balcazar, P., Oviedo-Rondón, E. O., Wineland, M. J., Pérez Serrano, M. and Brake, J. (2014). Effects of broiler breeder-feeding programme and feeder space change at photostimulation using maize-or wheat-based diets on eggshell properties and progeny bone development, *Br. Poult. Sci.* 55(3), 380–390. doi:10.1080/00071668.2014.921665.

Everaert, N., Kamers, B., Witters, A., De Smit, L., Debonne, M., Decuypere, E. and Bruggeman, V. (2007). Effect of four percent carbon dioxide during the second half of incubation on embryonic development, hatching parameters, and posthatch growth, *Poult. Sci.* 86(7), 1372–1379. doi:10.1093/ps/86.7.1372.

French, N. A. (1997). Modeling incubation temperature: The effects of incubator design, embryonic development, and egg size, *Poult. Sci.* 76(1), 124–133. doi:10.1093/ps/76.1.124.

French, N. A. (2009). The critical importance of incubation temperature, *Avian Biol. Res.* 2(1-2), 55–59. doi:10.3184/175815509X431812.

Gocsik, É., Silvera, A. M., Hansson, H., Saatkamp, H. W. and Blokhuis, H. J. (2017). Exploring the economic potential of reducing broiler lameness, *Br. Poult. Sci.* 58(4), 337–347. doi:10.1080/00071668.2017.1304530.

Goliomytis, M., Tsipouzian, T. and Hager-Theodorides, A. L. (2015). Effects of egg storage on hatchability, chick quality, performance and immunocompetence parameters of broiler chickens, *Poult. Sci.* 94(9), 2257–2265. doi:10.3382/ps/pev200.

Gouveia, C. H. A., Miranda-Rodrigues, M., Martins, G. M. and Neofiti-Papi, B. (2018). Thyroid hormone and skeletal development, *Vitam. Horm.* 106, 383–472. doi:10.1016/bs.vh.2017.06.002.

Granquist, E. G., Vasdal, G., De Jong, I. C. and Moe, R. O. (2019). Lameness and its relationship with health and production measures in broiler chickens, *Animal* 13(10), 2365–2372. doi:10.1017/S1751731119000466.

Groves, P. J. and Muir, W. I. (2014). A meta-analysis of experiments linking incubation conditions with subsequent leg weakness in broiler chickens, *PLoS ONE* 9(7), e102682. doi:10.1371/journal.pone.0102682.

Groves, P. J. and Muir, W. I. (2017). Earlier hatching time predisposes Cobb broiler chickens to tibial dyschondroplasia, *Animal* 11(1), 112–120. doi:10.1017/S1751731116001105.

Gualhanone, A., Furlan, R. L., Fernandez-Alarcon, M. F. and Macari, M. (2012). Effect of breeder age on eggshell thickness, surface temperature, hatchability and chick weight, *Braz. J. Poult. Sci.* 14, 9–14. doi:10.1590/S1516-635X2012000100002.

Guo, Y., Tang, H., Wang, X., Li, W., Wang, Y., Yan, F., Kang, X., Li, Z. and Han, R. (2019). Clinical assessment of growth performance, bone morphometry, bone quality, and serum indicators in broilers affected by valgus-varus deformity, *Poult. Sci.* 98(10), 4433–4440. doi:10.3382/ps/pez269.

Güz, B. C., Molenaar, R., De Jong, I. C., Kemp, B., van Krimpen, M. and van den Brand, H. (2020). Effects of eggshell temperature pattern during incubation on tibia characteristics of broiler chickens at slaughter age, *Poult. Sci.* 99(6), 3020–3029. doi:10.1016/j.psj.2019.12.042.

Güz, B. C., Molenaar, R., De Jong, I. C., Kemp, B., van Krimpen, M. and van den Brand, H. (2021). Effects of green light emitting diode light during incubation and dietary organic macro and trace minerals during rearing on tibia characteristics of broiler chickens at slaughter age, *Poult. Sci.* 100(2), 707–720. doi:10.1016/j.psj.2020.11.042.

Hamidu, J. A., Fasenko, G. M., Feddes, J. J. R., O'Dea, E. E., Ouellette, C. A., Wineland, M. J. and Christensen, V. L. (2007). The effect of broiler breeder genetic strain and parent flock age on eggshell conductance and embryonic metabolism, *Poult. Sci.* 86(11), 2420–2432. doi:10.3382/ps.2007-00265.

Hamidu, J. A., Fasenko, G. M., Guan, L., Barreda, D. R. and Feddes, J. J. (2011). Influence of parent flock age on embryonic metabolism in modern turkey strains, *Poult. Sci.* 90(2), 426–434. doi:10.3382/ps.2010-00967.

Hamidu, J. A., Torres, C. A., Johnson-Dahl, M. L. and Korver, D. R. (2018). Physiological response of broiler embryos to different incubator temperature profiles and maternal flock age during incubation. 1. Embryonic metabolism and day-old chick quality, *Poult. Sci.* 97(8), 2934–2946. doi:10.3382/ps/pey089.

Hammond, C. L., Simbi, B. H. and Stickland, N. C. (2007). In ovo temperature manipulation influences embryonic motility and growth of limb tissues in the chick (Gallus gallus), *J. Exp. Biol.* 210(15), 2667–2675. doi:10.1242/jeb.005751.

Hopcroft, R. L., Groves, P. J. and Muir, W. I. (2020). Changes to Cobb 500 chick characteristics, bone ash, and residual yolk mineral reserves during time spent in the hatcher, *Poult. Sci.* 99(4), 2176–2184. doi:10.1016/j.psj.2019.11.034.

Huang, S. C., Rehman, M. U., Lan, Y. F., Qiu, G., Zhang, H., Iqbal, M. K., Luo, H. Q., Mehmood, K., Zhang, L. H. and Li, J. K. (2017a). Tibial dyschondroplasia is highly associated with suppression of tibial angiogenesis through regulating the HIF-1α/VEGF/VEGFR signaling pathway in chickens, *Sci. Rep.* 7(1), 9089. doi:10.1038/s41598-017-09664-6.

Huang, S., Tong, X., Rehman, M. U., Wang, M., Zhang, L., Wang, L., Li, J. and Yang, S. (2017b). Oxygen supplementation ameliorates tibial development via stimulating vascularization in Tibetan chickens at high altitudes, *Int. J. Biol. Sci.* 13(12), 1547–1559. doi:10.7150/ijbs.22670.

Huang, S., Rehman, M. U., Qiu, G., Luo, H., Iqbal, M. K., Zhang, H., Mehmood, K. and Li, J. (2018). Tibial dyschondroplasia is closely related to suppression of expression of hypoxia-inducible factors 1α, 2α, and 3α in chickens, *J. Vet. Sci.* 19(1), 107–115. doi:10.4142/jvs.2018.19.1.107.

Kerschnitzki, M., Akiva, A., Shoham, A. B., Asscher, Y., Wagermaier, W., Fratzl, P., Addadi, L. and Weiner, S. (2016). Bone mineralization pathways during the rapid growth of embryonic chicken long bones, *J. Struct. Biol.* 195(1), 82–92. doi:10.1016/j.jsb.2016.04.011.

Leksrisompong, N., Romero-Sanchez, H., Plumstead, P. W., Brannan, K. E. and Brake, J. (2007). Broiler incubation. 1. Effect of elevated temperature during late incubation on body weight and organs of chicks, *Poult. Sci.* 86(12), 2685–2691. doi:10.3382/ps.2007-00170.

Li, J., Ma, Y., Zhang, L., Cai, C., Guo, Y., Zhang, Z., Li, D., Tian, Y., Kang, X., Han, R. and Jiang, R. (2022). Valgus-varus deformity induced abnormal tissue metabolism, inflammatory damage and apoptosis in broilers, *Br. Poult. Sci.* doi:10.1080/00071668.2022.2121640.

Liu, C., Zheng, W., Zhu, L., Tong, Q. and Li, D. (2022). Effect of elevated carbon dioxide on chicken eggs during the early and late incubation periods, *Animal* 16(4), 100499. doi:10.1016/j.animal.2022.100499.

Loh, B., Maier, I., Winar, A., Janke, O. and Tzschentke, B. (2004). Prenatal development of epigenetic adaptation processes in poultry: Changes in metabolic and neuronal thermoregulatory mechanisms, *Avian Poult. Biol. Rev.* 15(3), 119–128. doi:10.3184/147020604783637976.

Lourens, A., van den Brand, H., Heetkamp, M. J. W., Meijerhof, R. and Kemp, B. (2007). Effects of eggshell temperature and oxygen concentration on embryo growth and metabolism during incubation, *Poult. Sci.* 86(10), 2194–2199. doi:10.1093/ps/86.10.2194.

Lourens, A., van den Brand, H., Meijerhof, R. and Kemp, B. (2005). Effect of eggshell temperature during incubation on embryo development, hatchability, and posthatch development, *Poult. Sci.* 84(6), 914–920. doi:10.1093/ps/84.6.914.

Maatjens, C. M., Reijrink, I. A. M., Molenaar, R., van der Pol, C. W., Kemp, B. and van den Brand, H. (2014). Temperature and CO_2 during the hatching phase. I. Effects on chick quality and organ development, *Poult. Sci.* 93(3), 645–654. doi:10.3382/ps.2013-03490.

Maatjens, C. M., van Roovert-Reijrink, I. A. M., Engel, B., van der Pol, C. W., Kemp, B. and van den Brand, H. (2016). Temperature during the last week of incubation. I. Effects on hatching pattern and broiler chicken embryonic organ development, *Poult. Sci.* 95(4), 956–965. doi:10.3382/ps/pev447.

McNabb, F. M. A. (2006). Avian thyroid development and adaptive plasticity, *Gen. Comp. Endocrinol.* 147(2), 93–101. doi:10.1016/j.ygcen.2005.12.011.

Molenaar, R., van den Anker, I., Meijerhof, R., Kemp, B. and van den Brand, H. (2011a). Effect of eggshell temperature and oxygen concentration during incubation on the developmental and physiological status of broiler hatchlings in the perinatal period, *Poult. Sci.* 90(6), 1257–1266. doi:10.3382/ps.2010-00684.

Molenaar, R., Hulet, R., Meijerhof, R., Maatjens, C. M., Kemp, B. and van den Brand, H. (2011b). High eggshell temperatures during incubation decrease growth performance and increase the incidence of ascites in broiler chickens, *Poult. Sci.* 90(3), 624–632. doi:10.3382/ps.2010-00970.

Møller, A. P. (2006). A review of developmental instability, parasitism and disease: Infection, genetics and evolution, *Infect. Genet. Evol.* 6(2), 133–140. doi:10.1016/j.meegid.2005.03.005.

Møller, A. P., Sanotra, G. S. and Vestergaard, K. S. (1999). Developmental instability and light regime in chickens (*Gallus gallus*), *Anim. Behav.* 62(1), 57–71. doi:10.1016/S0168-1591(98)00213-5.

Monsoro-Burq, A. H. and Levin, M. (2018). Avian models and the study of invariant asymmetry: How the chicken and the egg taught us to tell right from left, *Int. J. Dev. Biol.* 62(1–2), 63–77. doi:10.1387/ijdb.180047ml.

Morita, V. S., Almeida, A. R., Junior, J. B. M., Vicentini, T. I., Zanirato, G. L. and Boleli, I. C. (2020). Neither altered incubation temperature during fetal development nor preferred rearing temperature improves leg bone characteristics of broilers, *J. Therm. Biol.* 93, 102726. doi:10.1016/j.jtherbio.2020.102726.

Mortola, J. P. and Labbè, K. (2005). Oxygen consumption of the chicken embryo: Interaction between temperature and oxygenation, *Respir. Physiol. Neurobiol.* 146(1), 97–106. doi:10.1016/j.resp.2004.10.011.

Muir, W. I. and Groves, P. J. (2018). Incubation and hatch management: Consequences for bone mineralization in Cobb 500 meat chickens, *Animal* 12(4), 794–801. doi:10.1017/S1751731117001938.

Muir, W. I. and Groves, P. J. (2019). The leg strength of two commercial strains of meat chicken subjected to different incubation profiles, *Animal* 13(7), 1489–1497. doi:10.1017/S1751731118002999.

Nangsuay, A., Meijerhof, R., van den Anker, I., Heetkamp, M. J. W., Kemp, B. and van den Brand, H. (2017). Effects of breeder age, strain, and eggshell temperature on nutrient metabolism of broiler embryos, *Poult. Sci.* 96(6), 1891–1900. doi:10.3382/ps/pew417.

Okur, N., Eratalar, S. A. and Yaman, A. (2017). Effect of high shell temperature and position in the incubator on embryonic mortality and hatchability of fertile broiler eggs, *Eur. Poult. Sci.* 81. doi:10.1399/eps.2017.206.

Olkowski, A., Wojnarowicz, C., Olkowski, B. and Laarvel, B. (2019). Cervical scoliosis and torticollis: A novel skeletal anomaly in broiler chickens, *Acta Vet. Scand.* 61(1), 47. doi:10.1186/s13028-019-0482-0.

Oviedo-Rondón, E. O. (2008). Predisposing factors that affect walking ability in turkeys and broilers, *34th Annual Carolina Poultr. Nutr. Conference*, Raleigh, NC, USA, 78–91.

Oviedo-Rondón, E. O., Halper, J., Wineland, M. J., Ferket, P., Cummock, J. and Kusovschi, A. (2010). Incubation conditions affect tendon development in broilers. *AVMA-AAAP Annual Meeting* (Vol. CD), Atlanta, GA.

Oviedo-Rondón, E. O., Lascelles, B. D. X., Arellano, C., Mente, P. L., Eusebio-Balcazar, P., Grimes, J. L. and Mitchell, A. D. (2017). Gait parameters in four strains of turkeys and correlations with bone strength, *Poult. Sci.* 96(7), 1989–2005. doi:10.3382/ps/pew502.

Oviedo-Rondón, E. O., Mente, P. L., Arellano, C., Lascelles, B. D. X. and Mitchell, A. (2018). Influence of gait on bone strength in turkeys with leg defects, *Poult. Sci.* 97(8), 2621–2637. doi:10.3382/ps/pey142.

Oviedo-Rondón, E. O., Small, J., Wineland, M. J., Christensen, V. L., Grimes, J. L., Funderburk, S. V. L., Ort, D. T. and Mann, K. M. (2008a). Effects of incubator temperature and oxygen concentration during the plateau stage of oxygen consumption on turkey embryo long bone development, *Poult. Sci.* 87(8), 1484–1492. doi:10.3382/ps.2007-00470.

Oviedo-Rondón, E. O., Small, J., Wineland, M. J., Christensen, V. L., Mozdziak, P. S., Koci, M. D., Funderburk, S. V. L., Ort, D. T. and Mann, K. M. (2008b). Broiler embryo bone development is influenced by incubator temperature, oxygen concentration and eggshell conductance at the plateau stage in oxygen consumption, *Br. Poult. Sci.* 49(6), 666–676. doi:10.1080/00071660802433149.

Oviedo-Rondón, E. O., Velleman, S. G. and Wineland, M. J. (2020). The role of incubation conditions in the onset of avian myopathies, *Front. Physiol.* 11, 545045. doi:10.3389/fphys.2020.545045.

Oviedo-Rondón, E. O. and Wineland, M. J. (2011). Incubation distress easily leads to splayed legs. World's poultry. Available at: http://www.poultryworld.net/poultry/incubation-distress-easily-leads-to-splayed-legs/ (accessed 6 January 2023).

Oviedo-Rondón, E. O., Wineland, M. J., Funderburk, S., Small, J., Cutchin, H. and Mann, M. (2009a). Incubation conditions affect leg health in large, high-yield broilers, *J. Appl. Poult. Res.* 18(3), 640–646. doi:10.3382/japr.2008-00127.

Oviedo-Rondón, E. O., Wineland, M. J., Small, J., Cutchin, H., McElroy, A., Barri, A. and Martin, S. (2009b). Effect of incubation temperatures and chick transportation conditions on bone development and leg health, *J. Appl. Poult. Res.* 18(4), 671–678. doi:10.3382/japr.2008-00135.

Özcan, S. E., Andriessens, S. and Berckmans, D. (2010). Computational study of the heat transfer of an avian egg in a tray, *Poult. Sci.* 89(4), 776–784. doi:10.3382/ps.2009-00230.

Oznurlu, Y., Sur, E., Ozaydin, T., Celik, I. and Uluisik, D. (2016). Histological and histochemical evaluations on the effects of high incubation temperature on the embryonic development of tibial growth plate in broiler chickens, *Microsc. Res. Tech.* 79(2), 106–110. doi:10.1002/jemt.22611.

Pines, M. and Reshef, R. (2015). Poultry bone development and bone disorders. In: Scanes, C. G. (Ed.), *Sturkie's Avian Physiology*. San Francisco: Academic Press, 367–377.

Radzak, K. N., Putnam, A. M., Tamura, K., Hetzler, R. K. and Stickley, C. D. (2017). Asymmetry between lower limbs during rested and fatigued state running gait in healthy individuals, *Gait Posture* 51, 268–274. doi:10.1016/j.gaitpost.2016.11.005.

Rahn, H. (1981). Gas exchange of avian eggs with special reference to turkey eggs, *Poult. Sci.* 60(9), 1971–1980. doi:10.3382/ps.0601971.

Rath, N. C. and Durairaj, V. (2022). Avian bone physiology and poultry bone disorders. In: Scanes, C. G. and Dridi, S. (Eds.), *Sturkie's Avian Physiology*. San Francisco: Academic Press, 529–543.

Salamon, A. (2020). Fertility and hatchability in goose eggs: A review, *Int. J. Poult. Sci.* 19(2), 51–65.

Sanotra, G. S., Lawson, L. G., Vestergaard, K. S. and Thomsen, M. G. (2001). Influence of stocking density on tonic immobility, lameness, and tibial dyschondroplasia in broilers, *J. Appl. Anim. Welf. Sci.* 4(1), 71–87. doi:10.1207/S15327604JAWS0401_4.

Shah, T. and Özkan, S. (2022). Effects of thermal manipulation and photoperiodic lighting during incubation on hatching performance, hatching time, chick quality and organ growth, *Ege Univ. Ziraat Fak. Derg.* 59(1), 17–31. doi:10.20289/zfdergi.938921.

Shao, Y. Y., Wang, L. and Ballock, R. T. (2006). Thyroid hormone and the growth plate, *Rev. Endocr. Metab. Disord.* 7(4), 265–271. doi:10.1007/s11154-006-9012-2.

Shim, M. Y. and Pesti, G. M. (2011). Effects of incubation temperature on the bone development of broilers, *Poult. Sci.* 90(9), 1867–1877. doi:10.3382/ps.2010-01242.

Skeffington, K. L., Beck, C., Itani, N., Niu, Y., Shaw, C. J. and Giussani, D. A. (2020). Hypertension programmed in adult hens by isolated effects of developmental hypoxia *in ovo*, *Hypertension* 76(2), 533–544. doi:10.1161/HYPERTENSIONAHA.120.15045.

Smith, F., Hu, D., Young, N. M., Lainoff, A. J., Jamniczky, H. A., Maltepe, E., Hallgrimsson, B. and Marcucio, R. S. (2013). The effect of hypoxia on facial shape variation and disease phenotypes in chicken embryos, *Dis. Model. Mech.* 6(4), 915–924. doi:10.1242/dmm.011064.

Sözcü, A. and İpek, A. (2015). Acute and chronic eggshell temperature manipulations during hatching term influence hatchability, broiler performance, and ascites incidence, *Poult. Sci.* 94(2), 319–327. doi:10.3382/ps/peu080.

Sözcü, A., İpek, A. and van den Brand, H. (2022). Eggshell temperature during early and late incubation affects embryo and hatchling development in broiler chicks, *Poult. Sci.* 101(10), 102054. doi:10.1016/j.psj.2022.102054.

Tazawa, H. and Whittow, G. C. (2000). Incubation physiology. In: Whittow, G. C. (Ed.), *Sturkie's Avian Physiology*. San Francisco: Academic Press, 617–634.

Teitelbaum, S. L. (2000). Bone resorption by osteoclasts, *Science* 289(5484), 1504–1508. doi:10.1126/science.289.5484.1504.

Tona, K., Bamelis, F., Coucke, W., Bruggeman, V. and Decuypere, E. (2001). Relationship between broiler breeder's age and egg weight loss and embryonic mortality during incubation in large-scale conditions, *J. Appl. Poult. Res.* 10(3), 221–227. doi:10.1093/japr/10.3.221.

Uddin, Z. and Hamidu, J. A. (2014). Prolong egg storage affects broiler breeder embryonic metabolism and chick quality, *J. Anim. Sci. Adv.*, 4(7), 973-977. doi:10.5455/jasa.20140713053547.

Van Brecht, A., Hens, H., Lemaire, J. L., Aerts, J. M., Degraeve, P. and Berckmans, D. (2005). Quantification of the heat exchange of chicken eggs, *Poult. Sci.* 84(3), 353-361. doi:10.1093/ps/84.3.353.

van der Eerden, B. C. J., Karperien, M. and Wit, J. M. (2003). Systemic and local regulation of the growth plate, *Endocr. Rev.* 24(6), 782–801. doi:10.1210/er.2002-0033.

van der Pol, C. W., van Roovert-Reijrink, I. A. M., Aalbers, G., Kemp, B. and van den Brand, H. (2017). Incubation lighting schedules and their interaction with matched or mismatched post hatch lighting schedules: Effects on broiler bone development and leg health at slaughter age, *Res. Vet. Sci.* 114, 416–422. doi:10.1016/j.rvsc.2017.07.013.

van der Pol, C. W., van Roovert-Reijrink, I. A. M., Gussekloo, S. W. S., Kranenbarg, S., Leon-Kloosterziel, K. M., van Eijk-Priester, M. H., Zeman, M., Kemp, B. and van den Brand, H. (2019a). Effects of lighting schedule during incubation of broiler chicken embryos on leg bone development at hatch and related physiological characteristics, *PLoS ONE* 14(8), e0221083. doi:10.1371/journal.pone.0221083.

van der Pol, C. W., van Roovert-Reijrink, I. A. M., Maatjens, C. M., Gussekloo, S. W. S., Kranenbarg, S., Wijnen, J., Pieters, R. P. M., Schipper, H., Kemp, B. and van den Brand, H. (2019b). Light-dark rhythms during incubation of broiler chicken embryos and their effects on embryonic and post hatch leg bone development, *PLoS ONE* 14(1), e0210886. doi:10.1371/journal.pone.0210886.

van der Pol, C. W., van Roovert-Reijrink, I. A. M., Maatjens, C. M., van den Anker, I., Kemp, B. and van den Brand, H. (2014). Effect of eggshell temperature throughout incubation on broiler hatchling leg bone development, *Poult. Sci.* 93(11), 2878–2883. doi:10.3382/ps.2014-04210.

Villagómez-Cortés, J. A., Guevara-Torres, B. L., Landin-Grandvallet, L. A. and Tirado-Madrid, A. (2021). Comparison of locomotion problems and its economic impact on Cobb and Ross broiler strains, *Vet. Sci. Res.* 3(2), 4126. doi:10.30564/vsr.v3i2.4126.

Visschedijk, A. H. J., Tazawa, H. and Piiper, J. (1985). Variability of shell conductance and gas exchange of chicken eggs, *Respir. Physiol.* 59(3), 339-345. doi:10.1016/0034-5687(85)90137-9.

Wang, D., Gilbert, J. R., Zhang, X., Zhao, B., Ker, D. F. E. and Cooper, G. M. (2020). Calvarial versus long bone: Implications for tailoring skeletal tissue engineering, *Tissue Eng. Part B Rev.* 26(1), 46-63. doi:10.1089/ten.TEB.2018.0353.

Werner, C., Wecke, C., Liebert, F. and Wicke, M. (2010). Increasing the incubation temperature between embryonic day 7 and 10 has no influence on the growth and slaughter characteristics as well as meat quality of broilers, *Animal* 4(5), 810-816. doi:10.1017/S1751731109991698.

Wideman Jr., R. F. (2016). Bacterial chondronecrosis with osteomyelitis and lameness in broilers: A review, *Poult. Sci.* 95(2), 325-344. doi:10.3382/ps/pev320.

Wijnen, H. J., Molenaar, R., van Roovert-Reijrink, I. A. M., van der Pol, C. W., Kemp, B. and van den Brand, H. (2020a). Effects of incubation temperature pattern on broiler performance, *Poult. Sci.* 99(8), 3897-3907. doi:10.1016/j.psj.2020.05.010.

Wijnen, H. J., van den Brand, H., Lammers, A., van Roovert-Reijrink, I. A. M., van der Pol, C. W., Kemp, B. and Molenaar, R. (2020b). Effects of eggshell temperature pattern during incubation on primary immune organ development and broiler immune response in later life, *Poult. Sci.* 99(12), 6619-6629. doi:10.1016/j.psj.2020.09.088.

Wijnen, H. J., van der Pol, C. W., Papanikolaou, A., Lammers, A., Kemp, B., van den Brand, H., Perricone, V., Matthijs, M. G. R. and Molenaar, R. (2022). Broiler resilience to colibacillosis is affected by incubation temperature and post-hatch feeding strategy, *Poult. Sci.* 101(10), 102092. doi:10.1016/j.psj.2022.102092.

Williams, B., Waddington, D., Murray, D. H. and Farquharson, C. (2004). Bone strength during growth: Influence of growth rate on cortical porosity and mineralization, *Calcif. Tissue Int.* 74(3), 236-245. doi:10.1007/s00223-002-2124-0.

Wineland, M. W., Christensen, V. L., Yildrum, I., Fairchild, B. D., Mann, K. M. and Ort, D. T. (2006a). Incubator temperature and oxygen concentration at the plateau stage in oxygen consumption affects intestinal maturation of broiler chicks, *Int. J. Poult. Sci.* 5(3), 229-240. doi:10.3923/ijps.2006.229.240.

Wineland, M. W., Christensen, V. L., Yildrum, I., Fairchild, B. D., Mann, K. M. and Ort, D. T. (2006b). Incubator environment interacts with genetic line of broiler at the plateau stage to affect embryo plasma thyroxine and triiodothyronine concentrations, *Int. J. Poult. Sci.* 5(8), 714-722. doi:10.3923/ijps.2006.714.722.

Yair, R., Cahaner, A., Uni, Z. and Shahar, R. (2017). Maternal and genetic effects on broiler bone properties during incubation period, *Poult. Sci.* 96(7), 2301-2311. doi:10.3382/ps/pex021.

Yair, R. and Uni, Z. (2011). Content and uptake of minerals in the yolk of broiler embryos during incubation and effect of nutrient enrichment, *Poult. Sci.* 90(7), 1523-1531. doi:10.3382/ps.2010-01283.

Yair, R., Uni, Z. and Shahar, R. (2012). Bone characteristics of late-term embryonic and hatchling broilers: Bone development under extreme growth rate, *Poult. Sci.* 91(10), 2614-2620. doi:10.3382/ps.2012-02244.

Yalçin, S., Molayoglu, H. B., Baka, M., Genin, O. and Pines, M. (2007). Effect of temperature during the incubation period on tibial growth plate chondrocyte differentiation and the incidence of tibial dyschondroplasia, *Poult. Sci.* 86(8), 1772-1783. doi:10.1093/ps/86.8.1772.

Yalçin, S., Özkan, S. and Shah, T. (2022). Incubation temperature and lighting: Effect on embryonic development, post-hatch growth, and adaptive response, *Front. Physiol.* 13, 899977. doi:10.3389/fphys.2022.899977.

Yalçin, S. and Siegel, P. B. (2003). Exposure to cold or heat during incubation on developmental stability of broiler embryos, *Poult. Sci.* 82(9), 1388–1392. doi:10.1093/ps/82.9.1388.

Yeboah, P. P., Konadu, L. A., Hamidu, J. A., Poku, E. A., Wakpal, D., Kudaya, P. Y., Dey, A. and Saddiq, S. M. (2019). Hatcheries hatching chicks contribute to poor development of day-old chick's base on biological and immunological assessment, *Vet. World* 12(11), 1849–1857.. doi:10.14202/vetworld.2019.1849-1857.

Part 3

Managing the hatching stage

Chapter 15

Managing the poultry hatcher environment

Roos Molenaar and Henry van den Brand, Wageningen University & Research, The Netherlands

1 Introduction

The hatcher phase comprises approximately the final 3 days of the incubation process and is an important period for the avian embryo in terms of maturation and emergence from the eggshell (Hamburger and Hamilton, 1951; Tong et al., 2013). The final days of the incubation process start in a conventional hatchery with candling and transferring fertile eggs from trays located in the setter to baskets in the hatcher in which chickens are able to emerge from their eggshell. Environmental factors during the final days of incubation have been found to affect survival and day-old chick quality as well as post-hatch performance (Leksrisompong et al., 2007; Maatjens et al., 2014a; Wijnen et al., 2020a). The most important environmental factors during the hatching phase appear to be temperature, relative humidity and gas concentrations, which can all be adjusted in most hatchers. Over the past decade, much research has been carried out related to the hatcher phase, as well as in relation to early feeding (van de Ven, 2012; Maatjens, 2016; De Jong et al., 2019; Souza da Silva et al., 2021).

It is important that environmental conditions during the final days of the incubation process optimize the chicken embryo's survival and development. An understanding of the embryonic developmental processes and environmental requirements to optimize these processes is essential to hatch a viable and strong chicken during artificial incubation. This chapter will discuss these two aspects of the hatching phase. First, the hatcher phase will be described from

http://dx.doi.org/10.19103/AS.2022.0118.18

an embryonic point of view, focusing on the developmental and physiological changes during the final days of the incubation process. Second, the hatcher phase will be described from an incubator point of view, discussing the influence of the main environmental conditions - temperature, relative humidity and CO_2 concentration - in relation to survival and chick quality at hatch and in later life. This chapter will focus mainly on broiler and layer chickens (*Gallus gallus domesticus*).

2 Embryonic phase

2.1 Development, physiology and metabolism

The final maturation of the chicken embryo and the hatching process take place in the last week of incubation (Hamburger and Hamilton, 1951; Tong et al., 2013). Rapid changes in development and physiology occur in the last 3 days of incubation when the embryo emerges from the eggshell (Yildirim and Aygün, 2005). The yolk is utilized by the yolk sac membrane up to a rate of 2.5 g/24 h during the final days of incubation (Molenaar et al., 2010b). The abundance of fatty acids in the yolk is used mainly as an energy substrate (Speake et al., 1993; De Oliveira et al., 2008; Givisiez et al., 2020). From day 18 of incubation, the residual yolk is retracted within the body cavity over a period of 2 days. Yolk utilization appears to shift in this period from the yolk sac membrane toward the yolk stalk, a narrow tube that connects the yolk sac with the small intestine of the embryo (van der Wagt et al., 2020).

At approximately day 19 of incubation, the initiation of the hatching process is triggered by hypoxic conditions as a result of oxygen (O_2) demand excelling O_2 availability (Haron et al., 2021). On the one hand, O_2 demand is high to sustain embryonic growth and metabolism during this stage of incubation (Dietz et al., 1998). On the other hand, O_2 availability reaches its maximum due to limitations in the diffusion capacity of the eggshell (Rahn and Ar, 1974). This is expressed by a plateau phase in O_2 consumption, occurring from approximately day 14 of incubation onward in broiler chicken embryos (Dietz et al., 1998). This hypoxic condition leads to the embryo pipping the air cell, referred to as internal pipping (IP) (Visschedijk, 1967b). Consequently, pulmonary ventilation is initiated and respiration via the chorioallantoic membrane (CAM) declines as a result of closure of the blood vessels (Rideout, 2012). Once IP has taken place, O_2 consumption and thereby heat production will increase, followed by a further increase after external pipping (EP) and emergence of the chicken from the eggshell (Dietz et al., 1998; Janke et al., 2004). In broiler chicken embryos, EP takes place between nine and 12 h after IP (Maatjens et al., 2014a), as a result of an increase in partial pressure of CO_2 (pCO_2) and a decrease in partial pressure of O_2 (pO_2) in the air cell (Rahn et al., 1974). Although the moment of EP can be different between avian species, the level of pCO_2 and pO_2 at which EP occurs appear quite similar (Rahn and Ar, 1974). Before EP, CO_2 concentration in the air cell rises to approximately 6% and O_2 concentration decreases to approximately 14% (Rahn, 1981; Christensen, 1995). The embryo pips the eggshell, using its pipping muscle in the neck and egg tooth at the tip of the beak (Brooks and Garrett, 1970). Thereafter, the embryo will break the eggshell further with its egg tooth, while turning clockwise in the egg, using its wing to rotate (Tong et al., 2013). When the majority of the eggshell cap is cut open, the chicken starts to stretch and push the eggshell with its legs to emerge from the eggshell (Burton and Tullett, 1985). Although the hatching process is energy-demanding and the chicken is wet and exhausted afterwards, within a few hours it is dry and able to walk and explore its surroundings. The total duration of the hatching process, from IP until emergence from the eggshell, varies between 20 h and 28 h in chicken embryos (Visschedijk, 1967a; Molenaar et al., 2010a; Maatjens et al., 2014a).

During the hatching process, the chicken embryo relies mainly on anaerobic glycolysis and glucose, which is stored as glycogen within the liver, muscles and yolk sac membrane (YSM), and is used as an energy substrate, especially in the muscles (De Oliveira et al., 2008; Givisiez et al., 2020). Because carbohydrates stores are very low in the freshly laid egg, gluconeogenesis is an important pathway during incubation to ensure that sufficient glucose is produced and stored as glycogen for the energy-demanding hatching process (De Oliveira et al., 2008). Embryonic mortality during the hatching process can be related to depletion of glycogen stores and a lack of readily available energy sources (Christensen et al., 1999; De Oliveira et al., 2008).

2.2 Hatching position

Another reason for embryonic mortality in the final days of incubation can be the incorrect hatching position. The chicken embryo starts to move into the correct position for hatching around day 14 of incubation and turns to its final position with the head toward the air cell and the beak under the right wing from day 18 of incubation onwards (Tong et al., 2013). If the chicken embryo is not in the correct position, or in a so-called malposition, it may fail to hatch. The six most common malpositions that are seen in unhatched eggs are head over the right wing, head under the left wing, head away from the air cell, leg over the head, upside down, or head between the legs (Fig. 1). Head over the right wing is not always lethal, but chicken embryos need to put more effort into emerging from the eggshell, resulting in weakness as well as red hocks and beaks. The other malpositions can result in moderate to high mortality as the chicken embryo will find it more difficult to pip and hatch from the eggshell. Malpositions can be a result of incorrect setting of eggs (e.g. air cell down),

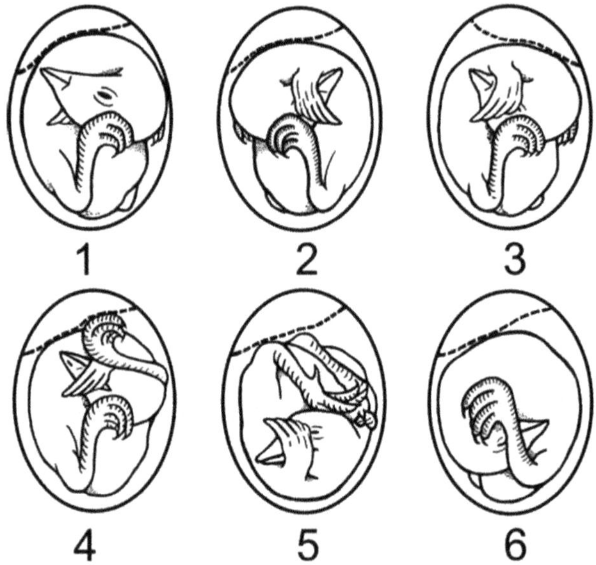

Figure 1 The six most common types of malpositions in chicken embryos: (1) Head over the right wing; (2) Head under the left wing; (3) Rotation away from air cell; (4) Leg over the head; (5) Upside down; (6) Head between the legs. Source: Adopted from Rideout (2012).

turning failures, or suboptimal environmental factors (Rideout, 2012) such as a high or low incubation temperature, which can negatively affect hatchability and day-old chick quality (Robertson, 1961). High numbers of unhatched eggs containing malpositioned embryos can relate to certain issues during artificial incubation and should be closely monitored.

2.3 Hatch time and hatch window

Hatch time is an important aspect of the hatcher phase and refers to the time in which a single chicken has hatched. Hatch window refers to the spread of hatch in a batch of eggs and is the duration between the first and the last hatched chicken (Decuypere et al., 2001). In chicken embryos, hatch time is around 20.5 days from the start of incubation and is influenced by different factors between and within batches of eggs such as breed, breeder flock age, egg characteristics, storage time, sex and incubation conditions (Tona et al., 2003; van Brecht et al., 2003; Careghi et al., 2005; Decuypere and Bruggeman, 2007; Tong et al., 2015b). These factors need to be considered when eggs are set. Chickens should not hatch too early in order to avoid remaining too long in the hatcher environment and becoming prone to dehydration and contamination (Tong et al., 2015a). However, all chickens should be hatched before processing.

In conventional hatcheries, chickens are generally processed after 510–512 h from the onset of incubation, which is just over 21 days, to allow as many chickens to hatch as possible.

Monitoring batches of eggs is important to evaluate the hatch time and can be done by counting the chickens a few times prior to pulling or by monitoring the humidity curve of the incubator (Romanini et al., 2013). Wing feather growth or the amount of meconium present on shell debris can also give an indication of the hatch time. Development of sensors and cameras, measuring temperature, vocalizations or movement may be potential techniques for future use (Bamelis et al., 2005; Romanini et al., 2013). Coordination of set and pull time and/or adjustments in incubation settings can be considered when hatch time is too early or too late.

The different hatch times of individual chickens within a batch of eggs follow a normal distribution and can be divided into three phases with chickens hatching early, mid, and late (Careghi et al., 2005). Chickens from these three phases differ in physiological traits at hatch as well as in performance in later life (van de Ven, 2012). van de Ven et al. (2011b) found no effect on body weight at hatch, but late-hatched chickens had higher blood glucose concentrations and higher relative organ weights compared to early and mid-hatched chickens when sampled at their biological age. This was probably caused by a higher yolk utilization in late-hatched chickens due to the longer incubation duration. In the post-hatch period, late-hatched chickens showed a lower growth rate up to 21 days of age and comparable results were found in other studies (Hager and Beane, 1983; Lamot et al., 2014). However, differences in post-hatch body weight may be partly explained by a time effect because chickens were weighed at their chronological age (= age calculated from pull time, around day 21.5 of incubation) instead of their biological age (= age calculated from true hatch time) (Careghi et al., 2005). Differences in biological age could be up to 24 h or even longer, depending on the hatch window duration.

Besides or because of differences in physiological and developmental traits between early, mid-term and late-hatched chickens, hatch time may also affect the behavior of chickens, but very few studies have been carried out on this topic. Løtvedt and Jensen (2014) found that hatching time affected behavior in a commercial laying hybrid breed, particularly in males. Compared to mid-term, early-hatched males showed a higher response to novel objects as well as slower partial learning in their early life, which may indicate they are more fearful. Stress coping may be different between early, mid-term and late-hatched chickens and may affect later-life behavior, such as feather pecking and performance in laying hens (de Haas et al., 2013). However, effects of hatch time on behavior in broiler chicken embryos and consequences of different coping styles and behavior requires further investigation.

The hatch window can take up to 48 h within batches of eggs (Decuypere et al., 2001). The hatch window is affected by several factors, including storage time and duration, breeder flock age, and temperature settings and distribution within an incubator (Tona et al., 2004; Careghi et al., 2005; Decuypere and Bruggeman, 2007). However, even when highly uniform conditions are applied to eggs of the same origin, the hatch window is often still found to be approximately 24 h (van de Ven et al., 2011a; Løtvedt and Jensen, 2014). This is likely the result of variations in factors that influence the hatch time of individual eggs, such as eggshell conductance, egg weight and laying sequence within a clutch of eggs (Visschedijk et al., 1985; Christensen, 1995; Decuypere et al., 2001; van de Ven, 2012). A small hatch window close to 24 h is desired, especially in conventional hatcheries, to prevent chickens from being held too long in the hatcher environment and thus becoming prone to dehydration or hatching too late to be processed. It is therefore important to monitor the hatch window with the techniques described above and resolve issues when the hatch window is too long.

Hatch time and hatch window might also be affected by communication among avian embryos. Chicken embryos begin to communicate with each other just before hatching (Tong et al., 2013). The first form of communication starts before IP by respiration movements (Rumpf and Tzchentke, 2010). Clicking sounds are produced simultaneously with respiration movements, probably by movement of cartilage tissue or fluid within the glottis (McCoshen and Thompson, 1968; Rumpf and Tzchentke, 2010). After IP, avian embryos communicate by species-specific vocalizations (Rumpf and Tzchentke, 2010) and calls of distress, contentment and brooding can be distinguished (Gottlieb and Vandenbergh, 1968). It has been found that clicking sounds especially contribute to hatch synchronization, resulting in a reduced hatch window. Artificial clicking sounds have been found to stimulate chicken embryos to hatch as well (Vince et al., 1970). However, in large-scale incubators, there is background noise of fans and engines (van de Ven et al., 2010) and if the noise level is above 80db, this may hamper communication among embryos and affect hatch time and hatch window duration (Rumpf and Tzchentke, 2010). This was confirmed by van de Ven et al. (2010), who found that broiler chickens incubated in silence hatched 1.5–7.2 h earlier than broiler chickens incubated with a background noise of either the mother hen or an artificial incubator (dB level not mentioned). The hatch window was longest (3–15 h increase) in chickens hatched with the background noise of an artificial incubator, suggesting that hampering embryonic communication may disturb hatch synchronization.

Donofre et al. (2020) investigated the effects of two levels of background noise of incubators (70 dB or 90 dB) in combination with species-specific vocalizations (present or absent) and found different results. The start of the hatching process was around six hours later in the treatment with the low noise level (70 db) and no species-specific vocalizations compared to all other treatments. No effect of treatment on the final hatch time or hatch window was found. Moreover, exposure to a high noise level of 90 dB increased hatchability compared to the noise level of 70 dB, and improved chicken quality traits in terms of navel and hock quality, but only when no species-specific vocalizations were applied. It was speculated that a loud rhythmic sound could stimulate chickens to hatch, but this may be the result of a stress reaction.

In summary, loud sounds during artificial incubation might affect hatchability and low sounds or silence may positively affect hatch synchronization and shorten the hatch window. The impact of different sound exposures and noise levels during incubation should be taken into account when assessing hatch time and hatch window duration and is an interesting field for further exploration.

3 Environmental settings within the hatching phase

The hatcher phase is a critical and sensitive period when the highest embryonic mortality is often seen (Rideout, 2012; Christensen, 2001). It is therefore important to manage environmental conditions during this period, to optimize not only hatchability but also day-old chick quality and post-hatch performance. During natural incubation, avian embryos appear to tolerate a wide range of temperatures, relative humidities and CO_2 concentrations (Freeman and Vince, 1974). However, it needs to be more strictly controlled during artificial incubation where eggs at a high density are incubated. Within an incubator, including a hatcher, environmental conditions vary among individual eggs, depending on the location of the egg relative to the heater, fan and nozzle, for example (French, 2001; Elibol and Brake, 2008b) and this is largely determined by the design of the incubator. The settings of the hatcher need to be adjusted to obtain the best hatchability and chick quality within batches of eggs with often differing origins. The three main environmental factors that can be controlled in conventional hatchers - temperature, relative humidity and CO_2 concentration - are discussed below. Effects of different factors on physiological and developmental changes of the chicken embryo and subsequent effects on hatchability, day-old chick quality and post-hatch performance are described. Confounding factors related to certain incubator settings are discussed where applicable to emphasize that not always one single environmental factor will change.

3.1 Temperature settings hatcher

Temperature during incubation is one of the main factors affecting hatchability, day-old chick quality and post-hatch performance (French, 1997; Christensen et al., 1999; Lourens et al., 2006). Comparing the temperature that the embryo experiences in natural versus artificial incubation needs to be done with great care as there is a major difference. With natural incubation, there is a temperature gradient across the egg due to heating by the brood patch on the one side and contact with the relatively cool nest material on the other side of the egg (Burke, 1925; Freeman and Vince, 1974). By changing the blood flow through the CAM, the avian embryo can influence its heat loss (Nichelmann and Tzschentke, 1999) and thereby also its temperature to some extent. With artificial incubation, eggs are often warmed or cooled with forced air and there is no temperature gradient present (Drent, 1970; Meijerhof and Van Beek, 1993). Embryos have, therefore, limited possibilities during artificial incubation to adjust their internal temperature, which explains the narrower temperature settings that are needed within the incubator.

There are two different ways to assess temperature settings during incubation. The first is by using the incubator temperature and the second is by using the embryo temperature (French, 2009). When the incubator temperature is used and set at a certain level, the temperature that the embryo experiences, or so-called embryo temperature, differs from the incubator temperature and is the result of air temperature, metabolic heat production and heat transfer from the egg to the environment (Meijerhof and Van Beek, 1993; Gualhanone et al., 2012; Hamidu et al., 2018). The difference between incubator and embryo temperature depends on multiple factors, such as breed, strain, embryonic age and air velocity within the incubator.

The second way to assess temperature settings is by using the embryo temperature. As this temperature is difficult to measure without invading the egg, eggshell temperature (EST) is often used as a non-destructive method to measure embryo temperature (Joseph et al., 2006). Operating incubators according to embryo temperatures or EST could help to optimize hatchability and chick quality (Lourens et al., 2005). Several studies have shown that an EST of 37.8°C (range from 37.5°C to 38.0°C) throughout incubation resulted in the best hatchability, day-old chick quality and post-hatch performance (Lourens et al., 2005; Leksrisompong et al., 2007; Wijnen et al., 2020a). This temperature range is comparable with EST measurements performed within the nest of the domestic fowl (Rahn et al., 1977). During the hatching process, however, it is difficult to maintain embryo temperature at this constant and uniform level as developmental stages of embryos vary (van der Pol et al., 2013), especially within large batches of eggs. Therefore, EST and air temperature are often controlled to maintain EST around 37.8°C until day 19 of incubation. Some studies have monitored the embryo temperature or EST during the final days of incubation. Romanini et al. (2013) monitored the EST of individually housed eggs in a small-scale incubator with a capacity of 300 eggs. They adjusted the incubator temperature to obtain an EST of around 37.8°C and were able to maintain the EST around this setpoint until hatch, when a drop in EST was found. This decrease in EST was used to assess hatch time. Hamidu et al. (2018) monitored the EST of individual eggs incubated in a small metabolic chamber at different incubation temperatures (ranging from 36.0°C to 37.5°C) from day 15 of incubation onward. They found that incubation temperature was highly correlated with EST and appeared to be influenced by strain, probably as a result of differences in metabolic rate. Measuring the core body temperature of chicken embryos is much harder. Holland et al. (1998) and Janke et al. (2002) used the temperature of the allantoic fluid or colonic temperature as a measure for deep body temperature in Muscovy ducks by placing small thermistor probes inside the egg. Individual eggs were incubated in a small metabolic chamber at an incubator temperature of 37.5°C and relative humidity of 70%. They showed that temperature increased, especially after IP, to a level of around 39.7°C.

EST and thereby core body temperature during the final days of incubation in large-scale incubators are likely to differ from results found in small-scale incubators as they will be the result of the incubator design and climate settings that affect temperature surrounding the eggs and heat transfer from the eggs to the environment. The egg and embryo characteristics within batches of eggs also affect metabolic heat production.

3.2 High temperatures

Prolonged exposure (more than 7 days of incubation) to high temperatures during artificial incubation has been shown to negatively affect hatchability, day-old chick quality and post-hatch performance (Gladys et al., 2000; Lourens et al., 2005; Leksrisompong et al., 2007; Molenaar et al., 2011; Wijnen et al., 2020a). As mentioned earlier, an EST of 37.8°C appears to give the best results in terms of hatchability, chick quality and post-hatch performance. Relatively small temperature deviations of 1°C above this optimum can have an impact (Zhang and Whittow, 1992; French, 1994; Lourens et al., 2005), even when only applied during the last few days of incubation (Maatjens et al., 2016b). High temperatures (EST > 38.9°C) are commonly found in conventional hatchers as a result of suboptimal cooling and air velocity during a period that embryonic heat production is high (French, 1997; Elibol and Brake, 2008a).

When embryo temperature increases (Fig. 2), the metabolic rate and O_2 requirement of the chicken embryo increases accordingly as they act mainly as a poikilotherm during the incubation period (Romijn and Lokhorst, 1955). Between day 7 and 14 of incubation, this is nicely illustrated by higher heat production and faster development at high (38.9°C) compared to control (37.8°C) EST (Lourens et al., 2007; Nangsuay et al., 2011). However, a plateau is reached in O_2 consumption and thereby heat production after approximately

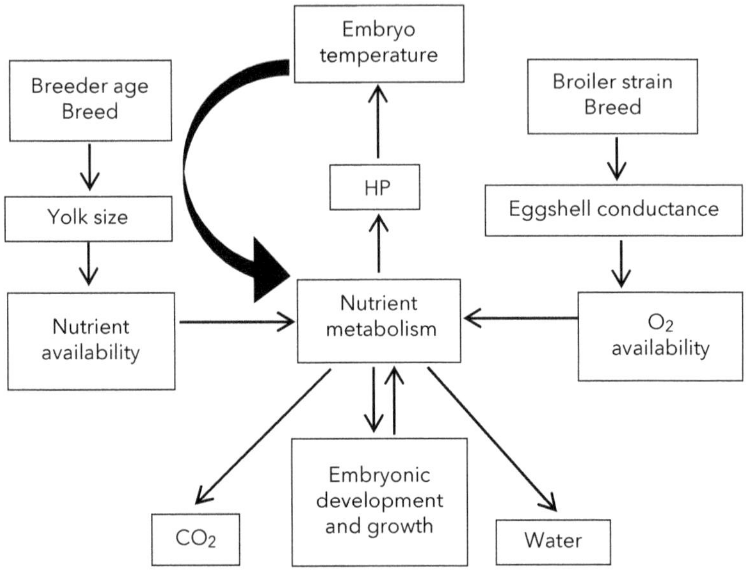

Figure 2 Overview of factors that affect embryonic development and growth during incubation; HP = heat production. Source: Adopted from Nangsuay (2016).

14 days of incubation in broiler chicken embryos (Lourens et al., 2006) and after approximately 16 days in layer chicken embryos (Nangsuay et al., 2015b; van den Brand et al., 2019) as O_2 availability is restricted by gas exchange across the eggshell and membranes (Wangensteen and Rahn, 1970). As a result, chicken embryos incubated at high EST reach hypoxic conditions sooner and start to IP and hatch earlier than at control EST. Maatjens et al. (2016a) showed a reduction of 5 h in hatch time when broiler eggs were incubated from day 17 of incubation onwards at a high EST of 38.9°C compared to a control EST of 37.8°C. Due to a reduction in hatch time, embryo development has been found to decrease at hatch, expressed by a lower yolk-free body mass (YFBM) and higher residual yolk (Lourens et al., 2005; Wineland et al., 2006; Leksrisompong et al., 2007). Moreover, this larger residual yolk needs to be retracted within the body cavity, which can lead to poorer and unhealed navels (Tona et al., 2005; Molenaar et al., 2010b). A poor navel condition can potentially increase cull and/or mortality rates in early life (Fasenko and O'Dea, 2008) as pathogens can more easily enter the chicken, causing yolk sac infections. Additionally, greater numbers of unsaleable second-grade chickens that were weak, pale, had short down and unhealed or black button navels were observed after incubation at high EST (Thommes and Just, 1964; Leksrisompong et al., 2007). Pale-colored feathers are probably due to poor yolk absorption (Leksrisompong et al., 2007).

Relative organ weights and functioning also appear to be impaired at high EST, and heart weight is consistently found to be lower (Wineland et al., 2006;

Leksrisompong et al., 2007; Maatjens et al., 2016a). This is possibly the result of a decreased amount of cell divisions (Romanoff, 1960; Leksrisompong et al., 2007) or changes in energy metabolism at high EST (Molenaar et al., 2010b; Molenaar et al., 2013). During the energy-demanding hatching process, readily available energy (i.e. stored as glycogen) is used for anaerobic metabolism that takes place mainly in the hatching and leg muscles (De Oliveira et al., 2008). However, energy metabolism appears to differ in embryos incubated at high EST, indicated by a lower hepatic glycogen concentration at IP and comparable at hatch at high EST (38.9°C) compared to control (37.8°C) EST (Maatjens et al., 2014b; Maatjens et al., 2017). A high EST of 38.9°C has been found to increase glucose oxidation and may therefore deplete glycogen stores further, probably as a result of the higher metabolic rate and increased anaerobic glycolysis (Christensen et al., 2007; Molenaar et al., 2013). When glycogen stores become limited or even depleted at high EST, alternative nutrients such as glucogenic amino acids may be used for energy instead of growth, which may (together with the earlier hatch time) contribute to a decrease in body development and alter energy status at hatch (Molenaar et al., 2013). In the worst case, when readily available energy becomes critically low during the hatching phase, the embryo may not be able to emerge from the eggshell (Beattie, 1964; Christensen et al., 1993). For example, Lourens et al. (2005) showed an increase of 6.3% of late embryonic mortality when EST was 38.9°C compared to 37.8°C from day 14 of incubation onwards. Furthermore, a higher incidence of the head between legs and head over wing malpositions was found at high EST and may contribute to lower hatchability (Molenaar et al., 2010b).

Additionally, effects of EST treatment may interact with other factors such as strain, breed and breeder flock age (Nangsuay et al., 2015a; Nangsuay et al., 2015b; Nangsuay et al., 2016; Nangsuay et al., 2017). For example, effects of high EST on embryonic development and energy status appear to be more pronounced in broiler compared to layer breeds, probably due to the higher metabolic rate and earlier O_2 shortage occurring in the second half of incubation in broiler chicken embryos (Molenaar et al., 2010a; Nangsuay et al., 2015b; van den Brand et al., 2019). The larger yolk size and nutrient availability, together with higher genetic growth potential, increase the metabolic rate and embryonic development of broiler compared to layer embryos (Nangsuay et al., 2015b). Broiler embryos may thereby experience an earlier and higher O_2 shortage, as indicated by the earlier start of the plateau phase in heat production (Lourens et al., 2007; Nangsuay et al., 2015b; van den Brand et al., 2019). Consequently, they may rely more on carbohydrate metabolism during the final days of incubation, as indicated by the lower hepatic glycogen stores found at hatch in broiler compared to layer embryos by Nangsuay et al. (2015b). Chick quality traits appear to be less affected by a high EST of 38.9°C in layers compared to broiler embryos. This was illustrated in the study by van den Brand et al. (2019) who incubated laying hen crossbreds of an old breeder flock at a high EST of 38.9°C compared to a control EST of 37.8°C from day 14.5 of incubation onward. Although lower relative heart weights were found when incubating at the high compared to control EST, no shorter incubation duration, higher incidence of red beaks and hocks, poorer navel condition or lower YFBM were found.

Moreover, O_2 availability is an important factor influencing the metabolic rate and development of chicken embryos and can interact with EST. Molenaar et al. (2010b) investigated the effects of two EST (37.8°C or 38.9°C) and three O_2 concentrations (17%, 21% or 25%) from day 7 to day 19 of incubation in a prime broiler breeder flock. They showed that a high EST of 38.9°C combined with a low O_2 concentration of 17% resulted in the lowest hatchability and YFBM and poorest navel condition. It was hypothesized that the lower hatchability was related to the poor development of the embryos, a decreased availability of energy (e.g. glycogen) for the hatching process, and/or the higher incidence of malpositions found.

Embryonic O_2 availability is furthermore influenced by eggshell conductance that affects the diffusion capacity of O_2 through the eggshell. Eggshell conductance can differ between breeds and strains. Eggs of a Cobb strain have, for example, a lower eggshell conductance than eggs of a Ross strain (Nangsuay et al., 2015a). As a probable result of this, embryonic heat production was found to be lower in a Cobb compared to Ross strain between day 15 and 18 of incubation and embryos appeared to come earlier in the plateau phase in a study by Nangsuay et al. (2015b). They incubated two breeder ages (young and old) of two broiler strains (Cobb and Ross) at two EST (37.8°C and 38.9°C). In addition to a lower embryonic heat production of Cobb compared to Ross embryos, a lower YFBM of 0.59 g (1.5%) was also found at hatch. These results indicated that the lower eggshell conductance may have contributed to reduced O_2 availability for Cobb compared to Ross chicken embryos, resulting in a lower metabolic rate in the second half of incubation and reduced body development at hatch.

No interactions were found between strain and EST in terms of chick quality traits or nutrient utilization in the study by Nangsuay et al. (2015b). This may be related to the relatively small difference in eggshell conductance between the two strains or the larger negative effect of the high EST. Studies investigating the interaction between embryonic O_2 availability and EST on hatchability and embryo development, particularly in the hatcher phase, are currently lacking.

Chickens that emerge from their eggshell can sometimes remain in the hatcher for quite long periods until they are pulled (>24 h), so it is important that they are maintained at the correct body temperature to prevent excessive dehydration. It is assumed that a body temperature between 40.0°C and 40.6°C is optimal for chickens after hatch (Maatjens et al., 2011). When the chicken's body temperature is maintained above 40.6°C, it starts to pant to remove heat by evaporative cooling and is prone to dehydration. Maatjens et al. (2011) showed that body weight loss in chickens of a young broiler breeder flock maintained at a body temperature above 41.0°C compared to chickens maintained between 40.0°C and 40.6°C was 2.5 g higher in the 24 h before pull time. This increase in body weight loss may negatively affect chick quality at hatch and in the post-hatch period.

A high EST in the final days of incubation has been found to negatively affect performance in later life of broiler chickens (Lourens et al., 2005; Hulet et al., 2007; Leksrisompong et al., 2009; Molenaar et al., 2011) and the severity of these negative effects is likely dependent on the duration and extent of the applied temperature treatment. These effects are further described in chapter 13.

3.3 Low temperatures

Although chicken embryos appear to be more affected by high than low temperatures (French, 1997; Romanini et al., 2015), low temperatures during the hatching phase can affect chicken embryos as well. It can be hypothesized that, by lowering the incubation temperature, the metabolic rate will be reduced, lowering O_2 demand and thereby potentially favoring fatty acid metabolism instead of carbohydrate metabolism. This could increase the survival rate and optimize embryonic development and energy status at hatch and in later life, especially for broiler chicken embryos that appear to experience a disbalance between their high metabolic rate and limitations in O_2 availability, particularly in the second half of incubation (van den Brand et al., 2019). This hypothesis appears to be largely true for the hatcher phase but not for later life, as shown by the few studies that have investigated the effects of low EST.

Yildirim and Yetisir (2004) found a higher late embryonic mortality (+5.3%) when chicken embryos of an old broiler breeder flock were incubated at an EST of 36.1°C compared to an EST of 37.7°C from day 17 of incubation onwards. Chick quality measurements were not performed in this study. Maatjens et al. (2016a, 2016b) found no effects on hatchability of fertile eggs and positive effects on chick quality traits when lower ESTs were applied during the final week of incubation. An EST of 35.6°C and 36.7°C from day 15 of incubation onward resulted in a higher YFBM (Δ = +1.31 g/+3.3% on average) and larger relative heart weight at hatch compared to an EST of 37.8°C. Hatch time was delayed by on average 16 h at an EST of 35.6°C and 4 h at an EST of 36.7°C compared to an EST of 37.8°C. The additional incubation time likely contributed to the more advanced development at hatch. Furthermore, hepatic glycogen stores at IP and hatch were higher at an EST of 35.6°C compared to the other two EST treatments, which may indicate that glucose oxidation was lower and possibly improved energy status at hatch. At day 7 of age, chickens incubated at an EST of 36.7°C showed the highest body weight compared to an EST of 35.6°C (Δ = +10 g/+5.0%) or 37.8°C (Δ = +6.8 g/+3.5%). No effects on mortality at day 7 were found. These results indicated that an EST lower than 37.8°C in the final week of incubation may positively affect development at hatch and possibly also in later life.

Wijnen et al. (2020a) studied the effects of low EST in the last week of incubation on post-hatch performance. Relative heart weight at hatch was increased at an EST of 36.7°C compared to an EST of 37.8°C, but YFBM at hatch did not differ. During the 40-day rearing period, body weight of broiler chickens incubated at an EST of 36.7°C compared to an EST of 37.8°C was lower. At day 40 of age, body weight after incubation at an EST of 36.7°C was 93 g (=3.0%) lower than after incubation at an EST of 37.8°C. Moreover, low EST was shown to negatively affect immune organ development at hatch and immune responses post-hatch and decreased several resilience parameters against a necrotic enteritis infection in later life (Wijnen et al., 2020b; Wijnen et al., 2021). This appears to indicate that although some chick quality traits can be positively affected by a lower EST during the hatching phase, performance and resilience in later life are negatively affected.

In summary, both high and low temperatures during the hatching phase that deviate more than 1°C from an EST of 37.8°C can affect hatchability, day-old chick quality and performance in later life. Maintaining and monitoring correct temperature settings and distributions within the hatcher is therefore of the utmost importance during artificial incubation.

3.4 Relative humidity settings in the hatcher

During the setter phase, relative humidity is important to obtain enough water loss to create a sufficient air cell for the start of the hatching process (Ar and Rahn, 1980; Barbosa et al., 2013). During the hatcher phase, relative humidity is suggested to play a role during the hatching process. It is often mentioned that relative humidity should not be too high so membranes and eggshell can be pipped by the embryo during the hatching process and relative humidity should not be too low to prevent dehydration of membranes and subsequent difficulties when hatching (Taylor, 1994; Wilson, 2004). However, it can be questioned whether these effects are only related to the actual relative humidity level or also accompanied by other confounded environmental factors when relative humidity levels become high or low within a hatcher. This is discussed further later in this chapter.

Very few studies have been carried out with regard to the effect of relative humidity during the hatcher phase on hatchability, day-old chick quality and post-hatch performance. Van der Pol et al. (2013) studied the effects of different relative humidity levels from embryonic day 2 onward, using eggs of a prime broiler breeder flock (47 weeks). They found that low relative humidity levels of between 30% and 35%, compared to normal relative humidity levels of between 50% and 55% when incubated at a constant EST of 37.8°C, increased egg weight loss, likely by decreasing allantoic fluid volume (Buhr, 1995). Late embryonic mortality was higher in the low relative humidity treatment and it was speculated that this was caused by the higher ventilation rate needed to obtain the lower relative humidity, negatively affecting the temperature distribution within the hatcher. However, this was not further expressed in chick quality characteristics at hatch. Neither chick quality and organ weights at hatch, nor body weight or feed conversion ratio until 4 days of age, were affected by the relative humidity treatment. This suggests that relative humidity levels during incubation appear to have minor effects as long as EST is maintained.

Bruzual et al. (2000) are one of the few to study the effects of relative humidity during the last 5 days of incubation on post-hatch performance. Eggs of a young broiler breeder flock (26 to 31 weeks) were incubated at either 43%, 53% or 63% of relative humidity from day 16 until day 21 of incubation and reared until 12 days of age. Hatchability, hatch or pull time, and body weight at hatch or pull time were not affected by relative humidity treatments. Comparable with the findings by Van der Pol et al. (2013), it was shown that body weight gain and FCR until 12 days of age did not differ between relative humidity treatments. Interestingly, mortality was higher in chickens incubated at 43% compared to 53% and 63% and there was a trend that this especially occurred at low compared to normal brooding temperatures. The reason for this effect is unknown.

It can be concluded that relative humidity levels in the hatcher itself appear to have minor effects on chick quality and post-hatch performance if other environmental factors are not largely influenced. Relative humidity levels during the hatcher phase are often realized by changes in ventilation rate or use of humidifiers and this can affect the temperature (distribution) and gas concentrations, as explained in more detail in chapter 12 and in other studies (Buhr, 1995; van der Pol et al., 2013). It is important to consider these confounding factors when effects of relative humidities are investigated, especially in large-scale incubators.

3.5 CO_2 concentration settings in the hatcher

Avian embryos appear to be able to tolerate high CO_2 concentrations or so-called hypercapnia in the second half of incubation especially (Taylor et al., 1956; Everaert et al., 2007; Burggren et al., 2015). In natural incubation, CO_2

concentrations up to 5% have been found in nests of domestic fowls at the end of incubation (Burke, 1925; Freeman and Vince, 1974). In large-scale hatchers, CO_2 concentration can differ considerably during the hatcher phase and is a result of incubator design and ventilation rate (van den Brand et al., 2021). High CO_2 concentrations are proposed in practice to induce IP and narrow the hatch window (Tona et al., 2013). However, studies related to high CO_2 concentrations have been inconclusive about this effect and results may have been affected by breeder flock characteristics, and/or level and application of CO_2 concentrations (Buys et al., 1998; Everaert et al., 2007; Everaert et al., 2010; Everaert et al., 2012; Tong et al., 2015b). Comparable with commercial practices, hypercapnia was obtained in some studies (Buys et al., 1998; Tong et al., 2015b) by lowering the ventilation rate. Other environmental factors, such as embryo temperature, relative humidity and O_2 concentration, were also likely affected and may have confounded the results (Maatjens et al., 2014a; van den Brand et al., 2021).

Maatjens et al. (2014a) therefore incubated eggs of a prime broiler breeder flock (41–45 weeks of age) from day 19 of incubation until 12 h after hatch at two CO_2 concentrations (0.2 or 1.0%), using CO_2 injection. Treatments were applied in combination with three EST treatments (low - 36.7°C , control - 37.8°C or high - 38.9°C). Interactions were mainly found between CO_2 concentration and EST treatment between IP and hatch on chick development, and differences appeared to be temporary in nature. Hatch time was not affected by CO_2 treatment, but embryos incubated at an EST of 37.8°C started IP 5 h earlier at 1.0% compared to 0.2%. At hatch (just after emergence of the eggshell), chick quality traits in terms of YFBM, residual yolk weight and relative heart, liver and lung weights were not affected by CO_2 concentrations. At 12 h post-hatch, relative heart and lung weight were found to be significantly higher for chickens incubated at high compared to low CO_2 concentration, and it was suggested that this may be caused by a higher respiration magnitude that has been found at higher CO_2 concentrations, requiring a higher output of these organs (Harada et al., 1985).

Van den Brand et al. (2021) incubated eggs of a prime breeder flock (37–45 weeks of age) from day 8 of incubation at 3 CO_2 concentration levels (0.1%, 0.4% and 0.8%) while maintaining EST at a normal (37.8°C) or high (38.9°C) level. O_2 concentration was maintained at 21.0%. No effects of CO_2 concentration on embryo development or chick quality traits in terms of hatch time, YFBM, residual yolk or relative organ weight were found at 6 h after hatching. Besides the effect on plasma glucose concentration at hatch, no effect of CO_2 concentration was found on plasma lactate and uric acid, or liver glycogen concentration at day 18 of incubation or at hatch. It was therefore concluded that effects of CO_2 concentrations at hatch were limited when other environmental conditions, such as EST, O_2 concentration and relative humidity, were maintained at a constant level.

It can be concluded from these studies that CO_2 concentrations up to 1% appear to have a limited effect in the hatcher phase on day-old chick quality, when other aspects such as egg temperature, relative humidity and O_2 concentration are kept constant. Minor effects of CO_2 concentration per se in the hatcher phase are to be expected on post-hatch performance.

4 Conclusion

During the hatcher phase, comprising approximately the last 3 days of incubation, the final maturation of the chicken embryo and the emergence of the eggshell takes place. In this phase, distinctive changes in embryonic development and physiology take place, which make it one of the most critical periods during incubation. Environmental conditions need to be maintained correctly to optimize hatchability, day-old chick quality and post-hatch performance and these are affected by incubator design and settings. The temperature that the embryo experiences is the most important environmental factor and small deviations from optimal can have a negative impact. Effects of relative humidity and CO_2 concentrations appear to be of minor importance and larger ranges can be applied without negative effects. However, confounded factors related to changes in relative humidity and CO_2 concentration should be considered, especially in large-scale incubators.

5 Where to look for further information

5.1 Further reading

- *The Avian Embryo* and *Biochemistry of the Avian Embryo* by Romanoff and Romanoff are beautiful thick reference books containing lots of information about prenatal development of avian embryos, measured in the older days.
- *'Development of the avian embryo'* by Freeman and Vince (1974) is a complete story of the physiology and behavior of avian embryos.
- Several PhD theses about incubation can be found online:
    - Setter and hatcher phase:
        - Lourens (2008) 'Embryo temperature during incubation: practice and theory'.
        - Molenaar (2010) 'Perinatal development and nutrient utilization in chickens – Effects of incubation conditions'.
        - Nangsuay (2016) 'Are all eggs equal? Embryonic development and nutrient metabolism in chicken eggs of different origins'.

- Hatcher phase:
  - Van de Ven (2012) 'Effects of hatching time and hatching system on broiler chick development'.
  - Maatjens (2016) 'Effects of temperature and CO_2 during late incubation on broiler chicken development'.
- Practical aspects of the hatcher phase can be found in:
  - 'Hatchery signals': Chapter 9, 'Hatching', and Chapter 10, 'Chick quality'.

5.2 Key journals and conferences

- *Poultry Science* and *World's Poultry Science Journal* publish the latest scientific studies related to incubation conditions, including the hatcher phase.
- World's Poultry Science Association organizes different annual meetings and conferences and strives to connect poultry science with the poultry industry. There are several working groups, and the Incubation and Fertility Research Group (IFRG) is targeting incubation and fertility research within avian species.

6 References

Ar, A. and Rahn, H. (1980). Water in the avian egg: Overall budget of incubation. *American Zoologist* 20(2): 373–384.

Bamelis, F., Kemps, B., Mertens, K., De Ketelaere, B., Decuypere, E. and DeBaerdemaeker, J. (2005). An automatic monitoring of the hatching process based on the noise of the hatching chicks. *Poultry Science* 84(7): 1101–1107.

Barbosa, V. M., Rocha, J. S. R., Pompeu, M. A., Martins, N. R. S., Baião, N. C., Lara, L. J. C., Batista, J. V. M. S. P. and Leite, R. C. (2013). The effects of relative humidity and turning in incubators machines on the incubation yield and chick performance. *World's Poultry Science Journal* 69(1): 89–98.

Beattie, J. (1964). The glycogen content of skeletal muscle, liver and heart in late chick embryos. *British Poultry Science* 5(3): 285–293.

Brooks, W. S. and Garrett, S. E. (1970). The mechanism of pipping in birds. *The Auk* 87(3): 458–466.

Bruzual, J. J., Peak, S. D., Brake, J. and Peebles, E. D. (2000). Effects of relative humidity during the last five days of incubation and brooding temperature on performance of broiler chicks from young broiler breeders. *Poultry Science* 79(10): 1385–1391.

Buhr, R. J. (1995). Incubation relative humidity effects on allantoic fluid volume and hatchability. *Poultry Science* 74(5): 874–884.

Burggren, W. W., Mueller, C. A. and Tazawa, H. (2015). Hypercapnic thresholds for embryonic acid-base metabolic compensation and hematological regulation during CO_2 challenges in layer and broiler chicken strains. *Respiratory Physiology and Neurobiology* 215: 1–12.

Burke, E. (1925). A study of incubation. *Bulletin* 178: 1–43.

Burton, F. G. and Tullett, S. G. (1985). Respiration of avian embryos. *Comparative Biochemistry and Physiology Part A* 82(4): 735–744.
Buys, N., Dewil, E., Gonzales, E. and Decuypere, E. (1998). Different CO_2 levels during incubation interact with hatching time and ascites susceptibility in two broiler lines selected for different growth rate. *Avian Pathology* 27(6): 605–612.
Careghi, C., Tona, K., Onagbesan, O., Buyse, J., Decuypere, E. and Bruggeman, V. (2005). The effects of the spread of hatch and interaction with delayed feed access after hatch on broiler performance until seven days of age. *Poultry Science* 84(8): 1314–1320.
Christensen, V. L. (1995). Factors affecting hatchability of turkey embryos. *Poultry and Avian Biology Reviews* 6: 71–82.
Christensen, V. L. (2001). Factors associated with early embryonic mortality. *World's Poultry Science Journal* 57(4): 359–372.
Christensen, V. L., Donaldson, W. E. and Nestor, K. E. (1993). Embryonic viability and metabolism in turkey line selected for egg production and growth. *Poultry Science* 72(5): 829–838.
Christensen, V. L., Donaldson, W. E. and Nestor, K. E. (1999). Length of the plateau phase and pipping stage of incubation affects the physiology and survival of turkeys. *British Poultry Science* 40(2): 297–303.
Christensen, V. L., Winelan, M. J., Grimes, J. L., Oviedo, E. O., Mozdziak, P. S., Ort, D. T. and Mann, K. M. (2007). Effect of incubator temperature and oxygen concentration at the plateau stage in oxygen consumption on turkey embryo muscle growth and development. *International Journal of Poultry Science* 6(6): 406–412.
de Haas, E. N., Kemp, B., Bolhuis, J. E., Groothuis, T. and Rodenburg, T. B. (2013). Fear, stress, and feather pecking in commercial white and brown laying hen parent-stock flocks and their relationships with production parameters. *Poultry Science* 92(9): 2259–2269.
De Jong, I. C., Gunnink, H., Van Hattum, T., van Riel, J. W., Raaijmakers, M. M. P., Zoet, E. S. and van den Brand, H. (2019). Comparison of performance, health and welfare aspects between commercially housed hatchery-hatched and on-farm hatched broiler flocks. *Animal* 13(6): 1269–1277.
De Oliveira, J. E., Uni, Z. and Ferket, P. R. (2008). Important metabolic pathways in poultry embryos prior to hatch. *World's Poultry Science Journal* 64(4): 488–499.
Decuypere, E. and Bruggeman, V. (2007). The endocrine interface of environmental and egg factors affecting chick quality. *Poultry Science* 86(5): 1037–1042.
Decuypere, E., Tona, K., Bruggeman, V. and Bamelis, F. (2001). The day-old chick: A crucial hinge between breeders and broilers. *World's Poultry Science Journal* 57(2): 127–138.
Dietz, M. W., van Kampen, M., van Griensven, M. J. M. and van Mourik, S. (1998). Daily energy budgets of avian embryos: The paradox of the plateau phase in egg metabolic rate. *Physiological Zoology* 71(2): 147–156.
Donofre, A. C., da Silva, I. J. O. and de Paula Ferreira, I. E. P. (2020). Sound exposure and its beneficial effects on embryonic growth and hatching of broiler chicks. *British Poultry Science* 61(1): 79–85.
Drent, R. (1970). Functional aspects of incubation in the herring gull. *Behaviour Suppl.* 1: 1–32.
Elibol, O. and Brake, J. (2008a). Effect of egg position during three and fourteen days of storage and turning frequency during subsequent incubation on hatchability of broiler hatching eggs. *Poultry Science* 87(6): 1237–1241.

Elibol, O. and Brake, J. (2008b). Effect of egg weight and position relative to incubator fan on broiler hatchability and chick quality. *Poultry Science* 87(9): 1913–1918.

Everaert, N., Debonne, M., Willemsen, H., Witters, A., Kamers, B., De Baerdemaeker, J., Decuypere, E. and Bruggeman, V. (2010). Interaction between ascites susceptibility and CO_2 during the second half of incubation of two broiler lines: Effect on embryonic development and hatching process. *British Poultry Science* 51(3): 335–343.

Everaert, N., Kamers, B., Witters, A., De Smit, L., Debonne, M., Decuypere, E. and Bruggeman, V. (2007). Effect of four percent carbon dioxide during the second half of incubation on embryonic development, hatching parameters, and posthatch growth. *Poultry Science* 86(7): 1372–1379.

Everaert, N., Willemsen, H., Debonne, M., Witters, A., Kamers, B., Darras, V. M., de Baerdemaeker, J., Decuypere, E. and Bruggeman, V. (2012). Interaction between ascites susceptibility and CO_2 during the second half of incubation of two broiler lines: The effect on post-hatch development and ascites mortality. *British Poultry Science* 53(2): 262–269.

Fasenko, G. M. and O'Dea, E. E. (2008). Evaluating broiler growth and mortality in chicks with minor navel conditions at hatching. *Poultry Science* 87(3): 594–597.

Freeman, B. M. and Vince, M. A. (1974). *Development of the Avian Embryo: A Behavioural and Physiological Study*. London, UK: Chapman and Hall.

French, N. A. (1994). Effect of incubation temperature on the gross pathology of turkey embryos. *British Poultry Science* 35(3): 363–371.

French, N. A. (1997). Modeling incubation temperature: The effects of incubator design, embryonic development, and egg size. *Poultry Science* 76(1): 124–133.

French, N. A. (2001). Temperature in a tunnel incubator - A long term study. *Avian and Poultry Biology Reviews* 12: 184–186.

French, N. A. (2009). The critical importance of incubation temperature. *Avian Biology Research* 2(1-2): 55–59.

Givisiez, P. E. N., Moreira Filho, A. L. B., Santos, M. R. B., Oliveira, H. B., Ferket, P. R., Oliveira, C. J. B. and Malheiros, R. D. (2020). Chicken embryo development: Metabolic and morphological basis for in ovo feeding technology. *Poultry Science* 99(12): 6774–6782.

Gladys, G., Hill, D., Meijerhof, R., Saleh, T. M. and Hulet, R. M. (2000). Effect of embryo temperature and age of breeder flock on broiler post-hatch performance. *Poultry Science* 79: S179.

Gottlieb, G. and Vandenbergh, J. G. (1968). Ontogeny of vocalization in duck and chick embryos. *Journal of Experimental Zoology* 168(3): 307–325.

Gualhanone, A., Furlan, R. L., Fernandez-Alarcon, M. F. and Macari, M. (2012). Effect of breeder age on eggshell thickness, surface temperature, hatchability and chick weigh. *Revista Brasileira de Ciência Avícola* 14(1): 9–14.

Hager, J. E. and Beane, W. L. (1983). Posthatch incubation time and early growth of broiler chickens. *Poultry Science* 62(2): 247–254.

Hamburger, V. and Hamilton, H. L. (1951). A series of normal stages in the development of the chick embryo. *Journal of Morphology* 88(1): 49–92.

Hamidu, J. A., Torres, C. A., Johnson-Dahl, M. L. and Korver, D. R. (2018). Physiological response of broiler embryos to different incubator temperature profiles and maternal flock age during incubation. 1. Embryonic metabolism and day-old chick quality. *Poultry Science* 97(8): 2934–2946.

Harada, Y., Kuno, M. and Wang, Y. Z. (1985). Differential effects of carbon dioxide and pH on central chemoreceptors in the rat in vitro. *The Journal of Physiology* 368: 679-693.

Haron, A., Ruzal, M., Shinder, D. and Druyan, S. (2021). Hypoxia during incubation and its effects on broiler's embryonic development. *Poultry Science* 100(3): 100951.

Holland, S., Hochel, J., Burgmeister, A., Janke, O. and Nichelmann, M. (1998). A method for measuring deep body temperature in avian embryos. *Journal of Thermal Biology* 23(2): 123-129.

Hulet, R., Gladys, G., Hill, D., Meijerhof, R. and El-Shiekh, T. (2007). Influence of egg shell embryonic incubation temperature and broiler breeder flock age on posthatch growth performance and carcass characteristics. *Poultry Science* 86(2): 408-412.

Janke, O., Tzschenktke, B. and Boerjan, M. (2004). Comparative investigations of heat production and body temperature in embryos of modern chicken breeds. *Avian and Poultry Biology Reviews* 15(3): 191-196.

Janke, O., Tzschentke, B., Hochel, J. and Nichelmann, M. (2002). Metabolic responses of chicken and Muscovy duck embryos to high incubation temperatures. *Comparative Biochemistry and Physiology. Part A, Molecular and Integrative Physiology* 131(4): 741-750.

Joseph, N. S., Lourens, A. and Moran Jr., E. T. (2006). The effects of suboptimal eggshell temperature during incubation on broiler chick quality, live performance, and further processing yield. *Poultry Science* 85(5): 932-938.

Lamot, D. M., van de Linde, I. B., Molenaar, R., van der Pol, C. W., Wijtten, P. J., Kemp, B. and van den Brand, H. (2014). Effects of moment of hatch and feed access on chicken development. *Poultry Science* 93(10): 2604-2614.

Leksrisompong, N., Romero-Sanchez, H., Plumstead, P. W., Brannan, K. E. and Brake, J. (2007). Broiler incubation. 1. Effect of elevated temperature during late incubation on body weight and organs of chicks. *Poultry Science* 86(12): 2685-2691.

Leksrisompong, N., Romero-Sanchez, H., Plumstead, P. W., Brannan, K. E., Yahav, S. and Brake, J. (2009). Broiler incubation. 2. Interaction of incubation and brooding temperatures on broiler chick feed consumption and growth. *Poultry Science* 88(6): 1321-1329.

Løtvedt, P. and Jensen, P. (2014). Effects of hatching time on behavior and weight development of chickens. *PLoS ONE* 9(7): e103040.

Lourens, A., Molenaar, R., van den Brand, H., Heetkamp, M. J., Meijerhof, R. and Kemp, B. (2006). Effect of egg size on heat production and the transition of energy from egg to hatchling. *Poultry Science* 85(4): 770-776.

Lourens, A., van den Brand, H., Heetkamp, M. J. W., Meijerhof, R. and Kemp, B. (2007). Effects of eggshell temperature and oxygen concentration on embryo growth and metabolism during incubation. *Poultry Science* 86(10): 2194-2199.

Lourens, A., van den Brand, H., Meijerhof, R. and Kemp, B. (2005). Effect of eggshell temperature during incubation on embryo development, hatchability, and posthatch development. *Poultry Science* 84(6): 914-920.

Maatjens, C. M. (2016). *Effects of temperature and CO_2 during late incubation on broiler chicken development*. PhD Thesis. Wageningen, The Netherlands: Wageningen University and Research.

Maatjens, C. M., Reijrink, I. A. M. and Molenaar, R. (2011). Effect of body temperature and relative humidity in the hatcher on body weight loss and chick length of the broiler chick. *Avian Biology Research* 4: 141-142.

Maatjens, C. M., van Roovert-Reijrink, I. A. M., Engel, B., van der Pol, C. W., Kemp, B. and van den Brand, H. (2016a). Temperature during the last week of incubation. I. Effects on hatching pattern and broiler chicken embryonic organ development. *Poultry Science* 95(4): 956-965.

Maatjens, C. M., van Roovert-Reijrink, I. A. M., van den Anker, I., Engel, B., van der Pol, C. W., Kemp, B. and van den Brand, H. (2016b). Temperature during the last week of incubation. II. Effects on first week broiler development and performance. *Poultry Science* 95(9): 2136–2144.

Maatjens, C. M., van Roovert-Reijrink, I. A. M., Engel, B., van der Pol, C. W., Kemp, B. and van den Brand, H. (2017). Temperature during the last week of incubation. III. Effects on chicken embryo physiology. *Poultry Science* 96(5): 1451–1458.

Maatjens, C. M., van Roovert-Reijrink, I. A. M., Molenaar, R., van der Pol, C. W., Kemp, B. and van den Brand, H. (2014a). Temperature and CO_2 during the hatching phase. I. Effects on chick quality and organ development. *Poultry Science* 93(3): 645–654.

Maatjens, C. M., van Roovert-Reijrink, I. A. M., van den Anker, I., Molenaar, R., van der Pol, C. W., Kemp, B. and van den Brand, H. (2014b). Temperature and CO_2 during the hatching phase. II. Effects on chicken embryo physiology. *Poultry Science* 93(3): 655–663.

McCoshen, J. A. and Thompson, R. P. (1968). A study of clicking and its source in some avian species. *Canadian Journal of Zoology* 46(2): 169–172.

Meijerhof, R. and Van Beek, G. (1993). Mathematical modelling of temperature and moisture loss of hatching eggs. *Journal of Theoretical Biology* 165(1): 27–41.

Molenaar, R., de Vries, S., van den Anker, I., Meijerhof, R., Kemp, B. and van den Brand, H. (2010a). Effect of eggshell temperature and a hole in the air cell on the perinatal development and physiology of layer hatchlings. *Poultry Science* 89(8): 1716–1723.

Molenaar, R., Meijerhof, R., van den Anker, I., Heetkamp, M. J., van den Borne, J. J., Kemp, B. and van den Brand, H. (2010b). Effect of eggshell temperature and oxygen concentration on survival rate and nutrient utilization in chicken embryos. *Poultry Science* 89(9): 2010–2021.

Molenaar, R., Hulet, R., Meijerhof, R., Maatjens, C. M., Kemp, B. and van den Brand, H. (2011). High eggshell temperatures during incubation decrease growth performance and increase the incidence of ascites in broiler chickens. *Poultry Science* 90(3): 624–632.

Molenaar, R., van den Borne, J. J., Hazejager, E., Kristensen, N. B., Heetkamp, M. J., Meijerhof, R., Kemp, B. and van den Brand, H. (2013). High environmental temperature increases glucose requirement in the developing chicken embryo. *PLoS ONE* 8(4): e59637.

Nangsuay, A. (2016). *Are all eggs equal? Embryonic development and nutrient metabolism in chicken eggs of different origins*. PhD Thesis. Wageningen, The Netherlands: Wageningen University and Research.

Nangsuay, A., Meijerhof, R., van den Anker, I., Heetkamp, M. J., Kemp, B. and van den Brand, H. (2015a). Development and nutrient metabolism of embryos from two modern broiler strains. *Poultry Science* 94(10): 2546–2554.

Nangsuay, A., Molenaar, R., Meijerhof, R., van den Anker, I., Heetkamp, M. J., Kemp, B. and van den Brand, H. (2015b). Differences in egg nutrient availability, development, and nutrient metabolism of broiler and layer embryos. *Poultry Science* 94(3): 415–423.

Nangsuay, A., Meijerhof, R., van den Anker, I., Heetkamp, M. J. W., Kemp, B. and van den Brand, H. (2017). Effects of breeder age, strain, and eggshell temperature on nutrient metabolism of broiler embryos. *Poultry Science* 96(6): 1891-1900.

Nangsuay, A., Meijerhof, R., van den Anker, I., Heetkamp, M. J. W., Morita, V. S., Kemp, B. and van den Brand, H. (2016). Effects of breeder age, broiler strain, and eggshell temperature on development and physiological status of embryos and hatchlings. *Poultry Science* 95(7): 1666-1679.

Nangsuay, A., Ruangpanit, Y., Meijerhof, R. and Attamangkune, S. (2011). Yolk absorption and embryo development of small and large eggs originating from young and old breeder hens. *Poultry Science* 90(11): 2648-2655.

Nichelmann, M. and Tzschentke, B. (1999). Thermoregulation in precocial avian embryos. *Ornis Fennica* 76: 177-187.

Rahn, H. (1981). Gas exchange of avian eggs with special reference to turkey eggs. *Poultry Science* 60(9): 1971-1980.

Rahn, H., Ackerman, R. A. and Paganelli, C. V. (1977). Humidity in the avian nest and egg water loss during incubation. *Physiological Zoology* 50(4): 269-283.

Rahn, H. and Ar, A. (1974). The avian egg: Incubation time and water loss. *The Condor* 76(2): 147-152.

Rahn, H., Paganelli, C. V. and Ar, A. (1974). The avian egg: Air cell gas tension, metabolism and incubation time. *Respiration Physiology* 22(3): 297-309.

Rideout, B. A. (2012). Investigating embryo deaths and hatching failure. *Veterinary Clinics of North America: Exotic Animal Practice* 15(2): 155-162.

Robertson, I. S. (1961). Studies of chick embryo orientation using x-rays. II. Malpositioned embryos and their subsequent hatchability. *British Poultry Science* 2: 49-58.

Romanini, C. E. B., Exadaktylos, V., Hong, S. W., Tong, Q., McGonnell, I., Demmers, T. G., Bergoug, H., Guinebretière, M., Eterradossi, N., Roulston, N., Verhelst, R., Bahr, C. and Berckmans, D. (2015). An insight into the heat and mass transfer mechanisms of eggshells hatching broiler chicks and its effects to the hatcher environment. *Journal of Thermal Biology* 48: 69-76.

Romanini, C. E. B., Exadaktylos, V., Tong, Q., McGonnel, I., Demmers, T. G., Bergoug, H., Eterradossi, N., Roulston, N., Garain, P., Bahr, C. and Berckmans, D. (2013). Monitoring the hatch time of individual chicken embryos. *Poultry Science* 92(2): 303-309.

Romanoff, A. L. (1960). *The Avian Embryo: Structure and Functional Development*. New York, NY: MacMillan.

Romijn, C. and Lokhorst, W. (1955). Chemical heat regulation in the chick embryo. *Poultry Science* 34(3): 649-654.

Rumpf, M. and Tzchentke, B. (2010). Perinatal acoustic communication in birds: Why do birds vocalize in the egg? *The Open Ornithology Journal* 3(1): 141-149.

Souza da Silva, C., Molenaar, R., Giersberg, M. F., Rodenburg, T. B., van Riel, J. W., De Baere, K., Van Dosselaer, I., Kemp, B., van den Brand, H. and de Jong, I. C. (2021). Day-old chicken quality and performance of broiler chickens from 3 different hatching systems. *Poultry Science* 100(3): 100953.

Speake, B. K., Noble, R. C. and McCartney, R. J. (1993). Tissue-specific changes in lipid composition and lipoprotein lipase activity during the development of the chick embryo. *Biochimica et Biophysica Acta* 1165(3): 263-270.

Taylor, G. (1994). Dry down and hatcher humidity. *Poultry Digest*: 18-30.

Taylor, L. W., Sjodin, R. A. and Gunns, C. A. (1956). The gaseous environment of the chick embryo in relation to its development and hatchability. *Poultry Science* 35(6): 1206–1215.

Thommes, R. C. and Just, J. J. (1964). Endocrine control of the yolk sac membrane glycogen levels in the developing chick embryo. *General and Comparative Endocrinology* 4: 614–623.

Tona, K., Bamelis, F., De Ketelaere, B., Bruggeman, V., Moraes, V. M., Buyse, J., Onagbesan, O. and Decuypere, E. (2003). Effects of egg storage time on spread of hatch, chick quality and chick juvenile growth. *Poultry Science* 82(5): 736–741.

Tona, K., Bruggeman, V., Onagbesan, O., Bamelis, F., Gbeassor, M., Mertens, K. and Decuypere, E. (2005). Day-old chick quality: Relationship to hatching egg quality, adequate incubation practice and prediction of broiler performance. *Avian and Poultry Biology Reviews* 16(2): 109–119.

Tona, K., Everaert, N., Willemsen, H., Gbeassor, M., Decuypere, E. and Buyse, J. (2013). Effects of interaction of incubator CO_2 levels and mixing hatching eggs of different embryo growth trajectory on embryo physiological and hatching parameters. *British Poultry Science* 54(4): 545–551.

Tona, K., Onagbesan, O., De Ketelaere, B., Decuypere, E. and Bruggeman, V. (2004). Effects of age of broiler breeders and egg storage on egg quality, hatchability, chick quality, chick weight, and chick posthatch growth to forty-two days. *Journal of Applied Poultry Research* 13(1): 10–18.

Tong, Q., Demmers, T., Romanini, C. E. B., Bergoug, H., Roulston, N., Exadaktylos, V., Bahr, C., Berckmans, D., Guinebretière, M., Eterradossi, N., Garain, P. and McGonnell, I. M. (2015a). Physiological status of broiler chicks at pulling time and the relationship to duration of holding period. *Animal* 9(7): 1181–1187.

Tong, Q., McGonnell, I. M., Roulston, N., Bergoug, H., Romanini, C. E., Garain, P., Eterradossi, N., Exadaktylos, V., Bahr, C., Berckmans, D. and Demmers, T. G. (2015b). Higher levels of CO_2 during late incubation alter the hatch time of chicken embryos. *British Poultry Science* 56(4): 503–509.

Tong, Q., Romanini, C. E., Exadaktylos, V., Bahr, C., Berckmans, D., Bergoug, H., Eterradossi, N., Roulston, N., Verhelst, R., McGonnell, I. M. and Demmers, T. (2013). Embryonic development and the physiological factors that coordinate hatching in domestic chickens. *Poultry Science* 92(3): 620–628.

van Brecht, A., Aerts, J. M., Degraeve, P. and Berckmans, D. (2003). Quantification and control of the spatiotemporal gradients of air speed and air temperature in an incubator. *Poultry Science* 82(11): 1677–1687.

van de Ven, L., Gosselink, M., van Wagenberg, A. V. and van den Brand, H. (2010). Hatching system affects broiler chick behaviorth European poultry conference. Tours, France, p. XIII.

van de Ven, L. J., Baller, L., van Wagenberg, A. V., Kemp, B. and van den Brand, H. (2011a). Effects of egg position during late incubation on hatching parameters and chick quality. *Poultry Science* 90(10): 2342–2347.

van de Ven, L. J., van Wagenberg, A. V., Debonne, M., Decuypere, E., Kemp, B. and van den Brand, H. (2011b). Hatching system and time effects on broiler physiology and posthatch growth. *Poultry Science* 90(6): 1267–1275.

van de Ven, L. J. F. (2012). *Effects of Hatching Time and Hatching System on Broiler Chick Development*. Wageningen, The Netherlands: Wageningen University & Research.

van den Brand, H., Meijerhof, R., Heetkamp, M. J. W., van den Anker, I., Ooms, M., Kemp, B. and Molenaar, R. (2021). Interaction between eggshell temperature and carbon dioxide concentration after day 8 of incubation on broiler chicken embryo development. *Animal* 15(6): 100223.

van den Brand, H., van de Kraats, S. J. F., Sözcü, A., Jöerissen, R., Heetkamp, M. J. W., van den Anker, I., Ooms, M. and Kemp, B. (2019). Both the rooster line and incubation temperature affect embryonic metabolism and hatchling quality in laying hen crossbreds. *Poultry Science* 98(6): 2632–2640.

van der Pol, C. W., van Roovert-Reijrink, I. A. M., Maatjens, C. M., van den Brand, H. and Molenaar, R. (2013). Effect of relative humidity during incubation at a set eggshell temperature and brooding temperature posthatch on embryonic mortality and chick quality. *Poultry Science* 92(8): 2145–2155.

van der Wagt, I., de Jong, I. C., Mitchell, M. A., Molenaar, R. and van den Brand, H. (2020). A review on yolk sac utilization in poultry. *Poultry Science* 99(4): 2162–2175.

Vince, M., Green, J. and Chinn, S. (1970). Acceleration of hatching in the domestic fowl. *British Poultry Science* 11(4): 483–488.

Visschedijk, A. H. J. (1967). The air space and embryonic respiration 2. The times of pipping and hatching as influenced by an artificially changed permeability of the shell over the air space. *British Poultry Science* 9: 185–196.

Visschedijk, A. H. J. (1968). The air space and embryonic respiration. 3. The balance between oxygen and carbon dioxide in the air space of the incubating chicken egg and its role in stimulating pipping. *British Poultry Science* 9(2): 197–210.

Visschedijk, A. H. J., Tazawa, H. and Piiper, J. (1985). Variability of shell conductance and gas exchange of chicken eggs. *Respiration Physiology* 59(3): 339–345.

Wangensteen, O. D. and Rahn, H. (1970–1971). Respiratory gas exchange by the avian embryo. *Respiration Physiology* 11(1): 31–45.

Wijnen, H. J., Molenaar, R., van Roovert-Reijrink, I. A. M., van der Pol, C. W., Kemp, B. and van den Brand, H. (2020a). Effects of incubation temperature pattern on broiler performance. *Poultry Science* 99(8): 3897–3907.

Wijnen, H. J., van den Brand, H., Lammers, A., van Roovert-Reijrink, I. A. M., van der Pol, C. W., Kemp, B. and Molenaar, R. (2020b). Effects of eggshell temperature pattern during incubation on primary immune organ development and broiler immune response in later life. *Poultry Science* 99(12): 6619–6629.

Wijnen, H. J., van der Pol, C. W., van Roovert-Reijrink, I. A. M., De Smet, J., Lammers, A., Kemp, B., van den Brand, H. and Molenaar, R. (2021). Low incubation temperature during late incubation and early feeding affect broiler resilience to necrotic enteritis in later life. *Frontiers in Veterinary Science* 8: 784869.

Wilson, H. R. (2004). *Hatchability Problem Analysis: University of Florida IFAS Extension CIR1112*. Gainesville: University of Florida.

Wineland, M. W., Christensen, V. L., Yildrum, I., Fairchild, B. D., Mann, K. M. and Ort, D. T. (2006). Incubator temperature and oxygen concentration at the plateau stage in oxygen consumption affects intestinal maturation of broiler chicks. *International Journal of Poultry Science* 5(3): 229–240.

Yildirim, Ī., and Aygün, A. (2005). Effects of high hatcher temperature application during plateau and pipping stages of incubation on hatching results in broiler breeder eggs. *Archive für Geflügelkunde* 69: 231–234.

Yildirim, I. and Yetisir, R. (2004). Effects of different hatcher temperatures on hatching traits of broiler embryos during the last five days of incubation. *South African Journal of Animal Science* 34: 211–216.

Zhang, Q. and Whittow, G. C. (1992). The effect of incubation temperature on oxygen consumption and organ growth in domestic-fowl embryos. *Journal of Thermal Biology* 17(6): 339–345.

Chapter 16

Alternative hatching systems for broilers

Henry van den Brand, Bas Kemp and Roos Molenaar, Wageningen University and Research, The Netherlands

1 Introduction

The day-old broiler chicken is a crucial hinge between the breeder and the broiler (Decuypere et al., 2001), and the quality of the day-old chicken appears to be related to later life health, welfare and performance (Tona et al., 2005; Willemsen et al., 2008; Van der Wagt et al., 2020). In the current conventional hatching system, two important aspects potentially reduce the quality of day-old chickens:

- Environmental conditions in hatchers and
- Hatch window.

Firstly, environmental conditions in the hatcher might be suboptimal regarding high temperature, high dust levels and use of disinfectants, which might negatively affect day-old chicken quality (Zulkifli et al., 1999; Leksrisompong et al., 2007; De Gouw et al., 2017). Secondly, chickens hatch over a hatch window of approximately 24–36 h (Careghi et al., 2005; Willemsen et al., 2010) and are removed from the hatchers only when the majority of chickens has

http://dx.doi.org/10.19103/AS.2022.0118.17

hatched. Chickens that hatch particularly early face a long period without feed and water before they are removed from the hatcher and placed in the broiler house. These two aspects may impair the health and welfare of chickens both at hatch and in later life. This has led to the development of alternative hatching systems, which differ on several aspects from the conventional hatching system. The aim of this chapter is

- to discuss the differences between the conventional and alternative hatching systems and
- to investigate the effects of different hatching systems on day-old chicken quality as well as on health, welfare and performance in later life.

2 Current alternative hatching systems

Alternative hatching systems can be divided into two categories:

- In the first category, chickens hatch in a hatchery and have feed (and water) available in the hatcher (hatchery fed).
- The second category consists of on-farm hatching systems in which, after candling at day 18 of incubation, fertile eggs are transported to the rearing farm where they can hatch.

Examples of the first category are:

- SmartStart (www.pasreform.com/en/smartstart) where, immediately after hatching, chickens have access to a semi-moist diet at two sides of the hatching basket. At pulling, chickens are handled and processed and transported in normal hatching baskets to the rearing farm.
- HatchCare (www.hatchtech.com/hatchery-products/hatchcare/; Wijnen et al., 2021) in which eggs are placed in a honeycomb-structured setter tray with holes above a hatching basket. After hatching, chickens fall through the holes into the hatching basket. Solid feed is available at two sides, whilst water is available via a drinking gutter at the two other sides. After pulling, chickens stay in the hatching basket during processing and handling and are transported in these baskets to the rearing farm.

Both systems can provide light during hatching in contrast to conventional hatchers (Archer et al., 2009).

Examples of on-farm hatching are:

- Patio (www.vencomaticgroup.com/product/poultry-production/patio-on-farm-hatching; Van de Ven et al., 2009). This is a multi-tier system, often in two rows with corridors at both sides. In this system, fertile eggs on setter trays arrive at day 19 of incubation and are placed on an elevated track system. After hatching, chickens fall on a belt with litter where they have access to feeders and drinkers.

- X-Treck (www.vencomaticgroup.com/product/poultry-production/x-treck-on-farm-hatching; De Jong et al., 2019; 2020) in which eggs on setter trays are placed on a suspended rail system. After hatching, chickens fall on a belt where they can dry. They then drop to the floor where they have access to regular feeding and drinking systems.
- One2Born (www.one2born.com/en/; Jessen et al., 2021a, b) in which eggs are placed in a specially developed cardboard hatch holder, which is placed on the litter in the barn. After hatching and drying, chickens move from the hatch holder to the litter where they have access to regular feeding and drinking systems.
- NestBorn (www.nestborn.eu/), in which eggs are placed directly on the litter by a specially developed egg-placing machine. After hatching, chickens have access to normal feeding and drinking systems.

The last three systems can be considered as in-barn systems because eggs hatch in a conventional barn for rearing chickens.

3 Differences between conventional and alternative hatching systems: environmental factors

An important difference between conventional and alternative hatching systems is the immediate access after hatching to feed (and water) in alternative systems, whereas this is delayed in the conventional hatching system. However, there are several other differences between both systems. Van de Ven (2012) has provided an overview of differences between a conventional hatching system and the Patio system. Table 1 shows these differences (with some additional variables) together with a comparison with the in-barn (X-Treck, One2Born and NestBorn), HatchCare and SmartStart systems. Data about all alternative hatching systems are provided by their manufacturers (the authors of this chapter are not responsible for this data). From Table 1, it can be concluded that differences between conventional hatchers and alternative hatching systems are considerable. Additionally, among alternative hatching systems, differences are present as well. The most important differences between conventional and alternative hatchings systems will be discussed below, including potential consequences for day-old chicken quality as well as health, welfare and performance in later life.

Table 1 Summary of differences between hatching systems, based on fast-growing broiler chickens

Variable	Conventional[a]	HatchCare[b]	SmartStart[c]	Patio[d]	X-Treck[d]	NestBorn[e]	One2Born[f]
Climate							
Air temperature, average, °C	38.3	35.0–36.7	36.1	33.6	32.9	34.0	34.0–35.0
Relative humidity, average, %	55.6	43.2	55.0	32.0	49.0	>30%	>35.0
Air velocity, average, m/s	0.57	Maximum 0.38		0.08	0.10	<0.10	<0.15
CO_2 level, average, ppm	~3500	Maximum 2500	~3000	~900	~900	Maximum 3000	Maximum 2500
Air volume per egg, dm^3	0.6–0.9	1.1	1.1	90–130	210	200	200
Chickens per m^2 at hatch	340–380	357	282	22	22	20	20
Feed and water access	Delayed	Immediately	Immediately	Immediately	Immediately	Immediately	Immediately
Egg position	Horizontal	Vertical	Horizontal	Vertical	Vertical	Vertical	Vertical
Other conditions							
Light	Off	On	On	On	On	On	On
Background noise, dB	85	80–90	85	56	56	50	50
Dust level during hatching	High	Medium	Medium	Low	Low	Low	Low

Disinfectants during hatching	Gas or spray	No, gas or drinking water	Gas or spray	No	No	No	No
Chick handling	Yes	Only removal of second grade	Yes	No	No	No	No
Second-grade chickens	Excluded	Excluded	Excluded	Included	Included	Included	Included
Transport of chickens	Yes	Yes	Yes	No	No	No	No

Data in the table is provided by representatives of the different manufacturers (except for conventional), and the authors of this chapter are not responsible for the exact values in this table. The representatives of all mentioned hatching system manufacturers agreed with publication of this data.

[a] Van de Ven (2012).
[b] Personal communication HatchTech.
[c] Personal communication PasReform.
[d] Personal communication Vencomatic.
[e] Personal communication NestBorn
[f] Personal communication One2Born.

3.1 Climate

An important potential difference between a conventional hatcher and on-farm hatching systems in particular is the ambient temperature. It should be emphasised that air temperatures provided in Table 1 are of less importance than the eggshell temperature (EST). The temperature in older conventional hatchers in particular can easily exceed the optimal EST of 37.8°C (Lourens et al., 2005), as a result of suboptimal cooling capacity or air velocity. In the final days of incubation in particular, embryonic metabolic heat production is high (up to 150 mW per egg at day 18 of incubation and even considerably higher after internal pipping; Van den Brand et al., 2015). Consequently, heat needs to be removed to maintain an optimal EST. A high EST during the hatching phase [>38.9°C; >day 14 of incubation; see also Chapter 15 of this book (by Molenaar et al.)] has been shown to negatively affect the chicken quality at hatch (Maatjens et al., 2014, 2016) and might negatively affect later life performance as well (Molenaar et al., 2011).

On-farm hatching can also result in variable EST (Souza da Silva et al., 2021). However, due to the much larger air volume per egg, the chance of overheating in on-farm hatching systems is potentially lower. In all systems (conventional or alternative), it is important to adjust the ambient temperature to maintain the EST at the desired level. Causes and consequences of suboptimal incubation temperature are discussed in more detail elsewhere in this book.

Air velocity plays an important role in maintaining an optimal and uniform EST, through convective heat loss and reducing the difference between air temperature and EST (Meijerhof and Van Beek, 1993). Where egg density is high, high air velocity is needed to maintain the EST. This requires forced ventilation in the hatcher (in both a conventional hatcher and in HatchCare and SmartStart systems). However, when the volume of air per egg is high (as in on-farm hatching systems), air velocity can theoretically be lower as long as ventilation around the eggs is sufficient. This is relevant for the three different in-barn systems.

In the X-Treck system, eggs are placed in a setter tray at considerable distance from the belt (14–17 cm), which allows sufficient air flow around the eggs. In the One2Born system, eggs are placed in a hatch holder on the litter, whereas in the NestBorn system, eggs are placed in the litter. In the last two systems, overheating can happen sooner than in the X-Treck system when air velocity across eggs is too low, resulting in too little convective heat loss. Maintaining the EST at the optimal level can be done by adjusting the temperature in the hatcher or in the barn or by adjusting the air velocity across the eggs.

Effects of relative humidity (RH), especially during the hatching phase, have been little studied. Bruzual et al. (2000) did not show large effects of RH in the last 5 days of incubation on day-old chicken quality. Van der Pol et al. (2013) also found little effect of different levels of RH (30–35% vs 55–60%) over the whole incubation period on day-old chicken quality. Consequently, it is not expected that differences in RH among the different hatching systems will result in significant differences in day-old chicken quality. However, in most cases, RH is confounded with or related to other environmental factors like temperature, meaning that the RH should be seen in relationship to other climatic factors.

The last climatic factor that might impact hatchability and day-old chicken quality is carbon dioxide (CO_2). A number of studies have been done on CO_2 in relation to hatchability and day-old chicken quality (see for overview Van den Brand et al., 2021). However, in almost all studies, effects of CO_2 were probably confounded with changes in RH or EST [see also Chapter 15 (Molenaar et al.) in this book]. Van den Brand et al. (2021) found that effects of CO_2 appeared to be marginal as long as the EST, RH and oxygen concentration were maintained. Earlier findings of impact of CO_2 on day-old chicken quality or later life health, welfare and performance appear to be due to other climate factors rather than CO_2 itself.

Table 1 indicates that variation in climate settings among hatching systems is considerable and that ambient temperature in alternative hatching systems appears to be lower than in the conventional hatching system. However, maintaining EST is key to guaranteeing optimal day-old chicken quality in all hatching systems. As long as EST is maintained, other climate factors appear to be less important.

3.2 Egg position

One difference between hatching systems is the position of the egg. In the conventional and SmartStart hatching systems, eggs are in a horizontal position, as in a nest, whereas in the other hatching systems, eggs are kept in a vertical position, with the blunt end up. Van de Ven et al. (2011a) investigated the effects of egg position during the hatching phase (from day 17 of incubation onwards; horizontal, vertical blunt end up and vertical blunt end down) and found limited effects on hatching time, minor effects on day-old chicken quality and no effects on hatchability. We are not aware of other studies on egg position during the hatching phase. It appears that egg position in the final phase of incubation is not of large importance for hatchability and day-old chicken quality.

3.3 Light

Another difference between the conventional hatching system and all alternative hatching systems is the absence of light in the former and its presence in the latter. Light has been shown to penetrate the eggshell of broiler eggs (Güz et al., 2021) and reach the embryo. Many studies on the effects of light during incubation on hatching characteristics, day-old chicken quality and later life behaviour have been done in the last decades, including the effects of light colour, light intensity, light schedule and light source [see also Chapter 11 (Rozenboim) in this book]. Tainika and Bayraktar (2022) concluded that exposure to light during the first 18 days of incubation stimulates embryonic development and increases hatchability compared to exposure to light during the whole incubation period. Additionally, a light schedule of 12 h light: 12 h darkness per day had a larger stimulatory effect on embryonic development than when eggs were exposed to light continuously. Finally, a too high light intensity appeared to negatively affect embryo development.

Based on this overview, it appears that exposure of eggs to light for only 3–4 days in alternative hatching systems will only have a minor effect on hatchability and day-old chicken quality. However, exposure to light during incubation has been shown to affect the behaviour of laying hens in particular in later life. Light exposure can affect brain lateralisation and consequently might affect (pecking) behaviour in later life. De Haas et al. (2021) reviewed the effects of light during incubation on physiological responses and concluded that, when laying hens were exposed to light during incubation, their ability to discriminate food improved, potentially lowering the risk of injurious pecking. Effects of light exposure during incubation on behaviour of broilers have been less investigated, but it can be hypothesised that effects might be comparable to laying hens, although broilers have a much shorter life span.

3.4 Noise

The level of noise during the hatching phase differs between the conventional hatching system, HatchCare and SmartStart, on the one hand, compared with on-farm hatching systems on the other hand. Noise in the hatcher might disturb vocal communication between chickens. Vocal communication between parents and embryos has been shown to be important for hatching synchronisation [Rumpf and Tzschentke, 2010; Dixon et al., 2016; Chapter 15 (Molenaar et al.) in this book]. Sound or noise during the hatching phase has been shown to affect the neurophysiology of the chicken embryo (e.g. Sanyal et al. 2013; Roy et al., 2014).

Forced-ventilated hatchers make a lot of noise (often >85 dB), meaning that they might disturb vocal communication among chickens and may consequently affect hatchability, hatching time and hatching synchronisation. Donofre et al. (2020) showed that exposure to background noise of a hatcher (90 dB) from the start of incubation onwards resulted in an earlier start of hatching (6 h) than when eggs were exposed to 70 dB, but only when no species-specific vocalisations were available. The hatch window was not affected by background noise. At a low noise level (70 dB), the presence of species-specific vocalisations from day 12 onwards resulted in a 4 h earlier start of hatching and better navel and hock quality in day-old chickens than when these vocalisations were absent. However, Tong et al. (2015) showed that species-specific vocalisations from day 10 of incubation onwards delayed the start of hatching, without affecting the average hatching time and the hatch window.

There has been little research on whether or not hatcher noise during the hatching phase influences vocal communication and hatching synchronisation among chicken embryos (as shown for e.g. Mallard ducks and pheasants; Persson and Andersson, 1999). However, it is possible that machine noise in conventional hatchers masks the potential benefits of species-specific vocalisations, as suggested by Donofre et al. (2020). Consequently, it can be speculated that on-farm hatching with lower noise levels might reduce the hatch window, as suggested by Van de Ven et al. (2010), though this has not been extensively investigated.

3.5 Dust and use of disinfectants

In the conventional hatching system, day-old chickens are commonly exposed to high levels of dust (Mitchell and Waltmann, 2003), high pathogen loads (Mitchell and Waltmann, 2003) and disinfectants (quite often formaldehyde) (Zulkifli et al., 1999). Dust originates from fluff of chickens during drying and can contain very small particles (<10 μm, named PM10), which are able to enter the trachea and lungs (Lai et al., 2012). De Gouw et al. (2017) observed high levels of PM10 concentrations in a conventional hatcher, which might stimulate later immune response in chickens (Lai et al., 2012). High concentrations of dust in the hatcher might therefore have negative effects on day-old chicken quality and later life performance, but the specific effects of dust have not been investigated to our knowledge.

Disinfectants, such as formaldehyde, are often used during the hatching phase to reduce bacterial load. Formaldehyde fumigation (maximum 23.5 ppm for 54 h) has been shown to cause lesions in the trachea of day-old chickens (Zulkifli et al., 1999), which might negatively affect later life health and performance. Wlazlo et al. (2020) showed a lower body weight (BW) at day 7 post-hatch for quails exposed to formaldehyde during the hatching phase (no concentration and duration provided) compared to a non-exposed control group (22.8 g vs 24.1 g), but this difference disappeared after day 7. However, overall survivability till day 7 of rearing was highest after formaldehyde treatment during the hatching phase compared to a negative control group (93.6% vs 88.4%), which is another important health and welfare aspect to consider.

Disinfectants cannot be applied in all alternative hatching systems. In the SmartStart and HatchCare systems, a gas (or spray) disinfectant can be used that is comparable with the conventional hatching system. In the HatchCare system, disinfection can also be given via the drinking water. Gas or spray disinfection aims at reduction of airborne contamination, whereas disinfection via drinking water aims to reduce bacterial or fungal growth in chickens. On-farm hatching systems can include disinfection via the drinking water, but this is not common practice. Drinking lines are often disinfected before hatching of eggs in the barn to ensure that day-old chickens have access to clean drinking water. In on-farm hatching systems, the air volume per egg is considerably higher than in the other systems, reducing the concentration of dust and pathogens on chicken level. Consequently, the use of disinfectants during the hatching phase is likely to be of less importance. Whether or not a disinfectant is used during the hatching phase depends on the hatchery or the farmer who, in turn, considers factors such as the origin of the eggs (e.g. breeder age or the inclusion of floor eggs) and (nest) hygiene protocols in making such a decision.

4 Differences between conventional and alternative hatching systems: management factors

4.1 Chicken handling and transport

In the conventional hatching system, chickens are removed from the hatching baskets (using an egg separator), moved over conveyor belts for selection, are vaccinated and manually sexed when necessary, counted and put in transport crates, after which they are moved to a holding room before transportation to the rearing farm. In the SmartStart system, chickens are handled in the same way or are manually removed from the hatching baskets, whereafter they are exposed to the same handling procedures. In the HatchCare system, only second-grade chickens are removed from the hatching cradles during processing, whereas first-grade chickens stay in the cradles. In the on-farm hatching systems, no chicken handling takes place at all. Chicken handling at the hatchery and transportation are both considered as potential stressors. They have been found to affect stress hormones such as corticosterone, behaviour and performance in laying hen chickens (Hedlund and Jensen, 2021; Hedlund et al., 2019, 2021). Giersberg et al. (2020a) investigated the effects of chicken-handling procedures in the hatchery more in detail, by varying the conveyor belt speed, drop height and acceleration from one conveyor belt to another. They concluded that broiler chicken health, welfare and behaviour were only affected to a limited extent by hatchery procedures, as long as the drop height (<280 mm) and conveyor belt speed (<27 m/min) were not too high.

Prolonged transport (4 h or 10 h) has been shown to be a fairly severe stressor for day-old broiler chickens, resulting in lower BW but less severe footpad dermatitis (FPD) at day 21 of age (Bergoug et al., 2013). Valros et al. (2008) showed that prolonged simulated transport (14 h) only showed minor effects on behaviour of laying hen chickens. Jacobs et al. (2017) showed that a prolonged transport duration (11 h) of day-old broiler chickens resulted in higher corticosterone concentrations and a tendency to lower glucose and lactate concentration upon arrival than after a transport duration of 1.5 h. In a comparable study, Jacobs et al. (2016) showed that prolonged transport (11 h) resulted in lower BW and yolk sac weight upon arrival (compared to a transport duration of 1.5 h), without affecting the chicken quality score (Tona et al., 2003), but performance or mortality during rearing did not differ between chickens of both transport durations. Hollemans et al. (2018) also showed that simulated transport for 3 h did not affect day-old chicken quality and later life performance. Varying results among studies might be related to factors such as poultry breed (layers vs broilers), hatchery, type of truck and road quality. However, hatchery handling and/or transport of day-old chickens, especially under suboptimal environmental conditions (especially temperature), might affect welfare and performance of chickens during and post-transport.

It is important to note that all these studies involved unfed chickens, making it unclear whether handling and particularly transport of fed chickens would produce comparable results. It is currently unclear whether or not chickens fed in the hatchery and transported afterwards without feed do experience negative effects upon placement in the rearing farm. Other factors (such as transport temperature) might also interfere in potential effects on health and welfare.

4.2 Removal of second-grade chickens

Second-grade chickens are normally removed in the hatchery during the selection process. Percentages of second-grade chickens vary among hatcheries and flocks. Van de Ven et al. (2012) found an average percentage of 1.15% second-grade chickens in both the conventional hatching and the Patio system. Second-grade chickens can be classified into six classes (physical anomaly, abnormal down, leg deformation, weak appearance, low-quality score and dead before examination) (Van de Ven et al., 2012). The mortality of second-grade chickens is high (when they are not removed from the flock), varying from 26.3% for the abnormal down class to 97.1% for the leg deformation class (Van de Ven et al., 2012). In on-farm hatching systems, second-grade chickens are not selected at the day of hatch, meaning that they will be present in the flock and selection should be done by the farmer in the first few days. In the case where second-grade chickens are counted as mortality, this may increase first-week and overall mortality figures at the broiler farm.

4.3 Time to feed and water access

Although more factors are involved (see above), the immediate post-hatch provision of feed and water is an important aspect of alternative hatching systems. In the last two decades, dozens of studies have been performed to study the effects of early feeding. In this chapter, only a few highlights will be discussed related specifically to the effects of solely early feed and water intake, but it will not cover the broader consequences of early feeding on chicken quality and later life performance. Recent reviews of specific aspects of early feeding highlight the main effects of early feeding of broiler chickens (De Jong et al., 2017; Taha-Abdelaziz et al., 2018; Jha et al., 2019).

It is little investigated how soon after hatching chickens eat. One of our students investigated how soon broiler chickens eat after hatching (Van Marwijk, unpublished results) and showed that most chickens (>85%) ingested the first feed within 150 min of emergence from the eggshell (Fig. 1a). For the first hatching chickens, it took longer to start eating than for the later hatching chickens. The latter appear to learn from their earlier hatched conspecifics and had their first feed sooner after hatching (Fig. 1b). Chickens originating from old breeder flocks (59 weeks) hatched later (on average 8 h) than chickens

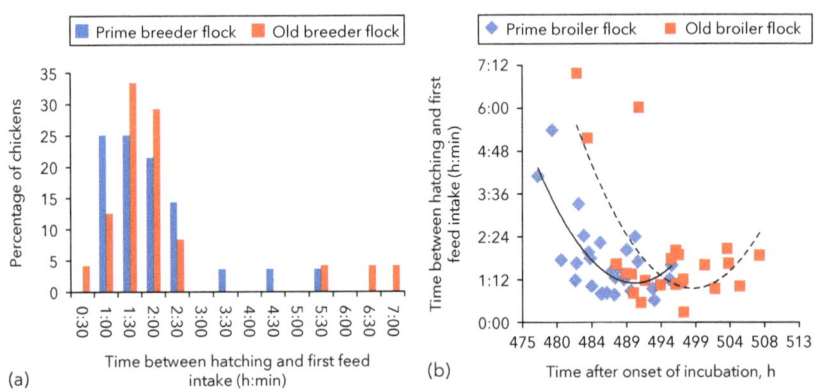

Figure 1 Effect of time between hatching and first feed intake and percentage of chickens eating (a) and the relationship between time after onset of incubation (hatching moment) and the time between hatching and first feed intake (b) in broiler chickens originating from a prime breeder flock (36 weeks) or an old breeder flock (59 weeks). Source: Van Marwijk (unpublished results).

originating from prime breeder flocks (36 weeks), but the interval between hatching and first ingestion of feed did not differ.

Another question is how much chickens eat immediately after hatching before being placed in the rearing farm. Very limited data on this topic is available. Cargill Animal Nutrition (unpublished results) showed that chickens eat approximately 0.1-0.2 g per chicken per hour in the first 2 h after hatching, independent of hatching time, and rest thereafter for approximately 15-20 h before they start eating again. Delayed-fed chickens (24 h) were eating considerably more after first provision of feed and continued eating thereafter (Fig. 2). Boyner et al. (2021) recently investigated eating behaviour in broiler chickens and showed that on average 5% of broiler chickens showed eating-related behaviour 25.4 h after hatching (29.5 h for early hatching chickens and 21.7 h for late hatching chickens). After approximately 30 h post-hatch, 50% of the chickens had a full crop filling.

These results demonstrate that feed intake in the first 24 h after hatching is quite limited. However, it appears that these small amounts of feed (and water) intake affect the metabolic rate of just-hatched chickens. Van Roovert-Reijrink et al. (2017) showed that the average metabolic heat production of immediately fed chickens was 521 mW/chicken at normal pulling time (day 21.5 after onset of incubation), whereas this was 340 mW for the unfed conspecifics (Fig. 3). This higher metabolic heat production is also reflected in a higher resistance to cold exposure in the first few days after placement in the rearing farm. Van

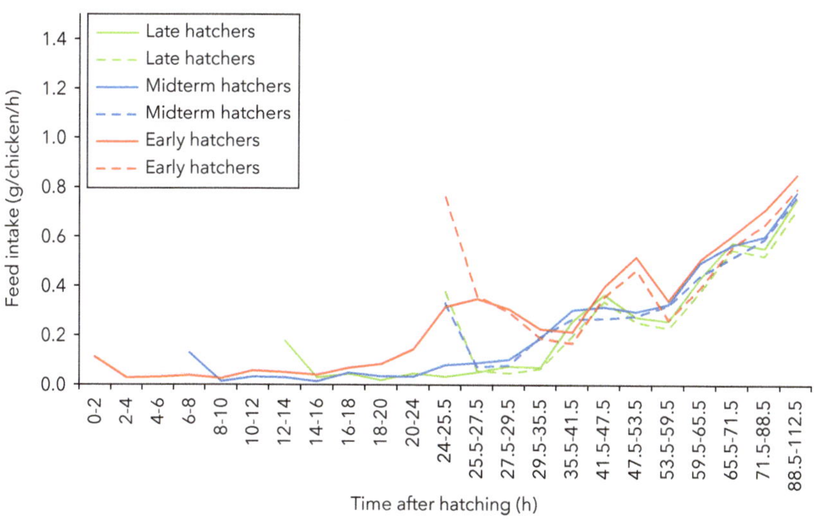

Figure 2 Relationship between time after hatching and feed intake for early, midterm and late hatching broiler chickens that are fed immediately after hatch (direct) or are withheld from feed for 24 h (delayed). Courtesy of Cargill Animal Nutrition (unpublished results).

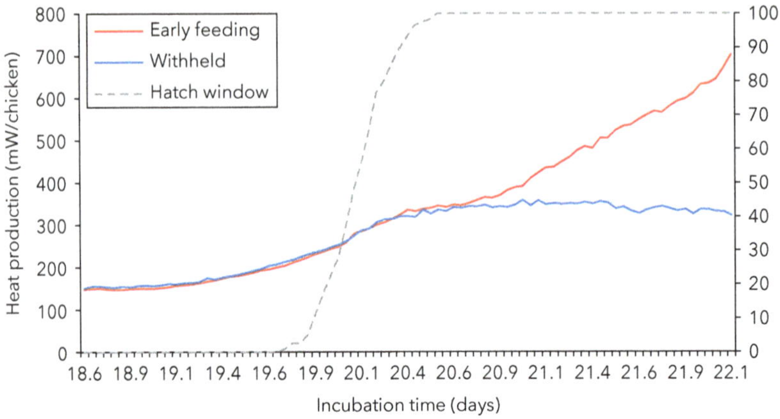

Figure 3 Relationship between time after start of incubation and heat production of immediately fed and unfed broiler chickens. Source: Van Roovert-Reijrink et al. (2017).

den Brand et al. (2010) demonstrated that immediately fed broiler chickens had a higher cloacal temperature before, immediately after and 30 min after a cold exposure (20°C for 30 min) at 2 or 3 days of age compared to their unfed conspecifics.

It is obvious that early hatching chickens have the most benefit from the availability of feed and water immediately after hatching, because they remain the longest in the conventional hatcher before pulling. This is for example expressed in interaction effects between hatching time and the moment of access to feed and water post-hatch (immediately or delayed) on day-old chicken characteristics, as found by Van de Ven et al. (2013), Lamot et al. (2014) and Deines et al. (2021a,b). The moment of hatching and spread of hatch should thus be taken into account when evaluating the effects of early feeding on chicken quality at placement, flock uniformity and later life health, welfare and performance (Van de Ven et al., 2011b) and, consequently, when hatching systems are evaluated.

5 Effects of alternative hatching systems on health, welfare and production

A number of studies have compared different alternative hatching systems with the conventional hatching system. This section discusses the most important results related to day-old chicken quality as well as later life health, welfare and performance. This section is based on studies using commercial, semi-commercial and experimental settings, as shown in Table 2. This section only considers studies that undertook a system comparison.

Table 2 Studies on the effects of different hatching systems of broiler chickens on day-old chicken quality and later life health, welfare and performance

Study	Conventional	HatchCare	Patio	X-Treck	One2Born	Other	Level[a]
Van de Ven et al. (2013)	X		X				Commercial
De Jong et al. (2019)	X			X			Commercial
De Jong et al. (2020)	X			X			Semi-commercial
Giersberg et al. (2020b)	X			X			Semi-commercial
Giersberg et al. (2021)	X	X		X			Semi-commercial
Jessen et al. (2021a,b)	X				X		Commercial
Lingens et al. (2021)	X					X[b]	Experimental
Souza da Silva et al. (2021)	X	X		X			Semi-commercial

[a] Commercial = large flocks at a commercial farm with barn as experimental unit; semi-commercial = flocks of <1500 chickens per pen with pen as experimental unit; experimental = flocks in small pens, with pen as experimental unit.
[b] System not indicated.

5.1 Hatchability and chicken quality

Souza da Silva (2021) found no difference in hatchability between the conventional, X-Treck and HatchCare systems, whereas Jessen et al. (2021b) found 0.5% higher hatchability in the One2Born compared to the conventional system. It appears that hatchability is not a point of concern in alternative hatching systems, as long as the temperature is maintained at the correct level (as discussed earlier).

Chicken quality can be expressed as the percentage of second-grade chickens, BW, yolk-free body mass (YFBM), navel score and organ development. Jessen et al. (2021b) found a lower percentage of second-grade chickens in the One2Born compared to the conventional hatching system (1.16% vs 1.61%). However, the selection of chickens was done by different persons, which means inter-observer variation cannot be ruled out. Jessen et al. (2021b) found that the quality of first-grade chickens was worse in the One2Born compared to the conventional system, based on the Pasgar score (Boerjan, 2002). This was mainly due to higher suboptimal navel scores as well as those for large yolk sac, impaired raising reflex and presence of red hocks. These chicken characteristics are typically related to a very high incubation temperature during the final days of incubation (Narinç and Aydemir, 2021), suggesting that in this particular study, eggs in the One2Born system experienced a too high EST. Worse navel quality and more red hocks were also found in the in-barn hatching system (X-Treck) by Souza da Silva et al. (2021) compared to the conventional and HatchCare system, with probably also EST as the explaining factor.

BW of chickens at placement was consistently higher in alternative hatching systems compared to the conventional hatching system (Van de Ven et al., 2012; De Jong et al., 2019, 2020; Jessen et al., 2021a; Lingens et al., 2021; Souza da Silva et al., 2021), which is most probably due to the feed and water intake in alternative hatching systems. This is also expressed in a higher YFBM (Van de Ven et al., 2012; De Jong et al., 2019, 2020; Souza da Silva et al., 2021). However, effects on residual yolk weight varied. De Jong et al. (2020) and Souza da Silva (2021) did not find that hatching system affected residual yolk weight, whereas De Jong et al. (2019) found a higher residual yolk weight at placement for the X-Treck system compared to the conventional system.

In parallel with YFBM, consistently higher stomach (gizzard plus proventriculus) and intestine weights (De Jong et al., 2019, 2020; Souza da Silva et al., 2021) (relative to YFBM) were found at placement in alternative hatching systems compared to the conventional hatching system. Hatching systems were found not to affect relative heart, liver, bursa and spleen weights. Altogether, this shows that alternative hatching systems result in further developed chickens at the moment that conventional hatched chickens normally are placed in the barn, which appears mainly due to earlier feed and water intake.

5.2 Mortality and performance

An important broiler chicken quality, health and welfare characteristic is first-week mortality. This can be difficult to study in alternative hatching systems because, in the HatchCare and SmartStart systems, second-grade chickens are removed before placement at the rearing farm, whereas in on-farm hatching systems, second-grade chickens are included and will be selected in the first days after placement. Including culling in the first-week, mortality can create a distorted picture (see also Section 4.2). When culling was not taken into account, first-week mortality varied among studies. Jessen et al. (2021b; 0.41% vs 0.99%, respectively) and Lingens et al. (2021; 1.08% vs 1.53%, respectively; not statistically tested) found lower first-week mortality in on-farm hatched chickens compared to conventionally hatched chickens. De Jong et al. (2020) found a tendency in the same direction (0.82% vs 1.15%, respectively). However, De Jong et al. (2019) and Souza da Silva et al (2021) did not find differences in first-week mortality between hatching systems. Total mortality (excluding selection) over the whole rearing period was lower in alternative hatching systems [De Jong et al., 2020; Jessen et al., 2021b; Lingens et al., 2021 (not statically tested)] than in the conventional hatching system or did not differ between hatching systems (De Jong et al., 2019; Souza da Silva et al., 2021).

BW at slaughter also showed differing results among studies. De Jong et al. (2019; 2020) did not find differences in BW at slaughter age, whereas Lingens et al. (2021) and Souza da Silva et al. (2021) found higher BWs at slaughter in alternative hatching systems compared to the conventional hatching system. This supports the findings of De Jong et al. (2017) who concluded based on an overview of studies that effects of early feeding on slaughter weight vary among studies, which might be due to factors such as broiler strain, age of the breeder flock and incubation conditions in different systems. Lamot (2017) extrapolated growing curves of early-fed and delayed-fed broiler chickens and concluded that differences in final BW could largely be explained by the later start of eating in delayed-fed chickens. So, it can be suggested that reaching the slaughter weight earlier in alternative hatching systems is an effect of the earlier start of feeding rather than an effect of changes in chicken development. This is also supported by the lack of effect of the hatching system on feed conversion ratio (De Jong et al., 2019; 2020; Souza da Silva et al., 2021).

5.3 Health and welfare

Important health aspects related to hatching systems are locomotion-related characteristics, such as gait score and FPD, and rejections at slaughter age. Prevalence of FPD did not differ between hatching systems (Jessen et al., 2021b), tended to be lower (De Jong et al., 2019, 2020) or was lower in alternative hatching systems (Giersberg et al., 2021; Lingens et al., 2021; Souza da Silva et al., 2021) compared to the conventional hatching system.

Because the prevalence of FPD is strongly related to litter quality (De Jong et al., 2014), litter quality in different systems has also been assessed. Some studies suggested that hatching systems had little effect on dry matter content of excreta (Lingens et al., 2021) or litter (Jessen et al., 2021b; Lingens et al., 2021). However, De Jong et al. (2020) found a higher litter dry matter content and De Jong et al. (2019) found a tendency to better litter quality after hatching in an on-farm hatching system than after hatching in a conventional hatching system. Because effects of alternative hatching systems on litter quality appear to be limited, it can be hypothesised that other factors might play a role in the prevalence of FPD. De Jong et al. (2012) showed that the hatchery from which chickens originated was a significant factor affecting the prevalence of FPD. Consequently, it cannot be ruled out that the different hatching systems might have influenced the prevalence of FPD without e.g. affecting intestinal integrity due to the provision of early feeding (e.g. Liu et al., 2020; Proszkowiec-Weglarz et al., 2020). Which hatchery-related factors might play a role and what physiological mechanisms are involved remain unclear. Despite the effects of hatching systems on FPD, no effects were found on gait of chickens (De Jong et al., 2019; 2020; Giersberg et al., 2021).

Jessen et al. (2021b) found a higher prevalence of breast blisters at slaughter in male chickens after hatching in a conventional system compared to the One2Born system, but this was not found in female chickens, whilst no effects on skin lesions and rejections were found at the slaughterhouse. The latter was confirmed by De Jong et al. (2019, 2020) and Giersberg et al. (2021). A last and important health and welfare aspect is the prevalence of breast myopathies at slaughter, such as wooden breast and white striping. Souza da Silva et al. (2021) demonstrated a higher prevalence of both breast myopathies at slaughter age after hatching in the HatchCare and X-Treck system compared to the conventional hatching system. Growth rate was higher in alternative hatching systems than in the conventional system, which might explain this higher prevalence since growth rate is an important predisposing factor for the occurrence of breast myopathies (for review see Petracci et al., 2019).

5.4 Behaviour

Behaviour of chickens hatched from different systems has been evaluated by home-pen observations (Giersberg et al., 2020b) and by several fear- and anxiety-related tests, such as the novel environment test (Giersberg et al., 2020b, 2021), the novel object test (Giersberg et al., 2020b, 2021; Jessen et al., 2021a), human approach test (Giersberg et al., 2020b, 2021; Jessen et al.,

2021a) and the tonic immobility test (Giersberg et al., 2021). Observations were done in day-old chickens and at different ages during the rearing period.

Home-pen behaviour hardly differed between the conventional hatching system and the X-Treck system at days 2, 5, 12, 20 and 36 post-hatch (Giersberg et al., 2020b). Results varied in fear- and anxiety-related tests. Giersberg et al. (2020b) showed that chickens hatched in the X-Treck system were more fearful and less active than conventionally hatched chickens; they suggested that this might be related to the higher intrinsic motivation to search for food or more exposure to humans or objects in the conventionally hatched chickens. However, a follow-up study (Giersberg et al., 2021) found no effect on fear- and anxiety-related behaviours between conventionally hatched chickens and chickens hatched in the X-Treck or HatchCare system. Finally, Jessen et al. (2021a) found a lower fear response towards humans in One2Born hatched chickens (day 1) than in conventionally hatched chickens based on a human approach test. However, this could not be confirmed in a novel object test. These varying results might have several reasons, such as broiler strain (fast vs slower growing), age of measuring (day 1 or later after hatching) or the type of test used. Incubation settings in the last couple of days before hatching may also have affected chicken quality and consequently the behaviour of the chickens. Because each study is performed at another place with different settings, it cannot be ruled out that findings are more related to the actual settings (e.g. temperature) of the different hatching systems than the hatching system itself.

6 Conclusion

Alternative hatching systems developed in the last 15 years differ in several ways from the conventional hatching system. An important difference is the immediate access to feed and water after hatching, but other (climate) aspects such as temperature play a role as well. Effects of different hatching systems on day-old chicken quality and health, welfare and performance vary and appear to be related to (suboptimal) settings in both the conventional and alternative hatching systems. This suggests that circumstances in each hatching system are of more importance than the hatching system itself.

7 References

Archer, G. S., Shivaprasad, H. L. and Mench, J. A. (2009). Effect of providing light during incubation on the health, productivity, and behavior of broiler chickens, *Poult. Sci.* 88(1):29–37.

Bergoug, H., Guinebretière, M., Tong, Q., Roulston, N., Romanini, C. E. B., Exadaktylos, V., Berckmans, D., Garain, P., Demmers, T. G. M., McGonnell, I. M., Bahr, C., Burel, C.,

Eterradossi, N. and Michel, V. (2013). Effect of transportation duration of 1-day-old chicks on postplacement production performances and pododermatitis of broilers up to slaughter age, *Poult. Sci.* 92(12):3300–3309.

Boerjan, M. (2002). Programs for single stage incubation and chick quality, *Av. Poult. Biol. Rev.* 13:237–238.

Boyner, M., Ivarsson, E., Andersson Franko, M. A., Rezaei, M. and Wall, H. (2021). Effect of hatching time on time to first feed intake, organ development, enzymatic activity and growth in broiler chicks on-farm, *Animal* 15(2):100083.

Bruzual, J. J., Peak, S. D., Brake, J. and Peebles, E. D. (2000). Effects of relative humidity during the last five days of incubation and brooding temperature on performance of broiler chicks from young broiler breeders, *Poult. Sci.* 79(10):1385–1391.

Careghi, C., Tona, K., Onagbesan, O., Buyse, J., Decuypere, E. and Bruggeman, V. (2005). The effects of the spread of hatch and interaction with delayed feed access after hatch on broiler performance until seven days of age, *Poult. Sci.* 84(8):1314–1320.

De Gouw, P., Van de Ven, L. J. F., Lourens, S., Kemp, B. and Van den Brand, H. (2017). Effects of dust, formaldehyde and delayed feeding on early postnatal development of broiler chickens, *Res. Vet. Sci.* 112:201–207.

De Haas, E. N., Newberry, R. C., Edgar, J., Riber, A. B., Estevez, I., Ferrante, V., Hernandez, C. E., Bjaer, J. B., Özkan, S., Dimitrov, I., Rodenburg, T. B. and Janczak, A. M. (2021). Prenatal and early postnatal behavioural programming in laying hens, with possible implications for the development of injurious pecking, *Front. Vet. Sci.* 8:678500.

De Jong, I. C., Gunnink, H. and Van Harn, J. (2014). Wet litter not only induces footpad dermatitis but also reduces overall welfare, technical performance, and carcass yield in broiler chickens, *J. Appl. Poult. Res.* 23(1):51–58.

De Jong, I. C., Gunnink, H., Van Hattum, T., Van Riel, J. W., Raaijmakers, M. M. P., Zoet, E. S. and Van den Brand, H. (2019). Comparison of performance, health and welfare aspects between commercially housed hatchery-hatched and on-farm hatched broiler flocks, *Animal* 13(6):1269–1277.

De Jong, I. C., Van Harn, J., Gunnink, H., Hindle, V. A. and Lourens, A. (2012). Footpad dermatitis in Dutch broiler flocks: Prevalence and factors of influence, *Poult. Sci.* 91(7):1569–1574.

De Jong, I. C., Van Hattum, T., Van Riel, J. W., De Baere, K., Kempen, I., Cardinaels, S. and Gunnink, H. (2020). Effects of on-farm and traditional hatching on welfare, health, and performance of broiler chickens, *Poult. Sci.* 99(10):4662–4671.

De Jong, I. C., Van Riel, J., Bracke, M. B. M. and Van den Brand, H. (2017). A 'meta-analysis' of effects of post-hatch food and water deprivation on development, performance and welfare of chickens, *PLoS ONE* 12(12):e0189350.

Decuypere, E., Tona, K., Bruggeman, V. and Bamelis, F. (2001). The day-old chick: A crucial hinge between breeders and broilers, *World Poult. Sci. J.* 57(2):127–138.

Deines, J. R., Clark, F. D., Yoho, D. E., Bramwell, R. K. and Rochell, S. J. (2021a). Effects of hatch window and nutrient access in the hatcher on performance and processing yields of broilers reared with equal hatch window representation, *Animals (Basel)* 11(5):1228.

Deines, J. R., Clark, F. D., Yoho, D. E., Bramwell, R. K. and Rochell, S. J. (2021b). Effects of hatch window and nutrient access in the hatcher on performance and processing yields of broilers reared according to time of hatch, *Poult. Sci.* 100(9):101295.

Dixon, L. M., Sparks, N. H. C. and Rutherford, K. M. D. (2016). Early experiences matter: A review of the effects of prenatal environment on offspring characteristics in poultry, *Poult. Sci.* 95(3):489–499.

Donofre, A. C., da Silva, I. J. O. and Ferreira, I. E. P. (2020). Sound exposure and its beneficial effects on embryonic growth and hatching of broiler chicks, *Br. Poult. Sci.* 61(1):79–85.

Giersberg, M. F., Molenaar, R., De Jong, I. C., Souza da Silva, C., Van den Brand, H., Kemp, B. and Rodenburg, T. B. (2021). Effects of hatching system on the welfare of broiler chickens in early and later life, *Poult. Sci.* 100(3):100946.

Giersberg, M. F., Molenaar, R., Pieters, R., Boyer, W. and Rodenburg, T. B. (2020a). Effects of drop height, convoyer belt speed, and acceleration on the welfare of broiler chickens in early and later life, *Poult. Sci.* 99(12):6293–6299.

Giersberg, M. F., Poolen, I., De Baere, K., Gunnink, H., Van Hattum, T., Van Riel, J. W. and De Jong, I. C. (2020b). Comparative assessment of general behaviour and fear-related responses in hatchery-hatched and on-farm hatched broiler chickens, *Appl. Anim. Behav. Sci.* 232:105100.

Güz, B. C., Molenaar, R., De Jong, I. C., Kemp, B., Van Krimpen, M. and Van den Brand, H. (2021). Effects of green LED light during incubation and dietary organic macro and trace minerals during rearing on tibia characteristics of broiler chickens at slaughter age, *Poult. Sci.* 100(2):707–720.

Hedlund, L. and Jensen, P. (2021). Incubation and hatching conditions of laying hen chicks explain a large part of the stress effects from commercial large-scale hatcheries, *Poult. Sci.* 100(1):1–8. doi:10.1016/j.psj.2020.10.015.

Hedlund, L., Palazon, T. and Jensen, P. (2021). Stress during commercial hatching processing induces long-time negative cognitive judgements bias in chickens, *Animals* 11(4):1083.

Hedlund, L., Whittle, R. and Jensen, P. (2019). Effects of commercial hatchery processing on short- and long-term stress responses in laying hens, *Sci. Rep.* 9(1):2367.

Hollemans, M. S., De Vries, S., Lammers, A. and Clouard, C. (2018). Effects of early nutrition and transport of 1-day-old chickens on production performance and fear response, *Poult. Sci.* 97(7):2534–2542.

Jacobs, L., Delezie, E., Duchateau, L., Goethals, K., Ampe, B., Lambrecht, E., Gellynck, X. and Tuyttens, F. A. M. (2016). Effect of post-hatch transportation duration and parental age on broiler chicken quality, welfare, and productivity, *Poult. Sci.* 95(9):1973–1979.

Jacobs, L., Delezie, E., Duchateau, L., Goethals, K., Ampe, B., Buyse, J. and Tuyttens, F. A. M. (2017). Impact of transportation duration on stress responses in day-old chicks from young and old breeders, *Res. Vet. Sci.* 112:172–176.

Jessen, C. T., Foldager, L. and Riber, A. B. (2021a). Effects of hatching on-farm on behaviour, first week performance, fear level and range use of organic broilers, *Appl. Anim. Behav. Sci.* 238:105319.

Jessen, C. T., Foldager, L. and Riber, A. B. (2021b). Effects of hatching on-farm on performance and welfare of organic broilers, *Poult. Sci.* 100(9):101292.

Jha, R., Singh, A. K., Yadav, S., Berrocoso, J. F. D. and Mishra, B. (2019). Early nutrition programming (in ovo and post-hatch feeding) as a strategy to modulate gut health of poultry, *Front. Vet. Sci.* 6:82.

Lai, H. T. L., Nieuwland, M. G. B., Aarnink, A. J. A., Kemp, B. and Parmentier, H. K. (2012). Effects of 2 size classes of intratracheally administered airborne dust particles on primary and secondary specific antibody responses and body weight gain of broilers: A pilot study on the effects of naturally occurring dust, *Poult. Sci.* 91(3):604–615.

Lamot, D. M. (2017). *First week nutrition for broiler chickens: Effects on growth, metabolic status, organ development, and carcass composition*. PhD Thesis, Wageningen University, Wageningen, The Netherlands.

Lamot, D. M., Van de Linde, I. B., Molenaar, R., Van der Pol, C. W., Wijtten, P. J. A., Kemp, B. and Van den Brand, H. (2014). Effects of moment of hatch and feed access on chicken development, *Poult. Sci.* 93(10):2604–2614.

Leksrisompong, N., Romero-Sanchez, H., Plumstead, P. W., Brannan, K. E. and Brake, J. (2007). Broiler incubation. 1. Effect of elevated temperature during late incubation on body weight and organs of chicks, *Poult. Sci.* 86(12):2685–2691.

Lingens, J. B., Abd El-Wahab, A., Ahmed, M. F. E., Schubrert, D. C., Sürie, C. and Visscher, C. (2021). Effects of early nutrition of hatched chicks on welfare and growth performance: A pilot study, *Animals (Basel)* 11(10):2888.

Liu, K., Jia, M. and Wong, E. A. (2020). Delayed access to feed affects broiler small intestinal morphology and goblet cell ontogeny, *Poult. Sci.* 99(11):5275–5285.

Lourens, A., Van den Brand, H., Meijerhof, R. and Kemp, B. (2005). Effect of eggshell temperature during incubation on embryo development, hatchability, and posthatch development, *Poult. Sci.* 84(6):914–920.

Maatjens, C. M., Reijrink, I. A. M., Molenaar, R., Van der Pol, C. W., Kemp, B. and Van den Brand, H. (2014). Temperature and CO2 during the hatching phase. I. Effects on chick quality and organ development, *Poult. Sci.* 93(3):645–654.

Maatjens, C. M., van Roovert-Reijrink, I. A., Engel, B., van der Pol, C. W., Kemp, B. and van den Brand, H. (2016). Temperature during the last week of incubation. I. Effects on hatching pattern and broiler chicken embryonic organ development, *Poult. Sci.* 95(4):956–965.

Meijerhof, R. and Van Beek, G. (1993). Mathematical modelling of temperature and moisture loss of hatching eggs, *J. Theor. Biol.* 165(1):27–41.

Mitchell, B. W. and Waltman, W. D. (2003). Reducing airborne pathogens and dust in commercial hatching cabinets with an electrostatic space charge system, *Av. Dis.* 47(2):247–253.

Molenaar, R., Hulet, R., Meijerhof, R., Maatjens, C. M., Kemp, B. and Van den Brand, H. (2011). High eggshell temperatures during incubation decrease growth performance and increase the incidence of ascites in broiler chickens, *Poult. Sci.* 90(3):624–632.

Narinç, D. and Aydemir, E. (2021). Chick quality: An overview of measurement techniques and influencing factors, *Worlds Poult. Sci. J.* 77(2):313–329.

Persson, I. and Andersson, G. (1999). Intraclutch hatch synchronization in pheasants and mallard ducks, *Ethology* 105(12):1087–1096.

Petracci, M., Soglia, F., Madruga, M., Carvalho, L., Ida, E. and Estevez, M. (2019). Wooden-breast, white striping, and spaghetti meat: Causes, consequences and consumer perception of emerging broiler meat abnormalities, *Compr. Rev. Food Sci. Food Saf.* 18(2):565–583.

Proszkowiec-Weglarz, M., Schreier, L. L., Kahl, S., Miska, K. B., Russell, B. and Elsasser, T. H. (2020). Effect of delayed feeding post-hatch on expression of tight junction- and gut barrier-related genes in the small intestine of broiler chickens during neonatal development, *Poult. Sci.* 99(10):4714–4729.

Roy, S., Nag, T. C., Upadhyay, A. D., Mathur, R. and Jain, S. (2014). Prenatal music stimulation facilitates the postnatal functional development of the auditory as well as visual system in chicks (*Gallus domesticus*), *J. Biosci.* 39(1):107–117.

Rumpf, M. and Tzschentke, B. (2010). Perinatal acoustic communication in birds: Why do birds vocalize in the egg?, *TOOENIJ* 3(1):141-149.

Sanyal, T., Kumar, V., Nag, T. C., Jain, S., Sreenivas, V. and Wadhwa, S. (2013). Prenatal loud music and noise: Differential impact on physiological arousal, hippocampal synaptogenesis and spatial behavior in one day-old chicks, *PLoS ONE* 8(7):e67347.

Souza da Silva, C., Molenaar, R., Giersberg, M. F., Rodenburg, T. B., Van Riel, J. W., De Baere, K., Van Dosselaer, I., Kemp, B., Van den Brand, H. and De Jong, I. C. (2021). Day-old chicken quality and performance of broiler chickens from 3 different hatching systems, *Poult. Sci.* 100(3):100953.

Taha-Abdelaziz, K., Hodgins, D. C., Lammers, A., Alkie, T. N. and Sharif, S. (2018). Effects of early feeding and dietary interventions on development of lymphoid organs and immune competence in neonatal chickens: A review, *Vet. Immunol. Immunopathol.* 201:1-11.

Tainika, B. and Bayraktar, Ö. H. (2022). Lighted incubation: Embryonic development, hatchability and hatching quality of broiler chicks, *World Poult. Sci. J.* 78(1):161-178. doi:10.1080/00439339.2022.1988806.

Tona, K., Bamelis, F., De Ketelaere, B., Bruggeman, V., Moraes, V. M. B., Buyse, J., Onagbesan, O. and Decuypere, E. (2003). Effects of egg storage time on spread of hatch, chick quality and chick juvenile growth, *Poult. Sci.* 82(5):736-741.

Tona, K., Bruggeman, V., Onagbesan, O., Bamelis, F., Gbeassor, M., Mertens, K. and Decuypere, E. (2005). Day-old chick quality: Relationship to hatching egg quality, adequate incubation practice and prediction of broiler performance, *Av. Poult. Biol. Rev.* 16(2):109-119.

Tong, Q., McGonnell, I. M., Romanini, C. E. B., Bergoug, H., Roulston, N., Exadaktylos, V., Berckmans, D., Bahr, C., Guinebretière, M., Eterradossi, N., Garain, P. and Demmers, T. (2015). Effect of species-specific sound stimulation on the development and hatching of broiler chicks, *Br. Poult. Sci.* 56(2):143-148.

Valros, A., Vuorenmaa, R. and Janczak, A. M. (2008). Effect of simulated long transport on behavioural characteristics in two strains of laying hen chicks, *Appl. Anim. Behav. Sci.* 109(1):58-67.

Van de Ven, L., Gosselink, M., Van Wagenberg, V., Kemp, B. and Van den Brand, H. (2010). Hatching system affects broiler chick behavior. In: *Proceedings of the 13th European Poultry Conference*, Tours, France.

Van de Ven, L. J. F. (2012). *Effects of hatching time and hatching system on broiler chick development*. PhD Thesis, Wageningen University, Wageningen, The Netherlands.

Van de Ven, L. J. F., Baller, L., Van Wagenberg, A. V., Kemp, B. and Van den Brand, H. (2011a). Effects of egg position during late incubation on hatching parameters and chick quality, *Poult. Sci.* 90(10):2342-2347.

Van de Ven, L. J. F., Van Wagenberg, A. V., Debonne, M., Decuypere, E., Kemp, B. and Van den Brand, H. (2011b). Hatching system and time effects on broiler physiology and post hatch growth, *Poult. Sci.* 90(6):1267-1275.

Van de Ven, L. J. F., Van Wagenberg, A. V., Decuypere, E., Kemp, B. and Van den Brand, H. (2013). Perinatal broiler physiology between hatching and chick collection in two hatching systems, *Poult. Sci.* 92(4):1050-1061.

Van de Ven, L. J. F., Van Wagenberg, A. V., Groot Koerkamp, P. W. G., Kemp, B. and Van den Brand, H. (2009). Effects of a combined hatching and brooding system on hatchability, chick weight and mortality in broilers, *Poult. Sci.* 88(11):2273-2279.

Van de Ven, L. J. F., Van Wagenberg, A. V., Uitdehaag, K. A., Groot Koerkamp, P. W. G., Kemp, B. and Van den Brand, H. (2012). Significance of chick quality score in broiler production, *Animal* 6(10):1677–1683.

Van den Brand, H., Heetkamp, M. J. W. and Kemp, B. (2015). Indirect calorimetry during incubation of hatching eggs. In: Gerrits, W. J. J. and Labussière, E. (Eds.), *Indirect Calorimetry: Techniques, Computations and Applications*. Wageningen Academic Publishers, Wageningen, the Netherlands, pp. 231–246.

Van den Brand, H., Meijerhof, R., Heetkamp, M. J. W., Van den Anker, I., Ooms, M., Kemp, B. and Molenaar, R. (2021). Interaction between eggshell temperature and carbon dioxide concentration after day 8 of incubation on broiler chicken embryo development, *Animal* 15(6):100223.

Van den Brand, H., Molenaar, R., Van der Star, I. and Meijerhof, R. (2010). Early feeding affects resistance against cold exposure in young broiler chickens, *Poult. Sci.* 89(4):716–720.

Van der Pol, C. W., Van Roovert-Reijrink, I. A. M., Maatjens, C. M., Van den Brand, H. and Molenaar, R. (2013). Effect of relative humidity during incubation at a set eggshell temperature and brooding temperature posthatch on embryonic mortality and chick quality, *Poult. Sci.* 92(8):2145–2155.

Van der Wagt, I., De Jong, I. C., Mitchell, M. A., Molenaar, R. and Van den Brand, H. (2020). A review on yolk sac utilization in poultry, *Poult. Sci.* 99(4):2162–2175.

Van Roovert-Reijrink, I.A.M., Van der Pol, C. W., Van den Brand, H., Priester, M. H. and Wijnen, H. J. (2017). Heat production of chicks with and without access to feed and water between hatching and pulling. Van Roovert-Reijrink, *Eur. Poult. Sci.* 81. doi:10.1399/eps.2017.204.

Wijnen, H. J., Van der Pol, C. W., Van Roovert-Reijrink, I. A. M., De Smet, J., Lammers, A., Kemp, B., Van den Brand, H. and Molenaar, R. (2021). Low incubation temperature during late incubation and early feeding affect broiler resilience to necrotic enteritis in later life, *Front. Vet. Sci.* 8:784869.

Willemsen, H., Debonne, M., Swennen, Q., Everaert, N., Careghi, C., Han, H., Bruggeman, V., Tona, K. and Decuypere, E. (2010). Delay in feed access and spread of hatch: Importance of early nutrition, *World Poult. Sci. J.* 66(2):177–188.

Willemsen, H., Everaert, N., Witters, A., De Smit, L., Debonne, M., Verschuere, F., Garain, P., Berckmans, D., Decuypere, E. and Bruggeman, V. (2008). Critical assessment of chick quality measurements as an indicator of posthatch performance, *Poult. Sci.* 87(11):2358–2366.

Wlazlo, L., Drabik, K., Al-Shammari, K. I. A., Batkowska, J., Nowakowicz-Debek, B. and Gryzinska, M. (2020). Use of reactive oxygen species (ozone, hydrogen peroxide) for disinfection of hatching eggs, *Poult. Sci.* 99(5):2478–2484.

Zulkifli, I., Fauziah, O., Omar, A. R., Shaipullizan, S. and Site Selina, A. H. (1999). Respiratory epithelium, production performance and behaviour of formaldehyde-exposed broiler chicks, *Vet. Res. Commun.* 23(2):91–99.

Chapter 17

Key issues in transportation of broiler and layer chicks

M. A. Mitchell, Scotland's Rural College (SRUC), UK

1 Introduction

According to the data from the Food and Agriculture Organization of the United Nations (FAO Stat, 2020), the global population of chickens (at any point) is approximately 30 billion. This figure includes all types of breeder birds, broiler chickens, laying hens and chicks. The broiler sector slaughters 71 billion broilers per annum (several crops per annum). It has been reported that the global population of laying hens stands at 6.1-7.6 billion (Fernyhough et al., 2020). Another study suggested that between 2000 and 2018, the number of laying hens in the world increased from 4.97 billion to 7.89 billion (Wu and Qin, 2019). The comparable figures for the EU are 6.4 billion broilers slaughtered per annum and a laying hen population of 456 million. According to the FAO data, total egg production in the world amounted to 1.577 billion units in 2019. The largest share of production (almost 64% of total production) belongs to Asian countries.

http://dx.doi.org/10.19103/AS.2022.0118.19

In the UK, 1.06 billion broiler chickens were slaughtered in 2020 (Shahbandeh, 2021). It is reasonable to conclude that these birds were transported to slaughter by road and that the corresponding newly hatched chicks were mainly delivered to the broiler farms by road transport. In the UK, 40.5-million-layer chicks were placed in 2020–2021 based on monthly averages and 39 million laying poults/pullets were placed per annum based on averaged weekly averages (GOV.UK 2021). Thus, whilst the newly hatched chicks of laying hens constitute as little as 4% of the UK chicks transported annually, the absolute volume of pullet chicks moved is very significant indeed.

Thus, in the UK, around 40 million laying hen newly hatched chicks may be transported from a hatchery to a growing farm and a substantial number of these birds will be relocated by road later in life to the laying farm. Inspection of the global production figures presented above indicates that global laying chick transportation may involve up to 7 billion laying chicks per annum. Whilst specific issues relating to transport may differ in accord with local conditions and procedures, vehicles and equipment and working practices, the more general problems for animal welfare, health and productivity will be of universal concern. A good deal of scientific research has been undertaken to provide insight into ways to improve the health and welfare of transported poultry whilst simultaneously optimising production efficiency. Many studies have addressed broiler transportation to slaughter and end-of-lay hen transportation to slaughter.

Whilst there have been significant research-based advances in improvements to procedures, practices and regulations and vehicle and container design and operation for poultry meat bird production (broiler chickens) and end-of-lay laying hens, far less attention has been paid to the transportation of pullet chicks and pullets. There have been a number of studies relating to the transportation of newly hatched broiler chickens examining welfare, health and production impacts of transportation, and it is generally accepted that the findings and recommendations from such research may be applied to newly hatched chicks from all major types of commercial production chickens.

2 Transportation of newly hatched chicks

In order to focus upon the transport of newly hatched or 'day-old chicks', it is at first necessary to define for transport and legal purposes the terms employed. Day-old chick means all poultry less than 72 h old, not yet fed as defined in point (19) of Article 2 of Delegated Regulation (EU) 2020/692.

All production and breeder birds must originally be placed on rearing farms following transportation from centralised hatcheries. The number of chicks transported at this stage must equal the final production figures plus those lost (mortalities and culls) during transport, rearing or production. The

Figure 1 Typical newly hatched chick boxes/trays/baskets/crates for commercial use.

birds are transferred from hatcheries within 1-3 days of their hatching. Chicks are transported mainly by road or in the case of breeder birds by air and road to these destinations. Chicks may be moved over relatively short distances in small numbers (e.g. pedigree birds, great grandparent lines) in high-specification containers and vehicles or over much greater distances in less sophisticated systems (see Fig. 1).

It may be proposed that the first journey is a potential, significant threat to the welfare and later production efficiency of all birds. This will be true of newly hatched chicks, including broiler chicks during transfer to rearing farms, and of pullet chicks transported to rearing farms and later intended for transfer to laying sheds for egg production. It is widely recognised that the husbandry of the birds during this period and the conditions under which they are maintained immediately prior to and after placement are vital in determining subsequent performance and health status (e.g. Decuypere et al., 2001; Langhout, 2001).

It is important to recognise the unique nature of the transportation of newly hatched chicks. No other species is transported in the immediate post-natal period (EFSA, 2011). Uniquely, chicks possess energy and water reserves in the form of the yolk sac and the chicks are sustained by this for a period after hatch. The frequently used term 'day-old chick' is a misnomer and may be misleading in relation to commercial chick transport, since some chicks may have been in a hatcher for some time after hatching and may thus be transported when approximately 24 h old. Thus, any debate addressing journey times for newly hatched chicks may thus focus upon the duration of the period in which the chick can be reasonably sustained by yolk sac reserves with no undue effect upon welfare or health.

3 Welfare issues affecting newly hatched chicks: pre-transport

There has been a strong consumer demand to take welfare into account in animal production, including during the complete production cycle for table eggs. This is particularly true in Europe and North America but increasingly around the world (Gautron et al., 2021). These authors concluded that 'consumers are increasingly aware of the production systems for laying hens and have strong opinions on what is favourable for a hen's welfare'. It is apparent that all aspects of the production process will be subject to increasing public concern and scrutiny and the development of new codes of practice and regulations and this will include the transportation of birds as well as the delivery of newly hatched chicks.

Historically, chick transportation from the hatchery to the farm is an issue that generally does not receive much attention. The main concern, in a commercial setting, is that mortality is kept to a minimum (Meijerhof, 1997).

The pre-transport for newly hatched chicks includes a number of procedures. After take-off, the post-hatch treatment includes sorting, grading, sexing in the case of laying chicks, counting, vaccination, waiting in the hatchery and, finally, transportation to the broiler house (Hammissou et al., 2019). All these factors can contribute to the stress imposed and will impact chick quality and welfare. There has always been a natural biological variation (24–48 h) in the time that chicks emerge from their eggshells (hatch time), even under optimal artificial incubation conditions. However, chicks are commonly removed from the hatchers at the same time (take-off). The length of the hatch window (time between early and late hatching chicks) will influence the duration without access to feed and water. Thus, a time window of 24–48 h exists between the first and last hatched chicks, depending on the eggs (e.g. size) and breeder flocks (e.g. age; Bautil et al., 2014; Decuypere et al., 2001); therefore, so-called 'day-old chicks' can in reality be up to 3 days old when they are transported (Jacobs, 2016). When chicks have been graded, sexed and vaccinated first-grade, they are mechanically (or sometimes manually) placed in plastic transportation crates typically with an area of 0.24 m^2, with circa 90 chicks per crate. Thereafter, the crates are stacked on carts, with spaces in between two stacks for ventilation, and placed in the holding room for up to 6 h, until the truck can be loaded (see Figs. 2–4).

Hedlund and Jensen (2022) investigated the impacts of the processes of 'commercial' incubation, hatching, handling/processing and transfer on newly

Figure 2 Commercial transport baskets or crates for newly hatched chicks. Comparison of possible basket/crate depths.

Figure 3 Newly hatched chicks in typical commercial transport baskets or crates.

hatched chicks and their subsequent development. Commercial layer chicks were compared to a control group hatched in a small incubator and handled gently and quietly in a quiet room without any processing and transport. The chicks were assessed at 4 days and at 1, 2, 3, 5, 8, 20 and 25 weeks of age. Feather pecking was studied at 15 weeks of age and damages to the feathers and injuries on the comb and wattle were assessed at 25 weeks of age. Commercially hatched chicks tended to show more feather pecking behaviour at 15 weeks of age compared to control chicks, although feather condition at 25 weeks of age showed the opposite pattern. Regarding production, commercially hatched chickens laid fewer and smaller eggs than the chicks hatched and handled under calm circumstances. From this experiment, it is concluded that the stressful experience in the commercial hatchery has

Figure 4 Typical newly hatched chicks' commercial vehicle transport space.

an overall negative effect on traits relevant to the industry, although it is not possible to determine which components of the commercial process had the most influence on chick well-being.

A proposed method of avoiding the rigours of transportation of newly hatched chicks is the transfer of incubated eggs to the farm rather than chicks. The hatching and brooding can thus take place at the destination farm. Both stakeholders and researchers have started exploring the potential to transport incubated eggs (Jacobs, 2016).

A preliminary study (de Jong et al., 2019) examined the behavioural and production consequences of on-farm hatching compared to traditional handling and transport of newly hatched chicks. On-farm hatching of broiler chicks using the X-treck system resulted in slightly better performance in the 1st week of life compared to flocks obtained from the hatchery, but no long-term effects on performance were found. With respect to broiler welfare, on-farm hatching resulted in less footpad dermatitis and a tendency for better litter quality. Which individual factors may have played a role in the difference in performance and welfare between the treatments cannot be determined from this evidence. This strategy is discussed in further detail elsewhere in this volume.

4 Welfare issues relating to transport of newly hatched chicks

The primary welfare concerns relating to the transportation of newly hatched chicks are similar to those identified for all livestock transport. Thus, the stress imposed by the physical environment is regarded as a central issue. The thermal micro-environment within the chick containers or boxes must be accurately controlled within tightly prescribed limits to avoid the imposition of heat stress and cold stress. The chicks must be placed carefully in the transport containers at stocking densities commensurate with the freedom to move and interact with other chicks but without compromising control of the thermal conditions or of the airflow through the boxes to allow adequate ventilation. The transport or journey times must be considered with attention to the period that the chicks will be kept without access to feed and water or early feeding strategies may be employed. The acceptable period without feeding or watering is based upon knowledge of the yolk sac reserves and their mobilisation and utilisation as described elsewhere. The chicks must also be protected from excessive vibrations, accelerations and impacts. Disturbing sounds and light stimuli should be minimised.

Transport of all livestock, including poultry, is regarded as a major source of stress and reduced welfare and a major cause of these problems is the thermal micro-environment in transit (Mitchell and Kettlewell, 1998; Cockram and Mitchell, 1999; Mitchell et al., 2000, 2001; Hunter et al., 2001; Kettlewell and Mitchell, 2001a, 2001b; Nilipour, 2002; Mitchell and Kettlewell, 2004a, 2004b, 2006). Many other factors may also contribute to transport stress such as handling, feed and water withdrawal, vibration, space restrictions upon behaviour, noise and pollutants (Mitchell and Kettlewell, 1998; Mitchell et al., 2000; Mitchell and Kettlewell, 2004a, 2004b, 2006). A number of previous studies have addressed some aspects of the transportation environments of day-old chicks but with a primary focus on minimising in-transit losses and maximising subsequent performance. It has been proposed that poor conditions in the transport of newly hatched chicks may have longer-term impacts on the health of production birds; for example transport conditions for newly hatched chicks have been reported as influencing the subsequent incidence of ascites and 'sudden death syndrome' (Maxwell and Robertson, 1998).

The Humane Society of the United States (HSUS) has described the many welfare concerns associated with the transport of newly hatched chicks

(HSUS, 2008, 2009) relating to the control of the thermal conditions at the level of the chicks and to variation of thermal condition across the load. When transporting day-old chicks, extreme temperature ranges may be detrimental to animal welfare, and, at excessively high and low temperatures, chicks may die in less than 1 h. Reduced ventilation may also adversely affect chicks. While good ventilation can help prevent high temperatures, it is not clear that current stacking configurations and spacing are adequate to promote proper ventilation.

According to the research available until recently, optimal conditions for day-old chick transport are a temperature range of 24–25°C and relative humidity (RH) of approximately 65% at a space allowance of 21–25 cm^2 per chick and a travel time from hatching no longer than 48 h, although there are some recommendations that this should be reduced to 12 h. These specifications are for the 'transport space'; i.e. the interior of the vehicle or the setting for the truck thermostat and for older designs of the vehicle, this will achieve an 'in-box' temperature at the level of the chicks of 30–32°C. Therefore, the temperature within the transport container should remain as close as possible to 31–32°C, though this may be difficult to monitor for all boxes and to control precisely in all locations.

These data are based upon vehicle ventilation rates and control systems found on typical chick transport vehicles and may be considered now to be out of date. With the advent of new vehicle designs and operation, the thermal specifications may have to be adjusted to match the new higher ventilation rates and improved internal air mixing (e.g. HatchTec systems). Indeed, the most recent Ross Broiler Handbook (Ross, 2018) states as follows:

1 Temperature should be adjusted so that the chick vent temperature is 39.4–40.5°C (103–105°F). Note that the required temperature control settings to achieve this chick vent temperature will vary between different vehicle designs.
2 Relative humidity (RH) should be maintained at a minimum value of 50%.
3 A minimum of 0.71 m^3 min^{-1} (25 ft^3 min^{-1}) of fresh air per 1000 chicks should be supplied. Greater ventilation rates may be required if the truck is not air-conditioned and ventilation is the only method available to cool the chicks.

HSUS (2008, 2009) has pointed out that temperatures within densely packed commercial containers may vary by several degrees if poor ventilation exists and that heterogenous distributions of thermal conditions have been demonstrated across the stacks of transport containers.

5 Assessing welfare of newly hatched chicks

The assessment of animal welfare has long attracted detailed examination and discussion (e.g. Broom, 2000, 2007). An accepted working definition of the welfare of an animal is 'the state of an individual in relation to its environment' (Broom, 1991). This definition is employed in conjunction with measures of the individuals' ability to cope with environmental challenges where difficulty in coping and failure to cope are regarded as indicators of poor welfare.

Potential welfare indicators include measures of physical health and condition and behavioural, physiological and neurophysiological responses. They can be used to infer not only the occurrence but also the intensity and duration of specific affective experiences that contribute to the animal's overall welfare state (Mellor and Beausoleil, 2015; Mellor, 2017; Beausoleil and Mellor, 2017).

It may be argued that assessing the well-being of chicks during a journey or appropriate experimental simulations by analysis of environmental variables and the concomitant physiological and metabolic responses will facilitate identification and definition of optimal transport conditions and practices. These will be consistent with the imposition of minimal physiological stress and as such will ensure optimum welfare and productivity in transit and during the immediate post-transport period and later development and growth phases.

6 Journey times and periods without feed and water

There have been a limited number of scientific publications relating specifically to this topic and it was from some of the older work that the original definition of a maximum journey time for newly hatched chicks of '24 hours as long as this is completed within 72 hours of hatching' was developed.

In many transported animals, the physiological challenges presented by the thermal conditions are compounded by extended periods without access to food or water. It has long been thought that the 1-day-old chick may be partially protected from such stresses by the presence of energy and water reserves in the yolk sac. Older studies proposed that yolk stores in the newly hatched chick constitute 18% of total body weight (BW) and contain approximately 2 g of lipid and 2.5 ml of water. In the absence of excessive thermoregulatory demands, this represents energy (75–80 kJ) and water supplies sufficient for 3 days without further provision of food and water (Freeman, 1984). More recent studies have proposed that in modern newly hatched chicks, high metabolic rate and rapid utilisation of resources in the first 24 h post-hatch coupled with delays in transit and placement result in poorer performance and health status throughout flock life (Tanaka and Xin, 1997; Xin and Lee, 1997; Hackl and Kaleta, 1997; Vieira and Moran, 1999; Bigot et al., 2003). However, the most recent studies suggest that the utilisation of yolk reserves is not greater in modern lines than in their genetic predecessors (van der Wagt et al., 2020; Aviagen, 2021).

It has been proposed that the major causes of in-transit and post-transport mortality and morbidity are dehydration and undernutrition (Xin and Lee, 1997). A suggested strategy to reduce metabolic depletion during extended transport is the exploitation of the reduction in metabolic rate in crated chicks in the dark (Tanaka and Xin, 1997; Mitchell, 2009).

The quantity and rate of use of metabolic reserves by basal metabolism, however, is clearly not the only factor which will influence chick survival during transportation. The prevailing micro-environment may impose thermoregulatory and metabolic demands upon the chicks which will require rapid mobilisation and utilisation of lipids or evaporation of water. Neonatal chicks do not possess fully developed effective homeothermic mechanisms (Lamoreux and Hutt, 1939; Freeman, 1965; Dunnington and Siegel, 1984) and consequently are vulnerable to the detrimental effects of thermal loads and fatigue and dehydration. In the immediate post-hatch chick, both body temperature and metabolic rate increase (Freeman, 1964); however, body temperature remains labile during exposure to suboptimal thermal environments (van der Hel et al., 1991). Thus, if the transportation environments are unduly hot or cold, then immaturity of thermoregulatory homeostasis, including inadequacy of lipid mobilisation or efficient evaporative heat loss during thermal polypnea, may result in stressful or life-threatening hypothermia or hyperthermia.

Until recently, there has been a recognised paucity of information relating to the scientific basis for prescribed transport times for newly hatched chicks. Current European legislation (EC 1/2005) prohibits journeys of greater than 24 h unless adequate food and water are provided. Such journeys should be completed within 72 h of hatching. Valros et al. (2008) examined the effects of journey times on the behaviour of newly hatched chicks from two strains of laying hens. The study assessed the effects of long transport durations on subsequent growth, the development of perching behaviour, fear of humans (measured as the duration of tonic immobility induced by manual restraint) and the willingness to compete for access to feed. Journey times of 4 and 14 h were compared in simulated transport trials. The results were inconclusive with no clear indication that long transport of newly hatched chicks has significant wide-reaching effects on behavioural development. Mitchell (2009) suggested that increased journey time was associated with adaptive physiological responses which may be primarily attributed to the duration without food; however, it was concluded that there was little evidence that prolonged transportation (with associated prolonged fasting) is more stressful for chicks from young or old breeders, compared to short transportation (with associated short fasting).

Tona et al. (2014) have addressed the issue that newly hatched chicks may remain without feed for about 24–48 h before they are placed on farms 'due to a series of logistic operations', including transportation. Those authors stress that the spread in hatching time can also mount up to 30–48 h for late versus early hatchers. The study examined the combined effects of hatching time with a delay in feed access of 48 h, starting from their hatch time (biological age). It is concluded that the spread of hatch influenced post-hatch performance, especially appetite and growth at least until day 5. Moreover, the delay in feed access interacted with the hatching time and caused adverse effects on the post-hatch performance.

Bergoug et al. (2013) investigated the effects of transportation duration on newly hatched chicks. Dehydration status, mortality, production performance and pododermatitis were assessed during the grow-out period. Transportation duration affected BW. Control groups had a significantly higher BW than chicks transported for 4 h and 10 h. This effect was temporary (absent after 21 days of age). No clear effect on haematocrit, feed uptake, feed conversion (FC) ratio or mortality was observed for birds transported up to 10 h. The decrease in weight in the 10-h transported group was associated with less severe pododermatitis. Chicks from older breeders also exhibited reduced mortality during the grow-out period.

Khosravinia (2015) examined the effects of transport distance upon weight loss (WL), residual yolk (RY) utilisation and serum biochemical parameters in newly hatched chicks. Journey distances (not times) were 0 km, 200 km, 400 km, 600 km, 800 km and 1000 km. Live weight decreased linearly by 0.42 g 100 km^{-1} and 0.48 g 100 km^{-1} of journey in male and female chicks, respectively. Yolk sac residual utilisation by 0.071 g 100 km^{-1} and 0.069 g 100 km^{-1} of travel in male and female chicks, respectively, over journey distances up to 800 km. Extended journeys from 800 km to 1000 km caused a greater decline of 0.51 g 100 km^{-1} and 0.58 g 100 km^{-1} in yolk weight in male and female chicks, respectively. On journeys of approximately 1000 km, serum glucose level was decreased by 40%.

Long-distance transport imposes greater physiological stress than shorter journey times. In one study (Jacobs et al., 2016), chicks from 60-week-old breeders were subjected to transportation of 1.5 h or 11 h duration. Sample chicks were randomly selected and assessed for 1-day-old chick quality, weighed and culled for yolk sac weighing. Mortality, average daily gain (ADG), BW, and FC of the experimental chicks were recorded until 41 days. Meat quality was measured for breast fillets. No interaction effect of parental age and transportation duration was found for any variables. BW and yolk sac weight at day 1 were lower for chicks transported 11 h than 1.5 h and for chicks from young versus old breeders. The effect of parental flock age on BW persisted until slaughter. Additionally, parental age positively affected ADG until slaughter. Chick quality was lower in chicks from old versus young breeders. Chick quality and productivity were not affected by transportation duration. Mortality and meat quality were not affected by either parental age or transportation duration. To conclude, no long-term detrimental effects were found from long post-hatch transportation in chicks from young or old parent flocks. Based on the results, it was suggested that 11 h post-hatch transportations under similar conditions will not impose long-term welfare or productivity risks.

De Jong et al. (2016) concluded, based on a thorough analysis of the existing literature, that post-hatch food and water deprivation for an average of 48 h (36–60 h) leads to lower BWs and higher total mortality in chickens up to 6 weeks of age. It was suggested that on the basis of these findings, it was not possible to state if the possible production changes and/or altered mortality constituted any prolonged risk to measurable bird welfare.

Interaction of transport duration and parental age upon physiological stress and chick quality has been examined by Jacobs et al. (2017). Chicks from young and old parent flocks were transported for either 1.5 h (short) or 11 h (long). Stress was assessed by measurement of plasma concentrations of corticosterone, lactate, glucose and thiobarbituric acid-reactive substances (TBARS). No interaction effects of parental age and transportation duration were found and no alterations to permissible journey times based on flock age are required. After long transportation, chicks showed increased corticosterone and lower TBARS levels compared to short transportation, reflecting the fundamental stress imposed by extended journey times.

Jacobs (2016) concluded from analysis of all available data that 11 h post-hatch transportations under normally employed conditions do not pose long-term risks for broiler welfare or productivity.

Some current concerns have arisen from an assertion that modern genetic lines of broiler chickens with high metabolic rates may deplete yolk sac reserves more quickly than their genetic predecessors and thus may be at greater risk of detrimental consequences. EFSA (2011) mentions Malik et al. (2010) who compared yolk sac reserve utilisation in fast- and slow-growing lines of chicks. The chicks were made to fast, and water was withheld for the first 48 h post-hatch or with the early provision of food and water. The study concluded that rapidly growing chicks do utilise yolk substrates more efficiently and quickly in the 48 h after hatch. However, as the slow-growing line studied represented laying hen chicks, it is not possible to extrapolate the findings in the context of yolk sac reserve utilisation in relation to transport. Barri (2008) presented for the first time the effects of altered embryonic incubation temperature and stress during transportation of newly hatched chicks on small intestine morphology, digestive organ development and expression of nutrient transporters mRNA in high-yield broiler chicks. These results contributed to the understanding of mechanisms by which either low or high temperatures, as compared to standard recommendations, during incubation and transportation can affect embryonic development and subsequent performance of broiler chicks.

Buzala et al. (2015) have reviewed the consequences of different growth rates as a result of long-term genetic selection on embryonic development and the metabolic rate of broilers and layers. The information presented that there might be significant differences in the metabolic demands between chicks of the two types of poultry which may have impacts on the utilisation of yolk sac reserves during early post-hatch handling and transport. These issues have been further addressed by Nangsuay et al. (2015). Thus, selection for production traits of broilers and layers leads to physiological differences, which may already be present during incubation and at hatch. Nangsuay et al. (2015) investigated the influence of strain (broiler versus layer) on egg nutrient availability, embryonic development and nutrient metabolism. A 0.24-g lower RY at the hatch of broiler embryos than for the layer embryos indicated that broiler embryos used more yolk and had a higher energy utilisation and energy deposition in yolk-free body mass. It was concluded that broiler and layer embryos have different embryonic development patterns, which affect energy utilisation and embryonic heat production. However, the embryos are equal in efficiency of converting the energy used to yolk-free body mass.

Van der Ven et al. (2013) have investigated the effects of hatching time on perinatal broiler physiology in two hatching systems, differing in conditions. A comparison was made between a conventional hatcher, where chicks are deprived of feed and water between hatching and the moment of chick pulling (d E21.5), and a patio system, in which the hatching and brooding phase are combined, and chicks have immediate post-hatch feed and water access. Climate conditions in the patio also differ with about 3°C lower temperature and 20% lower RH compared with conventional hatchers. At hatch, weights of the heart, lungs, stomach and intestine increased with hatching time, concurrent with a decrease in RY weight, regardless of the hatching system, indicating that later hatching chicks are more mature. Weights of the heart, liver, stomach and intestines were lower in hatcher than in patio chicks. Between hatch and E21.5, RY weight decreased, whereas organ weights increased in both fasted hatcher and fed patio chicks, but at a higher rate in the latter. Early feed and water access seem to enable early-hatching chicks to compensate for their apparent disadvantage in development at hatching, whereas chicks subjected to fasting show metabolic adaptations to preserve nutrients. Chick physiology at chick pulling time was shown to vary with time after hatching.

It is theoretically possible although unusual in commercial practice for newly hatched chicks to be held for more than 48 h prior to road transportation. The chicks may therefore experience long periods of fasting in commercial hatcheries. There are few publications describing in detail the physiological status of chicks in such situations. Tong et al. (2015) measured a range of stress responses and other biomarkers and observed that chicks may experience significant levels of stress during the holding period. Therefore, as newly hatched chicks are often subjected to other concurrent stressors (heat, dehydration, dust and deprivation of feed) in the first 24 h after hatching, extended holding may exacerbate the associated detrimental effects. As a consequence, the adverse effects during the critical early hours following hatch can result in an increased percentage of early mortality. It was concluded that shortening the hatch window and minimising the number of chicks that experience a long holding period before pulling may improve chick quality and physiological status, which may be due to unfavourable environmental conditions that include feed and water deprivation.

One study (Cardeal et al., 2021) assessed the effects of breeder age, placement time and the use of a pre-placement feed (PPF) on yolk sac absorption, duodenum villus height and crypt depth and performance of male broiler chicks. Early access to feed stimulated yolk sac absorption until 4 days of age. On the other hand, placement within 48 h negatively impaired the development of duodenum villus and crypt until 7 days of age for chicks from young broiler breeders, but this negative effect could be reduced by the use of PPF. These results suggest that the use of PPF inside transport boxes is recommended for chicks from young broiler breeders that will be submitted to delayed feed access in order to improve gastrointestinal tract development, mainly that of the duodenum mucosa. It should be noted that early feeding of chicks may later change the legal status of the birds as a 1-day-old chick is defined as poultry less than 72 h old, not yet fed as defined in point (19) of Article 2 of Delegated Regulation (EU) 2020/692. This may have implications for the application of transport regulations. Early feeding and PPF are discussed in further detail elsewhere in this volume.

The welfare concern relating to modern broiler chicks possibly having higher metabolic rates and more rapid depletion of yolk sac reserves compared to their immediate genetic predecessors (HSUS, 2008, 2009) may have resulted from proposals in an EFSA Opinion (2004) presented without any supporting published evidence. The statement offered in EFSA Opinion may have derived from one or two previous publications describing WL in post-hatch chicks *but* not differences or effects attributable to metabolic rate or differential genetic selection for performance. There are reports relating to the utilisation of yolk reserves, RY and other substrates in embryos and newly hatched chicks but the effects of genetic selection upon these measures are poorly characterised. As in the case of metabolic rate, methodological differences make many comparisons invalid and no comparisons within broiler lines are available that indicate any increase in demand or reserve mobilisation or depletion in higher performing modern lines. Thus, whilst it has been considered that the utilisation rate of yolk sac substrates and reserves in newly hatched chicks from modern commercial broiler lines might impact upon acceptable journey times for newly hatched chicks, review of all the available and reliable evidence (van der Wagt et al., 2020; Aviagen, 2021) suggests that this may not be the case and current European journey limits (EC 1/2005 and WATO 2006) are adequate to protect the chicks.

Novel husbandry systems have been developed, in which immediate provision of nutrition post-hatch or on-farm hatching is employed. In these novel systems, 1-day-old-chick handling and transport are minimised. This would avoid both fasting and post-hatch transportation (van Harn et al., 2014). In addition (Jacobs, 2016), the option to provide feed and water immediately after hatch in the hatcher has been explored by commercial companies (e.g. HatchTech, Veenendaal, the Netherlands). This is discussed in further detail elsewhere in this volume.

7 Thermal stress in transit

Many factors may contribute to transport stress such as handling, feed and water withdrawal, vibration, space restrictions upon behaviour, noise and pollutants (Weeks and Nicol, 2000; Mitchell et al., 2001b; Mitchell, 2004a and b, 2006). Prior to departure on their first journey, some chicks may have been in a hatcher for some time after hatching and may thus be transported when approximately 24+ h old. Any discussion (*vide infra*) on thermal conditions and journey times for newly hatched chicks must thus consider the duration of the period in which the chick can be reasonably sustained by yolk sac reserves with no undue effect upon welfare or health (EFSA, 2011; van der Wagt et al., 2020; Aviagen, 2021). It may be proposed that the improved design and operation of modern trucks (often air conditioned) for the transportation of newly hatched chicks reduces the risk of thermal stress; however, heat and cold stress are possible because of the following reasons:

1. Not all vehicles are air-conditioned (AC) and some journeys are undertaken on passively ventilated trucks.
2. Environmentally controlled vehicles will operate well in temperate countries but for many producers around the world, extreme weather conditions and inadequate technology create thermal problems.
3. Extremes of heat and cold in summer and winter can overwhelm technological solutions.
4. Poor ventilation flow control and dispersion leads to heterogeneous distributions of temperature and humidity in transport containers - hot spots and cold spots in the bio-load chapter.
5. AC can fail.

In many countries, typical newly hatched chick transport will take place on climate- controlled vehicles; however, on occasion, trucks with natural ventilation are used when transports are over small distances. The climate-controlled trucks regulate dry bulb temperature and air supply through defined vents. Many trucks can carry up to 100 000 chicks. Before transportation commences, recommendations by some companies require that trucks are preheated to 28–29°C and temperature is maintained at circa 25–27°C during transportation, which is monitored by the driver, sometimes with an alarm system when temperatures deviate from the preferred temperature (Jacobs, 2016). However, it should be emphasised that no standard newly hatched chick transportation trucks exist and that each hatchery or company is responsible for the development of its own trucks. There is a very wide variation in the duration of journeys undertaken upon these vehicles. Journey time can be lengthened when chicks for different producers are transported in the same truck, resulting in long transportation durations for chicks that are transported to the last farm.

7.1 Thermal comfort zones

Attempts to define thermal comfort zones for chicks in transit started seriously in 1984 when Kaltofen and Dijk (1984), at the Centre for Applied Poultry Science 'Het Spelderholt' in the Netherlands, investigated the effect of temperature and humidity on day-old chicks (Molenaar, 2012). Based on the behaviour of the chicks in varying conditions, a comfort zone was defined both for solitary chicks and for chicks placed with 50 in a box. At varying environmental conditions, the rectal temperature of the chicks was measured at 30–44°C and RH between 20% and 90%. Individual, solitary chicks did not show any abnormal behaviour at environmental temperatures of 32–34°C, representing the first definition of optimum temperatures for chick transport. Those chicks were able to maintain rectal temperatures at 39–40°C (assumed to be normal at the time). In other studies, the same group demonstrated that groups of chicks could maintain a relatively constant deep body temperature across a wider range of air temperatures than individual chicks by utilising behavioural thermoregulation. It was claimed that humidity hardly influenced the behaviour of newly hatched chicks. However, it was proposed that at RH values higher than 85%, the temperature comfort zone decreased by 1°C and at RH lower than 40%, the temperature comfort zone increased by 1°C. In order to maintain body temperature between 39°C and 40.5°C, the final recommendation was environmental temperature should be kept between 30.5°C and 36°C for solitary chicks and between 28°C and 30°C for chicks. It was emphasised that in the hatchery and during transportation, cold stress must be avoided. Ideally, all chicks should have body temperatures between 39.5°C and 40.5°C. The same study provided data on mean body temperature in groups over the whole post-hatch placement period and identified those periods in which the risk of thermal stress and hypothermia might be greatest (see Fig. 5).

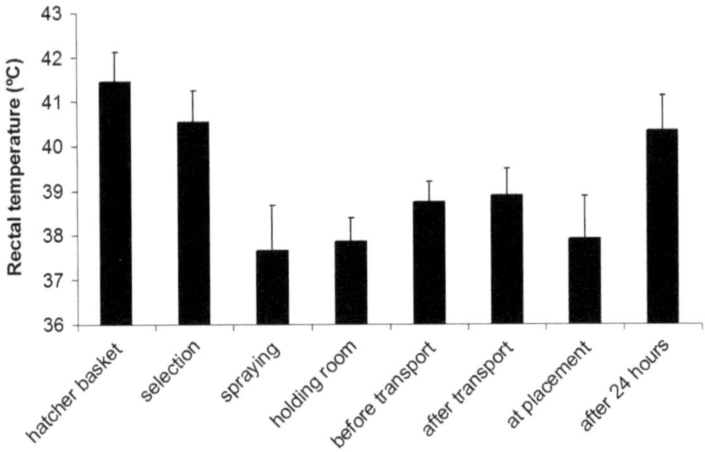

Figure 5 Rectal temperatures of newly hatched chicks from placement in the hatcher baskets until 24 h after arrival at the destination farm. Source: After Kaltofen and van Dijk (1984).

Thus, thermal stress may be the major source of welfare compromise during the transportation of newly hatched chicks. Whilst commercial breeders and producers have long recognised the necessity to maintain an appropriate thermal environment for chicks in transit (Tamlyn and Starr, 1986; Freiji, 1988; Laughlin, 1989; van der Hel and Henken, 1990; Qureshi, 1991; van der Hel et al, 1991), the conditions employed have been largely defined by empirical means and have been based upon minimisation of mortality rates during and following transport and efficient productivity during the subsequent rapid growth phase. In current practice, the recommended temperature for chick transport is 24–26°C (Ross Breeders, 1996; Meijerhof, 1997; Weeks and Nicol, 2000), although only the breeder companies' advice includes a recommendation of controlled humidity (75% at 24°C). An HSUS report (HSUS, 2008, 2009) concluded that 'extreme temperature ranges encountered during the transportation of one-day old chicks may be detrimental to welfare. At excessively high or low temperatures chicks may die within one hour. Reduced ventilation may also adversely affect chicks. While good ventilation can reduce the effects of high temperatures it is not clear in commercial practice if stacking configurations and spaces are adequate to promote proper ventilation'.

The problem of inadequate ventilation of chick boxes was emphasised and it was proposed that current container densities and ventilation should be reviewed, and further research should be undertaken to define the optimal practice.

Neonatal chicks do not possess fully developed, effective homeothermic mechanisms (Mitchell, 2009) and consequently are vulnerable to the detrimental effects of thermal loads, fatigue and dehydration. Both body temperature and metabolic rate increase immediately after hatching. However, body temperature remains labile during exposure to suboptimal thermal environments (Mitchell, 2009). Thus, if the transportation environments are unduly hot or cold, then immaturity of thermoregulatory homeostasis, including inadequacy of lipid mobilisation or efficient evaporative heat loss during thermal polypnea, may result in stressful or life-threatening hypothermia or hyperthermia. In addition, the accelerated rates of utilisation of energy and water reserves may result in premature dehydration. Freeman (1984), quoted in Mitchell (2009), has estimated that reserves may be completely exhausted in as little as 8–10 h at a temperature of 40°C.

It may therefore be proposed that, in order to simultaneously optimise survival, productivity and welfare of the newly hatched chick in transit, an effective strategy would be to match the thermal characteristics of the micro-environment to the biological requirements of the birds. Some previous studies have attempted to define thermoneutral or optimal environments for neonatal chicks on the basis of metabolic heat production and body temperature responses (Misson, 1976; Henken et al., 1989; Gates et al., 1989; van der Hel et al., 1991). In general, these studies did not measure other indicators of homeostatic effort which might better define the physiological impact of the thermal micro-environment. Xin and Harmon (1996) examined the effects of a range of temperatures and humidities (20–35°C and 40% to 17%) on day-old chicks by measuring metabolic rate and mortality. They concluded that optimum or thermoneutral conditions occurred at 30–32°C. Xin (1997) has also reported that chicks held at a constant 29°C do not exhibit different mortality or BW loss compared to birds exposed to as much as a 16°C cycling temperature around the same mean temperature. Transportation of 1-day-old chicks for 18 h at 25°C reduced the rate of subsequent growth to 45 days of age and some sustained biochemical changes compared with un-transported chicks (Pijarska et al., 2006). On the other hand, Valros et al. (2008) observed no indications of lasting effects of simulated transport upon behavioural development.

Mitchell et al. (1996) employed physiological stress modelling, measurement of metabolic rate and the concept of AET (Mitchell and Kettlewell, 1998; Mitchell et al., 2001a; Mitchell et al., 2001b; Mitchell and Kettlewell, 2004a, 2004b, 2006) to determine optimum transport thermal environments for newly hatched chicks. All measurements were performed on chicks in commercial transport containers in calorimeter chambers housed in controlled climate rooms. Temperatures of 20–35°C accompanied by RH of 50–65% and durations of exposure from 3 h to 12 h were employed. Metabolic heat productions ranged from 7.8±0.3 W kg^{-1} to 8.7±0.9 W kg^{-1} in close agreement with previously published values (van der Hel et al., 1991; Tanaka and Xin, 1997a). On the basis of minimal change in body temperature and minimal alterations in basal metabolic rate, hydration state, electrolyte balance, BW loss and plasma metabolite concentrations, an optimal temperature-humidity range of 24.5-25°C and 63-60% RH for the transport of chicks at commercial stocking density was identified. It was emphasised that these physiologically ideal conditions are very similar to those currently employed by commercial breeders and producers, although more recent studies have revised recommendations for the basis of control and the specifications (*vide infra*). The original studies also provided evidence that if the thermal micro-environment is appropriately controlled, then journey durations of at least 12 h are wholly acceptable. It was concluded that both productivity and welfare of newly hatched chicks in transit can be maintained by careful regulation of the temperature and water vapour density to these prescribed limits inside the transport containers. Further work on behavioural identification of the preferred thermal conditions for chicks in commercial transport simulations will provide further refinement and support to these strategies.

Thermoregulatory responses of newly hatched chicks to cold and heat challenges compared to a 'thermoneutral' environment have been studied in transport simulations by Viera et al. (2016). Chicks were exposed to the three thermal conditions - cold (21°C), thermoneutral (35°C) and hot (38°C) - for 1 h in commercial transport containers in an environmental chamber. The boxes were stacked as in commercial practice in order that any effect of box placement might be determined. Body weight, respiratory frequency, mean surface and cloacal temperature were monitored. The box placement had a negligible effect on the thermal comfort of birds. The mean surface and cloacal temperature responses were markedly affected by the cold treatment. It was suggested that longer simulated journey durations and higher temperatures might also affect the chicks' thermoregulatory responses.

Exposure to extreme thermal stress might increase chick mortality in transit; however, there appear to be few accurate scientifically based surveys yielding estimates of mortality during road transportation of newly hatched chicks. Commercial practice often involves calculation of 3-day and 7- or 8-day mortality reasoning that these values (total losses from birds transported) integrate the overall effects of the transport process and placement. On this basis, mortalities of 1-1.7% have been reported (Tanaka and Xin, 1997).

Yerpes et al. (2021) monitored the thermal micro-environments during chick transport in Spain and correlated the variability in journey length, hygrothermal conditions and carbon dioxide concentration in the boxes with effects upon chick WL, mortality and first-week mortality. The study examined 66 summer journeys of differing durations. The effects of other intrinsic factors on the chicks (e.g. chick number loaded, gender, breed, breeder flock age in weeks and the number of days eggs are stored) were examined. The results showed that chick WL was positively associated with journey duration and RH. No effect of environmental variables was found on mortality during transport. However, chick mortality during the first week of life was related to the number of day-old chicks loaded per journey, with the stocking density in the transport container and with chick gender. It was concluded that owing to the environmental heterogeneity during transport and the effect of the environment on chick weight during transport and mortality at the first week of life, there was an urgent need to refine the air-conditioning and ventilation systems of day-old chick transport towards a greater environmental homogeneity.

Despite the apparent success of the operation of modern designs of chick transport vehicles in terms of chick survival and subsequent productivity, few studies have actually characterised the thermal profiles in commercial chick transports and compared the findings to the recommended transport thermal micro-environments. Early studies by Quinn and Baker (1997) demonstrated thermal heterogeneity in chick transport trucks and made recommendations for improvements in ventilation regimes and system designs. Nazareno et al. (2015a) described heterogeneity of the distribution of thermal parameters in loads of air-conditioned trucks carrying eggs or newly hatched chicks. The study reported inadequate control of temperature and humidity conditions across the load and proposed that the thermal loads experienced by chicks on these trucks were not compatible with thermal comfort and as such caused welfare and production problems. In a subsequent study, Nazareno et al. (2015b) also examined the thermal profile of an air-conditioned truck with different levels of box placement during 1-day-old chick transport. Data were collected on 11 journeys carrying 630 chick boxes each and a total of 63 000 chicks. Thermal data included dry bulb temperature, RH and calculated specific enthalpy. The microclimatic conditions of the truck were not as per recommended values, which confirms a heterogeneous distribution of heat and moisture in the environment. The most stressful environment for birds inside the truck was located in front and at the centre of the truck, illustrating the failure to control airflow distribution within the transport space and chick containers.

A later study by Nazareno et al. (2016) described the thermal gradients between the container environment and the interior of chick boxes and mean surface temperature of chicks during transportation. Nine shipments were examined with different density boxes. Measurements included the mean surface temperature of chicks (using infrared thermography) and the container microclimate variables, i.e. air temperature, RH and specific enthalpy. The study confirmed a heterogeneous distribution of thermal micro-environments throughout the loads studied. It was proposed that some conditions were outside the thermal comfort zones for newly hatched chicks and that the associated heat stress might be detected by measurement of surface temperature.

Vieira et al. (2016) conducted a study to assess the thermoregulation of day-old chickens, placed in transport boxes, under simulated transport thermal conditions. Newly hatched chicks were subjected to three thermal treatments in transport containers. The box temperatures employed were 35°C + 60% RH (control), cold (15°C) + 60% RH and 40°C + 70% RH (hot). Body weight, respiratory frequency, mean surface and cloacal temperature were monitored. The box placement had a negligible effect on the thermal comfort of birds. The mean surface and cloacal temperature responses were markedly affected by the cold treatment. However, with an increasing exposure time, the hot treatment impacted negatively upon the chicks. This evidence tends to support the recommendations for thermal conditions currently prescribed by some sectors of the global industry; i.e. vehicle conditions should be maintained that will result in container temperatures of 31–32°C.

In subsequent studies, Vieira et al. (2019) assessed the variation of physiological responses and mortality of day-old chicks subjected to different thermal conditions and exposure times during simulated transport based upon the previous publications describing inappropriate thermal loads and heterogeneity of temperatures and humidity distributions within the loads on commercial vehicles. The physiological variables used in this trial were BW, respiratory rate, cloacal temperature, average surface temperature and gene expression of heat shock protein (HSP70). Similar heat loads were employed to those used in previous trials. There were small effects on BW loss associated with treatment, but the major effects were upon thermoregulatory responses (respiration rate) and deep body temperature. The chicks subjected to the heat treatment exhibited respiratory rates above 100 movements per minute, average cloacal temperatures above 44.7°C (approaching lethality) and surface temperatures above the comfort zone (greater than 39.6°C). The interpretation of exposures to these extreme thermal loads was to propose that transport of more than 3 h under extreme conditions would result in elevated mortality through severe thermal stress.

De Camargo et al. (2015) correlated the age of breeder flocks from which eggs were produced and holding periods post-hatch and thermal micro-environment with final chick quality. The ages of breeder flocks employed were 46, 54 and 64 weeks. The three holding periods prior to placement were 0 h, 14 h and 28 h. Monitored variables were the air temperature, RH and specific enthalpy. The response variables were body mass of chicks, cloacal temperature, average surface temperature and respiratory rate. The lairage time and the microclimatic conditions of the hatchery chick's room influenced the thermal comfort of the chicks. The best lairage time was 0 h. The highest percentage of WL due to the lairage time was observed in chicks from breeders with 54 weeks of age. It may be proposed that chicks from younger breeder flocks might better withstand conditions encountered in transit in the immediate post-hatch period and that the detrimental effects of the journey might be reduced if chicks are dispatched as soon as possible after hatch.

Hamissou Maman et al. (2019) considered the effects of deep body temperature in the first 12 h post-hatch on subsequent bird performance. Three groups of chicks were identified in the 12-h post-hatch handling period on the basis of deep body temperature: control, high- and low-temperature groups, with average body temperatures of 40°C, 42.6°C and 38.1°C, designated as control, high and low, respectively. Residual yolk sac weight was not affected by temperature group, whereas organ weights were significantly lower in the high body temperature group than in the control and low body temperature groups. Body weight was significantly lower at placement in chicks in the high-body-temperature group than in chicks in the control and low-body-temperature groups due to greater WL during the immediate post-hatch handling period. Feed conversion ratio (FCR) and mortality were numerically greater in the high body temperature chicks than in the control group, whereas FCR and the mortality in the low body temperature chicks were intermediate at 35 days. It is proposed that the study indicates that day-old chicks with high body temperatures (42.6°C) exhibited a greater percentage of BW loss due to dehydration and lower organ weights during the 12 h post-hatch handling period, which was followed by significantly poorer broiler performance. On the basis of these findings, it might be suggested that chicks with initially higher deep body temperatures might be more sensitive or vulnerable to potential heat stress in transit. It was suggested that this should be used in the screening of newly hatched chicks from high-value categories prior to long transports or on vehicles where control of the thermal micro-environment is not of the highest standard. However, the practicality of such a strategy is questionable and there is no clear option for how to deal with the chicks that might be 'screened out'.

7.2 Surface temperature measurement

Surface temperature measurements using thermal imaging (thermography/thermal cameras) can be employed to determine the thermal status (presence of heat or cold stress) in a number of species, including poultry (Rowe et al., 2019), with the appropriate calibrations. In addition, thermal imaging may be applied to assess other welfare impacts on live animals (Mitchell, 2013a, 2013b). Stress in hens has been quantified using thermal imaging (Herborn et al., 2015). Thermal comfort has been determined in broilers by thermography (Olawole and Atta, 2019). Sensible heat loss (Souza et al., 2019) and the impacts of airflow upon heat exchange (Akter et al., 2022) have been estimated by thermal imaging in broiler chickens. It is proposed that the technology and approach can be readily applied to the assessment of thermal comfort in chicks in transport containers at appropriate points in a journey.

7.3 Thermal micro-environments and ventilation

The ultimate determinants of the localised on-board vehicle (chick transporter) micro-environment are the prevailing climatic conditions, the addition of heat and water vapour to the load space from all sources, including the bio-load (chicks), and the ventilation rate and distribution. All these issues have been extensively addressed in relation to the transport of broiler chickens at slaughter age (e.g. Hoxey et al., 1996; Kettlewell et al., 2000), but the corresponding characteristics of chick transporters have received less detailed attention. Tanaka and Xin (1997) have investigated the effects of chick box structure and stacking configuration using simulations on the transport micro-environment. Modified box structures were compared with traditional designs. The results revealed that the measured ventilation rate under the then-current box structure and stacking arrangement (averaging 0.013 L s^{-1} per chick) seemed sufficient during cold weather, but it was considerably below the values recommended for mild to hot weather. Temperature lift in the traditional box design and configuration ranged from 3.4°C to 5.5°C. The improved structure and stacking configurations had an average airflow rate of 0.062 L s^{-1} per chick and temperature lifts were between 5.5°C and 9.9°C. It was suggested that the new structure and configurations were beneficial in potential heat stress conditions, but that excessive flow rate in some boxes may be detrimental in some circumstances and other measures should be taken to avoid cold stress.

Valuable measurements of heat and moisture production of chicks are available for calculation of vehicle ventilation requirements (Tanaka and Xin, 1997), but only the work of Quinn and Baker (1997) appears to have examined in detail the ventilation characteristics of commercial chick transporters. That work employed full-scale experimental determination of ventilation patterns and the prediction of the same by computational fluid dynamics (CFD). The important findings included the observation that the presence of the load of stacked chick boxes had a channelling effect upon the airflow through the load space with significant amounts of air bypassing the chick boxes and being recirculated. The implication of this ventilation regime for airflow in the chick containers was seen in the temperature distributions with peak temperatures occurring in the front central boxes and cooler air bypassing the load. In addition, cooler air entered from beneath the vehicle in the fully loaded configuration and reduced the flow through the load as well as potentially introducing exhaust fumes into the load space. That study emphasised three main points to optimise carriage of newly hatched chicks:

1 Thermal conditions should be controlled more effectively to improve chick welfare.
2 Air mixing is an important issue.

3 More effective ventilation systems should be developed (including containers) to ensure thermal homogeneity of the 'on-board' micro-environment and to optimise welfare.

It should be stressed that the design and operation of more modern vehicles, informed by this original research, have addressed these issues. Modern commercial vehicles have better controlled internal airflow, distributions and temperature and humidity control. This is an example of exploitation and extension of the fundamental research and developmental strategies applied so successfully to the optimisation of mechanical ventilation regimes for vehicles carrying other poultry and red meat species (Kettlewell and Mitchell, 2001a, 2001b). An example of the application of a high airflow ventilation system for chicks is the HatchTraveller vehicle (https://hatchtraveller.com/). The high airflow employed on these vehicles greatly improves overall ventilation and air mixing in the transport containers. As a consequence, micro-environment control is improved, but the enhanced removal of heat and water vapour means that the temperature set point for the vehicle transport space must be increased above normal recommendations. This is discussed in detail elsewhere.

8 Stocking density

In the UK, newly hatched broiler chicks are transported from commercial hatcheries to growing farms in custom-built vehicles with controlled environments. The chicks are carried in boxes, trays or drawers that are stacked on modules or trolleys. Typically, a vehicle might carry up to 60 000 chicks with around 100+ chicks per box. Current transport legislation (Regulation EC 1/2005) prescribes a stocking density or space allowance for chicks of 21–25 cm^2, which means generally 110 chicks would be placed in each 'standard'-type crate. Although coming into law as a Regulation in 2005, in force in 2007, the space allowances and thus stocking densities were derived from much older legislation (dating back to the early 1980s) and have not been revised in response to recent research findings and increasing scientific knowledge. The main purposes of regulating the space allowance are to avoid overcrowding, to ensure adequate ventilation and control of the internal hygrothermal micro-environment (temperature and humidity) matched to the chicks' heat and moisture production and airflow through the transport containers. It has been argued that with modern environmentally controlled vehicles with improved ventilation, it may be possible to design containers that will facilitate better control of the thermal environment and therefore safely carry more chicks per container. The advantages of such a system are carriage of a greater number of chicks per journey and therefore more efficient transport and delivery in relation to fuel and energy use and pollution/carbon footprint. Current industry practice and experience suggest that due to disparities in the design and structures of chick boxes employed in different European member states, it may be necessary to revise stocking densities and space allowances for chicks during road transport. This is based upon reports of chilling of chicks and decreased performance and increased mortalities in chicks transported at the upper end of the current space allowance range. This is attributed to the relatively low total metabolic heat production of chicks in each compartment of chick boxes with increased heights and therefore volumes. This must be a concern in relation to chick welfare, but more published data are required to support new recommendations. It might be proposed that in future, stocking densities or space allowances for chicks in containers might be defined better in terms of both surface area and volume rather than only surface area. The industry proposes to call for a reduction in the minimum space allowance from 21 cm^2 to 19 cm^2 per chick in order to avoid chilling in well-ventilated boxes or stacks in transport. It is unclear if increasing the recommended vehicle temperature and/or humidity would achieve the same objective and, therefore, further research was required in this area.

Control of the chick box thermal micro-environment by adequate ventilation is key to preventing thermal stress. Early studies have addressed these issues and proposed design improvements (Tanaka and Xin, 1997). More recently, boxes have been designed that have a much larger volume and head space (above the chicks) than their commercial predecessors (see Fig. 2). These boxes have been employed in commercial practice for several years with considerable success. Commercial results suggest that the boxes with higher volume/more head space are better ventilated and allow improved control of the micro-environment. An unpublished study (Mitchell et al., 2016) evaluated the potential benefits of the alternative chick box design. A total of 40 commercial journeys were studied consisting of 20 pairs of journeys, each member of the pair being undertaken at one of two different stocking densities (110 or 125 chicks per crate) during the summer and winter seasons. The same improved box design was used in all journeys. Journeys were paired using chicks from the same parent flocks and hatches. In each pair of trucks travelling to the same farm on the same day, one truck was stocked at 110 chicks per box. On the other journey (truck), chicks were carried at 125 chicks per box. All paired journeys were to the same farm and all farms employed in the study were similar distances from the hatchery. The two stocking densities sequence was alternated on successive days on journeys to the same farm. At this stage, there was no comparison of the improved box design with the older designs. Hygrothermal measures (dry bulb temperature [°C] and humidity [expressed as water vapour density gm^{-3}]) were made in a number of specified box locations in defined different stacks across the load thought to be representative of the extremes of local ventilation and airflow.

The animal-based measures were those associated with chick status. Deep body (rectal) temperature was measured before and after the whole transport period to indicate whether there is any evidence of heating or chilling during the transport period. In addition, chick BWs have been determined before and after transport to determine if any differences may indicate differential water loss and/or energy utilisation associated with stocking density. The internal vehicle thermal conditions were always controlled to the same set points.

A spatial distribution of thermal conditions in transport boxes was found at both stocking densities with upper boxes being warmer and more humid than the lower and the central stack on the vehicle being warmer and more humid. There were some spatial and temporal variations in the hygrothermal micro-environment during the journey. In-box temperatures and humidities were slightly higher at a stocking density of 125, but the differences were very small. The change in deep body temperature of chicks during the transport period for a stocking density of 110 across winter and summer conditions was -0.57 ± 0.02°C compared to -0.37 ± 0.02 for a stocking density of 125. For BW loss, the corresponding data were -9.25 ± 0.91 g (110) and -9.35 ± 0.88 g (125). It is concluded that for the design of box employed in these two studies stocking at 125 chicks per box compared to 110 chicks for journeys of the specified durations and type had no identifiable adverse effects. The deep body temperatures measured at each stage of the process and for both stocking densities may be considered 'normal' when compared to published values. It may be suggested that there is no environmental or bird measures evidence that either stocking density caused any problems or welfare issues. It was proposed that employing larger chick boxes with greater depth compared to conventional boxes might allow using a higher stocking density to underpin more efficient transport.

9 Exposure of chicks to acceleration and vibration

Donofre et al. (2014, 2017) investigated the potential effects of mechanical vibrations typical of live chick transportation on the stress responses of newly hatched chicks. In transport simulations, chicks were exposed to two levels of vibration (9.64 ms^{-2} and 15.19 ms^{-2}) for 2 h. The measured chick response variables included change in BW, respiratory rate and performance in the first week, compared to 160 control groups (without vibration). Chicks subjected to an acceleration of 15.19 ms^{-2} had increased respiratory rates and it was concluded that the vibrations may act as a potential stressor during the transport of these birds.

In an extension to these studies, the vibrational characteristics of chick transport vehicles that might influence bird well-being have been determined by Nazareno et al. (2015). Air-conditioned vehicles (container type) carrying a total of 63 000 chicks were used. The vibration levels (ms^{-2}) and shock vibrations were recorded using three triaxial accelerometers. It was concluded that the worst vibration levels occurred on poor roads and most shock vibrations occurred on asphalt roads. It was suggested that vibration and shocks might compromise chick welfare when transportation takes place on lower standard roads, but further research is required to clarify the measures required to minimise this risk.

10 Transport pathologies

Newly hatched chicks are frequently transported from the hatchery to the farm, including production, parent stock and grandsire parent stock chicks. It has been demonstrated that the associated transport stress can induce certain pathologies. However, the mechanism mediating the cellular and organ effects of transport stress is poorly understood. For example, transport conditions for 1-day-old chicks have been reported as influencing the subsequent incidence of ascites and 'sudden death syndrome' (Maxwell and Robertson, 1998), indicating an impact of transport upon the cardiovascular system.

Li et al. (2017) attempted to elucidate the mechanisms mediating possible transport-induced heart injury in newly hatched chicks. Chicks were transported for 2 h, 4 h and 8 h. The results demonstrated showed that transport stress increased the WL and the creatine kinase (CK) activity, disturbed the ionic (K^+, Ca^{2+}, Mg^{2+}) homeostasis and inhibited the ATPase (Na+-K+-ATPase, Ca^{2+}-ATPase, Mg^{2+}-ATPase and Ca^{2+}-Mg^{2+}-ATPase) activities, increased the ATP content and downregulated the gene expression levels of the ATPase-associated subunits in the heart. It was concluded that transport stress disturbed the ionic homeostasis in chick heart via modulating ion transporting ATPases and the transcriptions of the associated subunits, and ultimately induced WL and heart injury in chicks. In a further study, Sun et al. (2018) assessed the effects of transport stress on heart damage and the heat shock response (HSR) protection in newly hatched chicks exposed to 2 h, 4 h or 8 h of transport. Transport stress caused decreases in BW and increases in serum CK activity, nitric oxide (NO) content in heart tissue, cardiac nitric oxide syntheses (NOS) activity and NOS isoforms transcription. The mRNA expression of heat shock factors (HSFs, including HSF1-HSF3) and heat shock proteins (HSPs, including HSP25, HSP40, HSP47, HSP60, HSP70, HSP90 and HSP110) in the heart of 2 h transport-treated chicks was upregulated. After 8 h of transport stress in chicks, the transcription levels of the same HSPs and HSF2 were reduced in the heart. It was also found that the changes in the HSP60, HSP70 and HSP90 protein levels had similar tendencies. These results suggested that transport stress augmented NO generation by enhancing the activity of NOS and the transcription of NOS isoforms. Therefore, this study provides new evidence that transport stress induces heart damage in newly hatched chicks by blocking the cytoprotective HSR and augmenting NO production. In a complementary study, Ge et al. (2017) subjected chicks to transport and simulated transport for 2 h, 4h and 8 h and studied some adaptive responses to these treatments compared to control chicks. The mechanism examined was the activation of the nuclear factor-erythroid 2-related factor 2 (Nrf2)-induced antioxidant defence system. It was hypothesised that transport stress could cause oxidative stress in the cerebrum of newly hatched chicks by increasing lipid peroxidation and production of free radicals and decreasing the activities of antioxidant enzymes and the glutathione:oxidised glutathione ratio. It was demonstrated that transport stress activated the Nrf2 signalling pathway and triggered the transcription of antioxidant parameters. However, transport stress-induced cerebrum oxidative stress was not mitigated by activating the Nrf2 antioxidant defence response.

Cogburn et al. (2020) studied the immediate post-hatch period, in which newly hatched chicks were subjected to fasting for 4 h, 24 h or 48 h and then refed for 4 h, 24 h or 48 h and compared with a fully fed control group at each age (D1–D4). This may simulate the possible feed withholding that occurs during transportation in the first 72 h post-hatch. Transcriptional profiling of the liver during fasting/refeeding of newly hatched chicks revealed several highly expressed upstream regulators, which enable the metabolic switch from fasted (lipolytic/gluconeogenic) to fed or refed (lipogenic/thermogenic) states. This rapid homeorhetic shift of whole-body metabolism from a catabolic-fasting state to an anabolic-fed state appears precisely orchestrated by a small number of ligand-activated transcription factors that provide either a fasting-lipolytic state (PPARA, NR3C1, NFE2L2, SERTAD2, FOX01, NR0B1, RXR) or a fully fed and refed lipogenic/thermogenic state (THRSPA, SREBF2, PPARG, PPARD, JUN, ATF3, CTNNB1). THRSPA has emerged as the key transcriptional regulator that drives lipogenesis and thermogenesis in hatchling chicks, as shown here in fed and refed states. These findings may be indicative of the adaptive metabolic mechanisms that may be induced during the transportation of newly hatched chicks during the first 72 h post-hatch.

11 Guidance on good practice

> 'Welfare is top of the agenda for, politicians, food retailers, consumers and farmers alike. Its importance is such that all sectors of the poultry industry are subject to a plethora of welfare regulation embroiled in Legislation, Codes of Recommendation and Industry Codes of Practice' (Cruickshank, 2000).

The welfare of animals in transit is the subject of regulation and control across the world and it may be suggested that such legislation and controls on practices will increase in the future. The welfare of animals in transit in Europe is currently controlled by Regulation EC 1/2005. The European Commission has proposed that the efficacy of legislation will be improved by the adoption of 'guides to good practice'. Therefore, DG SANTE sponsored the production of Guides to Good Practice for cattle, sheep, pigs, horses and poultry in the European Union (2017) (http://www.animaltransportguides.eu/).

The poultry-specific guide was published in 2017 (Consortium of the Animal Transport Guides Project (2017) - 'Good practices for animal transport in the EU: poultry' (SANCO/2015/G3/SI2.701422) - but disappointingly little information was supplied concerning the transportation of newly hatched chicks beyond simple compliance with EC 1/2005. The major parts of the existing guidance and legislation relating to newly hatched chicks cover journey times/ feed and water withdrawal periods with industry guides providing specific recommendations for thermal micro-environments in the transport containers and in the vehicle. Such guides may also advise upon simple chick assessments to ensure welfare is optimal by inspection to determine cloacal temperature (range 40-40.5°C) and bird behaviour.

A wide range of commercial guidelines, handbooks, manuals and advisory publications for hatcheries, producers and livestock haulage companies are available relating to the preparation and transportation of newly hatched chicks to destination growing and production farms. These publications have been produced by poultry breeders, handing and transport companies, vehicle manufacturers, competent authorities and poultry business professional bodies. A selection is listed below which is probably incomplete but covers the main examples. Each guide exhibits similarities in making recommendations for optimum transport temperatures (vehicle and container), journey durations and chick inspection routines and nutritional and support strategies. Whilst specifics may differ, all recommendations are perhaps, unsurprisingly, generally very similar.

The major sources consulted have been Aviagen (2020), Hatchery Tips (2020), Barnett et al. (2008), Cazaban (2005), Cobb Hatchery Management Guide (2018, European Guide to Good Air Transport of Hatching Eggs and Day-old-chicks (2011), Heering (2020), Hendrix Genetics (2019), Management Guide (2020) Lohmann Tierzucht, Pas Reform (2012), Poultry Handling and Transportation Manual (2017), Poultry Handling and Transportation Manual (2017), Ross Broiler Management Handbook (2018), van der Linde (2018) and Veit Electronics (2021).

In the UK, the Farm Animal Welfare Committee (FAWC) advises Defra and the devolved administrations in Scotland and Wales on the welfare of farmed animals and any legislative or other changes that might be considered to improve standards of animal welfare. One of its chief activities is the preparation and presentation of formal reports and opinions on specific animal welfare issues. One such report was published in April 2019: 'Opinion on the Welfare of Animals during Transport' (FAWC, 2019). This report has identified a number of issues relating to poultry transportation, including the carriage of newly hatched chicks, and produced a number of recommendations for improving practices. The report suggested: 'Young (newly hatched) chicks are unable to effectively thermoregulate. Exposure to temperature stress, especially from long journeys, might be measurably demonstrated by rectal temperatures and/ or by weight loss'. Temperatures and humidity within the transport containers and within the transport space should be monitored and recorded. Of course, the vehicles currently employed for the transfer of newly hatched chicks to their sites of growing or production are effectively air-conditioned and as such routinely monitor on-board temperature which is used in automated control of the vehicle micro-environment.

Both the European Union and the UK Government are currently undertaking reviews, public consultations and discussions with industry and animal welfare organisations and NGOs with a view of developing improved and amended legislation for adoption in the near future (forecast 2023 and beyond). The review process in the UK was summarised in a FAWC Report (2019) which included a systematic scientific review (Defra project AW 0821 - 'A review of the evidence on welfare aspects of the transport of live animals'; Mitchell et al., 2018) examining the scientific evidence that might support or suggest possible improvements and amendments to animal transport regulation.

The mechanisms for the development of animal welfare science and policy and their translation into law (legislation) in the European Union have been extensively described and discussed by several authors in Hild and Schweitzer (2019). The parallel processes in the European Union, relating to animal transport, have moved on quickly. The Farm to Fork Strategy (2019) adopted by the Commission on 20 May 2020 announced that the Commission will, by the end of 2023, revise the animal welfare legislation to align it with the latest scientific evidence.

12 Conclusion

There is a considerable body of published evidence in relation to the transportation of poultry that might usefully inform policy, guidance and future legislation. Some of the issues have been addressed by the recently published Guides to Good Practice (DG SANTE, 2017), but many concerns remain to be resolved. The published research will facilitate the development of new policy, but there are also areas that might benefit from future research and which might be considered important research requirements. It has been considered that the utilisation of yolk sac substrates and reserves in newly hatched chicks from modern commercial broiler lines might impact acceptable journey times for newly hatched chicks. Review of all the available evidence suggests that this may not be the case and current journey limits are adequate to protect the chicks. In the case of newly hatched chicks, the maximum transport time is currently considered to be 24 h if the journey is undertaken in the first 72 h post-hatch. The available evidence suggests that this may still be appropriate based on available knowledge of yolk sac reserves and resource utilisation. There is a proposal in the UK that the maximum journey time for newly hatched chicks should be reduced to 21 h, but this is meeting with strong resistance. However, it may be appropriate to propose future research to specifically address this issue and to provide sound scientific evidence in support of the current position or to indicate that new guidelines or legislation should be developed.

It is well-recognised that newly hatched chicks are particularly vulnerable to cold stress in transport, but current industry practices and internal controls ensure that the risk of cold stress is minimised. There may still be some risk of exposure to heat or cold stress in transport containers due to poorly controlled ventilation regimes of 'chick boxes' (low or excessive local airflows). The risk of cold stress will be increased if chick space allowance is high or bird numbers are small. Hypothermia is the major risk to this category of birds. Heat stress in chicks will increase the rate of depletion of water and energy reserves and will pose an immediate risk to welfare and a long-term risk to productivity.

In keeping with the recommendations for broiler chickens, new legislation might prescribe temperature limits for newly hatched chick transport. The proposed values are supported by existing research and further investigation may not be required.

Early feeding of newly hatched chicks has been recommended also to improve welfare during long transport to farms. It is not clear from published research if the optimum method for delivery and composition of feed and water prior to farm placement has been established. Optimum vehicle temperatures for newly hatched chicks were proposed until recently to be 24–25°C with limits of 22-28°C and container temperatures of 30–32°C. The use of modern trailer designs with higher and more efficient ventilation systems requires revision of the recommended thermal envelopes. It is now proposed that the thermal micro-environment for newly hatched chicks must be controlled to ensure that chick rectal (vent) temperature is maintained at 39.4–40.5°C, a minimum RH of 50% within the containers and a ventilation rate minimum of 0.71 $m^3 min^{-1}$ (25 $ft^3 min^{-1}$) of fresh air per 1000 chicks should be supplied. Greater ventilation rates may be required if the truck is not air-conditioned and ventilation is the only method available to cool the chicks.

Adequate ventilation of chick vehicles and transport containers is vital to maintain welfare in transport. It is suggested that additional investigations might better characterise the ventilation of transport containers for newly hatched chicks and might examine the impact of different box designs and stocking densities or space allowances to further improve environmental control and chick health and welfare in transit. This area for newly hatched chick transport has been identified by the industry as priority. Thus, changes in chick box design will facilitate better ventilation and control of the thermal micro-environment will improve chick welfare. This might then allow changes in the permissible stocking density/space allowance for chicks. Most of the available research addresses the transport of broiler chicks, but it is proposed that the outcomes, principles, findings and recommendations can be safely applied to newly hatched chicks from other categories of poultry, including laying pullet chicks and breeder birds. It may be suggested that there may be some metabolic differences between these different groups of chicks but the magnitude of any disparity may be too small to merit specific recommendations or to justify adjustment to current recommendations.

A reasonable alternative approach to newly hatched chick transport is the delivery of hatching eggs to growing farms with on-farm hatching and brooding. There are several studies that have assessed the efficacy and practicality of this process and evidence suggests that this may provide an ideal solution to the welfare, health and productivity issues and concerns associated with road transportation of neonate chicks.

13 Where to look for further information

There is much current interest in revising and improving the guidelines and regulations governing the transport of all livestock. In Europe, EFSA have published Opinions on the Transport of Animal and in the UK, Defra and the devolved governments have undertaken reviews to provide the basis for improved legislation. These exercises, of course, must include consideration of the welfare of newly hatched chicks in transit. Such reviews often identify gaps in knowledge and future research requirements. Thus, the transport of newly hatched chicks may require further investigation. The currently available research suggests that journey time limits (based on yolk sac resource utilisation) are consistent with good chick welfare at this time. However, it might be proposed that new specific research addressing this assertion is merited. New guides to good practice and legislation might prescribe temperature limits for newly hatched chick transport. The proposed values appear to be by existing research, but further investigation may be required. In addition, changes in chick box design to facilitate better ventilation and control of the thermal micro-environment should be further investigated. This might improve chick welfare and may require changes in the permissible stocking density/ space allowance for chicks. This area should be studied as it offers a number of other potential benefits if proved appropriate. For further details of these and other issues, the FAWC Report (2019) and Review, the EFSA Opinions upon Animal Transport (2022) are recommended reading.

14 References

Akter, S., Cheng, B., West, D., Liu, Y., Qian, Y., Xiuguo Zou, X., Classen, J., Cordova, H., Oviedo, E. and Wang-Li, L. (2022). Impacts of air velocity treatments under summer condition: Part I–Heavy broiler's surface temperature response. *Animals: An Open Access Journal From MDPI* 12(3):328. https://doi.org/10.3390/ani12030328. Available at: www.mdpi.com/journal/animals.

Animal Transport Guides. Available at: http://www.animaltransportguides.eu/.

Aviagen. (2020). Hatchery. *TIPS 2020.* Available at: https://eu.aviagen.com/assets/Tech_Center/BB_Resources_Tools/Hatchery_Tips/HatcheryTips-EN.pdf.

Aviagen - Anon. (2021). Factors affecting chick comfort and liveability from hatcher to brooding house the poultry site. Available at: https://www.thepoultrysite.com/articles/factors-affecting-chick-comfort-and-liveability-from-hatcher-to-brooding-house (Vol. 2021).

Barnett, J. L., Edge, M. E., Thomson, L., Mackenzie, M., Sansom, G. and Kite, V. (2008). Manual for the hatchery and chick transport: National animal welfare standards for the chicken meat industry. *The Hatchery & Chick Transport*. National Animal Welfare Standards for the Chicken Meat Industry.

Barri, A. (2008). *Effects of incubation temperature and transportation stress on yolk utilization, small intestine development, and post-hatch performance of high-yield broiler chicks.* PhD Thesis. Submitted to the Faculty of the Virginia Polytechnic Institute and State University in partial fulfilment of the requirements for the Degree of Doctor of Philosophy In Animal and Poultry Sciences Copyright.

Bautil, A., Buyse, J. and Everaert, N. (2014). *Interactions between spread of hatch and delayed feed access with emphasis on the intestinal lipase activity*. MSc Dissertation. Leuven: KU Leuven.

Beausoleil, N. J., McKeegan, D. E. F. and Martin, J. E. (2021). Chapter 39: Avian welfare: Fundamental concepts and scientific assessment. In: Scanes, C. G. and Dridi, S. (Eds.), *Sturkie's Avian Physiology* (7th edn.). Academic Press. https://doi.org/10.1016/B978-0-12-819770-7.00013-X.

Beausoleil, N. J. and Mellor, D. J. (2017). Validating indicators of sheep welfare. In: Greyling, J. (Ed.), *Achieving Sustainable Production of Sheep*. Cambridge, UK: Burleigh Dodds Science Publishing, pp. 327–348.

Bergoug, H., Burel, C., Guinebretiere, M., Tong, Q., Roulston, N., Romanini, C. E. B., Exadaktylos, V., McGonnell, I. M., Demmers, T. G. M., Verhelst, R., Bahr, C., Berckmans, D. and Eterradossi, N. (2013a). Effect of pre-incubation and incubation conditions on hatchability, hatch time and hatch window, and effect of post-hatch handling on chick quality at placement. *World's Poultry Science Journal* 69(2):313–334.

Bergoug, H., Guinebretière, M., Tong, Q., Roulston, N., Romanini, C. E., Exadaktylos, V., Berckmans, D., Garain, P., Demmers, T. G., McGonnell, I. M., Bahr, C., Burel, C., Eterradossi, N. and Michel, V. (2013b). Effect of transportation duration of 1-day-old chicks on postplacement production performances and pododermatitis of broilers up to slaughter age. *Poultry Science* 92(12):3300–3309. http://doi.org/10.3382/ps.2013-03118.

Bigot, K., Mignon-Grasteau, S., Picard, M. and Tesseraud, S. (2003). Effects of delayed feed intake on body, intestine, and muscle development in neonate broilers. *Poultry Science* 82(5):781–788.

Broom, D. M. (1991). Animal welfare: concepts and measurement. *Journal of Animal Science*. 69, 4167–4175.

Broom, D. M. (2000). Welfare assessment and problem areas during handling and transport. In: Grandin, T. (Ed.), *Livestock Handling and Transport* (2nd edn.). Wallingford: CABI Publishing, pp. 43–61.

Broom, D. M. (2007). Quality of life means welfare: How is it related to other concepts and assessed? *Animal Welfare* 16(S1):45–53.

Buzala, M., Janicki, B. and Czarnecki, R. (2015). Consequences of different growth rates in broiler breeder and layer hens on embryogenesis, metabolism and metabolic rate: A review. *Poultry Science* 94(4):728–733 http://doi.org/10.3382/ps/pev015.

Cardeal, P. C., Araújo, I. C. S., Sousa, L. S., Melo, E. F., Carvalho, T. S. M., Triginelli, M. V., Pompeu, M. A. and Lara, L. J. C. (2021). Breeder age and post-hatch feed access influence physiology of residual yolk sac, intestinal development and performance of broiler chicks. *Livestock Science* 251:104617. https://doi.org/10.1016/j.livsci.2021.104617.

Cazaban, C. (2005). Day-old chick quality, CEVA Sante animale, la Ballastiere. *Hatchery Expertise on-Line Issue*, France 1(July):BP 126, 33501 Libourne Cedex.

Cobb. Home. Available at: cobb-vantress.com.

Cobb Hatchery Management Guide. (2018). Cobb animal welfare tips. Available at: https://www.cobb-vantress.com/.

Cockram, M. S. and Mitchell, M. A. (1999). Role of research in the formulation of 'rules' to protect the welfare of farm animals during road transportation. *BSAP Occasional Publication* 23:43–64. https://doi.org/10.1017/S0263967X00033243.

Cogburn, L. A., Trakooljul, N., Wang, X., Ellestad, L. E. and Porter, T. E. (2020). Transcriptome analyses of liver in newly hatched chicks during the metabolic perturbation of fasting and re-feeding reveals THRSPA as the key lipogenic transcription factor. *BMC Genomics* 21(1):109. https://doi.org/10.1186/s12864-020-6525-0.

Commission Implementing (31.3.2021), p. 11802021R0405 – EN – 05.03.2022 – 005.001 – 1. Regulation (EU) 2021/405 of 24 March 2021 laying down the lists of third countries or regions thereof authorised for the entry into the Union of certain animals and goods intended for human consumption in accordance with Regulation (EU) 2017/625 of the European Parliament and of the Council (Text with EEA relevance) (OJ L 114).

Cruickshank, G. (2000). Personal communication: Paper presented to Poultry World Industry Conference-Welfare Guideline Sponsorscopies available from original author upon request.

de Camargo, J. R., da Silva, I. J. O., Nazareno, A. C., Vieira, F. M. C., de Castro, A. C. and Dias, R. A. P. (2015). Qualidade de pintos em função do microclima, tempo de espera e idade de matrizes. Quality of chicks as a function of microclimate, lairage time and ages of breeder. *Revista Brasileira de Engenharia Agrícola* 19(11):1079–1085. Available at: http://www.agriambi.com.br. ISSN 1807-1929.

de Jong, I. C., Gunnink, H., van Hattum, T., van Riel, J. W., Raaijmakers, M. M. P., Zoet, E. S., van den Brand, H. and H. (2019). Comparison of performance, health and welfare aspects between commercially housed hatchery-hatched and on-farm hatched broiler flocks. *Animal: An International Journal of Animal Bioscience* 13(6):1269–1277. https://doi.org/10.1017/S1751731118002872.

de Jong, I. C., Van Riel, J., Lourens, A., Bracke, M. B. M. and Van den Brand, H. (2016). *Effects of Food and Water Deprivation in Newly Hatched Chickens. A Systematic Literature*

Review and Meta-Analysis. Wageningen Livestock Research, Report 999. This report can be downloaded for free at Available at: www.wur.nl/livestock-research http://doi.org/10.18174/401560 or at (under Wageningen Livestock Research publications).

Decuypere, E., Tona, K., Bruggeman, V. and Bamelis, F. (2001). The day-old chick: A crucial hinge between breeders and broilers. *World's Poultry Science Journal* 57(2):127–138.

Donofre, A. C., da Silva, I. J. O. and Nazareno, A. C. (2014). Mechanical vibrations: A stressor in the transport of chicken. *Revista Brasileira de Engenharia Agricola e Ambiental* 18(4):454–458.

Donofre, A. C., Silva, I. J. O., Nazareno, A. C. and de paula Ferreira , E. (2017). Mechanical vibrations in the transport of hatching eggs and the losses caused in the hatch and quality of broiler chicks. *Journal of Agricultural Engineering* XLVIII:593JAE, 36–41. https://doi.org/10.4081/jae.2017.593.

Dunnington, E. A. and Siegel, P. B. (1984). Thermoregulation in newly hatched chicks. *Poultry Science* 63(7):1303–1313.

EC 1/2005 (2005). Council regulation (EC) No 1/2005 of 22 December 2004 on the protection of animals during transport and related operations and amending Directives 64/432/EEC and 93/119/EC and Regulation (EC). *Official Journal of the European Union* 1255/97:L3/1.

EC Council. Regulation (EC) No 1/2005 on the protection of animals during transport and related operations and amending Directives 64/432/EEC and 93/119/EC and regulation (EC) (Vol. 1255/97).

EFSA (European Food Safety Authority). (2004a). Opinion of the scientific panel on animal health and welfare related to the welfare of animals during transport. *The EFSA Journal* 44:1–36.

EFSA (European Food Safety Authority). (2004b). Opinion of the Scientific Panel on Animal Health and Welfare (AHAW) related to the standards for the microclimate inside animal road transport vehicles. *The EFSA Journal* 122:1–25.

EFSA (European Food Safety Authority). (2005). *Opinion of the Scientific Panel on Animal Health and Welfare (AHAW) Related to the Welfare of Weaners and Rearing Pigs: Effects of Different Space Allowances and Floor*.

EFSA Journal (1666). 268, 1–19. EFSA Panel on Animal Health and Welfare (AHAW). (2010). Scientific opinion on the influence of genetic parameters on the welfare and the resistance to stress of commercial broilers. *EFSA Journal* 8(7).

ESFA, Q. (2020). 00834-EFSA-GMO-NL-2020-172_Summary.pdf. Art 29 - Request for scientific opinion on welfare: Transport of animals (EFSA-Q-2020-00481 & EFSA-Q-2020-00482); Evidence and supporting documents for EFSA-Q-2020-00482. Evidence Report, 2020.06 Welfare Farm2fork.

EFSA Journal. (2011). Statement on the use of animal-based measures to assess the welfare of animals. *EFSA Journal* 10(6):2767–2829. https://doi.org/10.2903/j.efsa.2012.2767.

EFSA Panel on Animal Health and Welfare (AHAW). (2011). Søren Saxmose Nielsen, Julio Alvarez, Dominique Joseph Bicout, Paolo Calistri, Elisabetta Canali, Julian Ashley Drewe, Bruno Garin-Bastuji, Jose Luis Gonzales Rojas, Christian Gortazar Schmidt, Mette Herskin, Virginie Michel, Miguel Ángel Miranda Chueca, Barbara Padalino, Helen Clare Roberts, Hans Spoolder, Karl Stahl, Arvo Viltrop, Christoph Winckler, Malcolm Mitchell, Leonardo James Vinco, Eva Voslarova, Denise Candiani, Olaf

Mosbach-Schulz, Yves Van der Stede and Antonio Velarde. *EFSA Journal* 20(9):7441. doi: 10.2903/j.efsa.2022.7441.

EPB-Association of the European Poultry Breeders. (2011). European guide to good air transport of hatching eggs and day-old-chicks. Available at: www.europeanpoultry breeders.eu.

europa.eu. (2019). Farm to fork strategy for a fair, healthy and environmentally-friendly food system. Available at: https://ec.europa.eu/food/horizontal-topics/farm-fork -strategy_enFork.europa.eu.

European Commission. (2020). Revision of the animal welfare legislation. Available at: https://ec.europa.eu/food/animals/animal-welfare/evaluations-and-impact -assessment/revision-animal-welfare-legislation_en.

European Food Safety Authority. (2011a). EFSA panel on animal health and welfare. Scientific opinion concerning the welfare of animals during transport. *EFSA Journal* 9(1). Available at: http://www.efsa.europa.eu/en/efsajournal/pub/1966.htm.

European Food Safety Authority. (2011b). Scientific opinion concerning the welfare of animals during transport. *EFSA Journal* 9(1). https://doi.org/10.2903/j.efsa.2011 .1966.

FAWC. (2019). Opinion on the Welfare of Animals During Transport April 2019 Farm Animal Welfare Committee Area 2D, Nobel House, 17 Smith Square, London SW1P 3JR. Available at: https://consult.defra.gov.uk/transforming-farm-animal -health-and-welfare-team/improvements-to-animal-welfare-in-transport/supporting _documents/fawcopiniononthewelfareofanimalsduringtransport.pdf.

FAWC. Opinion on the welfare of animals during transport. Available at: defra.gov.uk.

Fernyhough, M.,Nicol, C. J., van de Braak, T., Toscano, M. J. and Tønnessen, M. (2020). The ethics of laying hen genetics. Journal of Agricultural and Environmental Ethics 33(1):15–36.

Food and Agriculture Organization of the United Nations (FA). (December 2020). Meat market review, emerging trends and outlook.

Freeman, B. M. (1965). The relationship between oxygen consumption, body temperature and surface area in the hatching and young chick. *British Poultry Science* 6(1):67–72.

Freeman, B. M and Manning, A. C. (1984). Re-establishment of the stress response in Gallus. *Comparative Biochemistry and physiology. A, Comparative Physiology,* 78(2):267–270. DOI: 10.1016/0300-9629(84)90144-0 PMID: 6146446.

Freiji, M. (1988). Chicks need cool transport. *Poultry International (MISSET)* 4(3):18–19.

Gates, R. S., Turner, L. W., Turner, G. M. and Tuttle, J. W. (1989). Measured heat production in chicks. In: *Proceedings of the American Society of Agricultural Engineers: Paper No. 894513.*

Gautron, J., Réhault-Godbert, S., Van de Braak, T. G. H. and Dunn, I. C. (December 2021). Review: What are the challenges facing the table egg industry in the next decades and what can be done to address them? *Animal: An International Journal of Animal Bioscience* 15(Suppl. 1):100282.

Ge, J., Li, H., Sun, F., Li, X. N., Lin, J., Xia, J., Zhang, C. and Li, J. L. (2017). Transport stress-induced cerebrum oxidative stress is not mitigated by activating the Nrf2 antioxidant defense response in newly hatched chicks. *Journal of Animal Science* 95(7):287-2878. https://doi.org/10.2527/jas.2017.1559.

Giersberg, M. F., Poolen, I., de Baere, K., Gunnink, H., van Hattum, T., van Riel, J. W., Ingrid, C. and de Jong, I. C. (2020). Comparative assessment of general behaviour and fear-related responses in hatchery-hatched and on-farm hatched broiler chickens. *Applied Animal Behaviour Science* 232. Accessed November 2020. https://doi.org/10.1016/j.applanim.2020.105100.

Global Poultry Industry and Trends - Feed & Additive Magazine. Accessed January 23, 2022. Available at: feedandadditive.com.

Guides to Good Practices for Animal Transport in the EU: Consortium of the Animal Transport Guides Project. (2017). Good practices for animal transport in the EU: Poultry (SANCO/2015/G3/SI2.701422). Available at: http://animaltransportguides.eu/.

Hackl, G. and Kaleta, E. F. (1997). Methods for the estimation of the age of broilers and effects of water and feed deprivation within the first days of life. *Archiv fuer Gefluegelkunde* 61:280–286.

Hamissou Maman,A., Ozulu, S., Ucar, A. and Elibol, O. (2019). Effect of chick body temperature during post-hatch handling on broiler live performance. *Poultry Science* 98(1):244–250 http://doi.org/10.3382/ps/pey395.

Hatchability. Available at: https://www.hatchability.com/chick-temperature.html.

HatcheryTips, E. N. Available at: aviagen.com.pdf.

HatchTech, V. (2020). The Netherlands.

Hatchtraveller. Available at: https://hatchtraveller.com/.https://hatchtraveller.com/hatchery-products/.

Hedlund, L. and Jensen, P. (2022). Effects of stress during commercial hatching on growth, egg production and feather pecking in laying hens. *PLoS ONE* 17(1):e0262307. https://doi.org/ 10.1371/journal.pone.0262307.

Heering and van der Linde, G. (2018). Key issues for the successful transportation of chicks. International Hatchery Practice. Available at: www.positiveaction.info/pdfs/articles/hp30_5p21.pdf.

Heering and van der Linde, G. (2018). Key issues for the successful transportation of chicks. International Hatchery Practice Available at: www.positiveaction.info/pdfs/articles/hp30_5p21.pdf

Hendrix Genetics. (2019). *Hatchery Management Guide: Version L9180-2*. Boxmeer, The Netherlands: European Union.

Henken, A. M., van der Hel, W., Hanssen, E. J. M., Scheele, C. W. and van Kampen, M. (1989). The effect of dry and wet bulb temperature on heat production of newly hatched chicks. In: *Energy Metabolism of Farm Animals*. Wageningen, Netherlands, pp. 291–294.

Herborn, K. A., Graves, J. L., Jerem, P., Evans, N. P. and Nager. (2015). Skin temperature reveals the intensity of acute stress. *Physiology and Behavior* 225–30:152.

Hild, S. and Schweitzer,L. (2019). Animal welfare: From science to law, la fondation droit animal, éthique et sciences (LFDA). Available at: www.fondation-droit-animal.org.

Hoeven, E. (2021). On-farm hatching: More than just early feeding. *International Hatchery Practice* 35(4):7–9.

House of Commons Environment, Food and Rural Affairs Committee. (2021). Moving Animals Across Borders; First Report of Session 2021–22 Report, Together With Formal Minutes Relating to the Report Ordered by the House of Commons to Be Printed 23 September 2021 HC 79 Published on 30 September 2021 by Authority of the House of Commons.

Hoxey, R. P., Kettlewell, P. J., Meehan, A. M., Barker, C. J. and Yang, X. (1996). An investigation of the aerodynamic and ventilation characteristics of poultry transport vehicles: Part 1. *Journal of Agricultural Engineering Research* 65(1):77–83.

HSUS and The Humane Society of the United States. (2008). Welfare issues with transport of day-old chicks. *Impacts on Farm Animals*. Available at: https://www.wellbeingintlstudiesrepository.org/hsus_reps_impacts_on_animals/7 (Vol. 7).

HSUS and Humane Society of the United States. (2009). An HSUS report: Welfare issues with transport of day-old chicks. *Farm Animal Welfare Science and Research*. Available at: http://www.humanesociety.org/assets/pdfs/farm/welfiss_chick_transport.pdf.

Hunter, R. R., Mitchell, M. A. and Matheu, C. (2001). Mortality of broiler chickens in transit – correlation with the thermal micro-environment. In: Stowell, R. R., Buckling, R. and Bottcher, R. W. (Eds.), *Proceedings of the 6th International Livestock Environment Symposium*, Louisville, Kentucky, USA, pp. 542–549. (Quoted in EFSA 2004. AHAW Transport Report).

Jacobs, L. (2016). *Road to Better Welfare: Welfare of Broiler Chickens During Transportation*. Doctoral Thesis, Ghent University Doctor of Philosophy (PhD). ISBN: 978-94-6197-471-6.

Jacobs, L., Deleziea, E., Duchateau, L., Goethalsb, K., Ampea, B., Buyse, J. and Tuyttens, F. A. M. (2017). Impact of transportation duration on stress responses in day-old chicks from young and old breeders. *Research in Veterinary Science* 112:172–176. https://doi.org/10.1016/j.rvsc.2017.04.015.

Jacobs, L., Delezie, E., Duchateau, L., Goethals, K., Ampe, B., Lambrecht, E., Gellynck, X. and Tuyttens, F. A. M. (2016). Effect of post-hatch transportation duration and parental age on broiler chicken quality, welfare, and productivity. *Poultry Science* 95(9):1973–1979.

Jacobs, L., Delezie, E., Duchateau, L., Goethals, K. and Tuyttens, F. A. M. (2017). Broiler chickens dead on arrival: Associated risk factors and welfare indicators. *Poultry Science* 96(2):259–265. https://doi.org/10.3382/ps/pew353.

Jacobs, L., Delezie, E. and Tuyttens, F. A. M. (2015). Welltrans: Research on the welfare of broiler chickens during transport Available at: http://pure.ilvo.vlaanderen.be/portal/en/projects/onderzoek-naar-het-welzijn-bijtransport-van-vleeskippen(44e5094e-ea0a-4aba-94fb-faf7ec05e574).html.

Jessen, C. T., Foldager, L. and Riber, A. B. (2021). Effects of hatching on-farm on behaviour, first week performance, fear level and range use of organic broilers. Applied Animal Behaviour Science 238:105319.

Kaltofen, R. S. and Dijk, D. J. (1984). Dutch Report about the Resistance of Day-Old Chickens against Different Combination of Environmental Temperature and Relative Humidity. Centre for Applied Poultry Science 'Het Spelderholt', No. 183.0776P.

Kettlewell, P. J., Hoxey, R. P., Hampson, C. J., Green, N. R., Veale, B. M. and Mitchell, M. A. (2001). Design and operation of a prototype mechanical ventilation system for livestock transport vehicles. *Journal of Agricultural Engineering Research* 79(4):429–439.

Kettlewell, P. J. and Mitchell, M. A. (2001a). Comfortable ride: Concept 2000 provides climate control during poultry transport. *Resource: Engineering and Technology for a Sustainable World* 8:13–14.

Kettlewell, P. J. and Mitchell, M. A. (2001b). Mechanical ventilation: Improving the welfare of boiler chickens in transit. *Journal of the Royal Agricultural Society of England* 162:175–184.

Kettlewell, P. J., Hoxey, R. P. and Mitchell, M. A. (2000). Heat produced by broiler chickens in a commercial transport vehicle. *Journal of Agricultural Engineering Research* 75(3):315–326.

Khosravinia, H. (2015). Physiological responses of newly hatched broiler chicks to increasing journey distance during road transportation. *Italian Journal of Animal Science* 14(3):3964. https://doi.org/10.4081/ijas.2015.3964.

Lamoreux, W. F. and Hutt, F. B. (1939). Variability in body temperature in the normal chick. *Poultry Science* 18(1):70–75.

Langhout, P. J. (2001). Feeding during the first week of life. *International Poultry Production* 9(1):7–10.

Laughlin, K. F. (1989). Computerised customer care. *International Hatchery Practice* 4:43–45.

Li, Z. Y., Lin, J., Sun, F., Li, H., Xia, J., Li, X. N., Jing Ge, J., Zhang, C. and Li, J. L. (2017). Transport stress induces weight loss and heart injury in chicks: Disruption of ionic homeostasis via modulating ion transporting ATPases. *Oncotarget* 8(15):24142-24153 Available at: www.impactjournals.com/oncotarget/.

Malik, H. E. E., Ali, O. H. A., Elhadi, H. M. and Elzubeir, E. A. (2010). Residual yolk utilization in fast and slow-growing chicks, subjected to feed and water deprivation. *Asian Journal of Biological Sciences* 4(1):90–95.

Management Guide. (2020). Lohmann Tierzucht. Available at: https://www.hatchability.com/lohmann.pdflohmann.pdf; hatchability.com.

Maxwell, M. H. and Robertson, G. W. (1998). UK survey of broiler ascites and sudden death syndromes in 1993. *British Poultry Science* 39(2):203–215.

Meijerhof, R. (1997). The importance of egg and chick transportation. *World Poultry* 13(11):17–18.

Mellor, D. J. (2017). Operational details of the five domains model and its key applications to the assessment and management of animal welfare. *Animals: An Open Access Journal from MDPI* 7(8).

Mellor, D. J. and Beausoleil, N. J. (2015). Extending the 'Five Domains' model for animal welfare assessment to incorporate positive welfare states. *Animal Welfare* 24(3):241–253.

Misson, B. H. (1976). The effects of temperature and relative humidity on the thermoregulatory responses of grouped and isolated neonate chicks. *Journal of Agricultural Science* 86(1):35–43.

Mitchell, M. (2001). Prescriptive production issues - A UK/European perspective. *Poultry Science* 80(Suppl. 1):2.

Mitchell, M. A. (2002). Chick transport and welfare. In: Deeming, D. C. (Ed.), *Practical Aspects of Commercial Incubation in Poultry*, Ratite Conference Books, Lincolnshire, UK, pp. 59–64.

Mitchell, M. A. (2006). Using physiological models to define environmental control strategies. In: Gous, R. M., Morris, T. R. and Fisher, C. (Eds.), *Mechanistic Modelling in Pig and Poultry Production*. Wallingford, Oxfordshire, UK: CABI Publishing International, pp. 209–228.

Mitchell, M. A. (2009). Chick transportation and welfare. *Avian Biology Research* 2(1-2):99–105. http://doi.org/10.3184/175815509X431894.

Mitchell, M. A. (2013a). Thermal imaging in Physiology: Theoretical and practical considerations. In: Luzi, F., Mitchell, M., Nanni-Costa, L. and Raedelli, V. (Eds.), *Thermography: Current Status and Advances in Livestock Animals and in Veterinary Medicine* (Vol. 2013). Fondazione Iniziative Zooprofilattiche e Zootechniche, Brescia, Italy, pp. 47–68.

Mitchell, M. A. (2013b). Thermal imaging: Thermoregulation in relation to animal production and welfare. In: Luzi, F., Mitchell, M., Nanni-Costa, L. and Raedelli, V. (Eds.), *Thermography: Current Status and Advances in Livestock Animals and in Veterinary Medicine* (Vol. 2013). Fondazione Iniziative Zooprofilattiche e Zootechniche, Brescia, Italy, pp. 147–162.

Mitchell, M. A., Carlisle, A. J., Hunter, R. R. and Kettlewell, P. J. (2000). The responses of birds to transportation. In: *World Poultry Congress*, Montreal, Canada, 22nd–24th August, 2000, pp. 1–14.

Mitchell, M. A. and Kettlewell, P. J. (1998). Physiological stress and welfare of broiler chickens in transit: Solutions not problems! *Poultry Science* 77(12): 1803–1814.

Mitchell, M. A. and Kettlewell, P. J. (2004a). Transport and handling. In: *Measuring and Auditing Broiler Welfare, Editors C. A. Weeks and A. Butterworth*. Oxon: CAB International, pp. 145–160.

Mitchell, M. A. and Kettlewell, P. J. (2004b). Transport of chicks, pullets and spent hens. In: Perry, G. C. (Ed.), *Welfare of the Laying Hen.* Oxon: CAB International, pp. 345–360.

Mitchell, M. A., Kettlewell, P. J., Carlisle, A. J., Hunter, R. R. and Manning, T. (1996). Defining the optimum thermal environment for the transportation of one-day old chicks: Physiological responses during transport simulations. In: *British Poultry Science, Spring Meeting,* Scarborough, 20–21st March 1996 (Vol. 37), Wildlife Preservation Society of Australia/BSAP, UK, pp. S89–S91.

Mitchell, M. A., Kettlewell, P. J., Hunter, R. R. and Carlisle, A. J. (2001). Physiological stress modelling – Application to broiler transport. In: Stowell, R. R., Bucklin, R. and Bottcher, R. W. (Eds.), *Livestock Environment VI. Proceedings of the 6th International Symposium*, 21–23 May 2001, Louisville, Kentucky, USA, pp. 550–555. St. Joseph, MI: The American Society of Agricultural and Biological Engineers. Available at: www.asabe.org; https://doi.org/10.13031/2013.7116.

Mitchell, M. A. M., Kettlewell, P. J. and Martin, J. (2016). Unpublished findings - Commercial study.

Mitchell, M. A. M., Kettlewell, P. J. and Martin, J. (2019). A review of the evidence on welfare aspects of the transport of live animals, Defra Project AW0821, September 14th 2018, 323.

Molenaar, R. (2012). The importance of the brooding period XXIV World's Poultry Congress 5–9 August, Salvador, Bahia, Brazil.

Nangsuay, A., Molenaar, R., Meijerhof, R., van den Anker, I., Heetkamp, M. J., Kemp, B. and van den Brand, H. (2015). Differences in egg nutrient availability, development, and nutrient metabolism of broiler and layer embryos. *Poultry Science* 94(3):415–423. http://doi.org/10.3382/ps/pev007.

Nazareno, A. A., da Silva, I. J. O. and Donofre, A. C. (2016). Thermal gradients of container and mean surface temperature of broiler chicks transported on different shipments. *Journal of the Brazilian Association of Agricultural Engineering* 36(4):581–592. http://doi.org/10.1590/1809-4430-Eng.Agric.v36n4p581-592/2016.

Nazareno, A. C., da Silva, I. J. O., Donofre, A. C., Vieira, F. M. C. and Castro, A. Cd (2015). Vibrações e choques mecânicos em pintos de um dia transportados em diferentes estradas. Vibration and mechanical shocks in one day-old chicks transported in different roads. *Revista Brasileira de Engenharia Agrícola e Ambiental* 19(7):680–685. Available at: http://www.agriambi.com.br.

Nazareno, A. C., Silva, I. J. O. D., Vieira, F. M. C. and Santos, R. F. S. (2015a). Temperature mapping of trucks transporting fertile eggs and day-old chicks: Efficiency and/or acclimatization? *Revista Brasileira de Engenharia Agrícola e Ambiental* 19(2):134–139. http://doi.org/10.1590/1807-1929/agriambi.v19n2p134-139.

Nazareno, A. E., da Silva, I. J. O., Vieira, F. M. C., Santos, R. F. S. (2015b). One day-old chicks transport: Assessment of thermal profile in a tropical. *Revista Brasileira de Engenharia* 19(7):1807–1929. http://doi.org/10.1590/1807-1929/agriambi.v19n7p663-667.

Nilipour, A. H. (2002). Poultry in transit are a cause for concern. *World Poultry* 18(2):30–33.

Olawole, A. and Atta, A. T. (2019). State-of-the-art technologies for assessing thermal comfort of broiler chickens. *International Journal of Engineering Applied Sciences and Technology* 4(8):72–83.

Pijarska, I., Czech, A., Malec, H. and Tymczyna, L. (2006). Effect of road transportation of chicks on blood biochemical indices and productive results of broilers. *Medycyna Weterynaryjna* 62:408–410.

Poultry Service Association, Ontario Ministry of Agriculture, Food and Rural Affairs, Canadian Poultry & Egg Processors Council. (2017). Poultry handling and transportation manual. Available at: http://www.poultryserviceassociation.com/uploads/2/7/9/6/27967763/2017_poultry_handling_and_transportation_manual.pdf.

Poultry Service Association. (2017). poultry_handling_and_transportation_manual.pdf Available at: poultryserviceassociation.com.

Quinn, A. D. and Baker, C. J. (1997). An investigation of the ventilation of a day-old chick transport vehicle. *Journal of Wind Engineering and Industrial Aerodynamics* 67–68:305–311.

Qureshi, A. A. (1991). Losses due to dehydrated broiler chicks. *World Poultry (Misset)* 7(4):75–79.

Reform, P. (2012). Pas reform hatchery technologies maintaining the ideal climate for chick handling and transport. Available at: https://www.thepoultrysite.com/articles/maintaining-the-ideal-climate-for-chick-handling-and-transport.

Ross Broiler Management Handbook. (2018). Available at: https://en.aviagen.com/assets/Tech_Center/Ross_Broiler/Ross-BroilerHandbook2018-EN.pdf.

Ross Broiler Handbook. (2018). EN.pdf. Available at: aviagen.com.

Rowe, E., Marian Stamp Dawkins, M. S. and Gebhardt-Henrich, S. G. (2019). A systematic review of precision livestock farming in the poultry sector: Is technology focussed on improving bird welfare? *Animals: An Open Access Journal From MDPI* 9(9):614. https://doi.org/10.3390/ani9090614. Available at: www.mdpi.com/journal/animals.

Shahbandeh, M. (2021). Poultry industry in the United Kingdom - Statistics & facts. April 1st 2021. Available at: www.statista.com/topics/6102/poultry-in-the-united-kingdom/#dossierKeyfigures; https://www.statista.com/statistics/298514/annual-poultry-meat-production-in-the-united-kingdom-uk-by-type/.

Souza da Silva, C., Molenaar, R. and Giersberg, M. (2021). Day-old chicken quality and performance of broiler chickens from 3 different hatching systems. *Poultry Science* I(C):I. https://doi.org/10.1016/j.psj.2020.12.050.

Souza-Junior, J. B. F., El-Sabrout, K., de Arruda, A. M. V. and Costa, L. LdM. (2019). Estimating sensible heat loss in laying hens through thermal imaging. *Computers and Electronics in Agriculture* 166:105038. https://doi.org/10.1016/j.compag.2019.105038.

Sun, F., Zuo, Y. Z., Ge, J., Xia, J., Li, X. N., Lin, J., Zhang, C., Xu, H. L. and Li, J. L. (2018). Transport stress induces heart damage in newly hatched chicks via blocking the cytoprotective heat shock response and augmenting nitric oxide production. *Poultry Science* 97(8):2638–2646 http://doi.org/10.3382/ps/pey146.

Tamlyn, J. and Starr, J. R. (1986). Monitoring the environment during transportation of day olds. *International Hatchery Practice* 6(1):11–15.

Tanaka, A. and Xin, H. (1997). Effects of structural and stacking configuration of containers for transporting chicks in their microenvironment. *Transactions of the American Society of for Agricultural Engineers* 40(3):777–782.

Tona, K., Bamelis, F., De Ketelaere, B., Bruggeman, V., Moraes, V. M., Buyse, J., Onagbesan, O. and Decuypere, E. (2003). Effects of egg storage time on spread of hatch, chick quality, and chick juvenile growth. *Poultry Science* 82(5):736–741.

Tona, K., Bruggeman, V., Onagbesan, O., Bamelis, F., Gbeassor, M., Mertens, K. and Decuypere, E. (2015). Day-old chick quality: Relationship to hatching egg quality, adequate incubation practice and prediction of broiler performance. *Avian and Poultry Biology Reviews* 16:109–119.

Tona, K., Onagbesan, O., De Ketelaere, B., Decuypere, E. and Bruggeman, V. (2004). Effects of age of broiler breeders and egg storage on egg quality, hatchability, chick quality, chick weight, and chick posthatch growth to forty-two days. *Journal of Applied Poultry Research* 13(1):10–18.

Tong, Q., Demmers, T., Romanini, C. E., Bergoug, H., Roulston, N., Exadaktylos, V., Bahr, C., Berckmans, D., Guinebretière, M., Eterradossi, N., Garain, P. and McGonnell, I. M. (2015). Physiological status of broiler chicks at pulling time and the relationship to duration of holding period. *Animal: An International Journal of Animal Bioscience* 9(7):1181–1187. https://doi.org/10.1017/S1751731115000233.

UK, G. O. V. Monthly statistics on the activity of UK hatcheries and UK poultry slaughterhouses (data for October 2021). Accessed January 14, 2022. Available at: www.gov.uk.

Valros, A., Vuorenmaa, R. and Janczak, A. M. (2008). Effect of simulated long transportation on behavioural characteristics in two strains of laying hen chicks. *Applied Animal Behaviour Science* 109(1):58–67.

van de Ven, L. J. F., van Wagenberg, A. V., Decuypere, E., Kemp, B. and van den Brand, H. (2013). Perinatal broiler physiology between hatching and chick collection in 2 hatching systems. *Poultry Science* 92(4):1050–1061. http://doi.org/10.3382/ps.2012-02534.

van der Hel, W. and Henken, A. M. (1990). Give newly hatched chicks a safe flight. *Poultry International (Misset)* 6(1):8–9.

van der Hel, W., Verstegen, M. W. A., Henken, A. M. and Brandsma, H. A. (1991). The upper critical temperature in neonatal chicks. *Poultry Science* 70(9):1882–1887.

van der Linde, G. (2018). Key issues for the successful transportation of chicks. *International Hatchery Practice* Available at: www.positiveaction.info/pdfs/articles/hp30_5p21.pdf (Vol. 30 5). Heering, The Netherlands, pp. 20–23. And International Hatchery Practice 2015 Vol.29 No.6 pp.15, 17

van der Wagt, I., de Jong, I. C., Mitchell, M. A., Molenaar, R. and van den Brand, H. (2020). A review on yolk sac utilization in poultry. *Poultry Science* 99(4):2162–2175. https://doi.org/10.1016/j.psj.2019.11.041.

van Harn, J., Lourens, A. and Gunnink, H. (2014). Uitkomen kuikens in een traditionele vleeskuikenstal. *Wageningen UR Livestock Research*. Available at: http://edepot.wur.nl/292049.

Veit Electronics. (2021). Day-old chick vehicles with the ecotransport technology. Available at: https://www.veit.cz/day-old-chick-vehiclesOverview-VEIT Electronics.

Vieira, F. M. C., Groff, P. M., Silva, I. J. O., Nazareno, A. C., Godoy, T. F., Coutinho, L. L., Vieira, A. M. C. and Silva-Miranda, K. O. (2019). Impact of exposure time to harsh environments on physiology, mortality, and thermal comfort of day-old chickens in a simulated condition of transport. International Journal of Biometeorology 63(6):777–785.

Vieira, F. M. C., Silva, I. J. O., Nazareno, A. C., Faria, P. N. and Miranda, K. O. S. (2016). Termorregulação de pintos de um dia submetidos a ambiente térmico simulado de transporte (Thermoregulation of day-old chickens submitted to simulated transport thermal environment). *Arquivos Brasileiro de Medicina Veterinária e Zootecnia* 68(1):208–214. http://doi.org/10.1590/1678-4162-7871.

Vieira, S. L. (1999). Feeding the newly-hatched broiler chick. *World Poultry* 15:17–18.

Vieira, S. L. and Moran, E. T., Jr. (1999). Effects of egg of origin and chick post-hatch nutrition on broiler live performance and meat yields. *World's Poultry Science Journal* 55(2):125–142.

WATO. (2006). Statutory instruments 2006 No. 3260. Animals, England animal health the welfare of animals (Transport) (England) Order 2006 Made 5th December 2006; Coming into Force 5th January 2007.

WATO Guidance. (2006). Welfare of Animals During Transport Council Regulation (EC) No 1/2005 on the protection of animals during transport and related operations. The Welfare of Animals (Transport) (England) Order 2006 and parallel national legislation in Scotland, Wales and Northern Ireland. Guidance on implementation in the United Kingdom. Published by the Department for Environment, Food and Rural Affairs. February 2011 Originally published December 2006 Second edition February 2008 Third edition August 2010. Available at: www.defra.gov.uk.

Weeks, C. and Nicol, C. (2000). Poultry handling and transport. In: Grandin, T. (Ed.), *Livestock Handling and Transport* (2nd edn.). Wallingford UK: CAB International, pp. 363–384.

Weeks, C. A. (2014). In livestock handling and transport 4th (2014). In: Grandin Chapter, T. (Ed.), *20 Poultry Handling and Transport*, pp. 378–399. https://doi.org/10.1079/9781780643212.0000.

Xin, H. (1997). Mortality and body weight of breeder chicks as influenced by air temperature fluctuations. *Journal of Applied Poultry Research* 6(2):199–204.

Xin, H. and Harmon, J. D. (1996). Responses of group-housed neonatal chicks to post-hatch holding environment. *Transactions of the American Society of for Agricultural Engineers* 39(6):2249–2254.

Xin, H. and Lee, K. (1997). Physiological evaluation of chick morbidity during extended post-hatch holding. *Journal of Applied Poultry Research* 6(4):417–421.

Yerpes, M. P., Llonch, P. and Manteca, X. (2021). Effect of environmental conditions during transport on chick weight loss and mortality. *Poultry Science* 100(1):129–137. https://doi.org/10.1016/j.psj.2020.10.003.

Yuhuan, W. and Fu, Q. (2019). Analysis on production efficiency of laying hens in China–Based on the survey data of five provinces. *Journal of Agricultural Science* 11(9).

Index

www.ingramcontent.com/pod-product-compliance
Ingram Content Group UK Ltd.
Pitfield, Milton Keynes, MK11 3LW, UK
UKHW021824100926
8556IPUK00003B/144

9 781801 462525